Biologic Effects of Light 2001

Biologic Effects of Light 2001

Proceedings of a Symposium
Boston, Massachusetts
June 16-18, 2001

Editor

Michael F. Holick
Boston, Massachusetts, USA

KLUWER ACADEMIC PUBLISHERS
BOSTON/LONDON/DORDRECHT

Distributors for North, Central and South America:
Kluwer Academic Publishers
101 Philip Drive
Assinippi Park
Norwell, Massachusetts 02061 USA
Telephone (781) 871-6600
Fax (781) 681-9045
E-Mail: kluwer@wkap.com

Distributors for all other countries:
Kluwer Academic Publishers Group
Post Office Box 322
3300 AH Dordrecht, THE NETHERLANDS
Telephone 31 786 576 000
Fax 31 786 576 474
E-Mail: services@wkap.nl

Electronic Services <http://www.wkap.nl>

Library of Congress Cataloging-in-Publication Data

A C.I.P. Catalogue record for this book is available
from the Library of Congress.

Printed on acid-free paper.

Printed in the United States of America.

Table of Contents

Preface

As humans ventured into the twentieth century, the industrialized countries were confronted with the scourge of rickets. Although solariums were becoming common in the early 1900s and phototherapy was gaining popularity as a result of the awarding of a Nobel Prize to Finsen in 1903, it wasn't until 1921 when Hess and Unger demonstrated that rickets could be cured by exposure to sunlight that the healthful benefit of sun exposure appreciated. In 1941, Apperly (Cancer Research; 1:191-195, 1941) noted that the occasional increased risk of skin cancer was associated with a decreased risk of many other more common and serious cancers.

The alarming increase in the number of cases of skin cancer, especially melanoma, has caused great concern about the negative role of sunlight in health.

The Sixth International Arnold Rikli Symposium on the Biologic Effects of Light was held in Boston, Massachusetts from June 16th – 18th, 2001. The goal of this Symposium was to focus on the very popular practice of tanning either by sunlight or by artificial light sources and the overall impact this practice has on health and disease. The program was organized by members of the Scientific Advisory Committee and my co-chair emeritus, Professor Ernst G. Jung. The Program Committee organized an outstanding state-of-the-art program that was enthusiastically received by the participants. In the tradition of the Symposium, the general topics included a broad range of perspectives regarding the biologic effects of sunlight and artificial ultraviolet radiation. The participants were also given a great grandson's perspective about the life and contributions of Arnold Rikli. After seeing the Arnold Rikli prize posted on the internet, Dr. Arthur Rikli, a recent retiree from the United States Public Health Service, contacted Dr. Peter Bocionek and myself. He provided us with a wealth of information about his great grandfather. This was presented to the participants before awarding the Arnold Rikli prize to Dr. Rolfdieter Krause at the Symposium.

In the session on UV Measurement and Photobiological Evaluation state-of-the-art information was provided on photobiologically weighted irradiance exposures and their impact on ocular tissues.

The session on Suntanning and Pigmentation reviewed the role of DNA excision fragments in melanogenesis, the effectiveness of tanning procedures to provide photoprotection. The production and persistence of a UV induced tan was also described.

The session on Vitamin D and Bone Health reiterated the important role that vitamin D has on maximizing bone health. Emphasis was placed on the increased prevalence of vitamin D deficiency in both young and older adults and the role it plays not only in initiating and perpetuating osteoporosis, but also for causing osteomalacia and myopathy. A very interesting presentation on the survival of panther chameleons that are housed indoors also gave further insight into the broad range of biologic effects that sunlight may have on health and reproductive activity.

The session on Non-Calcemic Effects of Vitamin D further supported the new appreciation for vitamin D's function in non-calcium regulating tissues. Data was presented to suggest that the skin not only makes vitamin D, but also activates it similar to other tissues such as prostate, breast and colon. This newly appreciated autocrine/paracrine function of vitamin D in these tissues may be the explanation for the association of latitude with increased risk of dying of some of the more common cancers such as colon, breast and prostate. There was also evidence to suggest that pancreatic cancer and multiple myeloma may also be associated with increased latitude.

The session on Photoimmunology discussed the signal transduction of ultraviolet A radiation on human keratinocytes. New insights were presented on the effect of ultraviolet A radiation on mitochondrial oxidation and cellular damage. The effect of UVA 1-induced T cell apoptosis on clinical consequences and the effect of UVA-1 and UVB on HIV activation was discussed.

The Photomedicine and Photobiologic Aspects in Skin session reviewed the effect of light treatment in various skin diseases and the initial development of a personal UVB sensor that detects the radiation responsible for photolyzing 7-dehydrocholesterol to previtamin D3. Light emitting diodes that are lightweight was suggested as a portable phototherapy unit.

The session on Biologic Clocks reviewed the regulation of circadian clock genes and the molecular biology of the circadian clock. The role of melanopsin and opsins on circadian clock resetting was discussed.

The session on Circadian Rhythms reviewed the action spectrum for melatonin suppression and evidence was provided about novel circadian photoreceptor in the human eye. The entrainment of the non-24-hour circadian period of the human biologic clock was discussed.

I would especially like to thank Mr. Jorg Wolff of Wolff System Technology Corporation and Mr. Friedrich Wolff from the Institute Friedrich

Wolff A.G. for their financial support. I am grateful to Mr. Michael Stepp of Wolff System Technology Corporation and the Light Foundation and his staff for their help in organizing the Symposium. I am most grateful to Catherine St.Clair who not only worked with Intraworld Incentives, Inc. to help with the excellent accommodations for the participants, but who as managing editor spent an enormous amount of time and effort in organizing this symposium book.

I look forward to seeing you at the 7th International Arnold Rikli Symposium on the Biologic Effects of Light that will be held in 2003.

Michael F. Holick, Ph.D., M.D.

Arnold Rikli: A Great-Grandson's Perspective

Arthur E. Rikli, M.D., M.P.H.Professor Emeritus, *College of Medicine, Columbia MO.4003 Faurot Drive Columbia, MO.*

I INTRODUCTION

As a descendant of Arnold Rikli I would like to express my and my families appreciation to the Institute Friedrich Wolff AG for sponsoring the Arnold Rikli Prize and Symposium. I am impressed by the scientific quality of the research work that is documented in the Symposium report titled "Biologic Effects of Light 1998" Editors, Holick & Jung. Many different authors have documented the life history of Arnold Rikli. I find it difficult to comment objectively on his medical views because of my medical experience. In 1945, my first experience as a physician in the United States Public Health Service (USPHS) was in the Tuberculosis Control Program. The remainder of my career I was concerned with the prevention and treatment the major chronic diseases as heart disease, cancer, diabetes and kidney disease. I have always wondered to what extent surgical and medical treatment favorably altered the natural history of the tuberculosis and syphilis prior to drug therapy. These were the kinds of patients that were attracted to Arnold Rikli's health center located what is now called Bled, Slovenia. 1881 Arnold Rikli prepared a document describing the services provided in Bled. It was then called Veldes. I have selected from this document "Dr. Arnold Rikli's, Physico-Hydriatic Establishment" his description of his experience on which his "cure" is based. It is followed by the 10 principles of his "cure". This presents in great-grandfather's own words his experience and his views.

II PREFACE - Arnold Rikli

"Some thirty-six years ago, all the remedies from the apothecary's shop failing, I was dying with dysentery, when my life was saved through an advice I found in Dr. Munde's "Hydrotherapie." From inclination as well as from gratitude, I resolved upon devoting thence forth my energies to the study and practice of a method which had done so much good to myself and so many others, and which promised to strengthen and harden my naturally rather weak constitution. With Dr.Munde's excellent practical guide in hand, I gradually tried upon my own body all the different applications of the Watercure, and then commenced to practice upon others as opportunities offered. I soon discovered that the use of cold alone did not answer in every case, and that, with the aged, the feeble, and the chlorotic, the alternate use of heat did more service than that of cold water exclusively. Thus, adding the sunbath, the bed-steambath, the airbath and similar other applications to Priessnitz's system, I gradually formed that combined system of Physico-HydriaticTreatment, which has been practiced for the last twenty seven years, under the favorable influence of a mild sunny climate, the pure mountain air, and with the assistance of several powerful springs of the clearest and coldest water, on the beautiful little lake of Veldes, the "Pearl of the Julian Alps" (Fig. 1).

Figure 1. Photograph of Arnold Rikli 1823-1906

III GENERAL PRINCIPLES - Arnold Rikli

The physician cares for the patient, nature cures him. There is no direct healing power in medicines. Things which are poisons to the healthy, are also poisons to the sick. Anything indispensable for the preservation of health is a natural remedy for its restoration. Air, light, water, nourishment, and exercise are indispensable to health; they should always be considered the first remedies to be resorted to in disease. The worst place for a chronic patient to be cured is in his bed. Help yourselves and God will help you. Life is change of matter: have more care to remove waste matter from your system than to supply its loss. Waste matter retained in the system becomes poison. It is through the complete removal of the waste peccant matter and the invigoration of the nervous system that harmony in the functions of the organism is restored, the body as it were rejuvenated. No rejuvenation of the body can take place without the assistance of air, light, water, proper diet and exercise; we can not buy health in the apothecary's shop. Prof. Dr. med Zdenko Leventa l, Sulgenauweg 16, 3007 Bern, wrote in German a life history of Arnold Rikli titled: "Der Sonnendoctor - Arnold Rikli (1823- 1906) It is documented by 26 references. He gives credit to many people including my cousins Mathias and Heinrich Rikli, who live in Friedberg near Wangen. The English summary of his article follows. I have selected this life history summary of Arnold Rikli because others frequently quote it and not because I share his views.

IV SUMMARY - Prof. Dr. med Zdenko Leventa

Thanks to new data discovered in the archives of the Rikli family, it is now possible to rewrite the life-story of the "sun-doctor" Arnold Rikli, until now incomplete and containing wrong facts and interpretations. Rikli was the son of a distinguished dyer in Wangen an der Aare (Switzerland). Very early he developed interest for free life in nature as well as for nursing. Under the influence of some "natural therapy" authors, he decided at the age of 32 to abandon his father's profession and to create as therapeutic institution in Veldes (Bled) then in Austria, now in Yugoslavia. The main principles on his so-called "atmospheric treatment" have been, beside hydrotherapy, air- and sunbaths combined with exercise. A part of the patients lived in open-air cottages. Rikli was certainly the first healer in the 19th century who introduced sunbaths in the cure of chronic diseases and functional disorders. It is possible that Rikli's ideas, spread broadly over Central Europe, have initiated or fostered the development of what is the present-day's way of plain-air recreation and fitness, probably also of nudism. His practice of half-a-century in Slovenia and Trieste (Italy) have brought him great patient affluence, but only poor popularity and imitation by physicians. This could be also explained by his continuous conflicts with the profession and law-courts. Rikli's extremely odd and confused theories have been partly influenced by old "humoral physiology". Rikli was firmly opposed to all results of modern medical science and practice and considered academic medicine and physicians as the most dangerous enemies of mankind.

V. Commentary

Most of the information that I have accumulated regarding Arnold Rikli and his ancestors I have received from Mathias and Heinrich Rikli. We have visited their home at Friedbergon many occasions since 1964 when I was

Figure 2. Plaque commemorating Arnold Rikli located in a park in Bled, Slovenia. Site was visited in 1989 by Arthur Rikli and grandson, Todd Rikli.

assigned to Geneva, Switzerland by the USPHS (Fig. 2). Mathias lived in the house built for Arnold's father Abraham Friedrich Rikli (1824-1902). Heinrich lived in the classic Swiss farm/barn house where he raised his family and farmed Friedberg prior to building his own home. On one occasion when visiting with Mathias he showed me many family artifacts he had accumulated in his attic. They included a sitts bathtub used at Arnold’s center in Bled. He also offered to give me a bed-steambath invented by Arnold. Mathias' daughter is still living at Friedberg. Arnold's father was a "rotfarber". He was known for his use of red dye made from dyer's madder (Rubia tinctorum, a plant related to the woodruff) It was used to dye wool and other textiles. Arnold was born 1823 in Wangen A/Aare. A walled town located on the Aare River. Access to the highway is over a covered bridge. Arnold's birthplace, a stately mansion in the rustic Bern style, is still standing and bears a memorial plaque from the Swiss Association for Public Health. His birthplace was located north of the red dyeing business. This site was later replaced by a school. Arnold's ancestors can be traced back to Heinrich Rikli (1570-1654) Arnold had a summer home in Veldes and a winter home at

Rikliweg 18, Wolfsberg Austria. Arnold's great granddaughter, Margarethe Simack, is living in his winter home. She is the owner of a very successful pharmacy called the White Wolf Pharmacy. Her grandfather Oscar became responsible for operation of Arnold's health center in Bled following Arnold's death in 1906. We have visited her home and she has accompanied us on a visit to Bled. Essentially all of the buildings used by Arnold Rikli have been replaced. A Casino has replaced his primary location. A restaurant and bar carried his name. There is a memorial plaque located in a park. Rikli Strassa (street) ends at a large church. The museum located in a castle high above the city of Bled contains a display of the services provided by Arnold and pictures of his facilities. The Tourist Office documents still credit Arnold with discovering the health benefits of Bled and being responsible for encouraging tourists to come to Bled. They display a life-sized colored picture of Arnold in his bare feet, wearing loose fitting shirt and shorts, holding a mountain climbing staff and pointing the way (Fig. 3).

Figure 3. Painting of Arnold Rikli

Arnold's son Arthur came to the United States and married Margaret Oehl September 1st, 1854. They lived in Murdock, Nebraska. They had 12 children. My father, Arthur, became a physician. Following formal medical training in 1907 he continued his training in Europe. He visited his grandfather's health center the year following Arnold's death. Many of the descendants of grandfather Arthur have chosen or are choosing the health professions as doctors, dentists, nurses, technicians, physical therapists, dental hygienists and others. Dr. Roberta E. Rikli, Ph.D. serving as Chairman, Kinesiology and Health Promotion, at California State, Fullerton has become well known for her publications on the effects of exercise. Exercise was one of Arnold's major principles in addition to water, air, light and diet. Arnold's health interests have not only effected his family but many of his principles are still receiving increasing favorable emphasis in many places today.

Acknowledgement

Pictures were provided by the Arnold Rikli family archive.

UV Measurement and Photobiological Evaluation

MEASUREMENT OF PHOTOBIOLOGICALLY WEIGHTED IRRADIANCES AND EXPOSURES

Wolfgang Heering
Lighting Research Institute, University of Karlsruhe, Germany

INTRODUCTION

In order to predict and quantify photobiological effects that a distinct radiation will produce on human skin and eye or another photobiological receiver, it is more adequate to measure photobiologically weighted irradiances and exposures than physical radiant quantities as for instance integral UV irradiances. Actinic radiant quantities X_{act} are obtained by weighting spectral radiant quantities $X_{e\lambda}$ incident on the receiver with action spectra $s_{act}(\lambda)$ a lot of which are believed to be well known:

$$X_{act} = \int_0^\infty s_{act}(\lambda) X_{e\lambda}(\lambda) d\lambda \qquad (1)$$

The index "act" describes the actinic effect under consideration. In Figure 1 some action spectra of phototherapeutics are shown as examples. An

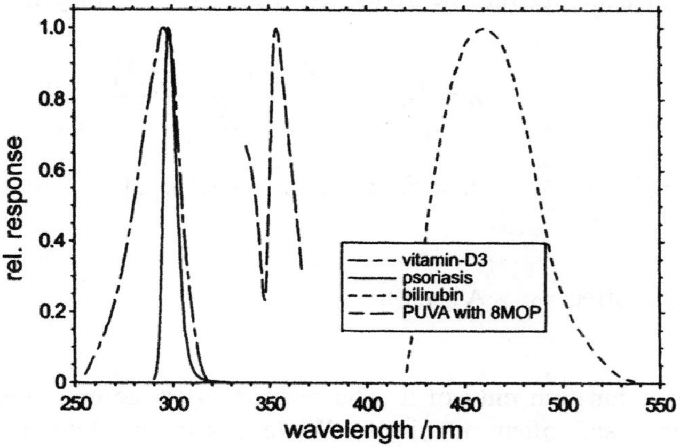

Figure 1. Action spectra of phototherapeutics

integrally measuring radiometer will measure the correct value of X_{act} if its relative spectral responsivity is perfectly matched to the action spectrum and if it is exactly calibrated. Imperfect calibration and matching as well as response to radiation outside of the action spectrum, deviations from cosine response, non-linearity, fatigue, dependencies on response on ambient temperature, modulation and polarization of radiation, non uniform sensitivity of the radiometer, offset and noise of the instrument etc. will produce uncertainties. This paper mainly concentrates on instrumental features which affect the quality of matching and presents figures to characterize the quality of matching to a target function.

However before, one should be aware that even if the actinic radiometer would measure the exact value of X_{act}, the measured actinic radiant quantity X_{act} might be not the right measure to quantify the actinic effect of the measured radiation. This is why actinic radiant quantities are based on a model of actinic effects which cannot be applied to each photobiological, photomedical and photochemical process. The conditions of the concept of actinic radiant quantities can be derived from the following model of actinic radiant effects.

MODELLING ACTINIC RADIANT EFFECTS

According to Kiefer [1] the process which radiation effects on a photobiological receiver occurs in four steps:

1. Absorption of radiation
2. Excitation of molecules, chromophores
3. Formation of photoproducts
4. Actinic effect

The number density n of photoproducts generated at the point x of the receiver by a monochromatic irradiance E(λ,x,t) there during the time t

$$n = \int_0^t \eta \cdot \left(-\frac{dE(\lambda,x,t)/(hc/\lambda)}{dx} \right)_{abs} dt \qquad (2)$$

with $$E = E_0(\lambda,t)\exp[-\sigma_a(\lambda)n_a x] \qquad (3)$$

gives an effect $$A = \mathrm{fct}(n) \qquad (4)$$

which is a function only of n. The relation (4) is definite, independent on wavelength and often non-linear. While n can mostly not be measured directly, the observable effect can be quantified, often as a threshold. From (1) and (2) it follows

$$n \propto H_{act} = s_{act}(\lambda) \cdot H(\lambda,x) \tag{5}$$

with
$$H(\lambda,x) = \int_0^t E(\lambda,x,t)dt \tag{6}$$

and
$$s_{act}(\lambda) = \frac{\eta(\lambda) \cdot \lambda \cdot \sigma_a(\lambda)}{\eta(\lambda_0) \cdot \lambda_0 \cdot \sigma_a(\lambda_0)} \tag{7}$$

$H(\lambda,x)$ is the radiant exposure and $s_{act}(\lambda)$ the so-called action spectrum, i.e. the spectral sensitivity of the photobiological receiver related to that at wavelength λ_0. $\sigma_a(\lambda)$ is the wavelength-dependent absorption cross section and $\eta(\lambda)$ the quantum yield, i.e. the number of photoproducts generated by one absorbed photon. The relation (5) implies the law of *Bunsen-Roscoe*, i.e. an actinic effect is determined by the actinic exposure H_{act}. So all physical, monochromatic radiant exposures applied at different wavelengths produce the same actinic effect if the actinic exposures are the same.

In practise, an actinic effect will not be produced by a quasi-monochromatic exposure but by a polychromatic one. According to *van Krefeld*, if the single effects at different wavelengths are independent on each other, i.e. no catalytic or hindering effects as photoaugmentation or photoreactivation appear, the actinic exposures by different spectral components of the incoming radiation can be added to an overall effect:

$$\begin{aligned} H_{act} &= \int H_{\lambda,act}(\lambda,x)d\lambda \\ &= \int s_{act}(\lambda) \cdot H_{\lambda}(\lambda,x)d\lambda \end{aligned} \tag{8}$$

An action spectrum is usually determined by measuring those quasi-monochromatic exposures $H_{th}(\lambda,x)$ at different wavelengths which produce the same measurable or just observable effect:

$$s_{act}(\lambda) \propto H_{th}^{-1}(\lambda,x) \tag{9}$$

The uncertainties of an action spectrum are quite large. They are by the uncertainties to reproduce and quantify the effect, the influence of other processes on the considered photoeffect, the uncertainty of measuring radiant exposures and by the influence of reversible photoreactions as for instance repair mechanisms in the DNA. Especially, if at certain wavelengths long exposure times are needed to get a distinct actinic effect, the uncertainties of $s_{act}(\lambda)$ are relatively large at these wavelengths. So, an actinic radiant quantity may be rather uncertain if it is generated by a relatively large radiant quantity of wavelength λ where $s_{act}(\lambda)$ is relatively small.

RADIOMETRY OF ACTINIC RADIANT QUANTITIES

Actinic irradiances, actinic radiances and actinic exposures – the focus is here on actinic irradiances – can be measured either by means of a spectroradiometer which scans spectral irradiances or spectral radiances to be weighted numerically by $s_{act}(\lambda)$ or by means of a broad-band actinic radiometer, a calibrated radiation detector the spectral responsivity of which is matched to $s_{act}(\lambda)$.

Spectroradiometric Measurements

The spectral method for the determination of actinic radiant quantities generally gives more exact results because no spectral matching to an action spectrum is needed. It is performed by a system of different components, i.e.

- input optics which couples part of the incident radiation into the spectral apparatus, defines the spectral radiant quantity to be measured and provides a correct illumination of the dispersing elements
- spectral apparatus which disperses the spectral components of the incoming radiation by time or by position
- detector which converts radiant power or energy at the output of the spectral apparatus into electrical signals
- signal-acquisition unit which processes and digitizes the electrical signals to be transmitted to a computer
- computer which computes the actinic radiant quantity from the measured spectral irradiances or radiances.

The entrance optics for measuring irradiances has first of all to provide for a cosine response to the radiations incident at very different angles. This is realized by a diffuser, a *Ulbricht* sphere or a diffusing plate in front of the input. The *Ulbricht* sphere is not very often used because its throughput is relatively small and dependent on wavelength and the most common coatings like $BaSO_4$, MgO_2 or PTFE are optically instable and/or get dirty with time. The mostly applied cosine adaption is by a diffusely transmitting plate of quartz glass or PTFE. However, transmitting diffusers give cosine responses which considerably depend on the wavelengths of the incoming radiation as it is demonstrated in Figure 2 on an example of a quartz glass diffuser [2].

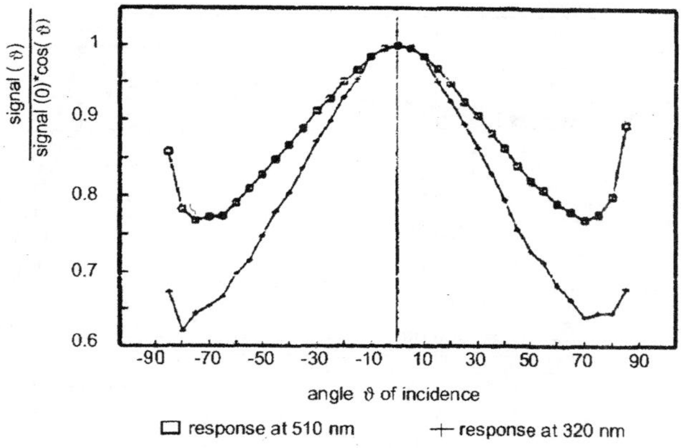

Figure 2. Radiometer signal with a transmitting quartz glass diffuser related to an ideal cosine response measured at different angles of incidence for two wavelengths [2]

We have produced a cosine response less dependent on wavelength by a roughened 10 mm thick PTFE plate which diffusely reflects parts of the incident radiation into a light guide coupled to the instrument (Figure 3). Figure 4 shows the relative deviations from cosine response at different wavelengths obtained by this method [3].

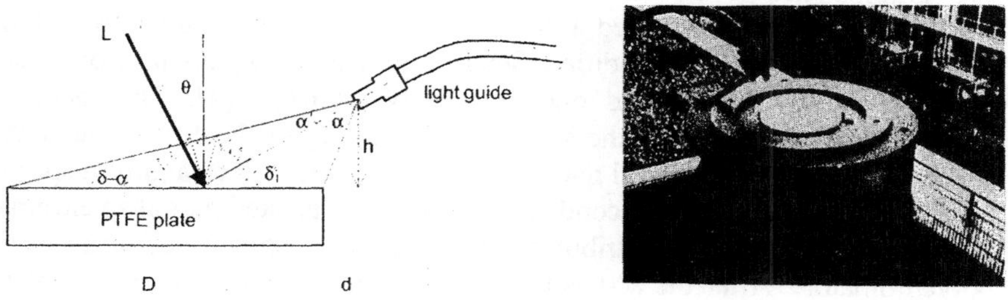

Figure 3. Entrance optics with a diffusely reflecting 10 mm thick PTFE plate and light guide coupling to the spectral apparatus

The incoming radiation is spectrally dispersed by means of a monochromator or a polychromator or filtered by system of interference filters. In order to measure the true spectral distribution of radiation, wavelength accuracy (Tol. ≤ 0.1 nm!), stray light reduction ($\leq 10^{-7}$!) and spectral resolution (FWHM ≤ 0.5 nm!) should be sufficiently high enough.

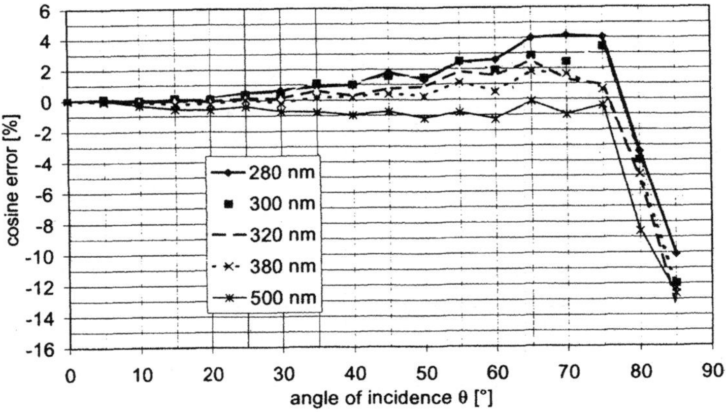

Figure 4. Relative deviations from cosine response with a reflecting PTFE plate for different wavelengths of the incoming radiation

The respective requirements depend on the shape of the action spectrum. This is normally achieved by the use of a precisely scanning double monochromator. However double monochromators are serially scanning instruments which usually need some minutes to scan a spectrum of a few hundreds of nanometers and which are not very robust because the rotation of two gratings has to be exactly synchronized. In order to overcome such problems we have constructed a new kind of double monochromator (Figure 5). The first stage of a modified Seya- Namioka arrangement works as a predisperser and the second one as a polychromator [4]. The incoming radiation gets dispersed by the step-scanned grating of the first stage with a spectral bandwidth of several nanometers and is then further highly dispersed by the fixed grating of the second stage so that the selected partial spectrum of the predisperser appears distributed on a part of the photodiode array of the polychromator. After the acquisition of a partial spectrum the grating of the predisperser is rotated to the next position so that the neighbored partial spectrum is incident on another part of the array. The partial spectra are picked up and scaled with different integration times depending on the irradiance level and are then numerically composed to one spectrum by a computer. This spectral procedure provides a short measuring time of only a few seconds for a complete UV (air) spectrum, small enough spectral bandwidth of less than 1 nm (0.6 nm) and a higher dynamic range than it would be possible with a usual photodiode array spectroradiometer. We have obtained a noise equivalent spectral irradiance of 30 $\mu W\ m^{-2}nm^{-1}$ at 300 nm

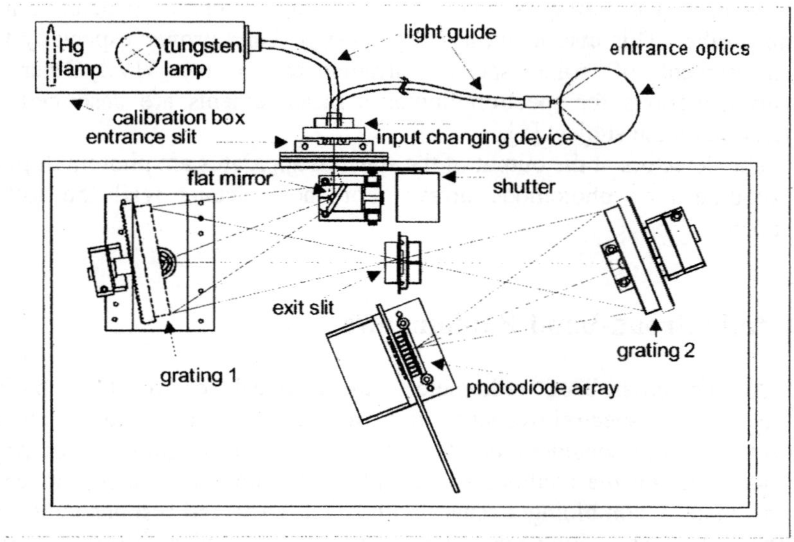

Figure 5. Optical arrangement of a fast measuring spectroradiometer with predisperser and photodiode array

for an exposure time of 10 s and an entrance optics like that of Figure 3. Furthermore the integral stray light suppression for broadband radiation is better than that of a single monochromator as it has been demonstrated by the measured spectral transmittances of a *Schott* WG 320 glass filter irradiated by a 150 W Xenon short arc lamp (Figure 6).

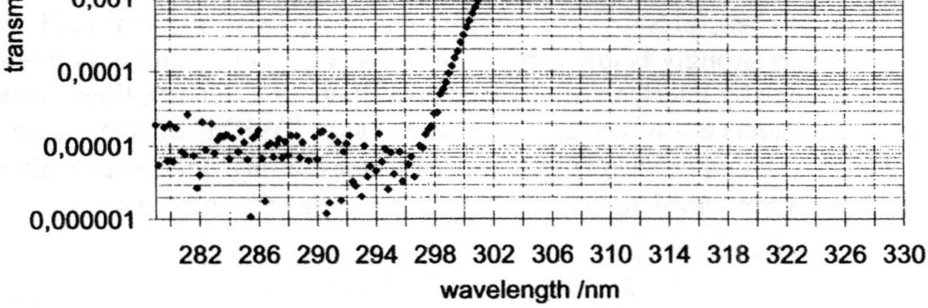

Figure 6. Integral stray light suppression proved by the measurement of spectral transmittances of a Schott WG 320 glass filter irradiated by a Xenon high pressure lamp

As only one grating is rotated the two stages cannot get detuned against one another. This instrument has been designed and proved especially for the measurement of solar spectral irradiances in the UV. Other usual instrumentations for spectroradiometric measurements are described in the literature on the subject [5].

The detectors at the output of the spectral apparatus are photomultipliers or photodiodes or photodiode arrays. For more details read the following chapter.

Actinic Broad-band Radiometers

Actinic broad-band radiometers are physical, chemical or biological detectors the spectral responsivities of which are matched to an action spectrum under consideration. The focus is here on the mostly used physical systems and on the methods and qualities of matching. Spectral matching is performed by combining a sensor with suitable spectral response with proper optical short pass, long pass or band pass filters and a diffuser in front of them. A simple physical broad-band radiometer with single filtering is shown in Figure 7.

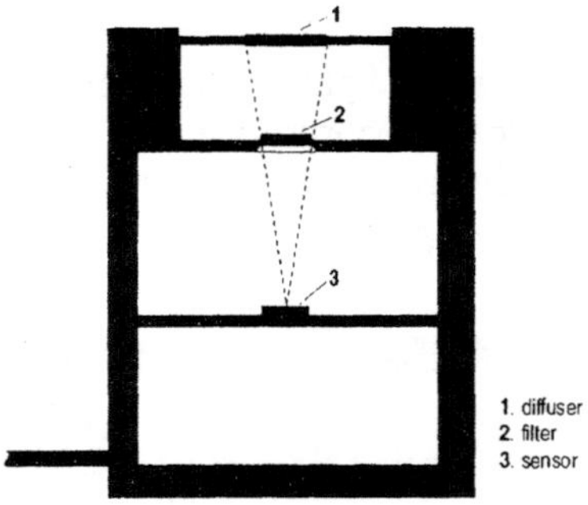

Figure 7. Actinic broad-band radiometer with single filtering

Filters of glass, plastics or gelatin in which pigments or metal oxides are solved absorb strongly below certain cut-off wavelengths. Band pass filtering is usually done by interference, MDI or ADI filters. Interference filters show a strong dependence on center wavelength and FWHM of the angle of incidence. The center wavelength λ_{max} of a transmitting interference filter is shifted to shorter wavelengths with increasing angle of incidence α:

$$\lambda_{max} / \lambda_{max,0} = \sqrt{(n^2 - \sin^2 \alpha) / n^2} \tag{10}$$

So the entrance optics of a filter radiometer has to be designed so that the filter transmits radiation always nearly at the same angle independently on the angular distribution of the incident radiation. Otherwise matching changes with the angle of incidence.

Also a temperature drift can have an influence on spectral matching. The center wavelength of ADI filters is shifted by about 0.01 – 0.03 nm/K and the cut-off wavelength of glasses by about 0.025 – 0.05 nm/K to longer wavelengths. Furthermore, the short circuit current of photodiodes depends on temperature too, as shown in Figure 8 on an example of GaAsP diodes. Intrinsic semiconductors begin to absorb at longer wavelengths when the band gap is lowered with increasing temperature.

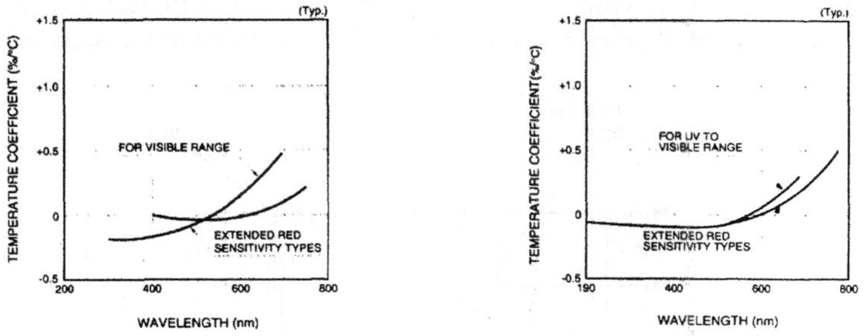

Figure 8. Temperature coefficients of GaAsP photodiodes, diffusion and Schottky types, as functions of temperature, according to Hamamatsu datasheet on photodiodes, p. 40-42 (1997)

Large exposures to short-wave UV will solarize glasses which are used as blocking filters or as substrates of interference filters or as fibers for light guides. This will shift and lower short-wave transmission, as it can be seen from Figure 9 [6].

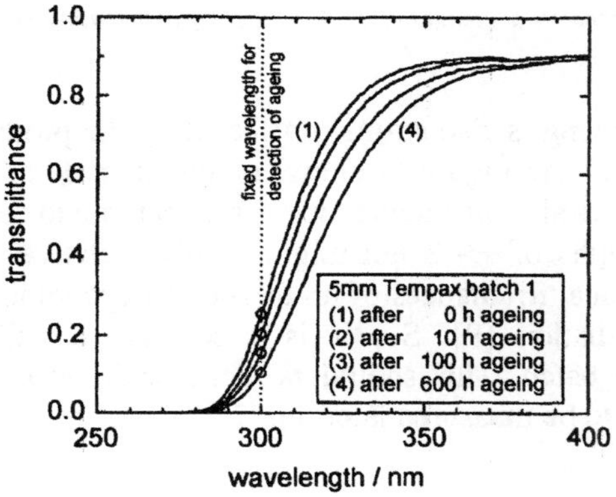

Figure 9. Shift of cut-off and decrease of transmittance of 5 mm borosilicate glass by an UV-B irradiance of about 30 W m^{-2} from a solar radiation simulator over hundreds of hours [6]

Solarization is less for fused silica glass than for silicate glass and smaller the purer the glass and the longer the wavelengths of exposure are. As it slows down with exposure time, a radiation pre-treatment may stabilize the UV transmittance. Strong UV-C exposures of $10^2 - 10^4$ J m^{-2} at wavelengths below 250 nm also considerably lower the spectral responsivities of sensors, especially those of diffused silicon p-n junction diodes and less those of Schottky diodes [7]. Radiation-attenuating means by aperture stops, (neutral density) filters or diffusers reduce such aging effects.

Spectral matching will also be influenced by humidity which penetrates into the dielectric layers of ADI interference filters. Figure 10 gives an order of the effect, the shift of center wavelength of transmission to larger values. It can be reduced by a densifying ion bombardment of the dielectric layers [8]. Additionally, water which gets solved in the porous material of a diffuser changes the spectral amounts of transmitted radiation onto the sensor.

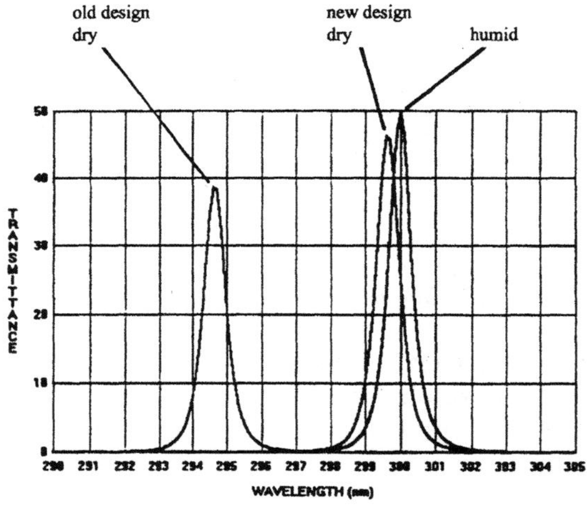

Figure 10. Shift of center wavelength of an ADI filter by penetrating humidity for different filter designs [8]

Spectral matching is also strongly affected by the production tolerances of interference filters. As Figure 11 shows on an example of an erythemal filter, the curves of spectral transmittance vary from sample to sample. Variations of cut-off wavelengths of +/- 3 nm might involve relative uncertainties of the measured erythemal irradiances of up to 100 % depending on the spectrum of the incident radiation [9]. So, it is necessary to calibrate each actinic radiometer of a batch. This should be done with respect to the illuminant which is mostly to be measured later on.

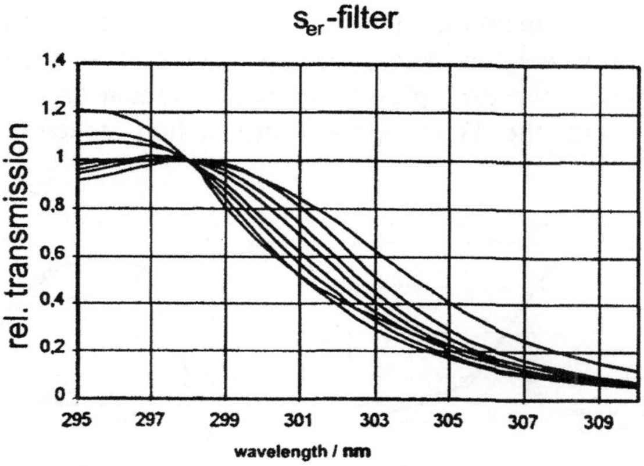

Figure 11: Measured spectral transmittances of interference filters of a batch to be used for filtering erythemal radiometers [9]

Last but not least a rather large uncertainty (≥5%) is introduced by short wave respectively long wave response of the actinic radiometer to radiation which is outside of the range of the action spectrum. Short pass or long pass filters can block radiation aside the action spectrum by many decades of magnitude.

A better spectral matching can be achieved by a multi-sensor system as shown in Figure 12. The sensors of a sensor array (photodiode array) are covered by different optical filters and the sensor signals (photocurrents fed

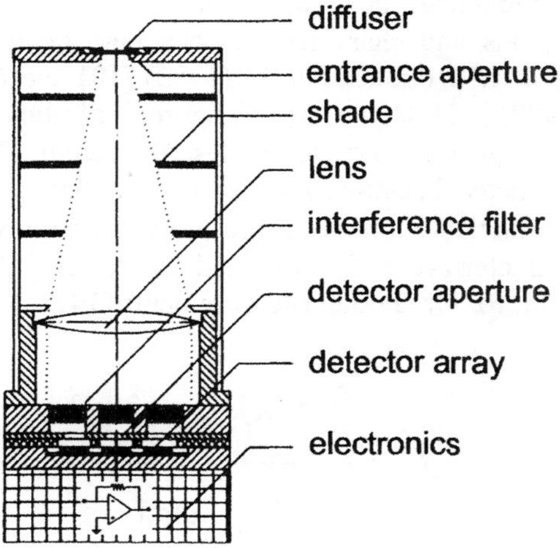

Figure 12. Multi-sensor actinic radiometer

over different resistances onto the summing point of an operational amplifier) are electronically weighted and added to a summed signal to give the right actinic response. Two examples of successful matching to target functions, the ACGIH and the Blue-Light-Hazard action spectra, are shown in Figure 13 a-b.

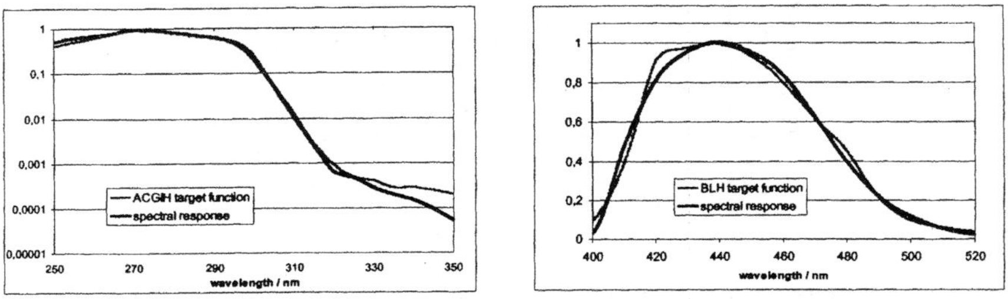

Figure 13 a-b. Spectral responsivities of two actinic detector heads compared to the ACGIH and BLH target functions, according to Gigahertz-Optik (Puchheim, Germany)

Characterizing Matching Quality

There are a lot of figures to characterize the performance of actinic broad-band radiometers, i.e. calibration uncertainty, matching to an action spectrum, short and long wave response, directional response, spatial response, non-linearity, temperature dependence on response, dependences on responsivity on modulation frequency and polarisation of radiation, fatigue and ageing. Methods, definitions and characteristics can here be used which are quite similar to those of CIE Publication 53 (1982) [10] and 69 (1987) [11] or to DIN 5032-6 (1997) [12]. However, the figure f_1 of photometry by which the relative degree of spectral mismatch is described cannot be simply applied to an actinic radiometer because neither the illuminant A is suitable as a reference nor the deviation of spectral responsivity from v(λ) is here relevant. The integral characterization f_1' proposed by Poppe [13] and the European Thematic Network for Ultraviolet Measurements [14]

$$f_1' = \frac{\int_0^\infty \left| s^*(\lambda) - s_{act}(\lambda) \right| d\lambda}{\int_0^\infty s_{act}(\lambda) d\lambda} \tag{11}$$

where
$$s^*(\lambda) = \frac{\int_0^\infty s_{act}(\lambda) S_{\lambda,C}(\lambda) d\lambda}{\int_0^\infty s(\lambda) S_{\lambda,C}(\lambda) d\lambda} s(\lambda) \tag{12}$$

is the normalized relative spectral responsivity, s(λ) the relative responsivity of the actinic radiometer and $S_{\lambda,C}$ the spectral distribution of the illuminant used for calibration, has the advantage that this definition is independent on the irradiating source but the disadvantage that it provides no real information on the uncertainty by mismatch for a given source. So, we think it is better to use

$$f_1(Z) = \frac{\int_0^\infty |s^*(\lambda) - s_{act}(\lambda)| S_{\lambda,Z}(\lambda) d\lambda}{\int_0^\infty s_{act}(\lambda) S_{\lambda,Z}(\lambda) d\lambda} \tag{13}$$

as a relative measure for mismatch when the actinic radiometer is irradiated by a source Z with spectral distribution $S_{\lambda,Z}(\lambda)$. For correcting the response of an actinic radiometer calibrated with a source C so that it gives the right result when irradiated by a source Z, the reading when measuring Z is to be multiplied with the reciprocal of the relative actinic responsivity

$$a(Z) = \frac{\int_0^\infty s(\lambda) S_{\lambda,Z}(\lambda) d\lambda}{\int_0^\infty s_{act}(\lambda) S_{\lambda,Z}(\lambda) d\lambda} \cdot \frac{\int_0^\infty s_{act}(\lambda) S_{\lambda,C}(\lambda) d\lambda}{\int_0^\infty s(\lambda) S_{\lambda,C}(\lambda) d\lambda} \tag{14}$$

Using tabulated test spectra as recommended by the European UV network [14] and shown in Figure 14 allows to calculate correction factors for different spectral distributions of radiations.

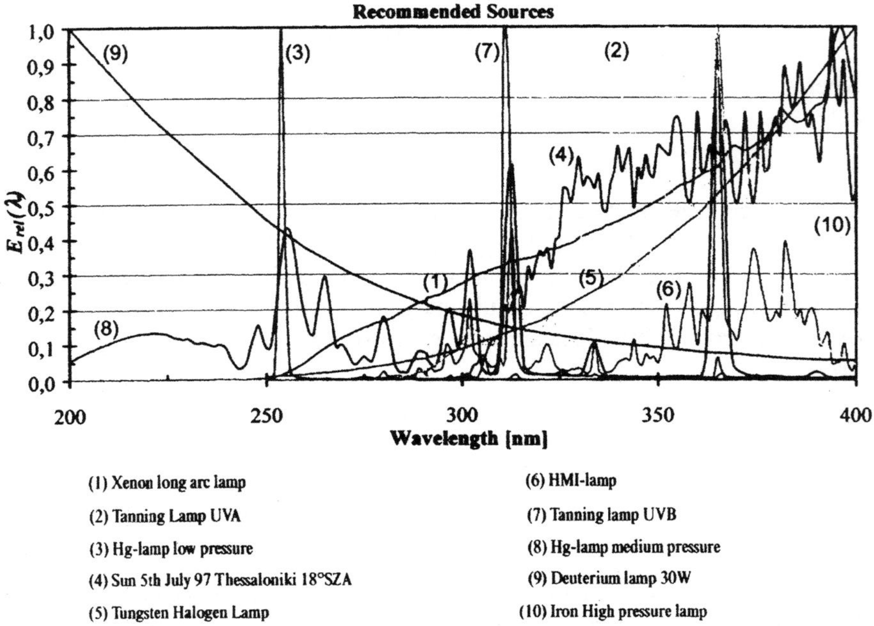

Figure 14. Test spectra for the evaluation of spectral matching and spectral correction of actinic broad-band radiometers

REFERENCES

[1] Kiefer, J., Methodik und Möglichkeiten der biologischen Aktionsspektroskopie, PTB-Opt **24**, 194-206 (1986)

[2] Seckmeyer, G. et al., Cosine error correction of spectral UV irradiances, Proc. SPIE **2049**, 140-151 (1993)

[3] Oppitz, C., Heering, W., Reflective optics for cosine-corrected irradiance measurements, Rev.Sc.Instr. **69**, 3487-3490 (1998)

[4] Oppitz, C., Ein neues Spektralradiometer zur Erfassung der solaren UV-B–Strahlung und der Ozongesamtsäule, Dissertation, Karlsruhe 1997

[5] Kostkowski, H.J., Reliable Spectroradiometry, Spectroradiometry Consulting, La Plata, Maryland 1997

[6] Döhring, T. et al., Spectral shaping of artificial UV-B irradiation for vegetation stress research, J.Plant Physiol. **148**, 115-119 (1996)

[7] Kuschnerus, P. et al., Characterization of photodiodes as transfer detector standards in the 120 nm to 600 nm spectral range, Metrologia **34** (1998)

[8] Kaiser, N. et al., Entwicklung hochstabiler dielektrischer Schichtsysteme zur Spektralmessung an der UV-B–Kante, 5. Statusseminar des Ozonforschungsprogramms, Bonn, 22.-23.05.1996

[9] Gugg-Helminger, A. et al., Problems with the classification of UV meters, UV News **5**, 15-17 (2000)

[10] CIE 53-1982, Methods of characterizing the performance of radiometers and photometers

[11] CIE 69-1987, Methods of characterizing illuminance meters and luminance meters

[12] DIN 5032-6 (1997), Lichtmessung: Photometer, Begriffe, Eigenschaften und deren Kennzeichnung

[13] Poppe, M., The measurement of actinic radiation, Proposal for a CIE Technical Report, 3. Draft (1993)

[14] UV News **6** (2000), Characterizing the performance of integral measuring UV meters, Part A, 1-26, and Part AA, 1-36

DETECTORS FOR UV MEASUREMENT – STATE OF THE ART AND NEW DEVELOPMENTS

Anton Gugg-Helminger
Gigahertz-Optik D-82178 Puchheim/Munich Germany

INTRODUCTION

Different types of detectors are used for the measurement of ultraviolet radiation including silicon, SiC, GaP and GaAsP Photodiodes, photocells and photomultipliers as well as biological and chemical detectors. Spectral responsivities, non-linearities, noise, dark signals and stabilities are traits of the sensors. For the measurement of irradiance and exposure an input optic is placed in front of the sensor, i.e. fiber bundles, transmitting or reflecting diffusers or an integrating sphere for cosine correction and/or an imaging system, which will influence the performance of UV radiometers. To match the spectral shape of a given actinic spectra, different correction filters are used. For best measurement results a spectral mismatch correction factor is needed. State-of-the-art integral UV radiometers are dependent on the consideration of all of these elements.

UV Radiometers

Typically a UV meter consists of four items: an input optic, filter set, light to signal transfer device and a read-out unit.

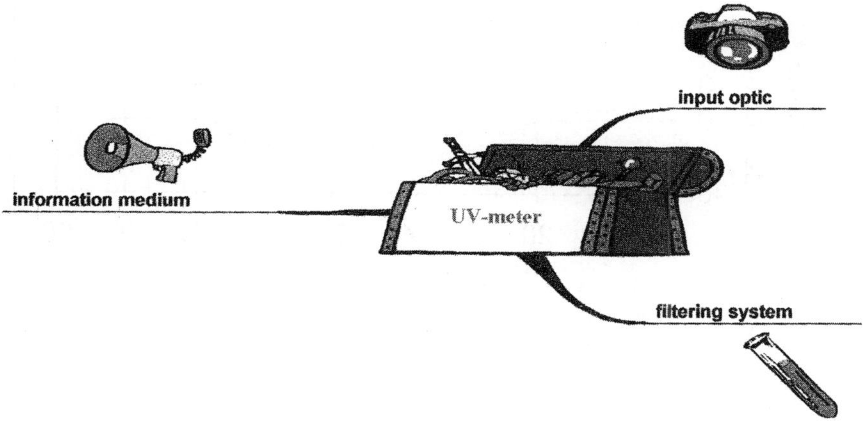

Figure 1. UV Radiometer

Input Optic

The input optic is made from different types of materials and designs.

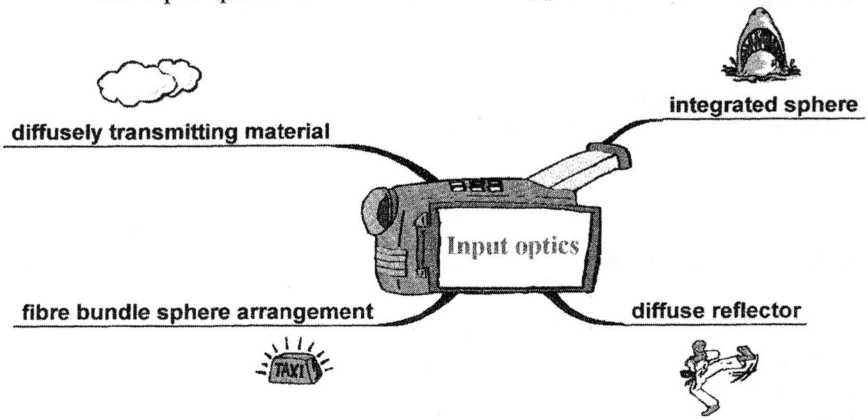

Figure 2. Input Optics

The integrating sphere is a well-known input device, which collects the entire incident light signal. Another type is a diffuse reflecting material or even a fiber bundle. The most commonly used input optic is the transmitting diffuser, which is small and can produce a good cosine correction. Good cosine correction is often neglected resulting in an additional problem when comparing measurement results. Up to 60 degrees, the cosine correction should have a maximum of 5% deviation from the ideal function.

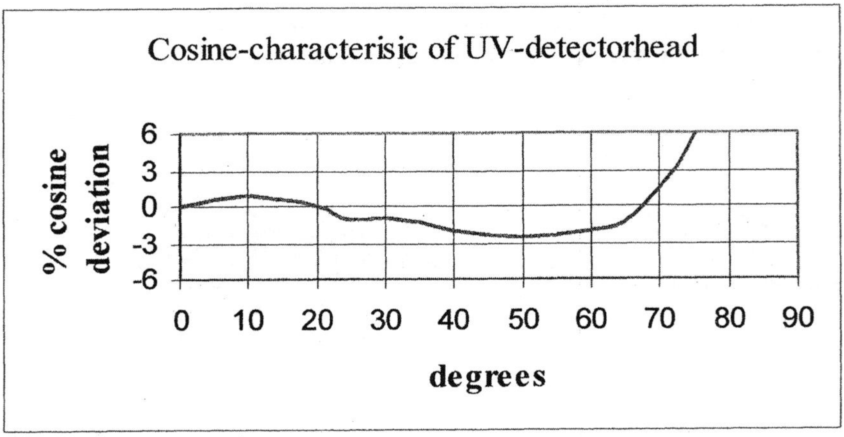

Figure 3. Cosine Characteristic

Filtering System

There are many different types of filtering systems in use. The choice of filter depends on the sensor selected for use.

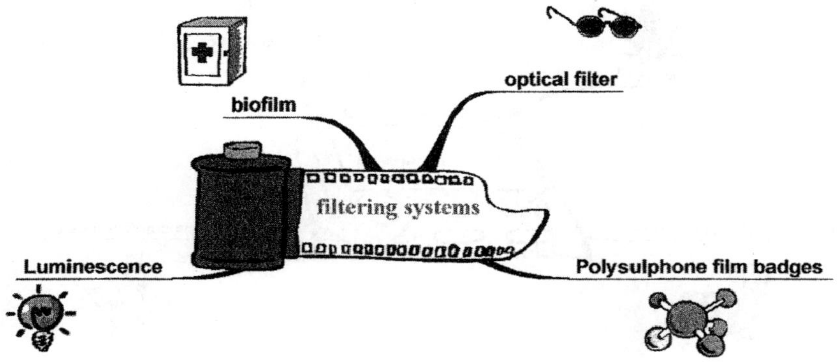

Figure 4. Filtering Systems

When using biologically effective elements, biofilm or polysulphone film badges may be chosen. Luminescence devices are used to change the incoming UV signal into other wavelengths (mostly into the visible). These "filters" are generally designed for a specific application. When trying to recreate a specific actinic spectral function, a set of different filters is Needed`.(1) Normally, color glasses and coated quartz plates are employed.

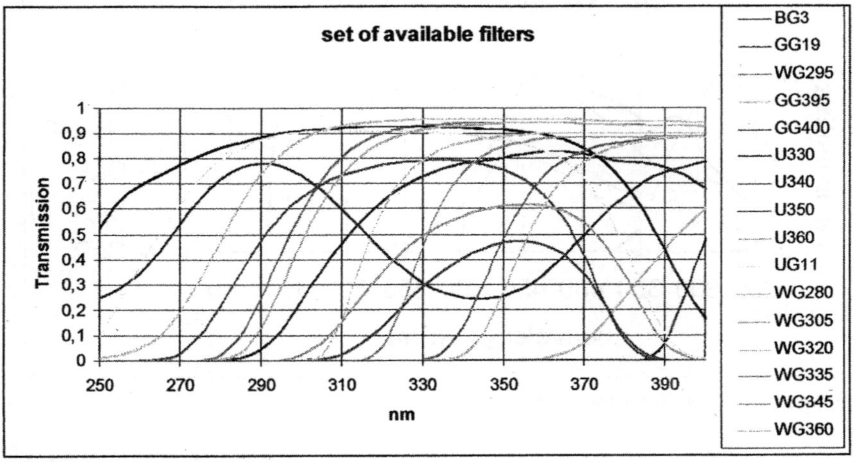

Figure 5.Set of Available Filters

Information Medium

Depending on the application, different units for changing the light into a readable signal are available.

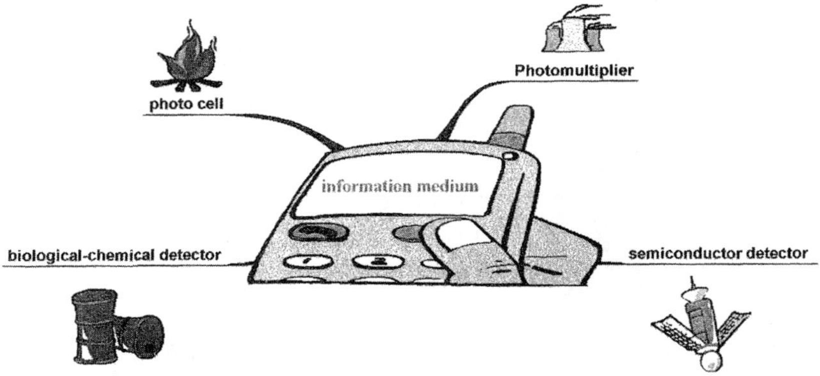

Figure 6. Information Medium

One of the oldest units in use is the photocell. For very low light levels photomultipliers are normally employed. Some actinic functions commonly rely on biological-chemical detectors. By far, most light measurement applications are covered by semiconductor detection units, which are available with different spectral responsivity functions in the UV-wavelength range.

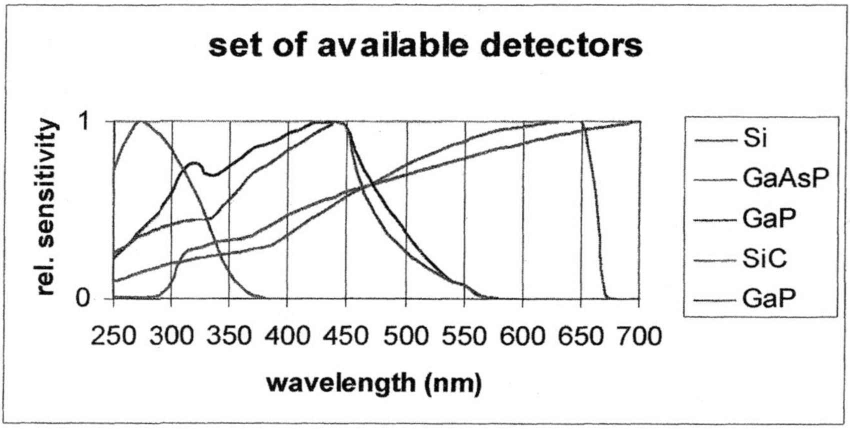

Figure 7. Semiconductor Detectors

In principle, it is possible to create the required and correct actinic functions with the different spectral sensitivities of these devices.

Other Effects of UV Radiometers

Some effects exist which should be realized when using UVdetectors. Depending on the semioconductor sensor unit, linearity is limited mostly to high signal levels. The sensors will become non-linear between 50 microamps and 2 milliamps. Noise and dark signal limit the low signal level sensitivity, which between 100 femtoamps and 100 picoamps.[2] However, this does not only depend on the sensor itself. The readout unit, bias voltage and the integrated circuit within the amplifier is often responsible for the minimum detection signal. All of these are fixed values in the UV meter but temperature is a variable, which will change the absolute values.
So the meter will have a temperature coefficient based on the materials used. It can be a positive or a negative or even a wavelength dependent temperature coefficient.

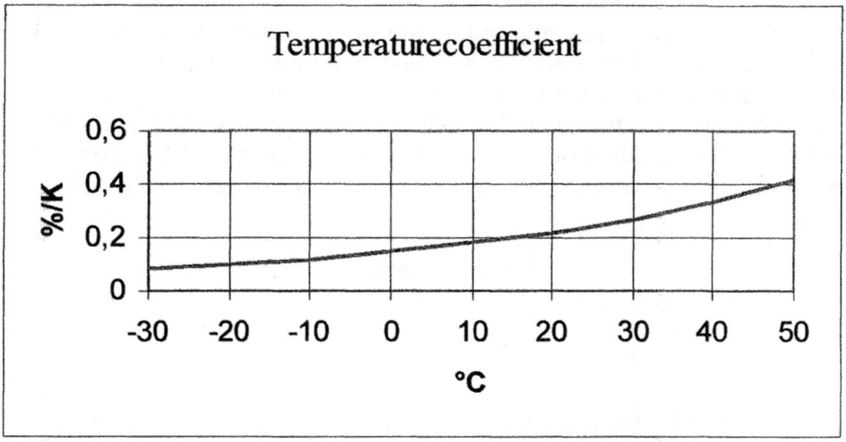

Figure 8. Temperature Coefficient of a Specific UV Detector

Spectral Wavelength Sensitivity

The most noticeable effect of all the possible uncertainties is the spectral matching of the detector compared to the desired function. Due to filter manufacturing limitations there will be never be 100% reproducibility. All optical glass filters will change either in their transmission or in the delivered thickness or both. Also, the interference principle may be used to create the desired shape. Shown in Figure 9, is a common batch of manufactured filters with a steep edge. The effects of a slight variation of only about 3 nm total in the shape of the filters are notably significant.

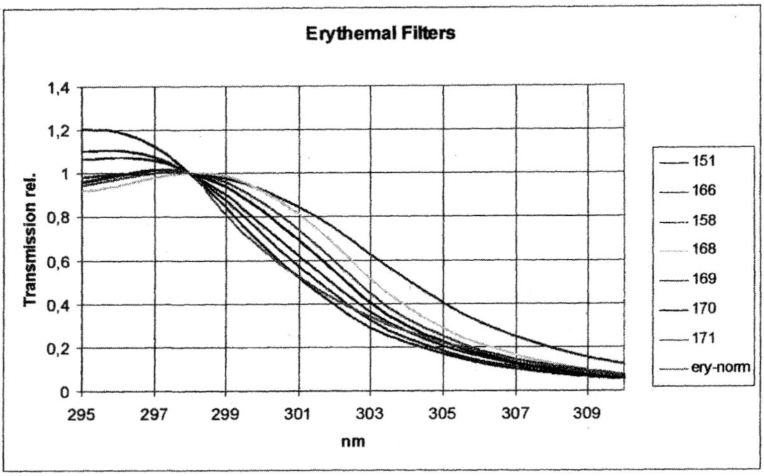

Figure 9. Batch of ErythemalFilters

In Figure 9., a linear scale of transmission versus wavelength is shown and all filters are within a 3 nm bandwidth. When looking at the logarithmic scale three things become very noticeable. First, one of the filter curves falls outside the others in the batch. Second, the transmission of one filter did not follow the desired curve below 1% of transmission and third, the shape of most of the filters seems to match the desired response curve very well.

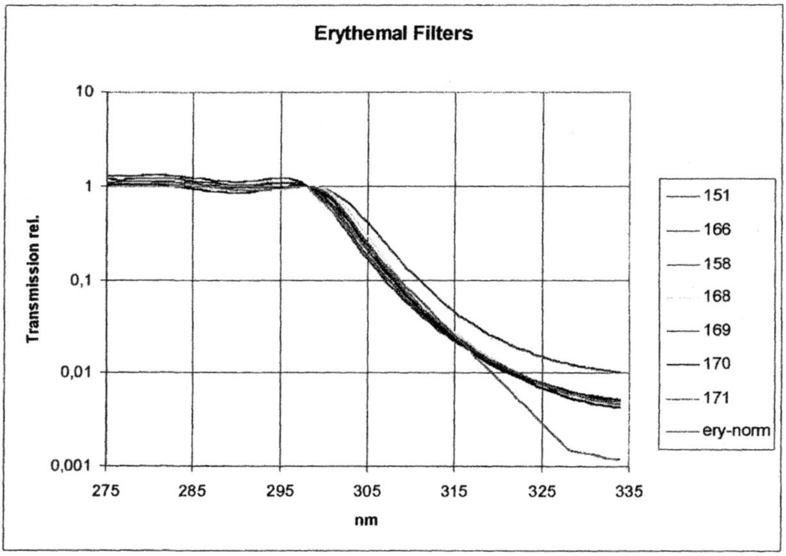

Figure 10. Same Batch of ErythemalFilters as inFigure 9. in Logarithmic Scale

The effects of the "good" batch with a 2 nm shift could be considered as negligible in light of other possible uncertainties. However, if we take into account some different light sources *(Figure 11)* [4] and calculate the expected uncertainty of the measurement, there is an additional uncertainty of 1.5 times or 50% away from the real value. *(Figure 12)*

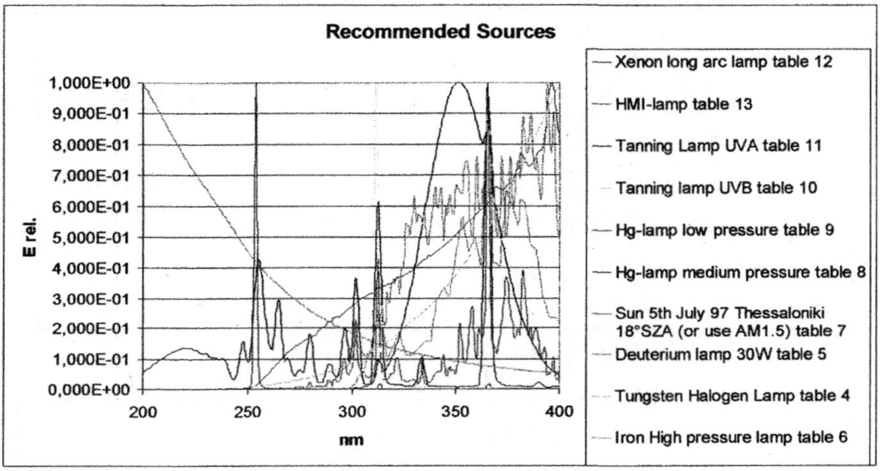

Figure 11. Recommended Light Sources

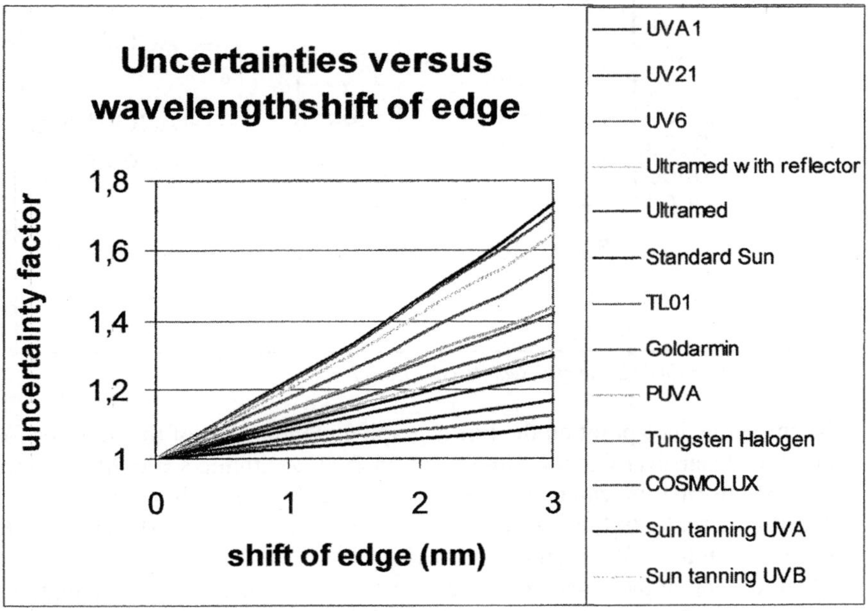

Figure 12. Wavelength Dependent Shift of the Edge of a batch of Erythemal Filters

A real detector matching the ACGIH function is shown in Figures 13 and 14, in both linear and logarithmic plots.

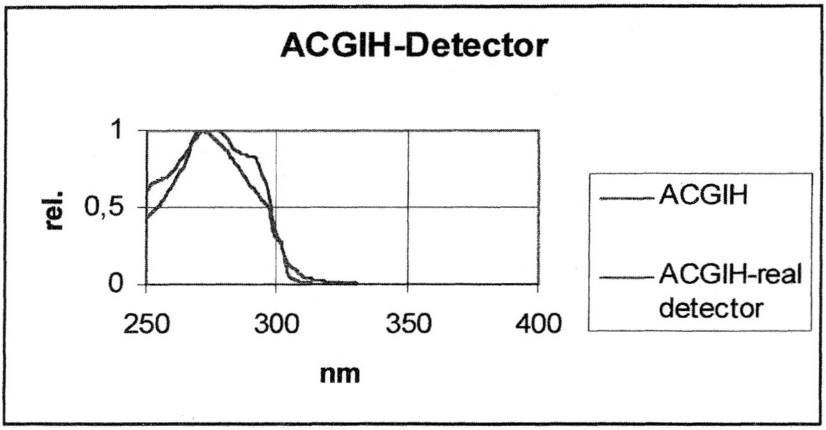

Figure 13. Linear Scale of a Real ACGIH Detector

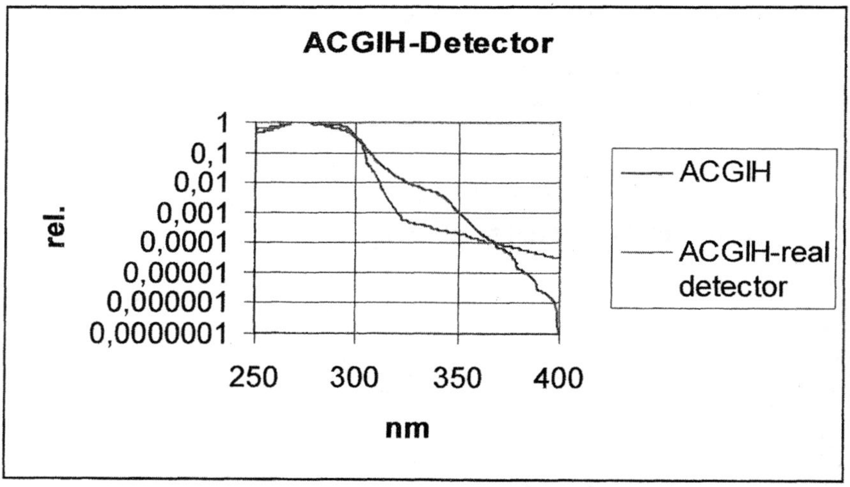

Figure 14. Logarithmic Scale of a Real ACGIH Detector

Compare the 50% point on the linear graph and the match of the detector for the critical sensitivity to the same point on the logarithmic scale. Much more detailed information can be seen about the outer band match on the logarithmic scale. In the specific cases of erythemal and ACGIH detectors the sensitivity in the UVA region is critical.

When looking at the linear scale the detector seems be a pretty good match. In the logarithmic scale a discrepancy in the UVA-region can be seen. Taking into account the many different possible light sources in use, a huge uncertainty in the measurement will occur.

When calculating for the detected ratio between UVA and UVB, a TL20 UVA light source for example, the ratio will be about 1% UVB.
When calculating the ratio between ACGIH weighted UVA and UVB for the same TL20 UVA light source, the ratio will be about 149% UVB!

This effect of the weighting function forces the spectral sensitivity below the 1% point to be taken into account.

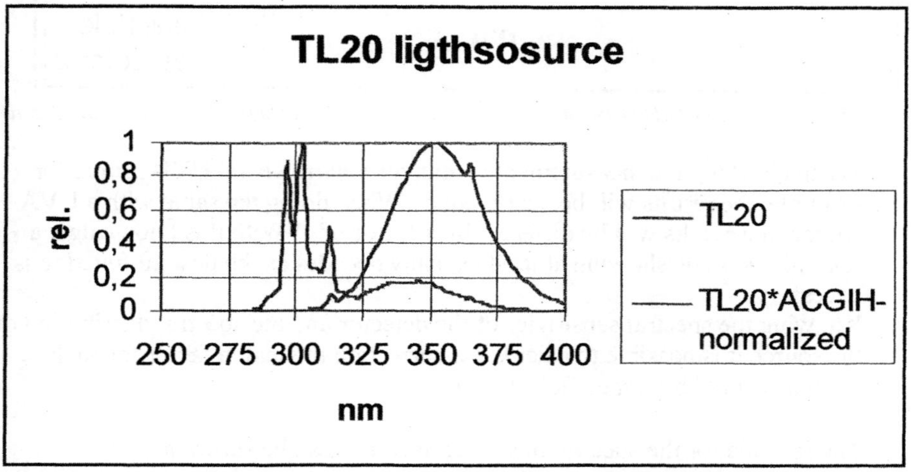

Figure 15. Comparison of Physical Function to Biological Weighted Function

When the logarithmic scale of the spectral sensitivity of a detector is not shown, the differences in readout can be shown by below example helping to understand the need to precisely know the spectral function.

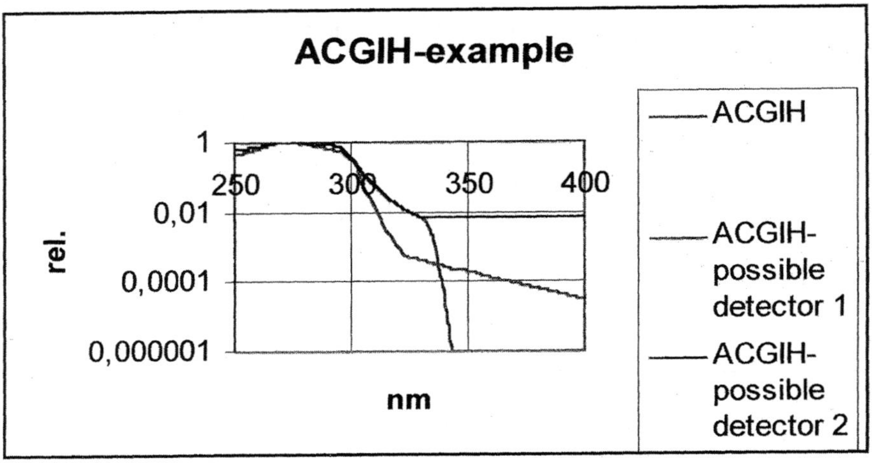

Figure 16. Real ACGIH Detector from Figure 13. with Flat and Steep Responses above 325 nm

When checking the measurement results for a narrowband UVB source for example, the results will be nearly equal. When doing the same with a UVA source, the results will be different by a factor of 4! With this knowledge, a logarithmic scale showing at least as many decades as the desired function is necessary.
Knowing the spectral sensitivity of the detector and the spectral distribution of the source it is possible to calculate a correction mismatch factor for each lamp measured by a specific detector.

The formula for the spectral mismatch factor [3] can be shown as:

$$a(Z_x) = \frac{\int_0^\infty E_{e,\lambda,Kal}(\lambda) \cdot s(\lambda)_{act,rel} \cdot d\lambda}{\int_0^\infty E_{e,\lambda,Kal}(\lambda) \cdot s(\lambda)_{rel} \cdot d\lambda} \cdot \frac{\int_0^\infty E_{e,\lambda,Z_x}(\lambda) \cdot s(\lambda)_{rel} \cdot d\lambda}{\int_0^\infty E_{e,\lambda,Z_x}(\lambda) \cdot s(\lambda)_{act,rel} \cdot d\lambda}$$

$a(Z_x)$ spectral mismatch correction factor for the test source Z_x
$E_{\lambda e,Kal}$ spectral irradiance of the calibration source at the reference plane
$s(\lambda)_{rel}$ relative spectral response of the radiometer head
$E_{e\lambda,Zx}$ spectral irradiance of the source Z_x at the reference plane
$s(\lambda)_{act,rel}$ relative spectral weighting function of specific actinic effect

$$Y = \frac{Y_{Z_x}}{a(Z_x)}$$

Y	correct value
Y_{Zx}	radiometer reading with source Z_x
$a(Z_x)$	according to above equation

When using these formulas and if all spectral functions are known, it will be easy to calculate correction factors for most of light sources.

This factor a(Z) will be an indicator of the quality of each detector head used for UV measurements.

Spectral mismatch correction factor:

$$f_1 = a(Z_x)-1$$

When using all these features properly, good UV detectors can be manufactured. The values which should be take into account are spectral mismatch, the short- and long wavelength response, the directional and the spatial response as the most sensitive units effecting measurement uncertainties

Properties of weighted function radiometers for uv irradiance measurement:

Properties	**Symbol**	**State of the Art Values**
Spectral mismatch correction factor	f_1	10%
Short wavelength response	u	5%
Long wavelength Response	r	5%
Directional response	f_2	3%
Linearity	f_3	0,2%
Display unit	f_4	0,1%
Fatigue	f_5	0,2%
Temperature	f_6	0,2%
Modulated radiation	f7	0,1%
Polarization	f8	0,1%
Spatial responsivity	f9	25%
Range change	f11	0,1%

Table 1. Properties of Weighted Functions of UV Radiometers for UV Irradiance Measurement

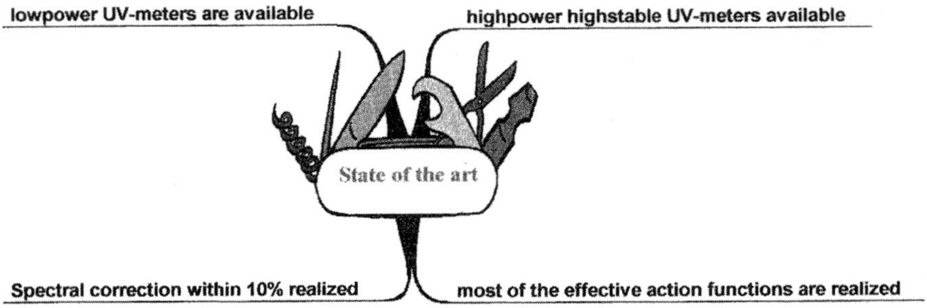

Figure 17 State of the Art Availability

As mentioned previously, possible additional uncertainties arise in the determination of UVA signal from narrow band UVB light sources. Therefore radiometers with two detector heads are required for proper measurement. One detector is for the UVB with an additional narrowband calibration value and a UVA detector.

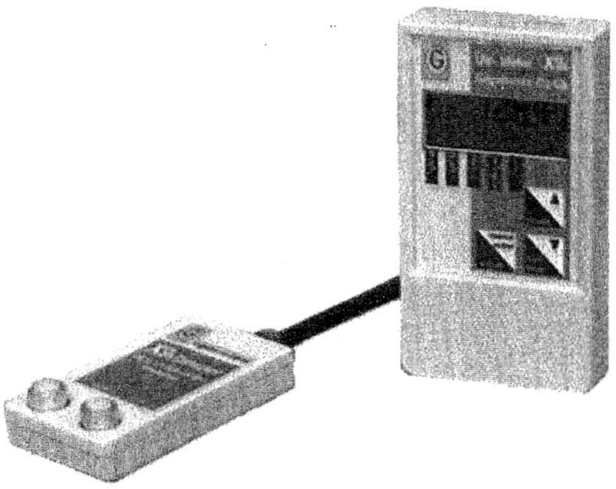

Figure 18. UVB/UVA Radiometer with Two Separate Detectors

With this "twin" radiometer the measured values in the target wavelength range are much more reliably interpreted than measurements made in only one of the wavelength regions.

The same arrangement is available with ACGIH and UVA detectors or erythemal and UVA detectors.
Any combination of these detectors is possible based on the UV source to be tested.

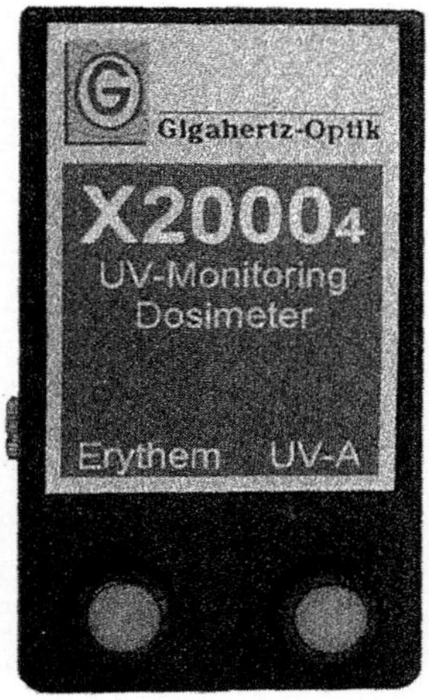

Figure 19 .Erythemal or ACGIH Detector Combined with UVA Detector

This unit can be used to check UVA irradiance to roughly calculate the erythemal or ACGIH effective irradiance in the UVA region. Using these results the value of the ACGIH sensor could easily be specified as a real value or as an artifact of the imperfectly matched UVA region.
Also, irradiance is logged, e.g. every second, to provide more information about time dependent irradiance as well as the dose.

Conclusions

Most of the action spectras are roughly realized. Beside this increase of low uncertainty devices broadband radiometric measurement inaccuracies can be improved with known spectral responsivity of the real detectorunit within the sensor sensitivity wavelenghtrange, known spectral distribution of the measured ligthsource, known calibration source, known of correct actinic function and as conclusion the calculated spectral mismatch correction factor for each recommended light source. Meaningful measurement required therefore a proper selection of an UV-meter and the relevancy of each specific measurement within the field of health hazard.

Referencies:

1. Schott Filter and Hoya Filter catalog 1999
2. Hamamatsu Photodiodes catalog 1999
3. UV news6, Characterizing the performance of integral measuring UV-meters, p. A-11
4. UV news6, Characterizing the performance of integral measuring UV-meters, p. AA-9

MEASUREMENT OF PHOTOBIOLOGICAL EXPOSURE OF OCULAR TISSUES

David H. Sliney, Ph.D.
US Army Center for Health Promotion and Preventive Medicine, Aberdeen Proving Ground, MD, USA

ABSTRACT

Because of a number of geometrical shading factors, such as the eyelids and brow ridge, the human eye is actually exposed to a very small fraction of incident ultraviolet radiation (UVR). The retina is exposed to visible light and some IR-A radiant energy within an imaged scene. These geometrical and imaging factors challenge the task of attempting to accurately measure the photobiologically significant exposure of the cornea, lens and retina to ultraviolet, visible and infrared radiation.

The lid opening varies with ambient scene luminance (brightness). It is possible to mathematically predict the opening of the lids and the angular field-of-view from our studies of lid opening. This aids in setting the field-of-view of measurement instruments designed to measure the UVR exposure dose to the cornea and lens. We have measured the UV exposure to the anterior segment of the eye as would occur when a person wears different types of sunglasses. The exposure is greatly affected by the type of sunglass frame and partially to the UV transmittance of the sunglass lenses. In some instances, UV exposures of some specific ocular tissues can actually equal or exceed those when not wearing sunglasses.

The light exposure to the retina is not at all uniform in outdoor daylight conditions. The central and superior regions of the retina receive much more light than the inferior retina. Thus instrumentation should simulate these geometrical factors.

Instrumentation designed to determine the photobiological dose to ocular tissues must therefore have acceptance field-of-views that mimic those of the human eye.

KEYWORDS: Eye, Ultraviolet Radiation, Radiometry, Light Measurement, Cataract,

DISCLAIMER: The opinions or assertions herein are those of the author and should not be construed as reflecting official positions of the US Department of the Army or Department of Defense.

INTRODUCTION

The previous presentations have explained the importance of many instrumentation characteristics. These include detector calibration, spatial response characteristics and spectral matching. The measurement of ocular exposure provides an excellent illustration of the great difficulties of providing an actual photobiological exposure dose for the complex organ of sight. As we shall see, the geometry of exposure becomes as important as spectral and temporal information. Ocular exposure to environmental ultraviolet radiation (UVR) has been of interest in terms of prescribing better eye protection, evaluating the potential hazards of environmental light exposure of the eye, and for setting human exposure limits for outdoor workers.

One objective of our laboratory at the US Army Center for Health Promotion and Preventive Medicine (USACHPPM) has been to determine the most effective protective measures for sunlight exposure to the eye and skin as well as to recommend human exposure limits to UVR and intense light. To that end, it has been necessary to characterize the primary ocular exposure to UVR and light, emphasizing the importance of geometrical exposure factors by measuring UVR and blue-light distributions in skylight under different environmental conditions. We have also studied the impact of lid opening upon optical radiation exposure dose to the cornea, lens and retina, and how scene luminance (brightness) affects the degree of lid opening and ocular exposure in outdoor subjects. These studies have shown the importance of good ocular dosimetry in any epidemiological study of the relationship between sunlight exposure and ocular disease.

In concert with the American Conference of Governmental Industrial Hygienists (ACGIH), we developed human exposure limits based upon an envelope action spectrum for a just-detectable photokeratitis (ACGIH, 1992; Sliney, 1972, 2000). These limits were carefully reviewed and adopted with slight modifications by the International Commission on Non-Ionizing Radiation Protection (ICNIRP, 1996; Duchene, 1991). This protection guideline applies to human exposure to solar UVR and to the UVR from artificial light sources. The exposure limit requires a spectral weighting by a hazard spectrum $S_{UV}(\lambda)$ for UVR incident upon the eye or skin. A number of field survey measuring instruments have been developed which employ detectors that match the $S_{UV}(\lambda)$ action spectrum as shown in Figure 1 (See Annex Table for listed values). However, as will be shown later, the geometry of the measurement is also of enormous importance when assessing risk of UV exposure to the eye. Outdoor safety assessments and epidemiologic studies can arrive at erroneous conclusions if measurements ignore geometrical factors and epidemiological assignments of exposure are seriously in error. A number of "reasonable" assumptions previously made by some epidemiologists regarding relative exposures have been shown to be false (Sliney, 1992; Sliney, 1998).

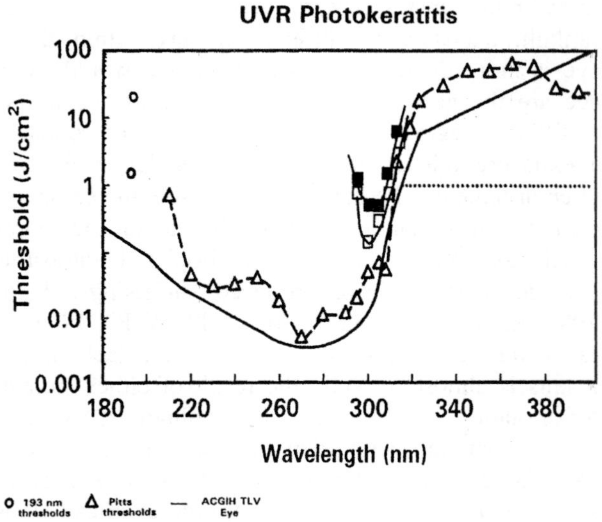

Figure 1. *The solid line shows the exposure limit, expressed as the reciprocal of the spectral response curve $S_{UV}(\lambda)$ compared to the photobiological action spectra (not normalized) for erythema, and photokeratitis expressed as thresholds. The exposure limits for ultraviolet radiation were extended into the UV-A region in 1985. Open triangles are the threshold data for photokeratitis. Solid squares are the data for permanent cataract, and open squares are for reversible cataract.*

Human exposure to sunlight, particularly the UVB component, is believed to be associated with a variety of eye disorders, including damage to the cornea, lens and retina (WHO, 1994). Photokeratitis (snowblindness) is clearly related to UVR exposure, whereas, cataracts (opacities of the lens) are the most frequently cited delayed consequence. Certain keratopathies (corneal degenerations) and pterygium (a fleshy growth on the conjunctiva) also are believed to result from excessive UVB exposures. Cataracts are a major cause of blindness in both developed and developing countries. However, the relative importances of different wavelengths in cataractogenesis, as well as the dose response curve, remain uncertain.

ENVIRONMENTAL EXPOSURE

The geographical variation in the incidence of age-related ocular changes such as presbyopia and cataracts and diseases such as pterygium and droplet keratopathies have led to theories pointing to sunlight, ultraviolet radiation (UVR) exposure and ambient temperature as potential etiological factors (WHO, 1995; Sasaki, 1999; Sliney, 1986). Some epidemiological evidence also points to an association of age-related macular degeneration (AMD) to sunlight (particularly blue-light) exposure (West, 1990??). The actual distribution of sunlight exposure of different tissues within the anterior segment of the eye is more difficult to assess than one would expect. Of greatest importance are the geometrical factors that influence the selective UVR exposure to different segments of the lens, cornea and retina. Studies show that sunlight exposure to local areas of the cornea, lens and retina varies greatly in different environments. Perhaps for this reason, the epidemiological studies of the potential role of environmental UVR in the development of ocular diseases such as cataract, pterygium, droplet keratopathies and age-related macular degeneration have produced surprisingly inconsistent findings (Dolin, 1997; Klein, 1992; and Taylor, 1989; WHO, 1995). All of these ocular diseases vary geographically; however, the lack of consistent epidemiological results is almost certainly the result of either incomplete or erroneous estimates of outdoor UVR exposure dose (Sliney, 1986, 1992, 1995, 1996, 2000, 2001). Geometrical factors dominate the determination of UVR exposure of the eye. The degree of lid opening limits ocular exposure to rays entering at angles near the horizon. Clouds redistribute overhead UVR to the horizon sky. Mountains, trees and building shield the eye from direct sky exposure. Most ground surfaces reflect only a small fraction of incident UVR. The result is that the highest UVR exposure appears to occur during light overcast where the horizon is visible and ground surface reflection is high. By contrast, exposure in a high mountain valley (lower ambient temperature) with green foliage results in a much lower ocular dose. Other findings of these studies show that retinal exposure to light and UVR in daylight occurs largely in the superior retina (Sliney, 1998).

Epidemiological Evidence

So why is it important to study the spectral and geometrical exposure of the human eye from solar UVR? It has long been suggested that the great latitudinal variation of "the world's most blinding disease," cataract (WHO, 1995), is strongly suggestive (as in the case of skin cancers) to result from UVR. WHO estimates that the worldwide prevalence of cataract exceeds 50 million (WHO, 1995). Indeed, if this dependence is examined along with a wide variety of laboratory studies, suggests that environmental factors surely plays a major role in the time of onset of lenticular opacities. However, most epidemiological evidence points to UVR in sunlight as a significant risk factor only in one type of cataract, i.e., cortical cataract (Taylor et al., 1988). The evidence for UVR as an etiological factor in at least two adverse changes in the cornea, i.e., droplet keratopathies and pterygium is much stronger (WHO, 1995; Taylor, 1999). Furthermore, the environmental studies of Sasaki (1999) have provided very important insights into the geographical variations in the incidence of cataract by examining the different types of cataract characteristic of different latitudes (Sasaki, 1999). Nuclear cataract was shown to be more common in the tropics; cortical cataracts, more common in mid-latitudes, and posterior subcapsular cataracts were not so clearly related to latitude. Despite this latitudinal variation, some other published epidemiological studies do not appear to show a relation between UVR and cataract. Nevertheless, a wide variety of scientific evidence, from laboratory studies of the UV photochemistry of lens proteins to a number of different animal exposure studies all provide support for the hypothesis that UVR should play a far greater role in cataractogenesis. Most age-related changes in the skin (from accelerated aging to skin cancer) have been conclusively shown to result from excessive exposure to solar UVR (or "sunlight exposure"). Although no one questions that UVR exposure produces the acute effects of "sunburn" (erythema) and snowblindness (photokeratitis), some have questioned whether pterygium and droplet keratopathies are clearly related to UVR exposure (Dolin, 1997; WHO, 1995). Even more under debate are theories that suggest that UVR and light may affect retinal diseases such as age-related macular degeneration (AMD) (WHO, 1995). A better resolution of these questions requires far better ocular dosimetry.

With the strong geographical variations in the incidence of nuclear cataract (Sasaki, 1999), it is surprising that most epidemiological studies show only weak or no apparent relation between UV or sunlight exposure and the incidence of nuclear cataract (Taylor, 1989). Perhaps it is necessary to consider the other important environmental parameter--environmental temperature. Both sunlight and ambient temperature have been cited as potential etiological factors in several age-related ocular diseases (Sliney, 1986). However, the combined role of these physical factors and their possible synergisms has generally not been carefully examined. These environmental co-factors, together with geometrical factors, have recently been examined to give some suggestions of the etiology of these ocular diseases (Sliney, 2001).

The Challenge of Measuring Actinic UVR in Sunlight

Both the *quantity* (irradiance) and *quality* (spectrum) of terrestrial ultraviolet radiation varies with the *solar zenith angle (Z)*, i.e., the angular position of the sun below the zenith (where $Z = 90^\circ -$ elevation angle above the horizon). The sun's position varies with time-of-day, the day of the year, and latitude. This variation is particularly striking in the UV-B spectral region (280-315 nm) because of the greater atmospheric attenuation along the direct atmospheric path. Stratospheric ozone absorption and molecular (Rayleigh) scattering by atmospheric N_2 and O_2 combine to attenuate the global UVR in this spectral region. However, in the troposphere, further absorption by air pollutants such as the oxides of nitrogen, sulfur-dioxide and ozone, and Mie scattering by water vapor in clouds and particulates can significantly add to the attenuation. Clouds and haze reduce the global UV (ground, horizontal) irradiance. However, haze and clouds redistribute the UVR, so that the UVR (i.e., the sky "brightness," or more correctly, radiance) of the horizon-sky can actually increase in comparison to a clear, blue-sky day. Since water vapor in clouds greatly attenuates infrared radiation, but does not significantly attenuate ground-level UVR, the warning sensation of heat to reduce the risk of sunburn can be absent on an overcast day. A light overcast, or light clouds scattered over a blue sky, do little to attenuate the UV global irradiance (unless a dense cloud lies directly over the sun), and many severe sunburns occur at the beach under such conditions. A light cloud cover may reduce the terrestrial global UVR (measured on a horizontal surface) to about half of that from a clear sky, although the horizon-sky UVR radiance does not decrease—and can even increase. Even under a heavy cloud-cover the scattered ultraviolet component of sunlight (termed the diffuse component," or "skylight") is seldom less than 10% of that under clear sky (Sliney, 1980). Only very heavy storm clouds can virtually eliminate terrestrial UV during summertime conditions. Although a higher global erythemal effective irradiance is measured at high altitude (Blumthaler, 1996), the atmospheric scatter (diffuse component) is less, and the horizon-sky UVR does not increase by climbing mountains (Sliney, 1986). All of these observations point to the complex geometrical factors that challenge any outdoor UVR measurements.

Since the eye receives most of its outdoor UVR exposure from ground reflectance, the proper measurement of reflected sunlight adds a further challenge. The ground reflection of solar UV-B radiation varies over 100-fold. Green grass reflects less than 1% and most artificial surfaces and rock normally reflects less than 10%. Two exceptions are ocean surf and white, gypsum sand, which reflect about 20-25%. Fresh snow reflects about 85%, thus producing "snow blindness." Open water reflects the entire sky (diffuse plus direct component) and can be of the order of 20%, although the direct, specular reflection from the sun's image is only about 2-3% of the incident, direct UV radiation. Table 1 provides more detailed values for ground reflectance

Terrain surfaces	Diffuse reflectance effective solar UV-B (ACGIH/ICNIRP spectral weighting)
Green mountain grassland	0.8 - 1.6 %
Dry grassland	2.0 - 3.7 %
Wooden boat dock	6.4 %
Black asphalt	5 - 9 %
Concrete pavement	8 - 12 %
Atlantic beach sand (dry)	15 - 18 %
Atlantic beach sand (wet)	7%
Sea foam (surf)	25 - 30 %
Dirty snow	59%
Fresh snow	88%

***Table 1**: Reflectance of ACGIH/ICNIRP-effective solar UV-B from terrain surfaces (Source: Sliney, 1983)*

Ocular exposure is far more affected than skin exposure by these geometrical factors. For an industrial UV source such as a welding arc, the cornea is shown to be more sensitive to UVR injury than the skin, but photokeratitis ("welder's flash" or "snow-blindness") seldom accompanies summer sunburn of the skin. This seeming paradox is explained by the fact that people do not look directly overhead when the sun is very hazardous to view, whilst most people may stare at the sun when it is comfortable to observe near the horizon. Fortunately, at sunset, the filtering of UVR and blue light by the atmosphere allows us to directly view the sun. When the solar elevation angle exceeds 10 degrees above the horizon, strong squinting is observed which effectively shields the cornea and retina from most direct exposure. These factors reduce the exposure of the cornea to a maximum 5% of that falling on the exposed top of the head. However, if the ground reflectance exceeds 15%, photokeratitis may be produced after a few hours exposure. If one were to ignore the squinting factor and proper instrument field-of-view, the photokeratitis threshold would be achieved in less than 15 minutes for midday summer sunlight!

When wearing sunglasses, the pupil dilates proportionally to the darkness of the sunglasses (Deaver, 1996). Coroneo et al (Coroneo, 1991; Coroneo, 1993) have shown that very oblique temporal rays can be refracted into the critical nasal equatorial region of the lens and this could explain the increased incidence of opacification originating in the nasal sector of the lens in cortical cataract. See Figure 2. The protective value of upper and lower lids, when they close down during squinting, determines the ocular UVR exposure dose in different environments. A brimmed hat or other headwear--associated with or without dark sunglasses--will modify greatly the UVR exposure dose. The geometrical factors that should be modeled by a radiometer now begin to appear as almost insurmountable. What can be done? Even geometrical positioning of the body greatly affects the solar exposure of human skin. The skin surfaces of the nose, ears, forehead and shoulders, can receive normally incident UV-B from the overhead sun, but behavior also influences this exposure (Diffey, 1977, 1979; Gies, 1995; Larko, 1983). The back always experiences greater UVR exposure thant the chest because of sun-avoidance behavior.

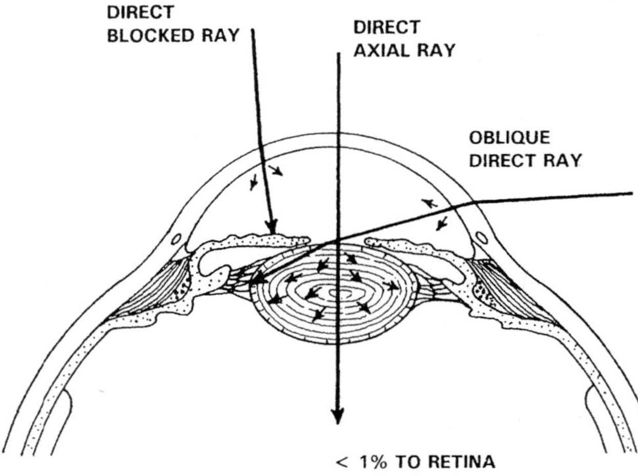

Figure 2. The Coroneo Effect. *Very oblique UV rays can reach the lens, even if spectacle or sunglass lenses block direct UVR.*

Biologically Relevant Measurements

With skin and ocular exposure so dependent upon geometry, the measurement challenge is not only to employ detectors with a spectral response that matches the action spectrum for biological risk, but also to design portable UV radiometric instruments ("safety meters") and personal dosimeters, such as film dosimeters that take into account at least some biologically relevant geometrical factors. Although spectroradiometers can be used to measure the biological hazard of fixed lamps and arcs, small meters or film badge dosimeters are generally needed to assess the UV risk in the workplace.

Two broad categories of detector systems have been developed and employed for the above health assessment goals. We have heard to this point in this symposium about **broadband electro-optic UVR detectors**. To assess the human health aspect of UVR exposure, UV-B or actinic detectors having a spectral response that mimics the UV photokeratitis action spectrum, which is effectively the ICNIRP/ACGIH UV hazard action spectrum $S_{UV}(\lambda)$, or the erythemal action spectrum. The spectral response of these UV "biometers" can be calibrated to provide an approximate measure of the erythemal effective irradiance or the UV hazard irradiance, since the action spectra are very similar within the solar spectrum from about 300 nm to at least 330 nm. In addition, a number of sensitive film dosimeters, which respond strongly to UV-B radiation with a small response in the UV-A. Fitting the appropriate action spectrum in the UV-A spectral region is important, since the UV-A in sunlight still contributes approximately 10-30 % of the total effective (summation) UV exposure dose.

Personal Ultraviolet Dosimeters

Although both fixed and portable broadband UV radiometers and spectroradiometers can be used to check and monitor ambient ultraviolet (UV) radiation either from the sun, open arcs or lamps, it is difficult to employ these instruments to measure the UV exposure of individuals. It is frequently interesting for epidemiological and occupational health studies to measure the localized UV exposures at several sites over the body, but most current radiometric instruments are too bulky and currently too costly for this use. Ideally, ultraviolet dosimeters designed for personal use should be compact, easy to read out, and not restrict normal movements of the wearers. The spectral sensitivity of the dosimeter ideally should match the action spectrum of the photobiological effect being monitored. The measured physical, chemical or biological change in the dosimeter should exhibit linearity with increasing UV exposure dose, and the dosimeter should show "photoaddition" (i.e., each wavelength acts independently, and a broad spectrum of radiation is the sum of the effects of all incident wavelengths) and not exhibit reciprocity failure (i.e., the measure of dose is independent of dose rate). Photoaddition can be difficult to achieve in photochemistry, in that a single chromophore (target molecule) would have to respond and yet respond as DNA or other biologically relevant molecules. The dosimeter response should not vary with changes in temperature and humidity; it should exhibit no 'dark response' under conditions of non-exposure; and it should show stability during its storage life. The dosimeter should not require laborious processing and should be easy to convert the physical, chemical or biological response to a measure of ultraviolet exposure dose. The dosimeter cost should be sufficiently low for extensive studies or even routine monitoring (Davies, 1976; Diffey, 1989;). To date, none of the available materials meets all of these ideals, but two types have been extensively investigated (polysulfone films and biological films). The biggest problem with most of these dosimeters is that they are not direct reading, and must be taken to a laboratory to read. However, several types of photosensitive exposure-rate cards have been developed that indicate approximate UVR dose-rate by a change in color scale; these have been given to the public for UV skin-cancer awareness campaigns.

Polysulfone Film Badge Dosimeters

To date, the most widely used UV film-badge dosimeter has been the thermoplastic film, polysulphone, as pioneered by Davies (1976). When polysulfone is exposed to UVR (particularly UVB), its UV absorption (generally calibrated at 330 nm) increases. This increase in absorbance has a useful linear-response range. Most typically, a 40- m thick piece of polysulphone film is mounted in a cardboard or plastic mount, and these are worn by the experimental subjects on anatomic sites of interest. Studies of

UV exposure have employed this type of film-badge dosimetry to estimate the solar exposure of children, or of adults taking part in different leisure pursuits or from different occupations—both indoors and outdoors. However, these are not used for routine worker exposure monitoring. The spectral response of polysulphone film tends to over-respond to UV-A compared to most action spectra of interest, but the collected data are claimed to be as reliable as to be to within ±15% (Diffey, 1989; Gies, 1995).

Biological Film Badges

Several biological molecules, viruses, and spores have been used in a variety of "bio-sensors." One of the most studied has been the bacteriophage T7, which has been used to monitor ambient UV radiation (Ronto, 1992, 1994), and when combined with an appropriate optical filter, a spectral response similar to the action spectrum for erythema in human skin can be achieved (Ronto, 1994). These types of dosimeters generally have spectral response functions that more closely follow the action spectrum for photokeratitis than erythema, since the living cells of the corneal epithelium face the environment without any shielding by the skin's stratum corneum. Intercomparisons of different types of bio-dosimeters have been made (Horneck, 1976; Munakata, 2000).

Direct Clinical Detection of UVR Exposure

The action spectrum and threshold for photokeratitis (snow-blindness) has been carefully studied, and was the basis of the action spectrum $S_{UV}(\lambda)$, hence, where individuals develop a threshold, just-detectable photokeratitis, the cornea itself is acting as a dosimeter. Sliney used this approach to estimate UV exposure to the crystalline lens or intra-ocular lens implant, and showed it to be an extremely small dose compared to that of the skin (Sliney, 1987).

Erythema has long been used as a measure of individual, acute exposure of the skin to UVR. The Minimum Erythemal Dose (MED) is a measure that varies with each individual, and this has been used to determine the action spectrum of erythema and actual exposure dose. To estimate chronic exposure, Diffey and colleagues used cutaneous microtopography to objectively measure the fine wrinkling of skin surfaces to indicate the level of sun-damaged skin. They hypothesized that this measure was a surrogate for skin cancer, and conducted controlled studies (Diffey, 1977, 1979, 1989) to show a good correlation. Dermatologists routinely examine benign markers of sun damage of skin on the face and on the back of the hands to judge the accelerated aging of the skin, and thereby estimate the risk for non-melanoma skin cancer.

Examples of Outdoor Measurements

There have been several research efforts to employ film-badge dosimeters on mannequins and on people to better understand the distribution of solar UVR exposure of different parts of the face and upper body. For example, Table 2 compares the mean percentage of ambient ultraviolet radiation relative to the top of the head ("vertex") received at different anatomical sites as measured on rotating mannequins and living subjects (in various outdoor activities). Not surprisingly, the shoulders generally receive the highest relative exposure (approximately two-thirds of the top of the global irradiance). The studies confirmed the large variation of exposures to different skin sites with differing postures and working activities. Because mannequins do not really behave as humans in sunlight, there were signifiant variations between the human and mannequin dosimetric studies. Several investigators have used UV film badges to measure solar UV radiation exposure on the face (Melville et al 1991; Rosenthal et al 1990; Holman et al 1983; Rosenthal et al 1991), and mannequins (Diffey et al 1977; Diffey et al 1979; Gies et al 1988; Gies et al 1992; Diffey & Cheeseman 1992; Airey et al 1995; Kimlin et al 1998). There was a divergence in the collected data reflecting factors such as season (affecting the range of solar zenith angles), the behavior of individuals and the influence of shading structures and vegetation. Some representative exposure values for different sites on the head are give in Table 2. As one would expect, there was considerable variation in relative exposure due to posture and head tilt. Airey et al (1995) showed the exposure on the nose relative to the horizontal dropped from 59% to 11% as the head tilted from 0° to 60° to normal. Hats (particularly with wide brims)and other headwear greatly affect exposure to the face(Diffey & Cheeseman 1992; Wong et al 1996); however, hats with shade the eyes, may actually lead to greater lid opening, rendering the eye more vulnerable to ground reflections (Sliney, 1999).

Site	Mannequin		Living subjects	
	Diffey et al 1977	Gies et al 1992	Holman et al 1983	Herlihy et al 1994
Cheek	31	24	15	13
Hand	50	-	24	29
Shoulder	75	94	66	43
Back	43	36	58	40
Chest	68	50	44	23
Thigh	33	-	16	25

Table 2 Sunlight exposure (relative to 100% on vertex) for rotating mannequins and living subjects engaged in outdoor activities (Source: Gies H.P., Roy C.R., Toomey S., MacLennan R., Watson M., 1995, Solar UVR exposures of three groups of outdoor workers on the Sunshine Coast, Queensland, Photochem. Photobiol., 62, 1015-1021).

Ocular Exposure Dosimetry

Very pertinent data could be given by UV-sensitive contact lens dosimeters (Sydenham, 1995), but most of the pertinent data will come from studies of lid opening conditions combined with directional field measurements and different environmental conditions. Also the ground reflection will enter in these evaluations. Tables 1 and 3 from Sliney (1995) provide information on the reflectance of ACGIH/ICNIRP-weighted solar UVB and the measured ACGIH/ICNIRP effective UVR from the sky with a 40° cone field-of-view. He effectively measured the UV radiance over a 40° averaging (acceptance) angle. This was later used to calculate ocular dose. Note that the relative effective UV irradiance near the horizon (i.e., within the eye's field of view) did not show big variations if the sky was visible and could increase with haze. Indeed, on an overcast day with the eyelids more open, the actual UV-B dose rate to the eye from the sky scatter can increase. Figure 3 shows a pair of Solar Light Biometers, one of which we modified to measure only the horizon sky. The ratio of the two measurements is not at all constant, and the horizon sky measurements show far less change with the hour of the day than do the global measurements.

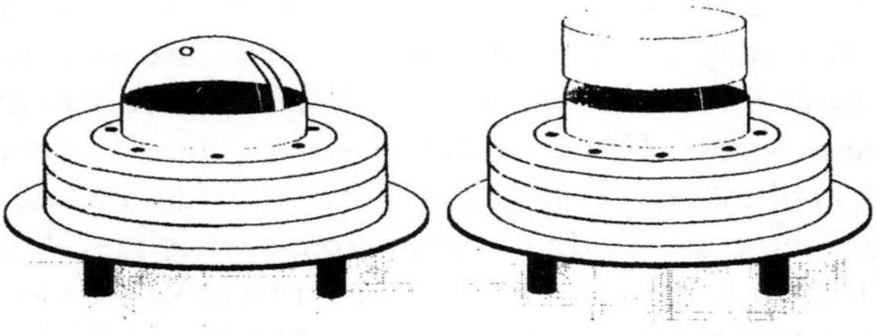

Figure 3. Two Solar Light, Inc. Biometers. One has been capped to measure only the horizon-sky UVR which exposes the eye.

Sky conditions location, elevation	Zenith reading ($\mu Wcm^{-2} sr^{-1}$)	Directly at sun ($\mu Wcm^{-2} sr^{-1}$)		Opposite sun ($\mu Wcm^{-2} sr^{-1}$)	Horizon sky ($\mu Wcm^{-2} sr^{-1}$)
Clear sky, dry, sea level	0.1	1.4	Z=70°	0.22	0.27
Clear sky, humid, sea level	0.27	4.1	Z=50°	0.27	0.24
Ground fog, sea level	0.04	0.19	Z=75°	0.04	0.03
Hazy humid, sea level	0.014	1.4	Z=70°	0.22	0.54
Cloudy bright, 700 m	0.54	0.44	Z=45°	0.27	0.05 (tree-line)
Hazy beach	0.54	0.60	Z=75°	0.54	0.60
Hazy beach	0.38	3.5	Z=40°	0.54	0.44
Clear mtn top 2750 m	0.54	1.6	Z=25°	0.82	0.08

***Table 3**: Measured ACGIH effective UVB from the sky with a 40° cone filed of view (Source: Sliney, 1983).*

The natural protection against overhead UVR afforded to the eye by the upper lid and brow ridge may be of little value for exposure to open-arc sources in the work place, since the source may frequently be within the normal horizontal line-of-sight. Hence greater eye protection is routinely required in industrial applications. Goggles and face shields are required when working around electric arc welding, which produces high levels of all UVR wavelengths. Close fitting facemasks with low transmittance to UVR, visible and infrared radiation are used for protection (Sliney and Wolbarsht, 1980).

Action-Spectrum Measurements

The actual published values of biological action spectra have a measurement uncertainty that also influences the final biological or health assessment of any physical measurement. The scientific community has been frequently puzzelled in the past by seemingly different action spectra determined for the same biological endpoint. There are a number of factors that influence this, but one physical factor that must be considered is the wavelength-sampling band. If this is a 5-10 nm bandwith, the action spectrum will be skewed (Sliney and Matthes, 1999). The apparent wavelength-shift from sampling in a region of rapidly changing source spectrum or rapidly changing action spectrum is illustrated in Figure 4.

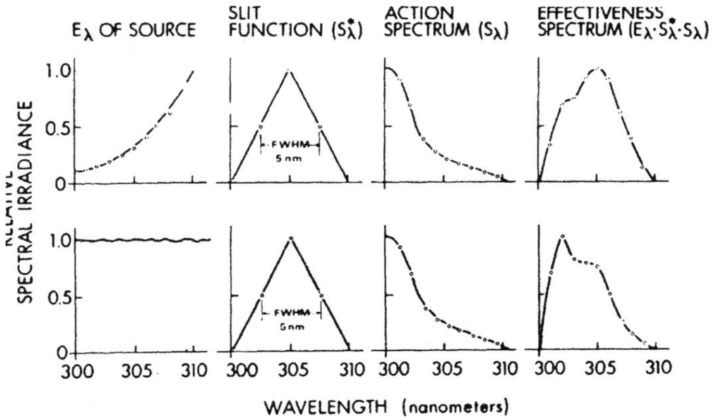

Figure 4. The effect of spectral bandwidth on the effective wavelength of emitted radiation. Note the spectral shift effect of using a xenon-arc monochromator where the spectrum is not changing with wavelength in the bottom panel.

The design of instrumentation intended to measure the photobiological dose to ocular tissues must take into account a number of geometrical shading factors, such as the eyelids and brow ridge, as well as behavioral aspects of vision. These geometrical and imaging factors challenge the task of attempting to accurately measure the photobiologically significant exposure of the cornea, lens and retina to ultraviolet, visible and infrared radiation. The human eye is actually exposed to a very small fraction of the global ultraviolet irradiance (diffuse plus direct radiation incident upon a horizontal surface). Therefore, the acceptance angle (field-of-view) should mimic those of the human eye. However, this acceptance angle varies with sky brightness. Thus, the instrument field-of-view must be adjustable in an instrument designed to really measure the UVR exposure dose to the cornea and lens. The UV exposure to the anterior segment of the eye can be measured in a mannequin fitted with UV detectors at the ocular positions, and measurements can be made with and without UV-absorbing spectacles, or as would occur when a person wears different types of sunglasses. The exposure is greatly affected by the type of sunglass frame and partially to the UV transmittance of the sunglass lenses. In some instances, UV exposures of some specific ocular tissues can actually equal or exceed those when not wearing sunglasses.

The retina is exposed to visible light and some IR-A radiant energy within an imaged scene. Although the lid opening varies with ambient scene luminance (brightness), it is possible to mathematically predict the opening of the lids and the angular field-of-view from studies of lid opening. The light exposure to the retina is not at all uniform in outdoor daylight conditions. The central and superior regions of the retina receive much more light than the inferior retina (Figure 3). Thus, instrumentation should simulate these geometrical factors.

References

1. ACGIH, Documentation for the Threshold Limit Values 4th Edn, Cincinnati: American Conference of Governmental Industrial Hygienists, 1992.
2. Airey DK, Wong JCF, Fleming RA, A comparison of human- and headform - based measurements of solar ultraviolet B dose, Photodermatol, Photoimmunol, Photomed, 1995; 11: 155-158.
3. American Conference of Governmental Industrial Hygienists (ACGIH) TLV's, Threshold Limit Values and Biological Exposure Indices for 1999, Cincinnati: ACGIH, 1999.
4. CIE, Erythemal reference action spectrum and standard erythemal dose, CIE Standard S007-1998, Vienna: CIE, 1998; also available as ISO 17166, 1999.
5. CIE, Standardization of the terms UV-A1, UV-A2 and UV-B, Report CIE-134/1, Vienna: CIE, 1999.
6. Challoner AVJ, Corless D, Davies A, et al, Personnel monitoring of exposure to ultraviolet radiation, Clin Exp Dermatol 1976; 1:175-179.
7. Cogan, DG, Kinsey, VE, Action Spectrum of Keratitis Produced by Ultraviolet Radiation Arch Ophthalmol 1946; 35:617-670.
8. Coroneo, MT, Müller-Stolzenburg, NW Ho, A, Peripheral light focusing by the anterior eye and the ophthalmohelioses, Ophthalmic Surg 1991; 22:705-711.

9. Coroneo, MT, Pterygium as an early indicator of ultraviolet insolation: an hypothesis, Brit J. Ophthalmol 1993;, 77:734-739.
10. Davis A, Deane GHW, Diffey BL. Possible dosimeter for ultraviolet radiation. Nature 1976; 261: 169-170.
11. Deaver DM, Davis J and Sliney DH, Vertical visual fields-of-view in outdoor daylight, Lasers Light Ophthalmol, 1996; 7(2/3):121-125.
12. Diffey BL. Ultraviolet radiation dosimetry with polysulphone film. In: Diffey, BL, ed. Radiation Measurement in Photobiology . London: Academic Press, 1989:135-139.
13. Diffey BL, Tate TJ, Davies A., Solar dosimetry of the face: the relationship of natural ultraviolet radiation exposure to basal cell carcinoma localisation, 1979, Phys Med Biol, 1979; 24:931-939.
14. Diffey BL. Kerwin M, Davies A, The anatomical distribution of sunlight, Br. J. Derm. 1977; 7:407-409.
15. Dolin, PJ, Ultraviolet radiation and cataract: A review of the epidemiological evidence, Br J Ophthalmol, 1994; 78:478-482 and reprinted in Optometry Today Nov/Dec 1997.
16. Duchêne, AS, Lakey, JRA, and Repacholi, MH, IRPA Guidelines on Protection against Non-Ionizing Radiation, New York: McMillan 1991.
17. Gies HP, Roy CR, Toomey S, MacLennan R, Watson M, Solar UVR exposures of three groups of outdoor workers on the Sunshine Coast, Queensland, Photochem. Photobiol., 1995; 62:1015-1021.
18. Ham, WT Jr., The photopathology and nature of the blue-light and near-UV retinal lesion produced by lasers and other optical sources In: Wolbarsht M L, ed, Laser Applications in Medicine and Biology, New York: Plenum Publishing Corp, 1983.
19. Ham, WT, Mueller, HA, Ruffolo, JJ, Guerry, D III, Guerry, RK, Action spectrum for retinal injury from near ultraviolet radiation in the aphakic monkey, Am J Ophthalmol, 1982; 93(3): 299-306.
20. Horneck G, Rettberg P, Rabbow E, Strauch W, Seckmeyer G, Facius R, Reitz G, Strauch K, Schott JU, Biological dosimetry of solar radiation for different simulated ozone column thickness. J. Photochem. Photobiol. B:Biol, 1996; 32:189-196.
21. International Commission on Non-Ionizing Radiation Protection (ICNIRP), Guidelines on UV Radiation Exposure Limits, Health Physics, 1996; 71(6):978 .
22. Klein, BEK, Klein, R, and Linton, KLP, Prevalence of age-related lens opacities in a population, the Beaver Dam Eye Study, Ophthalmology, 1992; 44(4):546-522.
23. Larko O, Diffey BL, Natural UV-B radiation received by people with outdoor, indoor, and mixed occupations and UV-B treatment of psoriasis. Clin Exp Dermatol, 1983; 8:279-285.
24. Munakata N, Kazadzis S, Bais A F, Hieda K, Rontó G, Rettberg P, Horneck G, Comparisons of spore dosimetry and spectral photometry of solar UV radiation at four sites in Japan and Europe, Photochem. Photobiol., 2000; 72:739-745.
25. Pitts, D.G., The human ultraviolet action spectrum Am J Optom Physiol Optics , 1974; 51(12):946-960.
26. Pitts, DG, Cullen, AP, Hacker, PD, Ocular effects of ultraviolet radiation from *295 to 365* nm, Invest Ophthalmol Vis Sci, 1977; 16(10): 932-939.
27. Quintern LE, Furusawa Y, Fukutsu K, Holtschmidt H. Characterization and application of UV detector spore films: the sensitivity curve of a new detector system provides good similarity to the action spectrum for UV-induced erythema in human skin. *J Photochem Photobiol* 1997; 37: 158-166.
28. Ringvold, A, Damage to the cornea epithelium caused by ultraviolet radiation, Acta Ophthalmologica, 1983; 61:898-907.
29. Rontó Gy, Gáspár S, Bérces A, Phage T7 in biological UV dose measurement. *J Photochem Photobiol* 1992; 12: 285-294.
30. Rontó Gy, Gáspár S, Gróf P, Bérces A, Gugolya Z. Ultraviolet dosimetry in outdoor measurements based on bacteriophage T7 as a biosensor. *Photochem Photobiol* 1994; 59: 209-214.

31. Sasaki K, Sasaki H, Kojima M, Shui YB, Hockwin O, Jonasson F, Cheng HM, Ono M, Katoh N. Epidemiological studies on UV-related cataract in climatically different countries. J Epidemiol (Japan), 1999; 9:S33-S38.
32. Sliney DH. Eye protective techniques for bright light. Ophthalmology 1983; 90(8):937-944.
33. Sliney DH, Estimating the solar ultraviolet radiation exposure to an intraocular lens implant, J Cataract Refract Surg 1987; 13:296-301.
34. Sliney DH, Ultraviolet radiation exposure criteria, Radiation Protection Dosimetry, 2000; 91(1-3):213-222.
35. Sliney, DH, Physical factors in cataractogenesis: ambient ultraviolet radiation and temperature, Invest Ophthalmol Vis Sci, 1986; 27(5):781-790.
36. Sliney DH, UV radiation ocular exposure dosimetry; J. Photochem. Photobiol. B: Biol 1995; 31: 69-77.
37. Sliney, DH, Krueger, RR, Trokel, SL, Rappaport, KD, Photokeratitis from 193-nm argon-fluoride laser radiation, Photochem Photobiol 1991; 53(6):739-744.
38. Sliney, DH, Geometrical gradients in the distribution of temperature and absorbed ultraviolet radiation in ocular tissues, Doc. Ophthalmol 2001 (In press).
39. Sliney, DH, The merits of an envelope action spectrum for ultraviolet radiation exposure criteria Amer Industr Hyg Assn J 1972; 33(10): 646-653.
40. Sliney, DH, and Matthes, R (Eds), The Measurement of Optical Radiation Hazards, ICNIRP Publication 6/98; CIE Publication CIE-x016-1998, ICNIRP: Munich and CIE: Vienna, 1999.
41. Sliney, DH, Wolbarsht, ML, Safety with Lasers and Other Optical Sources, New York: Plenum Publishing Corp, 1980.
42. Taylor H.R., Pterygium, Amsterdam: Kugler, 2000.
43. Taylor HR, West SK, Rosenthal FS, Munoz B, Newland HS, Abbey H, Emmett SK, Effect of ultraviolet radiation on cataract formation. New Engl J. Med 1989; 319:1429-1433.
44. World Health Organization, The Effects of Solar UV Radiation on the Eye, Report of an Informal Consultation, Geneva, August - September 1993, 1995 Publication WHO/PBL/EHG/94.1, Program for the Prevention of Blindness, Geneve: World Health Organization.
45. World Health Organization (WHO), Environmental health Criteria No. 160, Ultraviolet Radiation, Joint Publication of the United Nations Environmental Program, The International Commission on Non-Ionizing Radiation Protection and the World Health Organization, Geneva: WHO, 1994.
46. Wulf, H.C., Effects of ultraviolet radiation from the sun on the Inuit population, In: Petursdottir G, Sigurdsson SB, Karlsson MM, Axelsson J, eds, Circumpolar Health '93 Arctic Medical Research 1994; 53: 416-422.
47. Zuclich, JA, Ultraviolet-induced photochemical damage in ocular tissues, Health Physics 1989; 56(5):671-682.

ANNEX 1

APPLYING THE ULTRAVIOLET RADIATION EXPOSURE LIMITS

The ACGIH/ICNIRP EL for exposure to the eye and skin to UVR is 3 mJ/cm^2-effective, when the spectral irradiance E_8 at the eye or skin surface is mathematically weighted against the hazard sensitivity spectrum S(8) from 180 nm to 400 nm as follows:

$$E_{eff} = E\ E_8 ⚒ S(8) ⚒)8 \quad [1]$$

The permissible exposure duration, t_{max}, in seconds, to the spectrally weighted UVR is calculated by:

$$t_{max} = (3\ \mathrm{mJ/cm^2}) / E_{eff}\ (\mathrm{W/cm^2}) \quad [2]$$

In addition to the above requirement, the ocular exposure is also limited to 1 J/cm^2 for periods up to 1000 s (16.7 min) by ACGIH and to 30,000 s by ICNIRP. ACGIH limits the total irradiance to 1 mW/cm^2 for periods greater than 1,000 s. For this requirement, the total irradiance, E-uva, in the UV-A spectral region is summed from 315 nm to 400 nm:

$$E\text{-}uva = E\ E_8 ⚒)8 \quad [3]$$

where E_8 is the spectral irradiance in W/(cm^2-nm).

One way to express the ACGIH limit for the UV-A is: if the total irradiance exceeds the 8-hour criterion of 1 mW/cm^2, the maximum exposure must also be less than:

$$t_{max} = (1\ \mathrm{J/cm^2}) / E_{UVA}\ (\mathrm{W/cm^2}) \quad [4]$$

Equation [4] applies to all periods up to 8 h for the ICNIRP guidelines.

Table. UVR Hazard Spectral Weighting Function.

Wavelength (nm)	UVR Hazard Function $S_{UV}(\lambda)$
180	0.012
190	0.019
200	0.030
205	0.051
210	0.075
215	0.095
220	0.012
225	0.015
230	0.019
235	0.240
240	0.300
245	0.360
250	0.430
*254	0.500
255	0.520
260	0.650
265	0.810
270	1.000
275	0.960
280	0.880
285	0.770
290	0.640
295	0.540
300	0.300
305	0.060
310	0.014
315	0.003
320	0.001
325	0.0005
330	0.00041
335	0.00034
340	0.00028
345	0.00024
350	0.00020
355	0.00016
360	0.00013
365	0.00011
370	0.000093
375	0.000077
380	0.000064
385	0.000053
390	0.000044
395	0.000036
400	0.000030

*Low-pressure mercury germicidal lamp emission.

ANNEX 2
Summary of Useful Radiometric Quantities

The following radiometric quantities may be used in photobiology, photochemistry, photodermatology and illuminating science, and are briefly summarized here:

Irradiance (surface dose rate) and *radiant exposure* (surface dose) are units specifying power or energy incident upon a plane. These quantities are the most fundamental dose quantities used in all of photobiology. The units most commonly used are W/cm^2 and J/cm^2, respectively. 1 W = 1 J/s.

Fluence rate **and *fluence*** are used in some very sophisticated studies, where the internal surface dose with backscatter is included. These quantities are used correctly most often in theoretical studies of dose distribution and where photochemistry at the molecular level in tissue is enhanced as a result of multiple scattering events in tissue. Unfortunately, these terms are frequently misused to mean irradiance and radiant exposure because the units of W/cm^2 and J/cm^2 are the same.

Radiance (irradiance per solid angle) is an important quantity used by physicists in specifying a source. This quantity limits the ability of lenses and reflective optics in concentrating a light source. For, example, a xenon-arc lamp has a very high radiance and its energy can be focused to produce a very high irradiance on a target tissue. By contrast, a fluorescent lamp tube has a much lower radiance, and its energy cannot be focused to a high concentration. The units are $W/(cm^2 \cdot sr)$.

Radiant Intensity (power per solid angle) is used to indicate how collimated a light source really is. Although useful for specifying searchlights, it normally has very limited use in photobiology. The units are W/sr.

Spectral quantities (units per wavelength) are used for specifying the energy, power or irradiance per wavelength interval. When calculating a *photobiologically effective dose* the spectral quantity must be multiplied by the action spectrum. Examples: spectral radiant power, spectral irradiance, spectral radiant exposure, etc. The units for each quantity are modified by adding "per nanometer," e.g., W/cm^2 becomes $W/(cm^2 \cdot nm)$.

Photon (Quantum) quantities (units of photons) are used primarily in theoretical studies, and in photochemistry. In this case the radiant exposure is specified in photons/cm^2 and irradiance is specified in photon/(cm^2·s).

The following table summarizes useful radiometric terminology:

Table A1. Useful Radiometric Units[1,2]

Term	Symbol	Definition	Unit and abbreviation
Radiant Energy	Q	Energy emitted, transferred, or received in the form of radiation	joule (J)
Radiant Power	Φ	Radiant Energy per unit time	watt (W) defined as J/s
Radiant Exposure (Dose in Photobiology)	H	Energy per unit area incident upon a given surface	joules per square centimeter ($J \cdot cm^{-2}$)
Irradiance or Radiant Flux Density (Dose Rate in Photobiology)	E	Power per unit area incident upon a given surface	watts per square centimeter ($W \cdot cm^{-2}$)

Integrated Radiant Intensity	I_P	Radiant Energy emitted by a source per unit solid angle	joules per steradian ($J \cdot sr^{-1}$)
Radiant Intensity	I	Radiant Power emitted by a source per unit solid angle	watts per steradian ($W \cdot sr^{-1}$)
Integrated Radiance	L_P	Radiant Energy emitted by a source per unit solid angle per source area	joules per steradian per square centimeter ($J \cdot sr^{-1} \cdot cm^{-2}$)
Radiance[3]	L	Radiant Power emitted by a source per unit solid angle per source area	watts per steradian per square centimeter ($W \cdot sr^{-1} \cdot cm^{-2}$)
Optical Density	OD	A logarithmic expression for the attenuation produced by a medium $OD = -\log_{10}\left(\frac{\Phi}{\Phi}\right)$	unitless Φ_O is the incident power; Φ_L is the transmitted power

1. The units may be altered to refer to narrow spectral bands in which the term is preceded by the word *spectral* and the unit is then per wavelength interval and the symbol has a subscript λ. For example, spectral irradiance E_λ has units of $W \cdot m^{-2} \cdot m^{-1}$ or more often, $W \cdot cm^{-2} \cdot nm^{-1}$.

2. While the meter is the preferred unit of length, the centimeter is still the most commonly used unit of length for many of the terms

below and the nm or μm are most commonly used to express wavelength.

3. At the source $L = \dfrac{dI}{dA \bullet \cos\theta}$ and at a receptor $L = \dfrac{dE}{d\Omega \bullet \cos\theta}$.

UVR MEASUREMENT ERROR SOURCES: SPECTRAL RESPONSE OF FILTERED BROADBAND DETECTORS

Robert Angelo
Gigahertz-Optik, Inc., Newburyport, MA 01950-4009

INTRODUCTION

Over thirty potential sources of error can be identified in optical radiation measurement. Spectral error caused by a photodetector assembly's imperfect spectral 'shape' (responsivity versus wavelength), in relation to that of the light source to be measured, can be a major contributor to radiometric inaccuracy.

A brief overview of the optical radiation and spectral regions, using the CIE (Commission Internationale de l'Eclairage) defined UV-A spectral region, the concepts of spectral response and spectral error will be reviewed. An examination of two commercially available UV-A sensors' spectral responsivity and associated spectral error in relation to a UV-A fluorescent and filtered Xenon (ID65) light source will follow.

Optical Radiation

Optical radiation includes the wavelength range from approximately 200 nm to 10 micron (10,000 nm) (Figure 1). Below 200 nm, the atmosphere absorbs this radiation and above 10 micron the signal is known as heat, radio and microwaves.

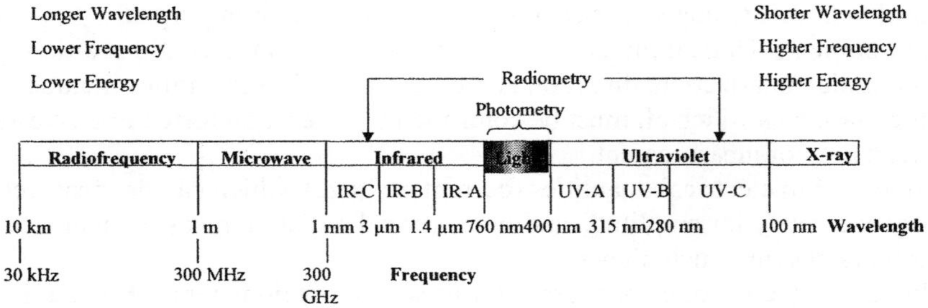

Figure 1. Electromagnetic Spectrum

Referring to the table below, the CIE has defined and named spectral bands within the total spectrum [1].

Table 1. CIE Vocabulary for Spectral Regions

CIE Vocabulary for Spectral Regions

Name	Wavelength Range
UV-C	100 to 280 nanometers
UV-B	280 to 315 nm
UV-A	**315 to 400 nm**
Visible	360-400 to 760-800 nm
IR-A*	780 to 1400 nm
IR-B	1.4 to 3.0 micron (mm)
IR-C**	3.0 m to 1 mm
* Also called 'near IR or NIR'	(1 micron = 1000 nm)
** Also called 'far IR'	

UVR Ultraviolet Radiation spans the spectral region from about 100 to 400 nm. The CIE has defined three spectral bands within the UV spectrum as UV-C, UV-B and UV-A.

Radiometry and Quality of Measurement

Radiometry deals with the measurement of optical radiation over the entire spectral range (see Figure 1). Below 200 nm, due to atmospheric absorption, absolute measurements are performed under vacuum or in purged environments. Optical radiation exists in frequency, wavelength, position, amplitude, polarization, direction and time. Each of these entities include other variables which all must be controlled or at least budgeted to arrive at an uncertainty of measurement.

Compared to electrical quantities (current, voltage), which can be measured very accurately for amplitude and time, optical radiation measurement inaccuracies run much higher.

Highest on the list of error sources for broadband radiometers are spectral error, spatial error and calibration.

Spectral Response and Error

Focusing on spectral error using the CIE defined UV-A bandpass, a radiometric sensor's spectral response or sensitivity versus wavelength should be 100% at each wavelength over the 315 to 400 nm bandpass since absolute rather than effective (no weighted spectral function) response is implied. Further, the ideal UV-A detector would have no response outside this range. A spectral plot illustrating this ideal sensor response for capturing total or absolute UV-A is presented in figure 2.

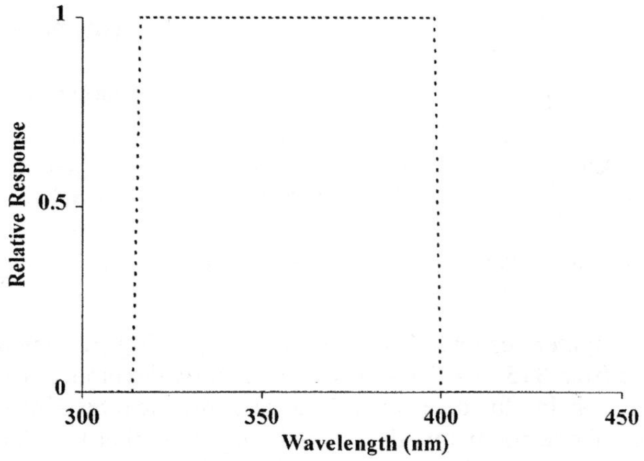

Figure 2. Ideal UV-A Spectral Response

A UV-A photodetector, correction filter and input optic device with a perfect match to this 'square wave' function does not exist and would be too expensive to produce so a compromise in the design is made by the manufacturer. An example of a typical UV-A detector's spectral response is shown in Figure 3. Due to the loss of response caused by the 'shoulders' of the Gaussian type response, the total area of the detector is –34% less than ideal. There is also a slight amount of response below 315 nm in the UV-B region which can adversely effect the measurement depending on the light source under test. Wavelengths above 400 nm are cut-off, but at a sacrifice of response from 390 to 400 nm.

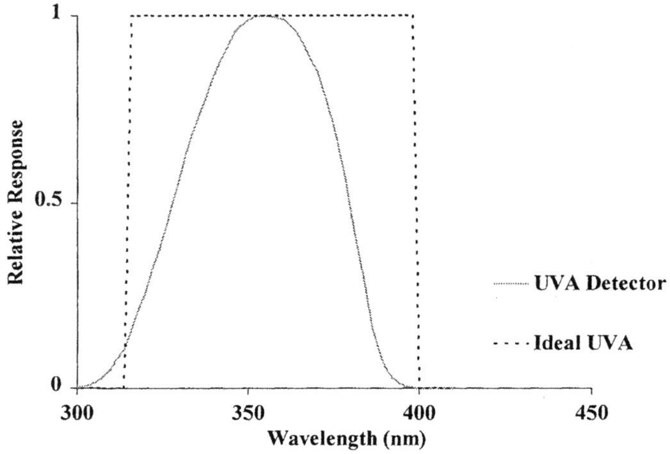

Figure 3. Typical UV-A Detector vs. Ideal Spectral Response

A new type computer designed UV-A detector /filter (Figure 4) with a near 'flat' response from 315 to 400 nm has been recently developed. Comparing total areas once again, this detector fall shorts of the ideal by –16%. There is some response above and below the desired range but other UV-A detectors with sharper cut-offs (not shown) can be selected based on the application.

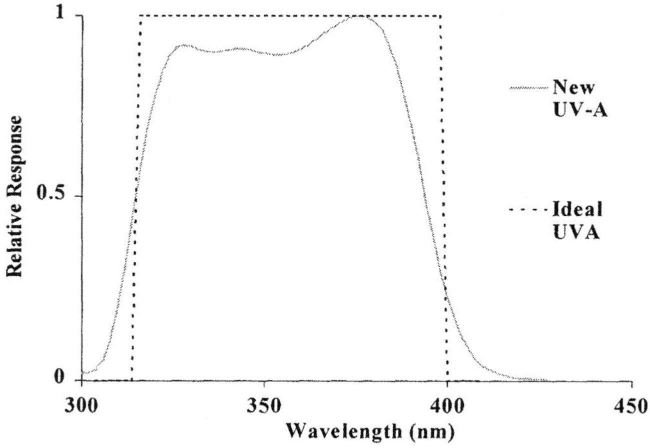

Figure 4. New UV-A Detector vs. Ideal Spectral Response

Determining and comparing total area by integrating under the curves for each wavelength across the desired bandpass is a quick and easy indicator of spectral error. Things get a bit more complicated when examining detector spectral error for a specific light source. The next graphic (Figure 5) presents a typical UV-A detector response, the ideal UV-A, a fluorescent 'blacklight' UV-A source and Xenon (ID65) type light source distribution.

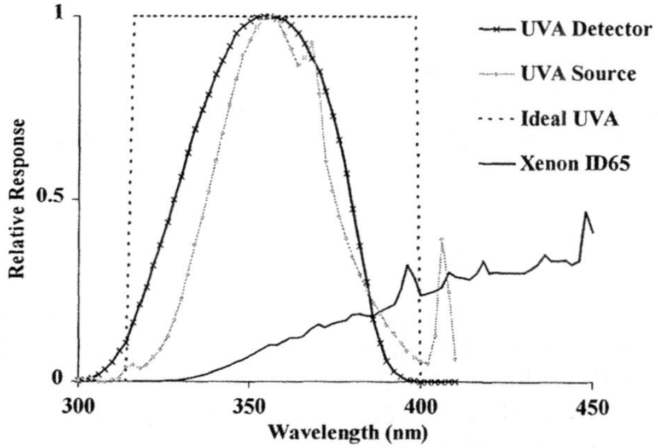

Figure 5. Typical UV-A Detector vs. Light Sources

Product integral calculations, discussed later, show the typical 'Gaussian' response detector would measure low by –18% for the UV-A 'blacklight' and –55% low for the Xenon (ID65) type sources. Since the spectral plot only shows wavelengths to 450 nm, it cannot be determined if any secondary response 'stray-light leak' occurs in the longer wavelengths. Without a blocking filter added to the bandpass correction filter, this can occur. If a silicon type photodiode, which is responsive out to 1100 nm, is used this stray-light reading will be added to the reading increasing the total spectral error.

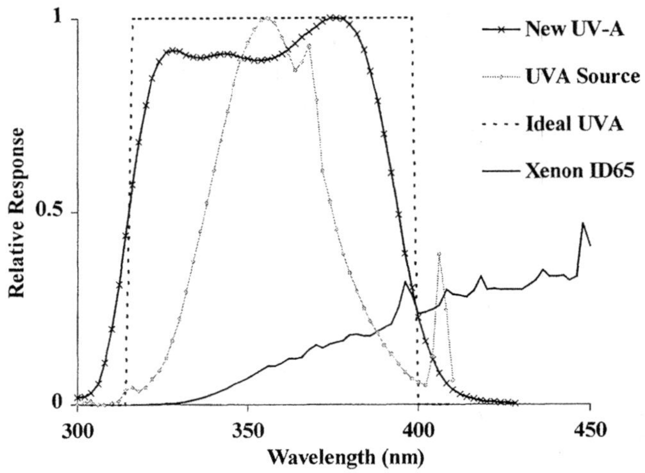

Figure 6. New UV-A Detector vs. Light Sources

The new type 'flat' UV-A detector shows improvement in measuring the true total UV-A output. This detector would read –9% low for the 'blacklight' source and –21.4% low for the Xenon source. The 'stray-light' error due to the detector response beyond 400 nm was calculated to be 5%. This error would be less for the UV-A source since there is much less distribution at this point. Since this detector employs a GaP Gallium Phosphide photodiode, which only responds out to 550 nm, a printout of the spectral response of this detector assembly out to 600 nm shows no secondary response (Figure 6).

Calculating and Correcting Spectral Error

If the relative spectral responses of the UV-A detector and source to be measured are known, product integrals can be performed to determine and correct for spectral error. The following equation will produce a correction factor which can be applied to the detector calibration for absolute spectral functions like the CIE defined UV-A, B or C where there is no effective weighting of the spectral response. With the factor applied, the detector would produce a close approximation of what a spectroradiometer would read on the same light source.

$$\text{Correction Factor} = \int_0^\infty H_{\lambda R} \bullet S_{\lambda R} \bullet d\lambda \div \int_{315}^{400} H_{\lambda R} \bullet d\lambda$$

$H_{\lambda R}$ = relative spectral response of detector

$S_{\lambda R}$ = relative spectral irradiance of source

Note the integration range from zero to infinity to account for all contributing radiation reaching the UV-A detector.
Another method to correct for spectral error is to calibrate the detector directly on the light source to be measured but this is not commonly done for practical reasons.
Normally the calibration source is simply stated on the certificate with a statement of uncertainty and/or other qualifying data.

Conclusions
Broadband radiometric measurement inaccuracies can be improved with new optical filter designs aided by computer technology as can be seen in Table 2.

Table 2. Comparison of Spectral Error of Typical and New Type UV-A Detector for Ideal UV-A Total Area and UV-A and Xenon Type Light Sources

	Ideal UV-A	UV-A Source	Xenon (ID65)
Typical Detector	-34%	-18%	-55%
New Detector	-16%	- 9%	-21.4%

However, due to the inevitable compromises that take place from the original design phase to manufacturing, corrections through calibration are needed to further improve accuracy. Currently the European Thematic Network and CIE Division 2 TC 2-47 are working on methods to characterize broadband UV radiometers similar to the way photometers are now. A good amount of the methodology deals with the spectral mismatch issue. More detailed information on correcting broadband detector calibrations for a specific light source is provided along with actual spectral distribution curves and datasets for many common light sources in use today.

References
1. CIE 17.4-1989 International Lighting Vocabulary

PULSED UVB IRRADIATION CONVERTS 7-DEHYDROCHOLESTEROL TO PREVITAMIN D_3 AND PHOTOPRODUCTS

*D. G. Changaris, **Z. Lu, T. Chen, M.F. Holick
**Synaptic Tan, Inc, Louisville, KY and ** Depts. Of Medicine, Dermatology, and Physiology, Boston University Medical Center, Boston, MA*

ABSTRACT

As a first step toward documenting efficacy of the Mark IV (Synaptic Tan, Inc., Louisville, KY, synaptictan.com) for clinical vitamin D_3 augmentation we sought to determine whether *in vitro* photoactivation of 7-dehydrocholesterol (7-DHC) occurs. The Mark IV provides uniform illumination (± 15%) over 0.5 m^2 at 60 cm utilizing non-spatial collimation with 20-100 μsec pulses of UVB/UVA. The optically coated reflectors serve as highly effective bandpass filters with output ranging over 270-370 nm with little light measured above and below this range. The device delivers 1 FDA corrected MED within 1-4 minutes depending on pulse rate and distance. Glass ampoules (total 27) containing 7-DHC in methanol were exposed to 0.5, 1, and 2 MED at 30, 45, and 60 cm. Assay for preVitamin D_3 (preD_3), Lumisterol (L_3), and Tachysterol (T_3) identified a dose-dependent linear relationship for the formation of preD_3 independent of distance. This confirms that "reciprocity" is preserved with 20-100 μsec pulses.

Both preD_3 and T_3 but not L_3 were detected above threshold levels in the 2 MED exposures at all distances. This data suggests that the Mark IV radiates sufficient ultraviolet energy to form in preD_3 humans sitting 30-60 cm from the device and provides the rational to justify human studies.

INTRODUCTION

Ultraviolet B initiates the transformation from 7-DHC to preD_3 which isomerizes thermally to vitamin D_3 (D_3). Lumisterol (L_3) and Tachysterol (T_3) also result from the cascade shown in Figure 1. The Mark IV (Figure 2) generates pulsed UVB/UVA with the wavelength distribution pattern shown in Figure 3. This paper represents a novel exploration to determine if pulsed light (ranging 20-100 μsec) can initiate the cascade that results in vitamin D_3 by converting 7-DHC to $preD_3$ with preservation of "reciprocity".

Peak and Peak reviewed the history of the concepts surrounding "reciprocity", a term that has come to mean that the absolute amount of energy determines the photochemical response, not the rate of delivery. However, "reciprocity" does not consider the possibility that short pulses may not follow this paradigm. Many photochemists use ultra-short pulses (femtoseconds to nanoseconds) to increase the photostability of fluorescence. Thus the ultra-short pulse may subvert "reciprocity" so central to photochemical reactions. Whether this failure of "reciprocity" extends into the microsecond pulse range has yet to be fully defined. Another question derives from the laws of heat transfer. The rate of heat transfer increases with temperature and depends on the ability of the material to transfer heat. A short pulse that raises the local temperature quickly will transfer more heat in less time than a continuous light. This suggest that in situations in which heat changes the conformation of a molecule to create a photoreactive state, pulsed light will be less efficient or more controllable as the light itself will not contribute to heat has efficiently as a continuous source. This can be critical where the light itself delivers significant infrared. Another situation becomes quite critical when heat results from the photochemical reaction that can be enhanced by the temperature. In certain biological settings added light may contribute heat that may potentiate toxic or unwanted secondary events.

The Mark IV provides a here-to-fore unavailable uniform illumination over several square feet (15% variance). Within the region used to expose the glass ampoules (about 12 square inches), the variance was less than 2%. The Mark IV has no detectable warmth at the distances used in this study. Due to the nature of the multiple reflections from the optical coating over a black surface, there emanates little light above 380nm from the Mark IV. This device delivers light in a novel manner that will identify whether 90 nm bandpass ultraviolet light follows reciprocity with 20-100μsec pulses. This study further addresses that issue whether this device will generate preVitamin D_3 *in vitro* as a basis to justify human clinical trials.

Methods

The Mark IV delivers ultraviolet light for clinical settings. Its design protects subjects from inadvertent electrical or heat burns. No part of the external device exceeds room temperature during use. Humans with little mobility and comfortably sit at 30-90 cm to expose selected surface areas. Due to the nature of the pulsing, the device was configured to deliver 1 instrumental MED (15.4 mj/cm^2 wavelength corrected per FDA guidelines) in approximately 3 minutes. Models may be configured to deliver one instrumental MED over 30-3000 seconds.

Glass ampoules containing 7-DHC were placed at 30, 45, and 60 cm from the center surface. Three glass ampoules were harvested after 0.5 MED, 1.0 MED and 2.0 MED for each distance and quickly frozen on dry ice and maintained in the dark until analyzed. The range variance of UVB measured with an IL 1700 was less than 2% over the glass ampoules during exposure. The photometry and FDA-wavelength-corrected MED was performed by Rapid Precision Technology Laboratories, Cordoba, Tennessee with an Optronics Laboratories OL754 system calibrated according to NIST standards. The measurement of chemicals in methanol was by HPLC.

Statistical Methods

The design of the study sought to study whether changes in dose and/or distance (exposure time increased linearly with distance). Analysis of variance with two readings of each of three samples for a given distance and dose of light. To confirm the preservation of reciprocity there should be no significant differences based within groups of similar dose at different distances or time. Within group variance was also reviewed for dose. For the light to be causative at initiating the photochemical reaction, doubling the dose of light should double the concentration of the photoactivated product as long as the ratio of product to substrate stays below 10%.

Efficiency and Longevity Studies

The light distribution was measured at 3-inch intervals using the IL 1700 (International Light Inc., Newburyport, MA) equipped with UVB detector and diffuser, calibrated to NIST standards. The line voltage and amperage was determined with a Fluke 23 Multimeter with standard test leads and a Fluke 80I-400 AC current probe. A test fixture utilizing the same electronic configuration was run continuously for 1240 hours. Output was determined using an IL 1700 similarly calibrated.

Results

The graph in Figure 4 shows that there is little if any photodegradation of 7-DHC to Lumisterol or Tachysterol. Figure 5 demonstrates visually that the within grouping of a given dose of light shows no significant changes based upon distance. Increasing the amount of light increased the amount of preVitamin D_3 in a linear fashion. Doubling the ultraviolet light dose doubled the formation of preD_3. Analysis of variance and regression analysis confirms these observations.

Energy consumption by the device was determined by measuring line voltage and current. The Mark IV consumes approximately 670 watts (5.99 amperes at 111 volts) to deliver 0.24 watts FA corrected UVA/UVB over 900 cm^2. These measurements were taken with the Mark IV calibrated to deliver 1 MED over 3 minutes. Due to the characteristics of the energy flux density, the dual reflector systems are stackable, extending the potential surface area to be covered by stacking another light-module pair within the device. Continuous discharging of the xenon lamps configured within the system design showed a 3% drop in output over 1240 hours.

Discussion

The Mark IV manufactured for UVB/UVA delivery generates enough UVB to form preVitamin D_3 from 7-DHC *in vitro* at clinically relevant distances while preserving reciprocity. The soft light illuminates complex surfaces as large as an average human with less that 15% variation. The device effects a Holtzman-like diffuse light with partial collimation with the practical result at doubling the distance from the light changes the amount of light delivered by 35% not 75% over these ranges. These characteristics favor the device for light activated therapies.

The optical coating of the Mark IV can be specified to reflect other bandwidths. Theoretically, any bandwidth from 10 nm to full spectrum can be selected. The efficiency of a single reflection over the desired bandwidth approaches 100%, above and below the desired band efficiencies fall to less than 30%. With multiple internal reflections the light output becomes quite specific to the desired band.

The near absence of light above 380 nm further enhances the clinical utility of the device. The patient perceives no heat and therefore does not perspire during the exposure interval. The device may allow an individual to expose back, chest, or abdomen and generate vitamin D to fulfill requirements for several days to weeks. Further study to document this potential utility can be justified from the current *in vitro* studies.

Pulsed light offers distinct advantages over continuous light for the dissipation of heat within biologic tissues. Heat transfers more efficiently when the heat difference is higher. With blood coursing through tissues it follows that more heat can be dissipated through pulsing than continuous light. With respect to the Mark IV, the absence of any significant infrared and the short exposures of specific bandwidth increase the potential uses of this device. With the ability to specify different bandwidths with different optical coatings, this device offers new avenues of clinical studies of potential treatments. It is more likely than not, that any photochemical reaction that depends on conformational changes generated by increasing temperature will be less likely to occur with pulsed light. The peculiar blend of efficiency (delivering a specified bandwidth) and inefficiency (delivering heat) of the Mark IV will allow for convenient, consistent, and soft light for accurate dosing in clinical settings.

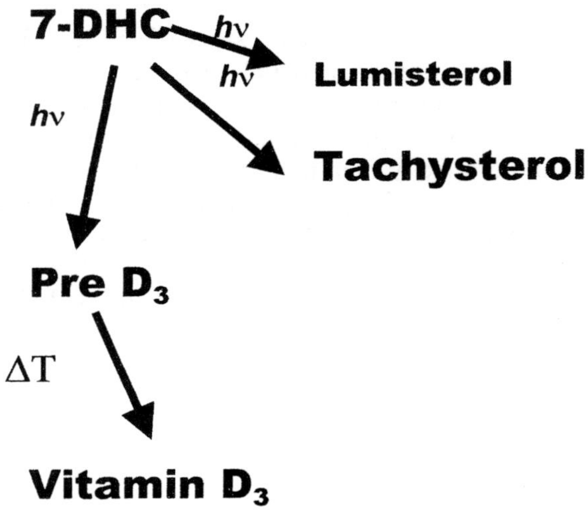

Figure 1. The vitamin D cascade begins with 7-dehydrocholesterol (7-DHC) and ends in the photo-products of pre Vitamin D_3 (Pre D_3), Lumisterol (L_3), and Tachysterol (T_3).

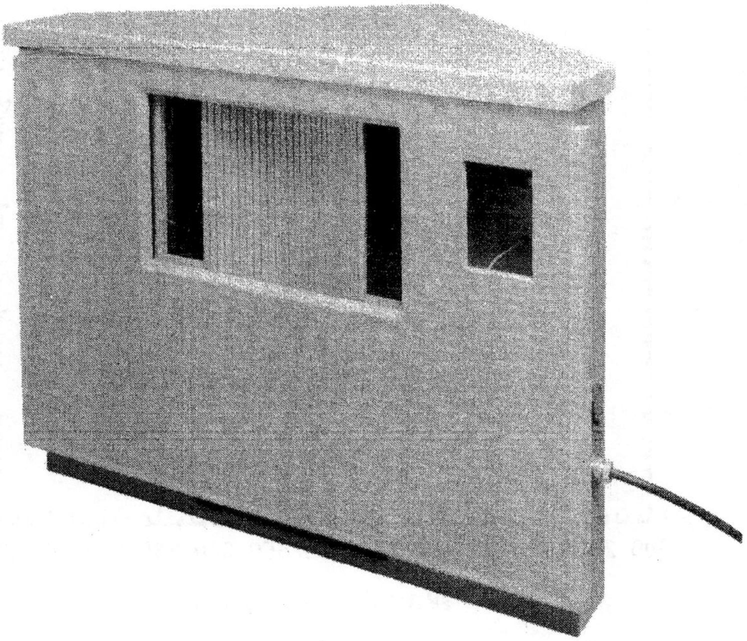

Figure 2. The Mark IV provides consistent ultraviolet over several square feet, minimizing the potential of injury from inadvertent touching of the lamp.

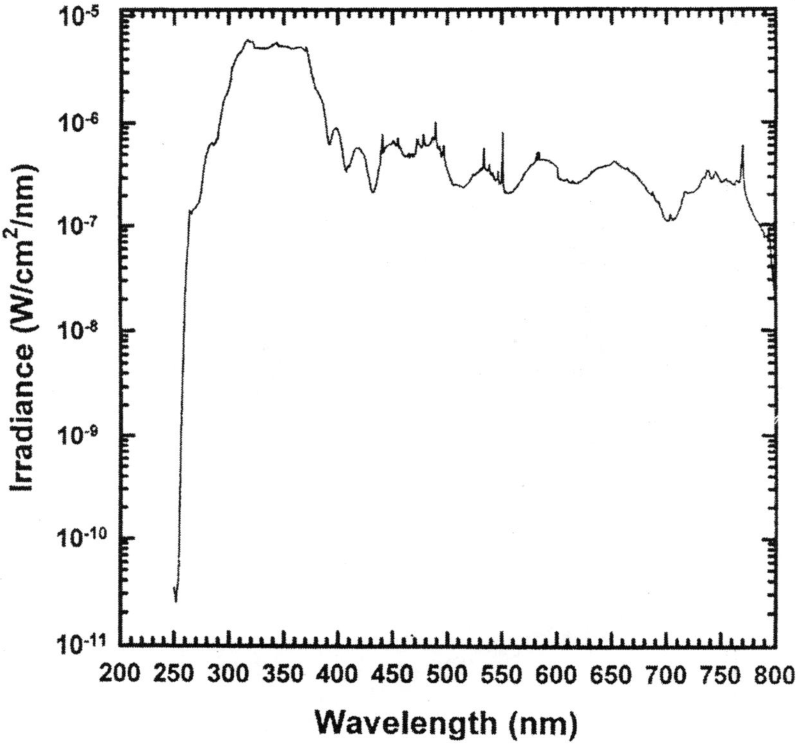

Figure 3. The spectral characteristics obtained with the Optronics Laboratories OL754 system showing light output ranging between 275 and 380 mμ . There is little output above this upper limit and no physically detectable heat.

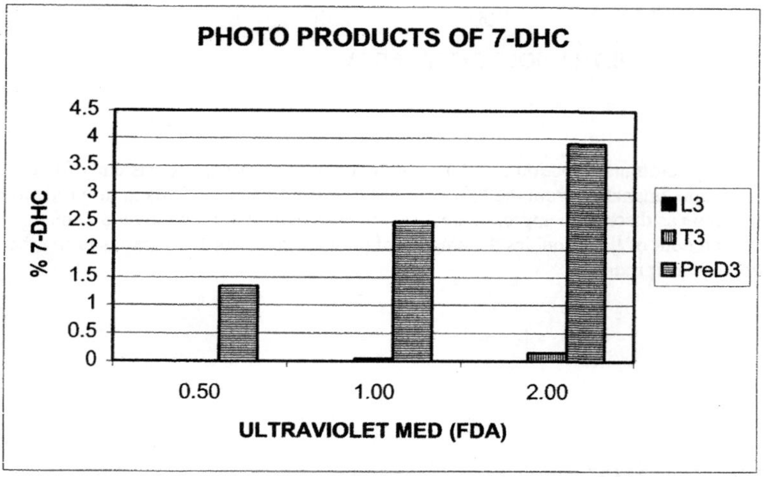

Figure 4. The Photoproducts of 7-DHC show minimal Tachysterol (T_3), no Lumisterol (L_3). There appears to be some isomerization of the samples by the spontaneous appearance of vitamin D_3.

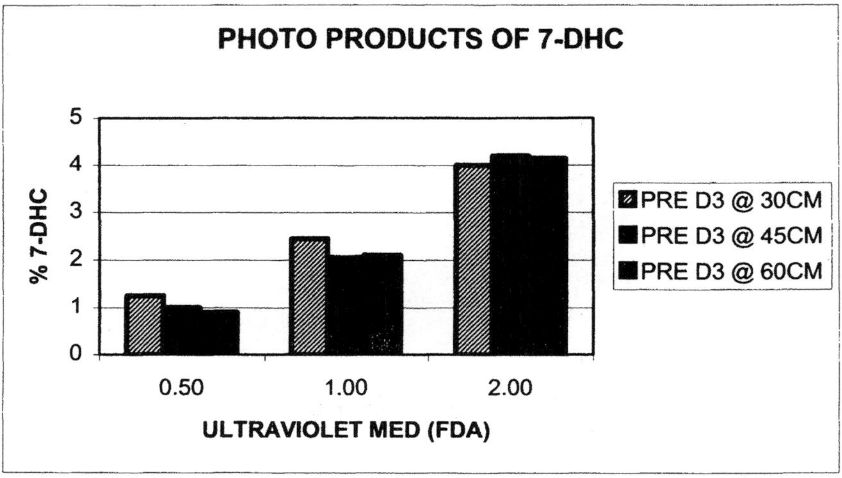

Figure 5. The phototransformation of 7-HDC to PreVitamin D_3 occurs consistently for a given dose of light emanating from the Mark IV, regardless of distance. This in turn reflects a variability of time as there is an approximate 40% increase in time at 60 cm over 30 cm. Doubling the amount of light doubles the amount of photoproduct, further demonstrating the confirmation of "reciprocity".

BIBLIOGRAPHY

Halliday, David and Resnick, Robert. *Physics Part I.* New York: John Wiley & Sons, Inc., 1966.

Holick, Michael F., *Vitamin D Molecular Biology, Physiology, and Clinical Applications.* Totowa, New Jersey: Humana Press, 1999.

Peak, J. G. and Peak, M. J. Lethality in Repair-Proficient Eserichia Coli after 365 nm Ultraviolet Irradiation is Dependent on Fluence Rate. Photochem Photobiol 36:103-105, 1982.

ASSESSMENT OF DESIGN CONFIGURATIONS FOR THE THERAPEUTIC USE OF DAYLIGHT

Phillip G. Mead, AIA.
College of Architecture, Texas Tech University, Lubbock, Texas.

BACKGROUND:

The relationship between light and architecture has been long and fruitful. Ancient to medieval temples were built in careful relationship to the motion of the sun in order to worship its existence, mark time and dramatize rituals. The ancient and Renaissance architects Vitruvius and Alberti both wrote how sunlight was to be placed in proper relationship in design. (1, 2) Not until the mid 19th century did medical design advocates like Florence Nightingale begin to promote the healthful qualities of light in houses and hospitals. Nightingale noticed that patients on the sunny side of a hospital had higher spirits than those located in darker areas. (3) This resulted in her promotion of the pavilion hospital plan that consisted of patient wings allowing ample light penetration and cross ventilation. (4) At the turn of the century it was found that sunlight could help cure tuberculosis resulting in hospitals with sunbathing balconies. This was also the dawn of modern architecture from which Frank Lloyd Wright and Le Corbusier began to promote the health qualities of light and air in their designs. The most celebrated therapeutic use of light in modern architecture occurred in 1927 with Richard Neutra's "Health House." Designed for Dr. Lovell, a staunch promoter of environmental therapies who wrote a weekly health column for the Los Angeles Times, the house allowed ample southeastern light through two story windows and provided multiple places for private sunbathing and exercise. But the incorporation of light in architecture for health reasons appears to fade after the wide spread application of sulfanilamide and penicillin in the 40's. In the mid 70's, energy conservation, and building practices further restricted window use in buildings. However, today, a new interest in healthy design is rekindled by the affects of sick building syndrome and daylight studies of increased worker/student performance. However, architecturally the therapeutic use of light for depression and sleep disorders has not been sufficiently investigated.

PROBLEM

Today's light level standards range from 300 lux for classrooms to 500 lux for offices. (5) Many home light levels are lower than 100 lux. For curbing depression and regularizing sleep, it is recommended that exposure to 1500 to 2500 lux for an hour a day is sufficient. (6) Thus, those who suffer sleep problems and seasonal affective disorder are not exposed to enough light. Turning up light levels could dramatically increase energy costs while spending time outdoors could be uncomfortable. Light boxes are commonly used, but are glaring, offer no view and emit a limited spectrum of light. A comfortable and productive alternative could be to configure building spaces and windows to harvest enough daylight to promote normal bodily functioning. This would be advantageous for a number of reasons. Window views are preferable because they offer a broader range of spatial sequences. (7) If scenes are of nature, a reduction in stress may result. Ulrich found that a view of trees by postoperative surgery patients may improve recovery time and lower doses of pain medication. (8) Unnecessary stress may also be induced due to the extreme light contrast between the light box and dim interiors, which may be the cause of headaches and eyestrain. Additionally, window light brings in the sky, which contains blue. In studies by Wolforth, his classroom color investigations found blue to lower blood pressure. (9)

HYPOTHESIS:

Indoor refuge areas that are small and expose many sides to daylight may produce enough light to regularize sleep and curb winter depression during cloudy December days in all of the lower 48 states.

METHODS:

From daylight availability tables found in Robbin's *Daylighting,* Seattle arguably has the lowest light levels in the lower 48 states due to its high percentage of cloud cover and low lux levels during December. Denver was chosen due to its mid level location in the US. In these cities, three architectural study designs were tested using the Graphic Daylight Design Method. The designs were predicated on 19th century Victorian towers and bay windows, which harvest a high amount of light within a small area. The first design is a modification of a bay window and contains a skylight. (Fig. 1) The second design is a bed located below a skylight. (Fig. 2) The third is a tower with windows and skylight. (Fig. 3)

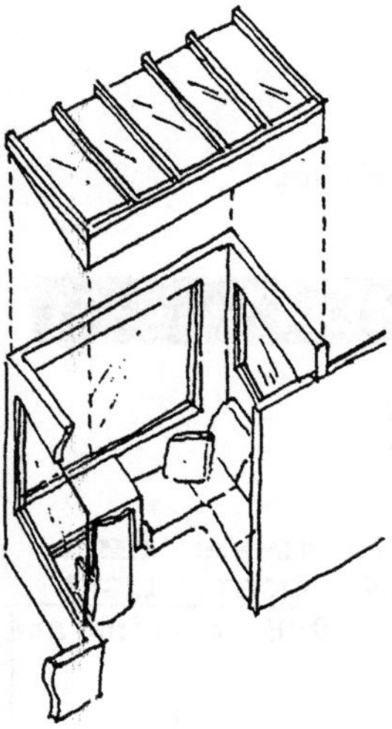

Figure 1. Peninsula Plan

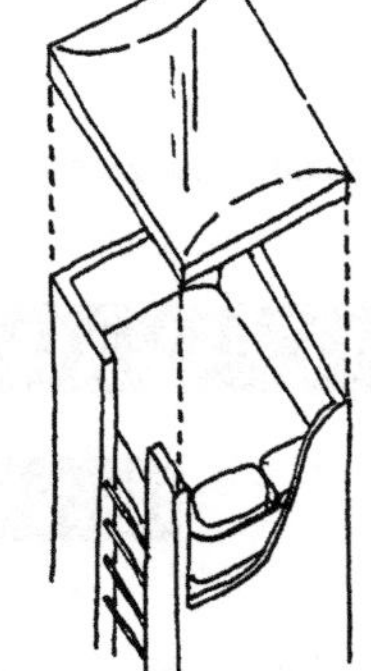

Figure 2. Skylight over Bed.

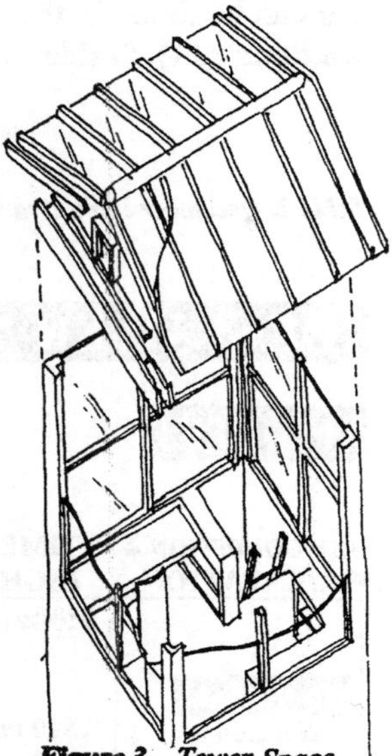

Figure 3. Tower Space

Graphic Daylight Design Method

Light measurements using the Graphic Daylight Design Method (GDDM) were used in this study to most nearly approximate light levels in Seattle and Denver. This formula takes the daylight factor (ratio of outdoor light to indoor light) and multiplies this figure by the outdoor light level. The daylight factor changes dramatically the further inside the space and is shown graphically by a series of light rings distributed from a window or skylight to a horizontal plane. In this study, daylight factors were used from footprints found in the book *Inside Out, Design Procedures for Passive Environmental technologies* .(10) In this study, the daylight factor was established at two feet six inches from a window (distance while sitting at a desk in front of a window). All desk and room orientations face north to avoid direct sunlight. The calculations are also considered "raw" because the method assumes interior surfaces are a non-reflective black. Therefore, internal and external reflectance, roof overhangs and the amount of dirt on windows were excluded. For this study, outside light conditions were taken from Seattle and Denver in the month of lowest light, December. Seattle and Denver light levels in December can be found in the tables provided in Robbin's *Daylighting, Design and Analysis*. (11) (Table 1)

Table 1. December Light Availability in Seattle and Denver at 9 and 11am.

DECEMBER LIGHT AVAILABILITY

SEATTLE

SKY CONDITION & ORIENTATION	TIME & LIGHT	TIME & LIGHT
	9:00am	**11:00am**
Cloudy Sky		
Horizon	1510 lux	2960 lux
Above	4190	6380
Clear Sky		
East	33,030 lux	23,540 lux
North	2490	4340
West	2590	5940
South	37,520	73,140
Above	9930 lux	26570 lux
Note: 36 % chance of sunshine.		

DENVER

SKY CONDITION & OREINTATION	TIME & LIGHT	TIME & LIGHT
	9:00 am	**11:00 am**
Cloudy Sky		
Horizion	3070 lux	5280 lux
Above	7090 lux	10,620
Clear Sky		
East	51,680 lux	29,850 lux
North	4190	6040
West	4320	8060
South	56,760	86,730
Above	21,040 lux	41,900 lux
Note: 77% chance of sunshine.		

Reference: *Daylighting, Design and Analysis,* Claude Robbins, New York, Van Nostrand Reinhold Co. 1986

Field Measurements for the Establishment of a Vertical Daylight Factor

Since the Graphic Daylight Design Method measures light levels on a horizontal work plane, this does not reflect light entering the eye from an outdoor scene. Therefore, a vertical plane daylight factor was established by measuring light levels from two feet six inches inside a window and dividing this by the measurements taken outside at the window. Through field measurements in summer and winter taken in Lubbock, Texas and Pocatello, Idaho, the ratio's ranged from 40 to 66 percent. A low average of 45% was established.

DATA:

The data found in this study could be considered conservative (with the exception of the skylight bed) since the Graphic Daylight Design method measures light on a horizontal surface, not the near vertical light levels that would be taken in when viewing outside. However, the horizontal measurements are relevant when a person is reading a book or working from a horizontal plane.

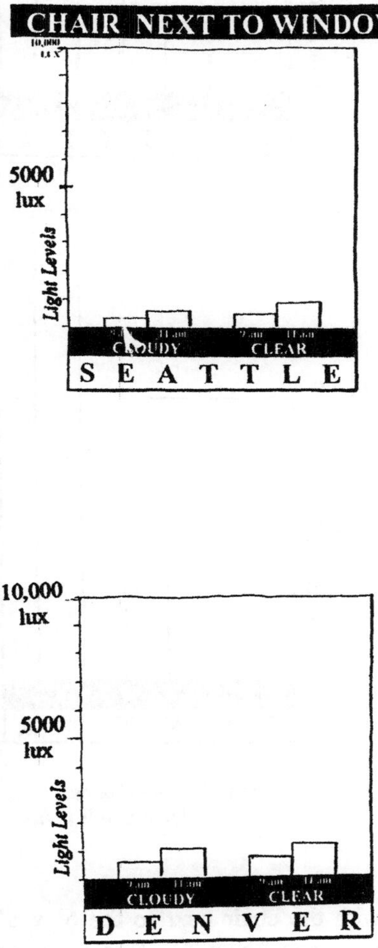

***Figure 4.* Chair Next to a Window. Light Levels in lux**

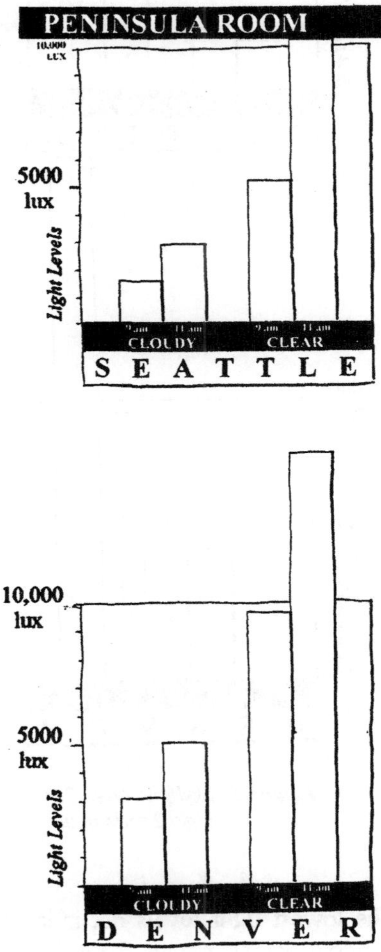

***Figure 5* Peninsula Room. Light Levels in lux**

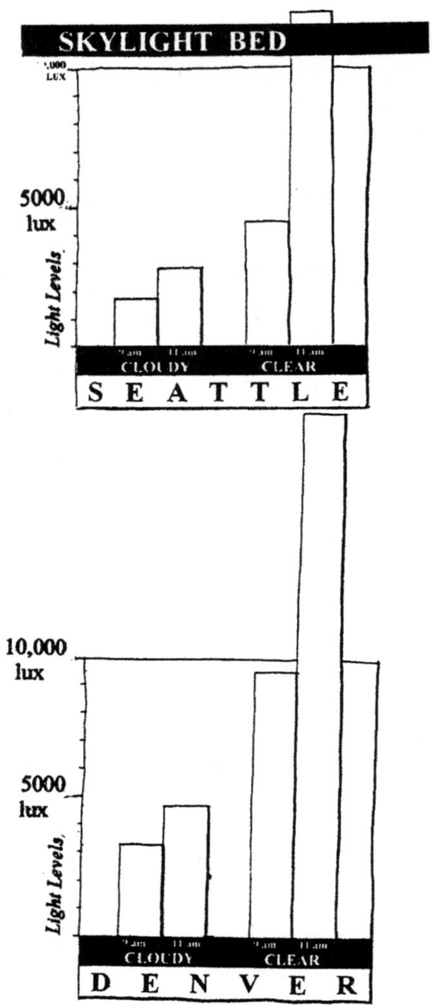

Figure 6 *Skylight Over Bed.*
Light Levels in lux

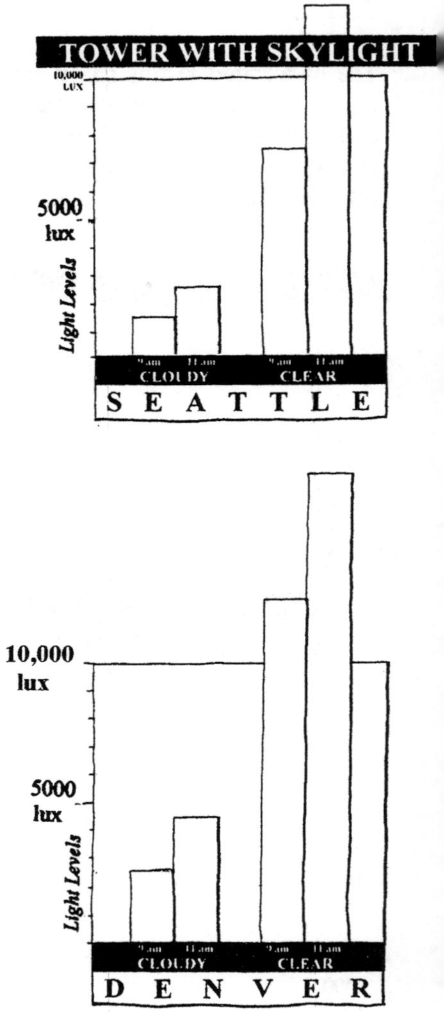

Figure 7 *Tower Space.*
Light Levels in lux

The lowest light levels occur in all cases with the chair next to the 6' x 6' window, which ranges from 300 to 1200 lux. (see figure 4) This supports Kripke's observation that sitting by a window in most cases does not have high enough lux values. (12) In contrast, the new designs on cloudy days at 9:00am in Seattle and Denver, range from 1400 lux to 3200 lux ...which is conceivably therapeutic. (Figs 5 - 7) 2500 lux is easily exceeded on cloudy Denver days at 11:00 am because all designs exceed 4300 lux. (Figs. 5 – 7)

Higher values are harvested at 9:00 am on clear Seattle days for all new designs range from 4300 to 7000 lux. (Figs. 5 – 7) On clear Denver days, 10,000 lux light box levels are approached or greatly exceeded with harvest levels ranging from 9000 to 18,000 lux. (Figs 5 - 7)

In examining light levels from windows compared to skylights, the skylights harvested from two to eight times more light than individual windows.

In analyzing the data, the percentage of clear days in both cities should be considered. In Seattle, there is slightly over 1/3 chance of sunshine from 9:00am to 4:00 pm in December while in Denver, there is nearly 80 percent chance of clear skies during the same time. (13)

CONCLUSION

For those suffering from depression and sleep disorders, the design solutions could strongly supplement rather than replace light box therapies. However, in Denver where blue skies predominate and the new designed spaces harvest at a minimum 9000 lux, artificial light might be replaced. On cloudy Denver days, bright artificial lighting could supplement daylighting. However, in all cases before 8:00am, artificial means would be needed. However in the spring and fall, the skylight bed could be useful in the early morning. Additionally, the skylight bed design could be used during lunch and mid-day breaks. The tower and peninsula room could be used while working, but it is unclear whether a computer screen could be comfortably used. Further study through ½ inch to full scale mock-ups need to be studied to see additional benefits and problems inherent in the proposals.

REFERENCES and NOTES:

1. Morgan Hickey, translator, *Vitruvius, The Ten Books on Architecture,* New York, Dover, 1914.
2. Rykwert Joseph. translator, *Leon Batsta Alberti, The Art of Building in Ten Books of Architecture* Cambridge Mass, MIT Press, 1988.
3. Nightingale, Florence *Notes on Nursing,* New York, Appleton, 1860.
4. Nightingale, Florence, *Notes on Hospitals,* 3rd edn, enl., London, Longman, Green, 1863
5. Stein, Benjamin and John Reynolds. *Mechanical and Electrical Equipment for Buildings* New York, John Wiley and Sons, 2000
6. 1500 to 2500 lux for an hour a day is recommended by Dr. Daniel Kripke in order to regularize sleep and curb depression. Kripke, Daniel. Phone interview by author, recorded on notepad, Lubbock Texas, May 25, 2001.
7. Ludlow, A.M. "The Function of Windows in Buildings" *Lighting Research and Technology 8,* 1976, pp. 57-68.
8. Ulrich, Roger. View through a window may influence recovery from surgery. *Science 224* April 1984, pp 420-21.
9. Wohlfarth, H. "The effects of color-psychodynamic environmental color and lighting modification of elementary schools on blood pressure and mood: a controlled study. *The International Journal of Biosocial Research 7,* 1985 no. 1 pp 9-16

10. Brown, G.Z. B. Haglud, J. Loveland, J. Reynolds and M. S. Ubbelohde. *Inside Out Design Procedures for Passive Environmental Technologies,* New York, John Wiley and Sons, 1992
11. Robbins, Claude. *Daylighting, Design and Analysis,* New York, Van Nostrand Reinhold, 1986
12. Kripke, Daniel, interview by author, recorded on notepad, San Diego, CA. June, 1999.
13 Robbins, 1986

VISUAL SIGNIFICANCE OF YELLOW OPTICAL FILTERS

Seymour Zigman, Boston University School of Medicine, Department of Ophthalmology Research Laboratory at Marine Biological Laboratory, Woods Hole, Massachusetts 02543

ABSTRACT

In this study, optical lenses were formulated to remove wavelengths shorter than 450 nm from various environmental light sources. They were tested for their ability to improve vision by preventing harmful and useless light from entering the eye, and to transmit nearly all visually useful wavelengths. A spectral colorimeter was used to determine the alterations of the transmission of mixed wavelengths of light caused by these yellow lenses. Further studies of how these lenses effect contrast vision in elderly and partially visually impaired human eyes were done.

The results have shown that the yellow lenses adjusted the wavelengths of light reaching the spectral colorimeter from various environmental light sources. They removed short wavelength light (<450 nm), but also transmitted from 500 to 600 nm very efficiently. Studies of contrast vision in elderly subjects and those with early cataracts (primarily) suggested that the yellow lenses improved contrast vision appreciably.

The findings show that an optical filter that can prevent short wavelength light from entering the eyes of elderly and visually compromised people has a positive influence on their vision.

INTRODUCTION

Yellow ocular lenses can enhance contrast vision by decreasing light-scattering anterior to the retina (Weale, 1989 ; Zigman, 1989, 1990,1992). Yellow optical filters may enhance vision by fine-tuning light wavelengths entering the eye into the maximum retinal spectral sensitivity region (i.e. 500-600-nm). Light-scattering and auto-fluorescence in the eye would be reduced by eliminating short wavelength light. The spectral radiance of common light sources was measured using a spectral colorimeter without and with a yellow filter, which prevents short wavelength blue and ultraviolet light from reaching the eye.

METHODS

A spectral colorimeter (Photo Research, Inc; Model PR 650) was used to detect the radiant energy derived from various light sources. A 2 mM thick yellow polycarbonate filter (SEE MORE®; cutoff filter < 450) was interposed between the light source and detector.

The data in Figure 1 was obtained using a VisTech Contrast Sensitivity (VCT S6500). Subjects sat a distance of 10 ft. The luminance, was 60 to 70 ft -L, from overhead cool-white 40W fluorescent lamps. The subjects, wearing their own corrective lenses, viewed sine wave displays (i.e. curved lines) in 2-1/2 " diameter circular areas. A series of such circular test areas with decreasing contrast was viewed across a top row from left to right by the subject. Rows of circles below the top row were of increasing spatial frequencies.

In some cases human contrast sensitivity was measured using a Pelli-Robson chart. This chart is composed of 2" black letters on a white background. The subject sat at 1 m from the chart so that the letters were at approximately 3° of angle, and the Snellen letter test acuity for these conditions would equate to 20.720. Room illumination was from cool-white fluorescent lamps with a radiance of approximately 100 cd/m^2. The conditions of the test are described in detail by Pelli, et al. 1988; Elliot, et al. 1991, and Zhang, et al. 1989.

In Table 1, column 1 the light sources are identified. The colorimeter was 30 cm from each source, except for the direct sky measurements. Column 2 indicates whether the measurements were made without or with the short wavelength filter. Column 3 presents the values for the total irradiance filter measured in Watts per steradian per meter square. Column 4 provides the wavelength maxima detected. The instrument also measured the radiance detected at the wavelength maxima as shown in column 5.

RESULTS

As shown in Table 1, emitted light with radiances predominantly at 400-500 nm was shifted to the 500-600 nm region by the yellow filter. This shift of radiant energy was into the region of maximal retinal spectral sensitivity.

Sources whose emission maxima were already in the 500-600 nm range was into the region were unaltered by the yellow filter.

Incandescent bulbs maximally at 780 nm. Their transmittance was unaltered by the filter. The total radiance and that at the wavelengths of maximum emission was reduced by the filter between 0 and 44%, The Phillips blue lamps reduced by 66% most radiances at the wavelength maxima were reduced by only 10%. The Phillips blue lamps and the blue sky radiances were reduced by approximately 35 to 40%.

DISCUSSION

The data suggest that a yellow filter that adjust the light entering the eye to the wavelengths of maximal sensitivity could improve contrast vision. Human subjects were tested to determine if this same yellow filter improves contrast vision and it was found that human contrast vision was improved by such filters (Zigman, 1990;1992). In another study (Figure 1), spatial contrast sensitivity measurements were made on elderly human subjects with or without common ocular lens deficiencies. Contrast function at high frequencies appeared to be enhanced in normal human subjects tested with a yellow filter, but only a trend and not a statistically significant difference were observed (Figure 1A). In subjects with varying degrees of cataract (i.e. prior to cataract surgery), a statistically significant improvement at high frequencies was observed (1B). In pseudophakic human eyes (i.e. cataractous lens removed and replaced with an intraocular lens implant), a significant improvement was observed, but only at the highest spatial frequencies (1C). Similar results were obtained using the Pelli-Robson chart (see Table 2).

From these observations we conclude that the yellow lenses used in this study, are capable of enhancing contrast vision by adjusting the light that can reach the retina to the region of maximal spectral sensitivity.

REFERENCES

1. Weale R. Aging and vision. Vis. Res 1986;26: 1507-1519.

2. Zigman S. Special features of the lens relative to the environment. Toxicity Res 1989;6:807-821.

3. Zigman S. Vision enhancement using a short wavelength light absorbing filter. Optom and Vis Res 1990;67:100 104.

4. Zigman S. Light filters to improve vision. Optom and Vis Sci 1992; 69:325-328.

5. Pelli DG., Robson JG.,Wilkins AM. The design of a new letter chart for measuring contrast sensitivity. Clin Vis Sci 1988;2:187-199.

6. Elliot DB., Bullimore MA., Bailey IL. Improving the relativity of the Pelli-Robson Contrast Sensitivity test. Clin Vis Sci 1991;5:471-475.

7. Zhang L., Pelli DG., Robson JG. The effects of luminance distance , and defocus on contrast sensitivity as measured by the Pelli-Robson chart. Invest Ophthalmol Vis Sci 1989;30:406.

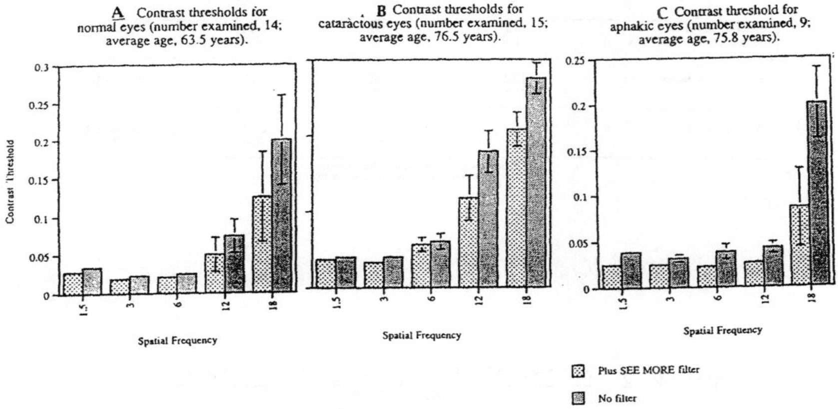

Figure 1. Contrast sensitivity measured by VisTech Chart a) Contrast threshold for normal eyes (number examined 14; average age 63,5 years). b) Contrast threshold for cataractous eyes (number examined 15: average age 76.5 years) c) Contrast threshold for aphakic eyes (number examined, 9; average age 75.8 years).
Dark bars = no filter; Light bars = use of filters.

TABLE 1 **EFFECT OF SEE MORE ® LENS FILTER ON WAVELENGTH AND RADIANCE**
by Seymour Zigman, Ph.D.

Lamp or Light Source	Minus or plus SeeMore Filter	Total Radiance (W/Sr/m²)	(λ -Max) wavelength of Maximum emission (nm)	Radiance at λ max	Change due to to SEE MORE® Filter
1. Sky through Window	-	10W/Sr/m²	412 nm	.0459	wavelength shift
	+	5.6	552	.0275	
2. X-Ray view box	-	7W/Sr/m²	544	.0536	9% lower
	+	4.6	544	.0481	
3. Reflection of Pelli Chart	-	0.433 W/Sr/m²	576	.00341	9.5% down
	+	0.348	576	.00325	
4. Phillips 60 W blue	-	16.5 W/Sr/m²	420	.044	wavelength shift
	+	5.6	544	.029	
5. Cool white 40 W lamps	-	7.3	544	.015	7% up
	+	8.5	544	.022	
6. 60 W incandescent bulb	-	10.7	780	.057	9% down
	+	9.4	780	.052	
7. Slide view box	-	6.3	436	.053	wavelength shift
	+	5.3	544	.053	

Measuring Device used: Photoresearch, Inc.
Spectral Colorimeter Model PR 650.

TABLE 2 MONOCULAR CONTRAST SENSITIVITY ON SUBJECTS WITH CATARACTS, RETINA DEGENERATION, PSEUDOPHAKIA*

Ocular Deficiency	Number	Average Age	Log Contrast Sensitivity No Filter	SM Filter	Corning Filter
Cataracts	15	77	1.15	1.31	1.13
Retinal Degeneration	12	73	1.23	1.33	1.23
Pseudophakia*	16	77	1.39	1.45	1.36

* No Lens-after Cataract Surgery

Suntanning and Pigmentation

THE ROLE OF DNA DAMAGE IN MELANOGENESIS: POTENTIAL ROLE FOR TELOMERES

Mark S. Eller*, Ina M. Hadshiew and Barbara A. Gilchrest
Boston University School of Medicine, Department of Dermatology, Boston, MA 02118

Introduction

Perhaps the best-studied and most widely recognized inducer of melanogenesis (tanning) in human skin is solar ultraviolet (UV) radiation. Although the regulation of melanogenesis is incompletely understood, evidence suggests that UV-induced DNA is at least one of the initial signals that stimulates melanogenesis in response to UV irradiation[1]. First, the action spectrum for the tanning response in human skin is essentially the same as that for the induction of the major DNA photoproducts[2,3]. Second, enhancing DNA repair by treatment of UV-irradiated melanocytic cells with the

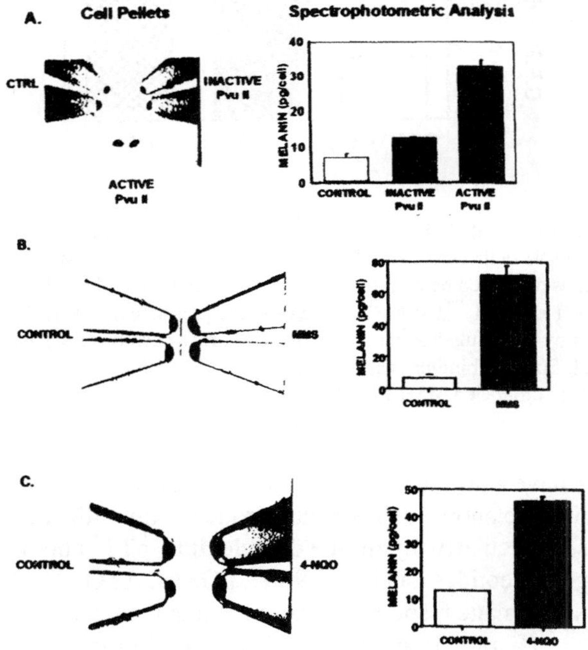

Figure 1. **Effect of DNA-damaging agents on pigmentation in S91 cells.**
Duplicate cultures of S91 cells were treated with the restriction enzyme *Pvu* II (A) or the DNA-damaging chemical agents MMS (B) and 4-NQO (C). After treatment, cells were cultured for 6 days, collected, and counted and an equal number of cells was pelleted for analysis. Melanin per cell was determined for each pellet. Averaged values of the duplicate cultures of each treatment are represented in the graphs. Because of the angle of centrifugation, only the thickest part of lightly pigmented cell pellets is visible, often making these pellets appear smaller. (Reprinted from Eller et al., ref 5)

prokaryotic DNA repair enzyme T4 endonuclease V approximately doubles the melanin content of these cells over a 4 day period as compared with irradiated cells treated with either heat- inactivated enzyme or diluent alone[4]. In addition, treatment of unirradiated S91 cells with DNA damaging agents such as methyl methanesulfonate (MMS), 4-nitroquinoline oxide (4-NQO) or the restriction endonuclease PvuII increases their melanin content (Figure 1). This results at least in part from the induction of the gene for tyrosinase, the rate-limiting enzyme in melanin synthesis[5], which is accompanied by an increase in the binding capacity for α-melanocyte stimulating hormone (α-MSH), a potent melanogenic hormone, to its cell surface receptor, as has been reported following UV-irradiation[6-9] (Figure 2). However, whether this induction of tyrosinase is dependent on α-MSH is unknown.

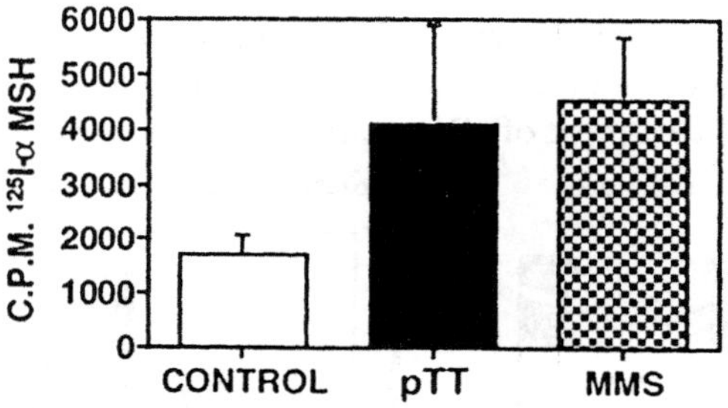

Figure 2. **Up-regulation of MSH binding to S91 cells by MMS and pTT.**
Duplicate cultures of S91 cells were treated with either diluent, DMEM, MMS, or thymidine dinucleotide (pTT), as described below, for 3 days. Cells were then collected and 1×10^3 cells were incubated with 1×10^{-9}M ^{125}I-α-MSH alone or in the presence of 5×10^{-6}M unlabeled α-MSH and processed by centrifugation through a 0.3 M sucrose cushion to separate un-bound from bound α-MSH. Specific binding was determined by subtracting ^{125}I counts detected in the presence of competing unlabeled MSH from those bound without competition. Data represent averages ± standard deviation of the duplicate cultures. (Reprinted from Eller et al., ref 5)

We and others have also shown that treatment of normal human skin cells, human and murine melanoma cells or intact rodent skin with specific small DNA fragments, particularly thymidine dinucleotide, pTT* and more recently other larger oligonucleotides (such as a 9mer, pGAGTATGAG), leads to the induction of DNA damage responses. These include increased melanogenesis[5,10,12](Figure 3), inhibition of cell proliferation[11-13], enhanced DNA repair capacity[13,14], and transient immunosuppression[15]. These responses are accomplished at least in part through induction of p53 and p53-regulated genes such as p21[13,14]. As with DNA damage, the increase in pigmentation in S91 cells treated with pTT is preceded by an increase in the level of

* Previously designated pTpT; in the oligonucleotide designation used here, "p" refers to a 5' phosphate group; all other bases are joined by phosphodiester linkages.

tyrosinase mRNA and protein[5,10]. Furthermore, treatment of these cells with pTT, as with DNA damaging drugs or restriction enzymes, increases the binding of α-MSH to the surface of these cells[5] (Figure 2) and increases their

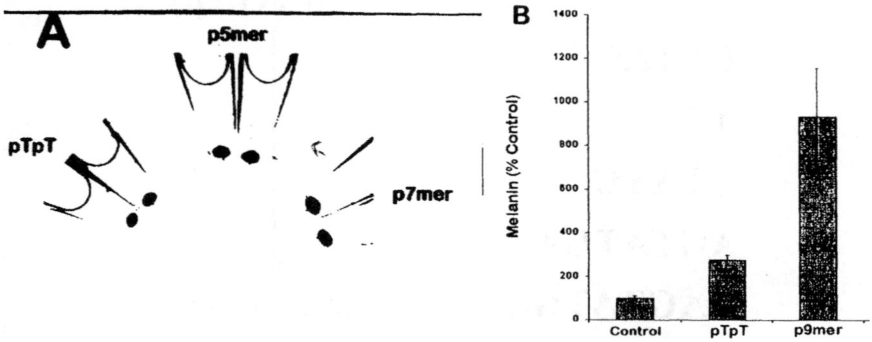

Figure 3. **Multiple oligonucleotides induce melanogenesis.**
Duplicate cultures of S91 cells were treated with 100μM oligonucleotide or diluent as a control. (A) Cell pellets of an equal number of cells from duplicate cultures of one representative experiment. (B) Melanin content (OD 495 nm) graphed as percent of diluent-treated controls (mean ± SD) for 3 independent experiments. (From Hadshiew et al., ref 11)

melanogenic response to α-MSH (Figure 4). More recently, using melanoma cells permanently transfected with a dominant-negative mutant form of p53 or the same cell line expressing wild type p53 alone, we have also shown that tyrosinase mRNA is induced by p53 activation in response to UV irradiation or treatment with the oligonucleotides[16,17]. Cells expressing the dominant-negative mutant form of p53 fail to induce tyrosinase in response to UV or pTT treatment. Current studies address whether p53 directly affects tyrosinase gene expression or whether it acts through an intermediary gene product. For example, whether pTT affects the expression of the α-MSH receptor through p53 is not known.

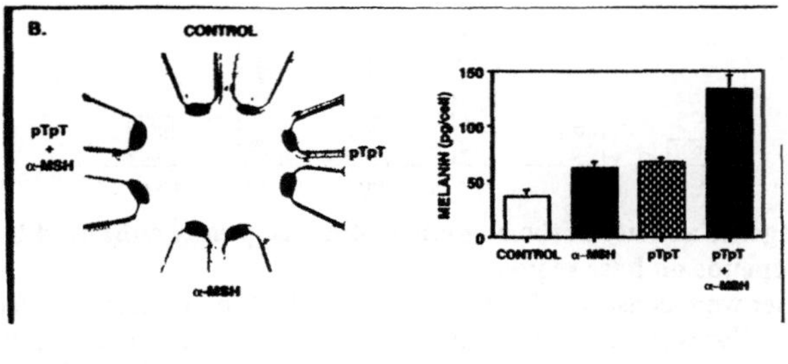

Figure 4. **Effect of MSH on pigmentation of pTpT-treated cells.**
Cultures of S91 cells were treated with pTpT as described above and cultured for 3 days. Duplicate plates of cells were then treated with 1 x 10^{-7}M α-MSH or diluent for an additional 3 days. Cells were then collected and an equal number of cells of each condition were pelleted for melanin analysis.

The melanogenic activity of these oligonucleotides generally increases with an increase in their molecular size, although sequence also has a major influence on this activity (Figure 5).

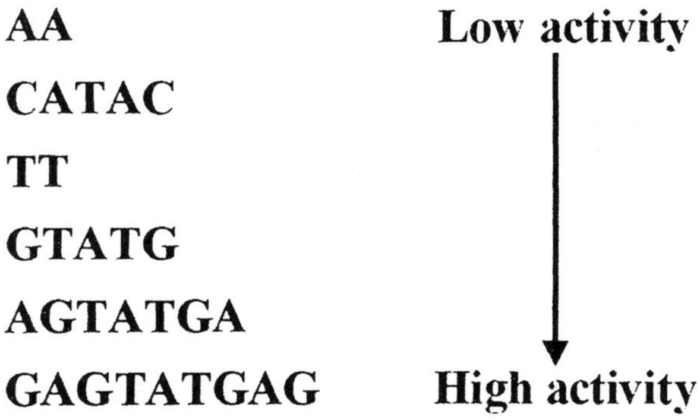

Figure 5. **Melanogenic activity of oligonucleotides depends on size and sequence.** Multiple oligonucleotides are arranged in order of their melanogenic activity. Generally, the activity increases with molecular size, but sequence also has a major influence.

For example, the dinucleotide pTT induces a 7-fold increase in melanin content in S91 cells[5,10], whereas the deoxyadenine dinucleotide pA\, the complementary sequence, is inactive[10]. Also, the 5-nucleotide sequence pGTATG (a truncated form of the 9mer described above), is a more potent stimulator than pTT whereas pCATAC, an oligonucleotide of identical size, is

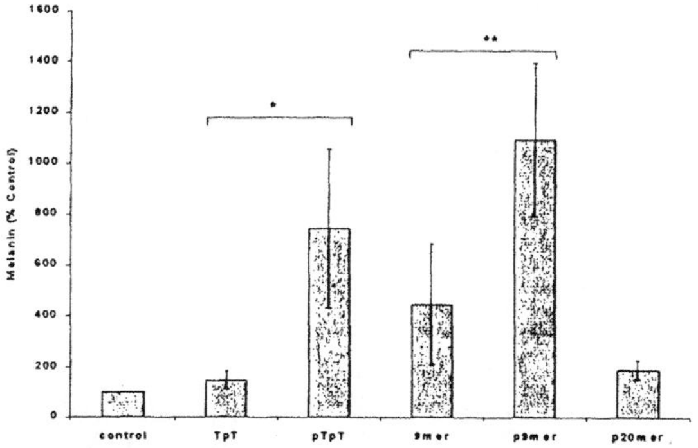

Figure 6. **Melanogenic activity of the oligonucleotides is greatly enhanced by a 5' phosphate and depends on base sequence.**
pTpT and the p9mer were constructed with and without the 5' phosphate and were added to triplicate S91 cell cultures at a concentration of 100 μM. In addition, a p20mer was also tested at 100 μM. The values were calculated as percent diluent-treated control. The graphed values represent three independent experiments. *p<0.004, **p<0.03, two-tailed student's t-test. (From Hadshiew et al., ref 11)

completely inactive[11]. In addition, the largest oligonucleotide tested to date, a 20mer composed of 5 repeats of GATC, fails to stimulate pigmentation above the level of control, diluent-treated cells (Figure 6)[11]. A phosphate on the 5' end of the oligonucleotide is critical for these oligonucleotides to be internalized by the cells, as demonstrated by confocal microscopy analysis using fluorescently labeled oligonucleotides, (Figure 7) [11]. Whether it is also separately necessary for their melanogenic activity is difficult to assess (Figure 6).

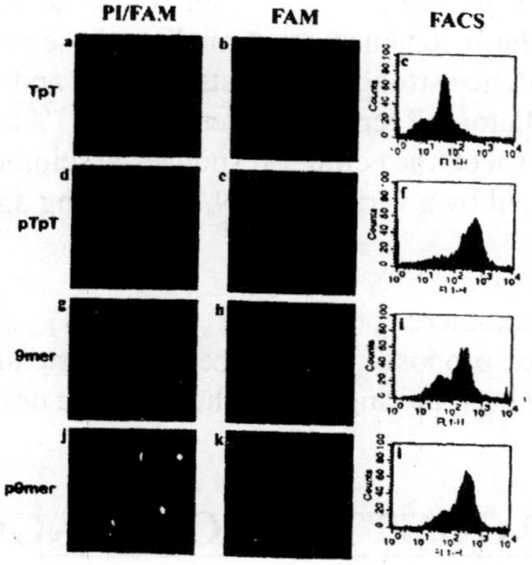

Figure 7. **Confocal microscopy and fluorescent activated cell sorter (FACS) analysis demonstrated FAM-labeled oligonucleotides accumulate intracellularly.**
Cultures of S91 cells were treated with FAM-labeled oligonucleotides for 4 hours and then processed for confocal microscopy or FACS analysis. Images a, d, g and j show both propidium iodide (red) and FAM-oligonucleotide (green) fluorescence. Co-localization of the two fluorescent molecules is assigned a yellow color by the computer program. Images b, e h and k show only the green FAM fluorescence. FACS analysis of FAM-labeled oligonucleotide-treated cells is presented in c, f, i and l. (From Hadshiew et al., ref 11)

Similarly, confocal analysis suggested that the more active oligonucleotides accumulate preferentially in the cell nucleus, compared to the less active pTT (Figure 7). These data strongly imply that these oligonucleotides act intracellularly, most likely in the cell nucleus.

The observation that these oligonucleotides stimulate multiple responses also stimulated by DNA damage suggests that they are detected within the cell as damaged DNA or as some consequence of this damage and in this way, trigger these responses[1]. Recent findings linking the disruption of telomeres with DNA damage-like responses have provided evidence for a more detailed and provocative hypothesis[18].

Most mammalian cells, including melanocytes, have a tightly regulated program of replicative senescence suggested to be a fundamental defense against cancer[19]. Cell senescence is controlled in large part by the length of telomeres, tandem repeats of the DNA sequence TTAGGG, several thousand base pairs long[20]. In germline cells and most cancer cells, immortality is associated with maintenance of telomere length by the reverse transcriptase subunit of the enzyme telomerase, TERT, that adds repeats of TTAGGG to the 3' ends of chromosomes[22-23]. TERT is generally not expressed in normal somatic cells[20], so telomeres become shorter with each round of DNA replication and, when critically shortened, trigger either replicative senescence or massive genomic instability and eventually cell death[24]. Although the molecular mechanisms that mediate these responses are not fully understood, evidence strongly suggests that p53 and the p53-regulated genes are major mediators of replicative senescence[25-27] and cell death by apoptosis[31]. In this respect, the cellular responses to telomere shortening are similar to those induced by a variety of DNA damaging agents, such as UV irradiation[28].

The 3' end of each telomere consists of a ...TTAGGG... single-stranded overhang that has been proposed to stabilize a loop structure at the chromosome ends by base-pairing within the telomere double helix[29] (Figure 8).

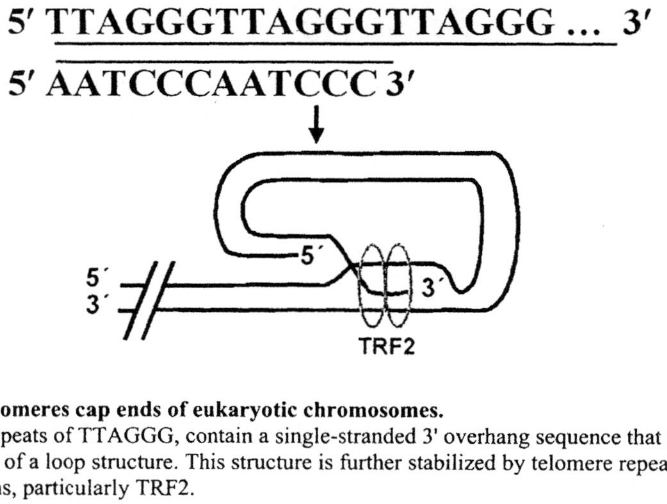

***Figure 8.* Telomeres cap ends of eukaryotic chromosomes.**
Telomeres, repeats of TTAGGG, contain a single-stranded 3' overhang sequence that facilitates the formation of a loop structure. This structure is further stabilized by telomere repeat factor (TRF) proteins, particularly TRF2.

In addition, mammalian telomeres are associated with two proteins, telomere repeat factors 1 and 2 (TRF1 and 2), which regulate telomere length and integrity respectively[30]. Disruption of the telomere-TRF structure by expression of a dominant negative mutant version of TRF2 ($TRF2^{DN}$) results in the exposure and degradation of the 3' telomere overhang, as well as the induction of p53 and p53-dependent apoptosis in certain cell types[31] (Figure 9).

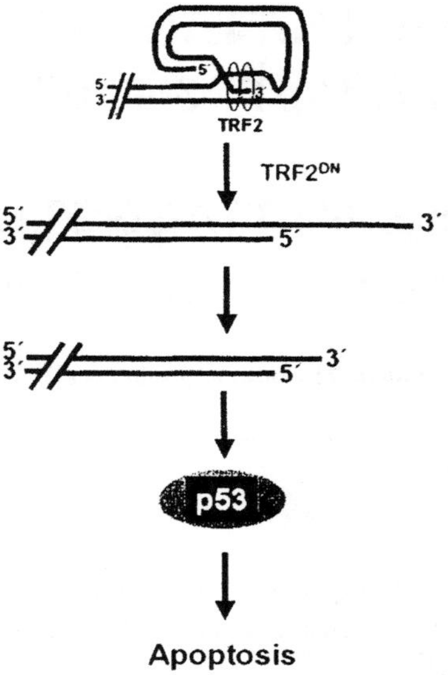

***Figure 9.* Telomere disruption leads to p53 induction of apoptosis.**
Disruption of the telomere loop structure by expression of a dominant negative mutant form of TRF2 leads to the exposure of the 3' overhang and induction of p53 and p53-dependent apoptosis in certain cells types[31].

Recently, Blackburn[32] proposed that telomere function is governed not simply by length, but also by an equilibrium between a "capped", or silent, state and an "uncapped" state that triggers cell cycle arrest and other DNA damage responses. The exact molecular nature of these two states and how they initiate DNA damage responses have not been described. Recently, however, Baumann and Cech reported a single-stranded telomere-binding protein in yeast and human cells and proposed that this protein caps and protects telomere ends[33].

Inspection of the base sequence of the active, melanogenic, oligonucleotides reveals that these DNAs show a substantial homology to the telomere 3' overhang repeat sequence, TTAGGG (Figure 10), ranging from having one-third to two-thirds identity. Furthermore, oligonucleotide sequences that did not induce the DNA damage responses, such as pAA and pCATAC , are not represented in this telomere sequence.

In melanocytes, a differentiated phenotype is characterized by increased melanogenesis, increased expression of the micropthalmia transcription factor and a cell cycle arrest resulting from decreased activity of the cyclin-dependent kinases which mediate progression through the cell cycle[34]. These

Overhang: 5′ ..TTAGGGTTAGGG.. 3′
pTT: 5′ pTT 3′ 5′ pTT 3′

Overhang: 5′ ..TTAGGGTTAGGG.. 3′
p9mer: 5′ pGAGTATGAG 3′

***Figure 10.* Melanogenic oligonucleotides and telomere 3' overhang show sequence homology.**
PTT and the 9mer oligonucleotide, both active stimulators of melanogenesis are compared to the telomere 3' overhang.

characteristics can be induced not only by telomere shortening[20], but also by oxidative[35] and genotoxic stress[28], certain oncogenic stimuli[36,37] and exposure to high levels of cyclic AMP[34].

Because critical telomere shortening during aging[38], experimental telomere disruption by expression of $TRF2^{DN}$ [31] and treatment with these DNA oligonucleotides[13,14] all lead to p53 induction, DNA damage responses and differentiation, and because all the active but none of the inactive oligonucleotides studied[10,11,15] have partial sequence homology to the telomere overhang, we hypothesize that exposure of single-stranded telomeric DNA is a primary physiologic signal that leads to p53 induction and subsequent DNA damage responses such as cell senescence, apoptosis or melanogenesis, depending on the cell type and signal intensity. Furthermore, we speculate that the DNA oligonucleotides mimic the exposed telomeric DNA in the cell nucleus and thus trigger the same responses in the absence of DNA damage or telomere disruption.

Together, these data suggest a model whereby exposure of the single-stranded 3' telomeric DNA is a critical signal for the induction of DNA damage responses such as cell cycle arrest/replicative senescence, a more differentiated phenotype (such as enhanced melanogenesis) or apoptosis, depending on cell type and/or signal intensity. Exposure of this DNA could occur during normal DNA replication or from destabilization of the telomere loop structure due to critical telomere shortening or from DNA damage, such as thymine dimers formed as a result of UV irradiation. In this respect, telomeres in mammalian cells may not only serve as a measure of replicative age but also of genotoxic stress and may contribute to the generation of DNA damage responses. Additional experiments using inhibitors of these telomere responses will further define their role in constitutive as well as UV-induced pigmentation.

References

1 Gilchrest, B.A., and Eller, M. S. DNA photodamage stimulates melanogenesis and other photoprotective responses. *J Investig Dermatol Symp Proc* 1999; 4: 35-40.

2 Parrish, J.A., Jaenicke, K.F., Anderson, R.R. Erythema and melanogenesis action spectra of normal human skin. *Photochem Photobiol* 1982; 36: 187-91.

3 Freeman, S.E., Hacham, H., Gange, RW., et al. Wavelength dependence of pyrimidine dimer formation in DNA of human skin irradiated in situ with ultraviolet light. *Proc Natl Acad Sci U S A* 1989; 86; 5605-9.

4 Gilchrest, B.A., Zhai, S., Eller, M.S., et al. Treatment of human melanocytes and S91 melanoma cells with the DNA repair enzyme T4 endonuclease V enhances melanogenesis after ultraviolet irradiation. *J Invest Dermatol* 1993; 101: 666-72.

5 Eller, M.S., Ostrom, K., Gilchrest, B.A. DNA damage enhances melanogenesis. *Proc Natl Acad Sci USA*, 1996; 93:1087-1092.

6. Hadley, M.E.,Levine, N. Pigmentation and Pigmentary Disorders, ed. Levine, N. CRC, Boca Raton, FL 1993: 95-114.

7. Fuller, B.B., Lunsford, J.B., Iman D.S., et al. Alpha-melanocyte-stimulating hormone regulation of tyrosinase in Cloudman S-91 mouse melanoma cell cultures. *J Biol Chem.* 1987; 262: 4024-4033.

8. Donatien, P.D., Hunt, G., Pieron, C. The expression of functional MSH receptors on cultured human melanocytes. *Arch Dermatol Res.* 1992; 284:424-426.

9. Bolognia, J., Murray, M., Pawelek, J. UVB-induced melanogenesis may be mediated through the MSH-receptor system. *J Invest Dermatol.* 1989; 92: 651-656.

10. Eller, M.S., Yaar, M., Gilchrest, B.A. DNA damage and melanogenesis. *Nature* 1994; 372: 413-414.

11. Hadshiew, I.M., Eller, M.S., Gasparro, F.P., Gilchrest, B.A. Stimulation of melanogenesis by DNA oligonucleotides: effect of size, sequence and 5' phosphorylation. *J Dermatol Sci* 2001; 25: 127-138.

12. Pedeux,R, Al-Irani, N., Mateau, C., et al. Thymidine dinucleotides induce S phase cell cycle arrest in addition to increased melanogenesis in human melanocytes. *J Invest Dermatol* 1998; 111:472-477.

13. Eller, M.S., Maeda, T., Magnoni, C., et al. Enhancement of DNA repair in human skin cells by thymidine dinucleotides: Evidence for a p53-mediated mammalian SOS response. *Proc Natl Acad Sci USA* 1997; 94:12627-12632.

14. Maeda, T, Eller, MS, Hedayati, M, et al. Enhanced repair of benzo(a)pyrene-induced DNA damage in human cells treated with thymidine dinucleotides. *Mutat Res* 1999; 433:137-145.

15. Cruz, P.D .Jr, Leverkus, M., Dougherty, I., et al. Thymidine dinucleotides inhibit contact hypersensitivity and activate the gene for tumor necrosis factor α. *J Invest Dermatol* 2000; 114:253-258.

16. Khlgatian, M., Asawanonda, P., Eller, M.S. et al.Tyrosinase expression is regulated by p53. *J Invest Dermatol* 1999; 112: 548.

17. Khlgatian, M.K., Hadshiew, I.M., Asawanonda, P. et al.Tyrosianse gene expression is regulated by p53. *J Invest Dermatol*, (In press).

18. Eller, M.S. Hadshiew I.M., Puri N. et al. The single-stranded telomeric DNA induces DNA damage responses. *J Invest Dermatol* 2000;114:756.

19. Campisi, J. Replicative senescence: an old lives tale? *Cell* 1996; 84:497-500.

20. Greider, C.W. Telomere length regulation. *Annu Rev Biochem* 1996; 65:337-365.

21. Feng, J., Funk, W.D., Wang, S.S., et al. The RNA component of human telomerase. *Science* 1995; 269:1236-1241.

22. Harrington, L., McPhail, T., Mar, V., et al. A mammalian telomerase-associated protein. *Science* 1997; 275:973-977.

23. Nakamura, T.M., Morin, G.B., Chapman, K.B., et al. Telomerase catalytic subunit homologs from fission yeast and human. *Science* 1997; 277:955-957.

24. de Lange, T. Telomeres and senescence: ending the debate. *Science* 1998; 279:334-335.

25. Dimri, G.P., Itahana, K., Acosta, M., Campisi, J. Regulation of a senescence checkpoint response by the E2F1 transcription factor and $p14^{ARF}$ tumor suppressor. *Mol Cell Biol* 2000; 20:273-285.

26. Atadja, P., Wong, H., Garkavtsev, I. Increased activity of p53 in senescing fibroblasts. *Proc Natl Acad Sci U S A* 1995; 92: 8348-8352.

27. Campisi, J., Dimri, G.P., Hara, E. Control of replicative senescence, In E. Scneider and J Rowe (ed.) Handbook of the biology of aging, 4th ed. Academic Press, New York, NY. 1996; 121-149.

28. DiLeonardo, A., Linke, S.P., Clarkin, K., Wahl, G.M. DNA damage triggers a prolonged p53-dependent GI arrest and long-term induction of Cip1 in normal human fibroblasts. *Genes Dev.* 1994; 8: 2540-2551.

29. Griffith, J.D. Comeau, L., Rosenfeld, S., et al. Mammalian telomeres end in a large duplex loop. *Cell* 1999; 97:503-514.

30. van Steensel, B., Smogorewska, A., de Lange, T. TRF2 protects telomeres from end-to-end fusions. *Cell* 1998; 92:401-413.

31. Karlseder J., Broccoli D., Dai Y., et al. p53- and ATM-dependent apoptosis induced by telomeres lacking TRF2. *Science* 1999; 283:1321-1325.

32. Blackburn, E. Telomere states and cell fates. *Nature* 2000; 408: 53-56.

33. Baumann, P., Cech, T.R. Pot1 the putative telomere end-binding protein in fission yeast and humans. *Science* 2001; 292: 1171-1175.

34. Haddad, M.M., Xu, W., Schwahn, D.J., Liao, F., et al. Activation of a cAMP pathway and induction of melanogenesis correlate with association of p16(INK4) and p27(KIP1) to

CDKs, loss of E2F-binding activity, and premature senescence of human melanocytes. *Exp Cell Res* 1999; 253: 561-72.

35. Chen, Q.M., Bartholomew, J.C., Campisi, J., et al. Molecular analysis of H2O2-induced senescent-like growth arrest in normal human fibroblasts: p53 and Rb control G1 arrest but not cell replication. *Biochem J* 1998; 332: 43-50.

36. Serrano, M., Lin, A.W., McCurrach, M.E., et al. Oncogenic ras provokes premature cell senescence associated with accumulation of p53 and p16INK4a. *Cell* 1997; 88: 593-602.

37. Zhu, J.,Woods, D., McMahon, M., Bishop, J.M. Senescence of human fibroblasts induced by oncogenic Raf. *Genes Dev*. 1998; 12: 2997-3007.

38. Stein, G.H., Drullinger, L.F., Soulard, A. and Dulic, V. Differential roles for cyclin-dependent kinase inhibitors p21 and p16 in the mechanisms of senescence and differentiation in human fibroblasts. *Mol. Cell. Biol.* 1999; 19: 2109-2117.

39. Dimri, G.P., Lee, X., Basile, G., et al. A biomarker that identifies senescent human cells in culture and in aging skin in vivo. *Proc Natl Acad Sci U S A* 1995; 92: 9363-9367.

HOW MUCH PHOTOPROTECTION DOES A TAN AFFORD?

Antony R Young
Department of Environmental Dermatology
St John's Institute of Dermatology
Guy's, King's and St Thomas' School of Medicine
King's College London
St Thomas' Hospital
London SE1 7EH

SUMMARY

Pigmentation, whether constitutive (i.e. base skin colour) or facultative (induced by UVR), is considered to have photoprotective properties in human skin. Much of the evidence for this is epidemiological and is based on the inverse relationship between skin cancer risk and skin pigmentation. Some studies have evaluated the photoprotective role of facultative pigmentation using erythema and epidermal DNA photodamage caused by a single UVR exposure. In general, pigmentation-related protection factors against these endpoints have been shown to be about 2-3 in different white skin types and were independent of the level of pigmentation. In vitro studies suggest that epidermal pigmentation may be the mammalian equivalent of a bacterial SOS response. We have shown no difference in human epidermal DNA repair capacity in different skin types after a single exposure to solar simulating radiation (SSR). However, repeat SSR exposure studies suggest that DNA repair may be induced in skin types IV that tan well, but not in skin types II that tan poorly. One interpretation of these studies is that pigmentation ability is an indicator of inducible DNA repair capacity and it is this, rather the photoprotective properties of pigment per se, this is important in the prevention of skin cancer.

1. INTRODUCTION

Human skin may be classified according to its constitutive colour and its ability to undergo melanogenesis (tanning) in response to UVR exposure. These properties are often described as constitutive and facultative pigmentation respectively and, along with sensitivity to sunburn, are the basis of skin phototype as summarized in Table 1. The adverse acute and long-term effects of solar exposure are well established and, in general, are inversely related to skin type, with skin type I being the most sensitive and skin type VI being the most resistant. It is widely assumed that sensitivity to UVR is directly related to pigmentation or tanning ability and this assumption is primarily based on epidemiological evidence that shows that skin cancers are much less common in people who tan well (e.g. skin types III and IV) or who have high levels of constitutive pigmentation (e.g. skin types V and VI). Furthermore, studies comparing dark skinned peoples with related albinos show the latter to have a higher incidence of photodamage (McFadden, 1961) and non-melanoma skin cancer (Lookingbill et al, 1995). However, comparisons based on pigmentation alone may fail to take into account other factors that may be important in UVR sensitivity, such as DNA repair capacity and sensitivity to UVR-induced suppression that is widely believed to be important in human skin cancer (Euvrard et al, 1997). For example, studies by Kelly et al (2000) have shown that on previously unexposed white buttock skin, cell mediated immunity is more readily suppressed in skin types I/II when compared with skin types III/IV. These differences were highly significant and the observations held good whether solar simulated radiation (SSR) doses were expressed in terms of individual MED or in J/cm^2. Such data strongly suggest inherent differences in UVR sensitivity that are not obviously related to erythema or pigmentation.

Skin Type	**Susceptibility to Sunburn**	**Constitutive Skin Colour**	**Facultative Tanning Ability**	**Susceptibility To Skin Cancer**
I	high	white	very poor	high
II	high	white	Poor	high
III	moderate	white	good	moderate
IV	low	olive	very good	low
V	very low	brown	very good	very low
VI	very low	black	very good	very low

Table 1 Classification of human skin types with respect to relative response to acute and long-term solar exposure

This chapter reviews the evidence for the photoprotective properties of UVR-induced pigmentation against the acute effects of UVR and will focus primarily on human studies in vivo that have been done with environmentally relevant sources of UVR.

2.PHOTOPROTECTION BY MELANIN

Photodermatological literature often states that human pigmentation is photoprotective, without qualitative and quantitative comment. For example, there is often a lack of differentiation between constitutive and facultative pigmentation and information is often lacking on the level of protection that is provided by pigmentation. This view generally assumes that the primary function of melanin is that of an optical filter. Whilst photoprotection may be considered to be a passive physical process, e.g. the attenuation of UVR as is the case with conventional sunscreens, it may also be considered as an active enzymatic process, e.g. as a means by which DNA repair is enhanced or reactive oxygen species (ROS) are inactivated. For example, in vitro chimeric epidermal reconstructs with melanocytes from one skin type added to keratinocyte cultures of a different skin type suggest keratinocyte/melanocyte interaction with both cell types regulating antioxidant defence in a skin type dependent way (Bessou-Touya et al, 1998).

2.1. Erythema

Photoprotection by melanin has been approached using ex vivo and in vivo experimental techniques. The former have relied on UVR transmission measurements. Kaidbey et al (1979) used spectroscopic and biological techniques, in which cadaver skin was used to attenuate UVR, came to the conclusion that black skin attenuates 5 times as much UVR as white skin.

Cripps (1981) assessed the protection factor afforded by a Wisconsin (~45°N) summer tan (obtained over 3.5 months) in skin types II, III and IV by comparing the solar simulating radiation (SSR) MED on tanned and untanned buttock skin. Protection factors were 2.4 ± 0.65 (SD), 2.45 ± 0.5 and 2.1 ± 0.22 respectively with a mean of 2.33 ± 0.05. Although not stated, it is presumed that tanning was according to skin type and it seems that better tans did not afford better protection factors against erythema. Melanogenesis and stratum corneum thickening occur during the normal tanning response. Some studies have attempted to delineate between the photoprotective properties of each. Kaidbey and Kligman (1978) induced pigmentation by repeated UVA exposure without appreciable thickening of the stratum corneum. Skin type was not specified but was probably III/IV. Gniadecka et al (1996) assessed the photoprotective role of the stratum corneum in vitiligenous and normal adjacent skin (skin type not specified) that had had no UVR exposure for 3 months. Sensitivity to erythema was compared with the thickness of the stratum corneum and the viable epidermis as well as pigmentation assessed by a reflectance device. Overall, the authors concluded that the stratum corneum accounted for almost 2/3 of the photoprotection of normal skin and therefore was more important than pigmentation and that the thickness of the viable epidermis was not important in photoprotection.

Comparisons of tanned and untanned skin showed that a tan afforded a protection factor of about 2-3 against erythema induced by solar simulating radiation. Sheehan et al (1998) induced tanning in skin types II and III with repeated sub-erythemal exposure to SSR. Photoprotection was assessed by assessing the MED on untanned and tanned sites, included tanned sites that had had the stratum corneum removed by tape-stripping. As expected, the tanning response in skin type III was greater than in skin type II but there was no significant difference in the level of photoprotection with a protection factor of about 2 in both skin types. Removal of the stratum corneum generally reduced the level of photoprotection by about 20%, which suggested that the stratum corneum was relatively unimportant in the tanning response. However, the degree of photoprotection could not be correlated with the level of the tan. In a comparable later study but without the tape-stripping component, Sheehan et al (2001a) used a similar protocol in skin types II and IV. Despite the difference in the tanning response, the protection factor against erythema, assessed by MED determination, was again of the order of 2 with no difference between the skin types (Sheehan et al, 2001a). Overall, the data of Sheehan et al (1998, 2001a) in skin types II, III and IV show that the degree of photoprotection against erythema was very similar but could not be correlated with the level of tan, as presumed to be the case with the field study of Cripps (1981). This suggests that other, as yet undefined, inducible forms of photoprotection may be in operation and that the level of melanin may be of less importance.

2.2 DNA Photodamage

The effects of UVR on DNA include the formation of dipyrimidine lesions such as cyclobutane pyrimidine dimers (CPD) and pyrimidine (6-4) pyrimidine photoproducts (6-4). Human studies have shown that CPD and (6-4) are readily induced with sub-erythemal exposure of SSR, UVB and UVA (Young et al, 1996, 1998). There is increasing evidence that epidermal DNA is a major chromophore for many of the acute and long-term effects of solar exposure (Young, 1999). Photodamage to DNA may result in gene mutation, e.g. p53, that is thought to be the initial step in non-melanoma skin cancer. However, as stated above, DNA is also considered to be the chromophore for erythema and melanogenesis.

Furthermore there is also considerable evidence that DNA photodamage, especially CPD, initiates many of the immunological effects of UVR (Young, 1999). Given the significance of DNA photodamage it is not surprising that living organisms have developed highly effective DNA repair mechanisms. Studies in our laboratory and by others have shown that, after a single exposure of SSR, (6-4) repair in human skin in situ is relatively rapid and that CPD repair is much slower with many lesions persisting for at least 24 hours (Young et al, 1996, Bykov et al, 1999). Such slow repair also means that lesion induction may be cumulative if skin is exposed to UVR the following day, as is the case in "real life". There is also evidence that cytosine containing lesions (C=C, C=T) are repaired more rapidly than those with thymine only (T=T) (Bykov et al, 1999, Xu et al, 2000). Apart from the DNA repair response, it might also be expected that photoprotection by melanin would include protection against DNA photodamage in situ.

One study compared the SSR-induction of T=T in white and black skin ex vivo and came to the conclusion that constitutive pigmentation afforded DNA protection factors of 2-4 (Strickland et al, 1988). Bykov et al (2000) reported an inverse relationship between constitutive pigmentation in white-skinned people and UVB-induced epidermal DNA photodamage. Some studies have assessed the ability of UVR-induced pigmentation in white skin to protect epidermal cells from CPD by a subsequent challenge dose of UVR. Gange et al (1985) reported that UVB and UVA induced tans, in people who tanned well, resulted in protection factors of about 2 against the induction of epidermal CPD by UVB. Sheehan et al (2001a) treated skin types II and IV with 0.65MED SSR for two weeks and exposed SSR-treated and untreated skin to 2MED SSR one week after the last tanning treatment. An analysis of the data, taking into account CPD caused by the tanning treatment, showed that pigmentation was associated with a DNA protection factor of about 2-3. Surprisingly, despite the superior tan in skin type IV, the level of photoprotection was not significantly higher than skin type II. Assessment of photoprotection against erythema was also made and the protection factors for erythema and CPD were virtually identical, as might be expected in DNA is the chromophore for erythema. These data are in contrast to that of Gange et al (1985) who, as stated above, reported that comparable UVB and UVA induced tans gave comparable protection against UVB-induced CPD, but that only the UVB tan resulted in protection against UUB-induced erythema with a protection factor of 3. These data suggest a lack of relationship between DNA photodamage and erythema.

In a study of skin explants taken from habitually sun-exposed skin of two people of skin type III, Kobayashi et al (1998) showed an inverse relationship between the level of melanin in supranuclear caps and UVB-induced DNA photodamage in the corresponding keratinocytes. The authors showed DNA protection factors ranging from just over 1 to about 5.

Barker et al (1995) compared the UVB dose-responses for CPD in vitro in human melanocytes of different skin type. Melanocytes from skin types IV-VI showed much more tyrosinase activity and melanin than those from skin type I-II. However, the dose-response data show a relatively modest difference in CPD from a single UVB exposure that suggests a protection factor of less than 2. Melanocytes and adjacent basal layer keratinocytes in untanned human skin in situ show similar sensitivity to the induction of CPD by UVB and UVA suggesting no inherent differences in sensitivity to DNA photodamage (Young et al, 1998). In combination, the data from these two studies would suggest that melanocytes have no great advantage over keratinocytes in terms of photoprotection from DNA damage.

Studies on reconstructed human skin with and without melanocytes from skin type II-III showed that the presence of melanocytes had no effect on the induction of CPD or (6-4) (Cario-Andre et al, 2000). This perhaps is not surprising, as melanin was not detected in the preparations. However, the presence of melanocytes inhibited the formation of apoptotic sunburn cells (SBC) that are considered to be a means of eliminating keratinocytes with mutagenic and therefore carcinogenic potential. In theory, a reduction of SBC is likely to enhance skin cancer risk. These data suggest that melanocytes might influence the effects of UVR in ways that are unrelated to melanin production. However, recent studies in our laboratory have not shown any skin type (I-IV) dependent differences in SBC formation on untanned skin (Sheehan et al, 2001b).

Overall, several human studies suggest that the acute photoprotection afforded by constitutive and induced pigmentation against DNA photodamage in keratinocytes and melanocytes and erythema is equivalent to wearing a sunscreen with a sun protection factor (SPF) of 2-3. The comparable results with DNA photodamage and erythema provide additional evidence that DNA is an important chromophore for erythema. Protection factors of 2-3 should result in a 50-60% reduction of biologically effective dose. Such a reduction, if maintained over long periods, would be significant in terms of long-term risk of skin cancer. However, the maintenance of a tan requires repeated exposure to UVR, either from the sun or an artificial source and it is likely that this is associated with the accumulation of epidermal DNA photodamage (Sheehan et al, 2001a) that may minimize the benefits of the protection. There are conflicting data about the role of a tan in affording protection against malignant melanoma in woman with Weinstock et al (1991) suggesting that a tan may afford protection and Holly et al (1995) reporting no benefit from a tan.

2.3. Facultative Tanning and Erythema may be Indicators of DNA Repair Status

The lack of excision repair of epidermal DNA is believed to be the major reason for the exceptionally high and early incidence of skin cancer in xeroderma pigmentosum (XP) patients. Compensation of this repair deficiency by the topical application of liposome encapsulated T4N5 in an 18 month study has been shown to significantly reduce the incidence of actinic kerstoses (Yarosh et al, 2001) in XP patients even though the effect on this topical treatment on epidermal DNA repair, at least for T=T, in XP patients in situ is relatively modest (Yarosh et al, 1996).

Our laboratory has shown that repeated daily sub-erythemal exposure to SSR in skin types I and II results in a cumulative and persistent erythema with no tanning in skin type I and minimal tanning in skin type II (Young et al, 1991, Sheehan et al, 1998). In contrast, similar exposure protocols, based on given daily MED exposures, in skin types III and IV show transient erythema and a good tanning response, whether assessed histologically, clinically or by reflectance measurements (Young et al, 1991, Sheehan et al, 2001a). Such protocols also show an increase in the number of cell layers in the stratum corneum that was independent of skin type (Young et al, 1991). Excision repair of epidermal DNA from a single UVR exposure has been shown to have large inter-person variation that is independent of skin type (Xu et al, 2000) and this is also the case in our laboratory (unpublished). However, we have also compared the effects of repeated sub-erythemal (0.65MED) on DNA repair after SSR exposure in skin types II and IV (Sheehan et al, 2001a). In these studies, the daily SSR dose received by skin type IV was approximately twice that received by skin type II when expressed in physical units, i.e. J/cm2. An assessment of epidermal DNA photodamage at the end of the two week tanning protocol showed a higher level of DNA damage in skin type IV, presumably as a direct result of the higher daily SSR dose received. However, a similar assessment made one week after the last SSR exposure showed a non-significant reduction in DNA damage in skin type II and a highly significant reduction in skin type IV. These data support the hypothesis proposed by Gilchrest et al (1993) that DNA repair is an inducible response comparable to the SOS response in bacteria. They also suggest the tanning response may be better regarded as a clinical indicator of DNA repair capacity rather than a primary method of photoprotection by optical screening of UVR. Furthermore, they are consistent with the hypothesis that DNA is the chromophore for erythema and that DNA photolesions initiate and, if not readily repaired, maintain the persistent erythema response as seen in skin types I and II. In this context it should be noted that XP patients often have a persistent erythema response but also seem to be capable of tanning, although it is certain that systematic studies in XP patients will not have been done for erthical reasons.

Additional human studies will need to be done to confirm the hypothesis that tanning ability is an indicator of inducible epidermal DNA repair capacity and that this may be more important that the optical screening, or other, properties of induced melanin which have been shown to be rather limited or remain speculative. Such a hypothesis might be extended to propose that the decreasing susceptibility to skin cancer in white skin types I through to IV is due to increasing DNA repair inducibility rather than any inherent photoprotective properties of melanin per se. Quite apart from inducible DNA repair, it is also possible that antioxidant, immunological and other defences may also be inducible and that these may account for the adaptive properties observed in skin after repeated UVR exposure. The differences that we have observed in DNA repair capacity between acute and repeat exposure stress the need to do "real life" studies in human skin in vivo.

3. CONCLUSIONS

There is no doubt that constitutive pigmentation or the ability to tan reduces the adverse acute and long term effects of solar UVR exposure. It is has been widely assumed that this is mainly due to the optical screening properties of melanin per se. However, there is increasing evidence that skin type related UVR responsivity, that is believed to be associated with melanogenesis, may be more complex than previously thought. In particular, it seems possible that the ability to tan is a reflection of DNA repair capacity. Furthermore, studies in our laboratory in skin types II, III and IV have shown modest but comparable levels of photoprotection against acute DNA photodamage/erythema despite clear differences in the degree of pigmentation induced by repeated exposure to SSR. These data suggest the possibility of other defence mechanisms that may be as important as melanogenesis.

REFERENCES

Barker D., Dixon K., Medrano E.E., Smalara D., Mitchell S.I.D., Babcock G., Abdel-Malek Z.A. Comparison of the responses of human melanocytes with different melanin contents to ultraviolet B irradiation. Canc Res 1995; 55:4041-46

Bessou-Touya S., Picardo M., Maresca V., Surleve-Bazeille J.E., Pain C., Taieb A. Chimeric human epidermal reconstructs to study the role of melanocytes and keratinocytes in pigmentation and photoprotection. J Invest Derm 1998; 111:1103-08

Bykov V.J., Sheehan J.M., Hemminki K., Young A.R. In situ repair of cyclobutane pyrimidine dimers and 6-4 photoproducts in human skin exposed to solar simulating radiation. J Invest Derm 1999; 112: 326-31

Bykov V.J., Marcusson J.A., Hemminki K. Effect of constitutional pigmentation on ultraviolet B-induced DNA damage in fair-skinned people. J Invest Derm 2000; 114: 40-3

Cario-Andre M., Bessou S., Gontier E., Maresca V., Picardo M., Taieb A. The reconstructed epidermis with melanocytes: a new tool to study pigmentation and photoprotection. Cell Mol Biol 1999; 45:931-42

Cripps D.J. Natural and artificial photoprotection. J Invest Derm 1981; 77:154-7

Euvrard S., Kanitakis J., Pouteil-Noble C., Claudy A., Touraine J.L. Skin cancers in organ transplant recipients. Annal Trans 1997; 2:28-32

Gange R.W. Blackett A.D., Matzinger E.A., Sutherland B.M., Kochevar I.E. Comparative protection efficiency of UVA- and UVB-induced tans against erythema and formation of endonuclease-sensitive sites in DNA by UVB in human skin. J Invest Derm 1985; 85:362-64

Gilchrest B.A., Zhai S., Eller M.S., Yarosh D.B., Yaar M. Treatment of human melanocytes and S91 melanoma cells with the DNA repair enzyme T4 endonuclease V enhances melanogenesis after ultraviolet irradiation. J Invest Derm 1993; 101:666-72

Gniadecka M., Wulf H.C., Mortensen N.N., Poulsen T. Photoprotection in vitiligo and normal skin. A quantitative assessment of the role of stratum corneum, viable epidermis and pigmentation. Acta Dermato-Venereologica 1996; 76:429-32

Holly E.A., Aston D.A., Cress R.D., Ahn D.K., Kristiansen J.J. Cutaneous melanoma in women. I. Exposure to sunlight, ability to tan, and other risk factors related to ultraviolet light. Amer J Epidem 1995; 141:923-33

Kaidbey K.H., Kligman A.M. Sunburn protection by longwave ultraviolet radiation-induced pigmentation. Arch Derm 1978; 114:46-8

Kaidbey K.H., Agin P.P., Sayre R.M., Kligman A.M. Photoprotection by melanin--a comparison of black and Caucasian skin. J Amer Acad Derm 1979; 1:249-60

Kelly D.A., Young A.R., McGregor J.M., Seed P.T., Potten C.S., Walker S.L. Sensitivity to sunburn is associated with susceptibility to ultraviolet radiation-induced suppression of cutaneous cell-mediated immunity. J Exp Med 2000; 191:561-66

Kobayashi N., Nakagawa A., Muramatsu T., Yamashina Y., Shirai T., Hashimoto M.W., Ishigaki Y., Ohnishi T., Mori T. Supranuclear melanin caps reduce ultraviolet induced DNA photoproducts in human epidermis. J Invest Derm 1998; 110:806-10

Lookingbill D.P., Lookingbill G.L., Leppard. Actinic damage and skin cancer in albinos in northern Tanzania: findings in 164 patients enrolled in an outreach skin care program. J Amer Acad Derm 1995; 32:653-58

McFadden A.W. Skin disease in the Cunan Indians. Arch Derm 1961; 84:1013-23

Sheehan J.M., C.S. Potten C.S., Young A.R. Tanning in human skin types II and III offers modest photoprotection against erythema. Photochem Photobiol 1998; 68: 588-92

J.M. Sheehan, N. Cragg, A.R. Young, C.S. Potten, Tanning affords the same protection against DNA photodamage & erythema in human skin types II and IV but stimulates faster DNA repair in skin type IV. J Invest Derm (submitted 2001a)

Sheehan J.M., Young A.R. Sunburn Cell reaction in different skin types. In preparation (2001b)

Strickland P.T., Bruze M., Creasey J. Cyclobuta-dithymidine induction by solar-simulating UV radiation in human skin, I. Protection by constitutive pigmentation. Photo-Derm 1988; 5:166-9

Weinstock M.A., Colditz G.A., Willett W.C., Stampfer M.J., Bronstein B.R., Mihm M.C.J., Speizer F.E. Melanoma and the sun: the effect of swimsuits and a "healthy" tan on the risk of nonfamilial malignant melanoma in women. Amer J Epidem 1991; 134: 462-70

Xu G., Snellman E., Bykov V.J., Jansen C.T., Hemminki K. Effect of age on the formation and repair of UV photoproducts in human skin in situ. Mut Res 2000; 459: 195-02

Yarosh D., Klein J., O'Connor A., Hawk J.L.M., Rafal E., Wolf P. Reduction of skin cancer in Xeroderma Pigmentosum: randomized, double blind, placebo controlled study of topically applied DNA repair enzyme (T4 endonuclease V) in liposomes. Lancet 2001; 357: 926-29

Yarosh D., Klein J., Kibitel J., Alas L., O'Connor A., Cummings B., Grob D., Gerstein D., Gilchrest B.A., Ichihashi M., Ogoshi M., Ueda M., Fernandez V., Chadwick C., Potten C.S., Proby C.M., Young A.R., Hawk J.L. Enzyme therapy of xeroderma pigmentosum: safety and efficacy testing of T4N5 liposome lotion containing a prokaryotic DNA repair enzyme. Photoderm Photoimmun Photomed 1996; 12:122-30

Young A.R., Potten C.S., Chadwick C.A., Murphy G.M., Hawk J.L., Cohen A.J. Photoprotection and 5-MOP photochemoprotection from UVR-induced DNA damage in humans: the role of skin type. J Invest Derm 1991; 97:942-48

Young A.R., Chadwick C.A., Harrison G.I., Hawk J.L., Nikaido O., Potten C.S. The in situ repair kinetics of epidermal thymine dimers and 6-4 photoproducts in human skin types I and II. J Invest Derm 1996; 106: 1307-13

Young A.R. Potten C.S., Nikaido O., Parsons P.G., Boenders J., Ramsden J.M., Chadwick C.A. Human melanocytes and keratinocytes exposed to UVB or UVA in vivo show comparable levels of thymine dimers. J Invest Derm 1998; 111:936-40

Young, Antony R. "The Molecular and Genetic Effects of Ultraviolet Radiation Exposure on Skin Cells." In Photodermatology, John L.M. Hawk, ed. Arnold, London, 1999.

Ziegler A., Leffell D.J., Kunala S., Sharma H.W., Gailani M., Simon J.A., Halperin A.J., Baden H.P., Shapiro P.E., Bale A.E. Mutation hotspots due to sunlight in the p53 gene of nonmelanoma skin cancers. Proc Nat Acad Sci USA 1993; 90:4216-20

PRODUCTION AND PERSISTENCE OF UV-INDUCED TAN

Sharon A. Miller, Janusz Z. Beer, Nelson T. Lao, Barbara Z. Zmudzka
Food and Drug Administration, Rockville, MD 20850

INTRODUCTION

Background

Approximately 1,000,000 individuals patronize indoor tanning salons in the U.S. on any given day (1). It was estimated that in 1999 there were approximately 27 million visits to tanning salons in the U.S. (2). A survey of users in the U.K. reported that a typical use pattern is 20 visits per year, but that some avid tanners make 100 visits or more per year (3). Additionally, although not well quantified, large numbers of people use tanning lamps in the home. This could result in a large proportion of an individual's total cumulative UV exposure being contributed by artificial UV sources.

The carcinogenic and photoaging properties of UV radiation have been proven in numerous studies. These effects become apparent predominantly later in life. Approximately 1,300,000 new cases of skin cancer are diagnosed each year in the U.S., and approximately 10,000 deaths were attributed to skin cancer in 1999 (4). The photoaging effects of chronic UV exposure have also been well documented (5, 6). The knowledge of these harmful effects prompted development of safety and performance standards for the use of artificial tanning devices. The U.S. relies on the FDA Performance Standard for Sunlamp Products of 1985 (7). It is used in conjunction with a Policy Letter of 1986 providing the criteria for manufacturers use in establishing the maximum exposure time and exposure schedule (8). The intention of these documents is to provide for reliable control of exposures and to limit acute and delayed damage. Together, they specify information to be provided to the user, limits on short wavelength radiation, and requirements for protective eyewear in addition to requirements for the maximum timer interval and a recommended exposure schedule. In Europe, many countries adhere to the International Electrotechnical Commission (IEC) standard for UV- and infrared-emitting appliances (9), though some countries have additional, more stringent requirements for products marketed in their countries. Currently attempts are being made to update both the FDA and IEC standards and to harmonize them.

Dynamics of Tanning

It is fairly well established that the full manifestation of an UV-induced tan requires several days. When a single exposure to either an UVB source or solar radiation was used, several researchers noted that melanogenesis required 3-7 days (10-14).

Contrary to results with UVB-emitting sources, when pure UVA lamps are used, the pigmentation is visible immediately (i.e., immediate pigment darkening, or IPD). The color resulting from this process is reported to have a bluish-gray tone that partially fades within approximately 2 h and then remains stable from 4 h to 7 days (11, 15). This stable component of the IPD still contains a grayish hue and is termed "persistent pigment darkening". This is seen as distinct from delayed tanning (neomelanization) which has a brown or dark yellow hue and a delayed onset of at least 3 days (15).

Current Recommended Exposure Schedules

As mentioned above, FDA uses a policy letter as a companion document to the FDA standard (8). This policy letter provides criteria for the development of the maximum recommended exposure time and exposure schedules for tanning equipment. Such information is required on the label and must be provided by the manufacturer to the user. In the Policy Letter, FDA uses a minimal erythema dose (MED) value of 156 J/m^2-eff (MED_{FDA}) and provides a specific action spectrum for erythema. Both the MED_{FDA} and the FDA erythema action spectrum differ slightly from those used by the IEC (200 J/m^2-eff and the CIE erythema action (16)). Some of the changes being considered in the FDA Performance Standard for Sunlamp Products are to adopt the CIE action spectrum for erythema and to change the MED to 200 J/m^2-eff.

The FDA policy recommends that the first few exposures (3 times the first week) be limited to 0.75 MED. After the first week of exposures, the policy allows for gradual increasing of exposures up to 4 MED_{FDA} per session. The number of recommended sessions per week after the first week is not specified. After "maximal tanning has occurred", the FDA policy letter recommends "biweekly or weekly exposures of up to 4 MEDs" (8). However, based on exposure schedules that are displayed on the exterior surface of some sunbeds, it seems that the FDA-recommended exposure schedule is sometimes misinterpreted by the industry to allow 3 tanning sessions per week, at 4 MED_{FDA} for tanning induction and maintenance. Such an exposure regimen, carried out for a full year, may result in an accumulation of a total of 75,000 J/m^2-eff. This is 3 times the average annual UV exposure received from the sun in the U.S. (25,000 J/m^2) (17, 18). From the viewpoint of the desired effect, this is most likely excessive considering the dynamics of tanning.

The IEC standard does not recommend an exposure schedule, per se. However, it does recommend that the first exposure should not exceed 100 J/m^2-eff, which would be equivalent to 0.5 MED at 200 J/m^2-eff. In addition, this standard recommends a limit on annual exposures to tanning equipment of 15 kJ/m^2-eff (weighted with the CIE erythemal action spectrum). This would be equivalent to approximately 38 tanning sessions, at 2 MED (at 200 J/m^2 per MED) per session.

The scientific basis for all these recommendations is rather limited, and many elements of the standards are based primarily on the opinions of experts, rather than on the results of rigorous research. This is particularly true for the exposure schedules. In one of the few studies of repeated exposures, Kaidbey (10) demonstrated that – after induction of a tan – tan maintenance was achieved with 1 MED per week to solar simulated radiation. However, he noted that statistical evaluation of the data was not possible with only seven subjects. Caswell (19) recently completed a study that evaluated the tanning effectiveness of the FDA-recommended exposure schedule for sunlamps. The exposures were administered at a minimum of 48-h intervals and subjects were exposed for 8 weeks. His data demonstrated that approximately 2 weeks were required before a significant tan could be visualized or measured. Follow-up data from this study demonstrated that the tan persisted for at least 10 weeks after the cessation of exposure (20). Thus, although not rigorously proven, to date, there are data to indicate that a tan can be maintained with significantly lower doses than are currently being used. Therefore, there is a need to develop scientifically-based exposure schedules that could reduce the cumulative exposure of sunlamp users without compromising the desired effect. With recent progress in methodology, it is possible to quantitatively explore UV responses of human skin and to use the results of such explorations to achieve this goal.

OUR STUDY

We are conducting a study to compare the physical and biological methods for testing and standardization of UV response in human skin. This is not a study of tanning, nevertheless it provides insight into the tanning process in a variety of human subjects. Our study is conducted on volunteers representing a broad range of phototypes and different racial/ethnic origins. The subjects are grouped according to their predicted photosensitivity and racial/ethnic origin (African-American or Black; American Indian or Alaska Native; Asian;

Hispanic or Latino; Native Hawaiian or Pacific Islander; White). To date, we have completed evaluations of 76 subjects. In this paper, we use examples of data for Asians, Blacks and Whites. At this stage in the study, we have sufficient representation (6 Asians, 24 Blacks, 36 Whites) in these 3 categories.

Methods

UV Measurements

The UV radiation (250 – 400 nm) is measured in units of spectral irradiance (W/cm^2-nm) at 1-nm increments with an Optronic Laboratories, Inc. (Orlando, Florida) Model 754 portable spectroradiometer. This system employs a photomultiplier tube detector and a double-grating monochromator to minimize stray light. The input optics consists of a spectralon-coated integrating sphere. For irradiation of individual subjects, before and after the exposure, the amount of UV radiation delivered to each 2 x 2 cm area is checked with a special low profile detector that is coupled to a radiometer (International Light, Newburyport, Massachusetts) and adjustments to the exposure time of each area are made to achieve the desired dose.

UV Exposures

An array of 8 Kodacel-filtered FS lamps are used as the source of UV radiation. This source emits approximately 40% UVB and 60% UVA radiation, with a broad peak at 313 nm. The use distance (approx. 22 cm) is chosen to deliver 100 J/m^2-eff (weighted with the CIE action spectrum for erythema) per minute. For the purposes of this chapter, the units of J/m^2-eff will be referred to as erythemal effective units (EEU) to indicate that they are a measure of erythemal-effective dose. Seven 2x2-cm dorsal skin areas are exposed to increasing UV doses (0.5 to 4.0 $MED_{predicted}$). Visual and instrumental observations are performed 1, 2, 8, and 16 days later. One day after exposure, the MED - defined as the lowest dose that produces pink erythema with at least one discernible border - is determined. Eight days after exposure, the minimal tanning dose (MTD) – defined as the lowest dose that produces a light brown tan – is determined.

Assessments of erythema/pigmentation

Visual

Visual evaluations of erythema are made at 24 h and 48 h after exposure, using the grading scales below:

For Erythema:

0	No reaction
0.5	Observer indecisive
1	Minimal perceptible erythema, faint pink, no borders
2	Pink erythema, at least one border defined
2.5	Pink erythema, two or more borders defined
3	Red
4	Fiery red
5	Violaceous red

The MED is defined in our study as a grade of 2 on the above erythema scale.

For pigmentation, visual evaluations are made at 8 and 16 days after exposure, using the grading scale below:

0	No difference from unexposed control
0.5	Observer indecisive
1	Pigmentation discernible through pressed glass
2	Light brown tan
3	Moderately brown
4	Dark brown
5	Black

The MTD is defined in our study as a grade of 2 on the above pigmentation scale.

Instrumental

Instrumental measurements are made at every session, i.e. on days 1, 2, 8 and 16 after exposure. A DiaStron Erythema/Melanin Index meter (DiaStron, Andover, U.K.) and a Minolta CM-2002 spectrophotometer (Minolta Corporation, Ramsey, New Jersey) are used for these evaluations.

The DiaStron meter measures the reflectance at 3 wavelengths: 546, 632 and 905 nm (Full Width Half Maximum < 9 nm at each wavelength) and displays the erythema or melanin index.

In this instrument, the erythema index (EI) is defined as:

$$EI = \log_{10}(R_{632\,nm}/R_{546\,nm}) * 1000$$

and the melanin index (MI) as:

$$MI = \log_{10}(R_{905\,nm}/R_{632\,nm}) * 1000$$

The EI and MI are reported as the difference between the respective mean index of each exposed area and the respective mean index of an unexposed, adjacent skin site.

The Minolta CM-2002 measures the diffuse reflectance (DR) from 400 to 700 nm at 10-nm increments. Erythema by DR is calculated as follows using the readings at 580 nm (maximal absorption of hemoglobin) and 630 nm:

$$\text{Erythema by DR} = \Delta AA_{580\text{ nm}} - \Delta AA_{630\text{ nm}}$$

Where $\Delta AA = \log(DR_{control}/DR_{UV})$, AA = Apparent Absorbance

The change in melanin is calculated from the slope of the AA between 630 and 700 nm as described in Kollias and Baqer (21).

Results and Discussion

Visual Observations

For White subjects, the MED ranged from 150 to 500 EEU, for Asian subjects, the MED ranged from 300 to 674 EEU, for Black subjects, the MED ranged from 400 to 1175 EEU.

Following observations by Parrish (12), Pathak (11), and others, we based our evaluation of MTD on assessments at day 8. Significant pigmentation was not noted until then, (we made no observations between day 2 and day 8) except in a few dark-skinned subjects that demonstrated IPD. The MTD in White subjects ranged from 272 to 1000 EEU, in Asian subjects, it ranged from 350 to 693 EEU, and in Black subjects it ranged from 430 to 898 EEU.

Instrumental Observations - DiaStron

The absolute values of MI for unexposed areas exhibited a very large range among our subjects. For the White subjects, they ranged from –134 to –27, for the Asian subjects, they ranged from –101 to 4, and for the Black subjects, they ranged from –3 to 220, indicating large differences in constitutive pigmentation, even within the same racial group. The differences between the MI of unexposed skin vs. that of the area noted as the MTD, were typically very low for the White subjects, ranging from –14 to 34. For the Asian subjects, the MI increase ranged from 2 to 37. For the Black subjects, the MI increase ranged from –9 to 76, and was frequently more than 100 at doses higher than the MTD.

Appearance and Disappearance of Tan

Results with the DiaStron instrument indicated that, in general, the tan reached its peak at day 8. However, these results were confounded in the cases where there was still visible erythema at day 8, usually seen in subjects with little constitutive pigmentation. In this situation, the DiaStron indicated similar MIs at both day 8 and day 16, or maximum MI at day 16. Examination of the raw data indicates that in subjects with insignificant melanin, the detection capability (signal/noise) of this instrument is severely challenged. This was also seen in dark-skinned subjects at high doses, where erythema remained visible for 16 days.

Figures 1-3 illustrate the change in MI with time for 3 different subjects, 1 White (S46), 1 Asian (S16) and 1 Black (S24). In the case of S46, there is little change in the MI, except for the highest dose, which was 2.1 MED. In the case of S16, an increase in MI was seen for all doses greater than 1.0 MED. The MI appears to be still increasing at day 18, except for the area that received 4.0 MEDs; this was likely due to peeling in that area, as was noted in our records. In the case of S24, there is a significant increase in MI, peaking at day 8, for all doses greater than 1.0 MED.

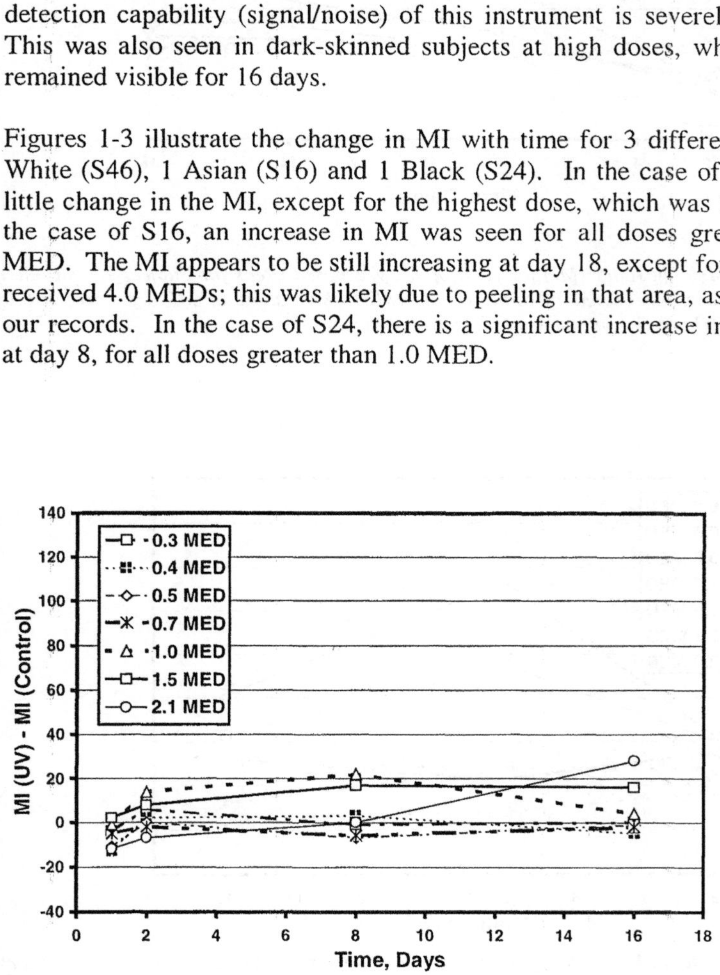

Figure 1. The Melanin Index (MI) for each of 7 UV-exposed areas for Subject 46, a White subject. UV doses are noted in terms of the MED. All values of MI are corrected by subtracting the MI of adjacent, unexposed control areas.

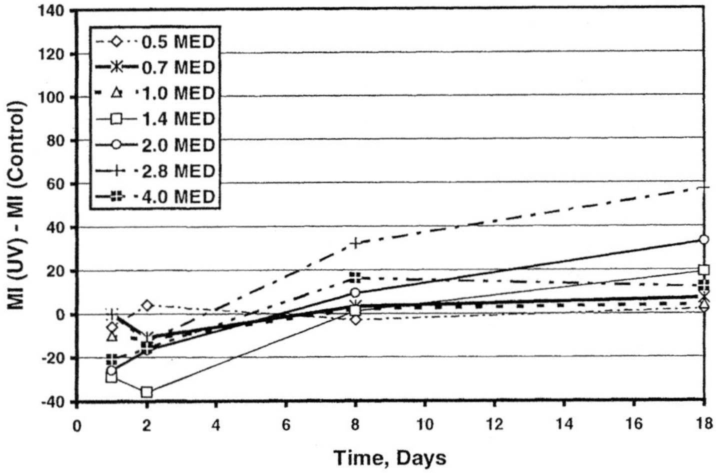

Figure 2. The Melanin Index (MI) for each of 7 UV-exposed areas for Subject 16, an Asian subject. UV doses are noted in terms of the MED. All values of MI are corrected by subtracting the MI of adjacent, unexposed control areas.

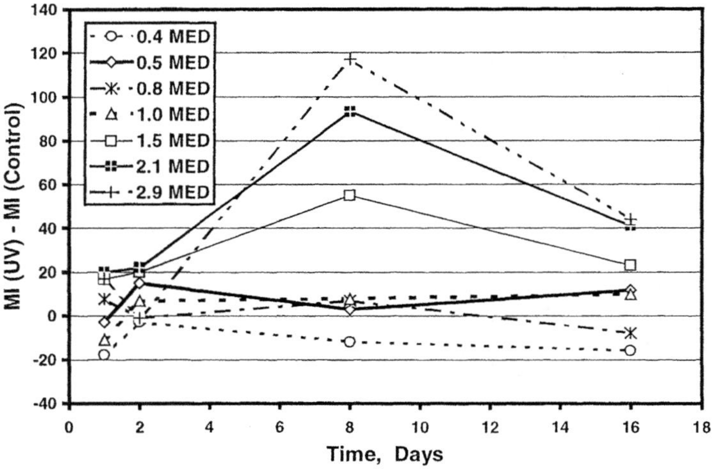

Figure 3. The Melanin Index (MI) for each of 7 UV-exposed areas for Subject 24, a Black subject. UV doses are noted in terms of the MED. All values of MI are corrected by subtracting the MI of adjacent, unexposed control areas.

Our final observations and measurements 16 days (18 days in the case of S16) after exposure indicate that the UV-induced tan can persist for a relatively long time. We also have limited observations indicating that in some subjects, an UV-induced tan may persist for many months (data not shown). This confirms the observations of Caswell (20).

Relationship between Erythema and Pigmentation

The amount of UV energy needed to produce a tan depends on UV sensitivity (MED). However, the ratio of minimal tanning dose to minimal erythema dose (MTD/MED) varies with UV sensitivity. This ratio may be >1 for sensitive skin and <1 for resistant skin. As early as 1983, Pathak and Fanselow reported this variation for skin types II-IV. They noted that the MTD is equal to the MED for individuals of skin type II, while the MTD is less than the MED for skin types III and IV, when full spectrum solar radiation was used. In our study, the MTD was typically 2 times the MED in White subjects (which includes skin types II), approximately equal to the MED in Asian subjects and less than the MED in Black subjects. In a few extreme cases of White subjects, the MTD was almost 3 times the MED. Thus, in White subjects, no tan was produced unless significant erythema was first produced. The differences between our results and those of Pathak and Fanselow can most likely be explained by the difference in spectrum used, and the lower proportion of UVA in our light source. The ratio of MTD to MED for 73 subjects is illustrated in Figure 4.

Two Components of Tan – Pigmentation and Erythema

Caswell (19) noted that the redness of the skin (as measured by a* with a Minolta Chromameter) was significant at visit 7 (7 exposures within 15 days) and increased slightly thereafter, stabilizing after 26 days. Our instrumental measurements demonstrate that 8 days after exposure, the tanned skin areas are still affected by erythema - to different degrees in different subjects. This can be readily seen in the spectral data collected with the Minolta spectrometer. In Figures 5 -7, the increased concentration of hemoglobin in the skin can be clearly seen. The two peaks at 540 nm and 580 nm, characteristic of the in vitro absorption spectrum of hemoglobin are readily apparent. In the examples shown the increased hemoglobin is still apparent at day 8, but is subsiding at day 18. The increase in melanin, represented by an increase in apparent absorbance in the 630 – 700 nm wavelength range is clearly visible by day 8 and persists to day 18.

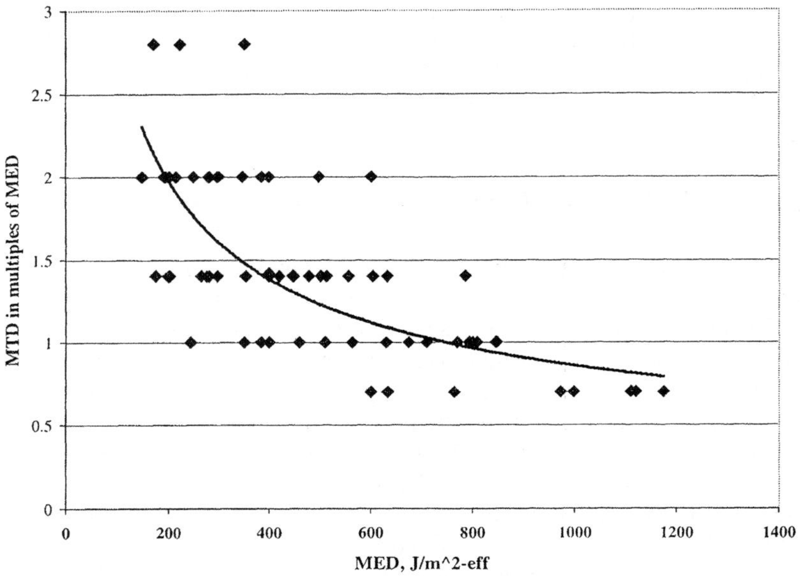

Figure 4. The Minimal Tanning Dose – expressed in multiples of MED – vs. MED. Data are shown for 73 subjects.

Conclusions

In our study, we used FS lamps which are primarily a UVB radiation source. Thus, our observations on tanning cannot fully applicable to other UV sources, especially the UVA-rich sources. It is well know that the efficiency of the tanning doses depends on the emission spectrum of the exposure source. Bech-Thomsen et al. (22) evaluated the tanning efficiency of sub-erythemagenic exposures to 6 different sunlamps and showed that the most significant tanning was produced by those sources with the largest proportion of UVA radiation. The tan produced by UVA sources was apparent within a few days. The sources with mainly UVB emission produced no discernible change in pigmentation, even after 4 weeks. Our experience suggests that pigmentation would have likely been observed after several days, if higher doses of UVB had been given in this study. Thus, source spectrum and exposure dose are important variables in the tanning process.

In general, fair-skinned individuals require higher doses in multiples of MED to produce a tan than do darker-skinned individuals. In fact, even a fraction of MED can produce a tan in dark-skinned UV-resistant persons, while much more than one MED may be needed to produce a tan in fair-skinned, UV-sensitive persons. This indicates that we can expect less deleterious side effects of the tanning process in dark than in fair skin. This is unfortunate, since the tan would be relatively more beneficial for the more UV-sensitive population. All these facts indicate that, new recommended tanning schedules should be tailored to different UV-sensitivities/tanning proficiencies.

Erythema is a primary effect of UV exposure. Thus, it is much easier (and probably quite common) to produce a "tan" that consists of some increased pigmentation and an increased amount of blood in the skin than to induce real melanogenesis. This makes it possible to satisfy the desire of many users of tanning equipment for a rapid effect. Such procedures should be discouraged. Observational evaluation of the tan is insufficient. Takiwaki and Serup (23) noted that the measurement of erythema in areas that have significantly different levels of pigmentation is problematic with simple meters such as the DiaStron. The development of a simple and reliable technique that would allow evaluation of the quality of UV-induced skin color changes ("tan") would be helpful.

Literature data and our observations show that both erythema and pigmentation can be quite persistent, depending on the UV dose and the melanocompetence of the individual. Thus, one way to reduce an individual's cumulative UV exposure would be to limit the frequency of exposures to the minimum required to maintain the tan. With new instruments available to measure the color of skin quantitatively, this can now be evaluated in a rigorous manner.

In summary, new recommended tanning schedules should be developed when minimal UV doses needed to produce and maintain a tan are known for different classes of tanners. Many observations indicate that such schedules may require considerably lower cumulative UV exposures than the current schedules. Development of such schedules should be followed by aggressive educational efforts addressed to the regulators, the industry, and the tanners. In particular, the latter group should be convincingly informed that the rapid production of a true tan is not possible and that attempts to produce such a tan involves the additional risk of erythema, especially in the most UV-sensitive skin types.

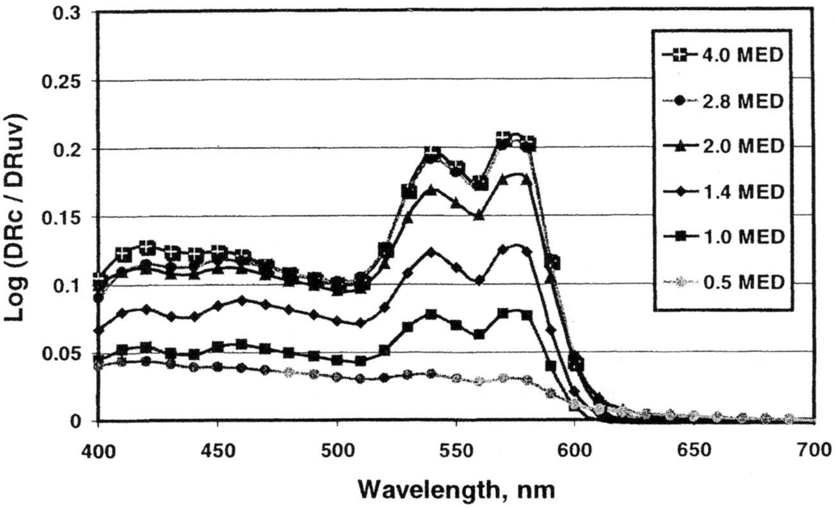

Figure 5. Diffuse reflectance spectrum of Subject 16 (Asian) at day 1.

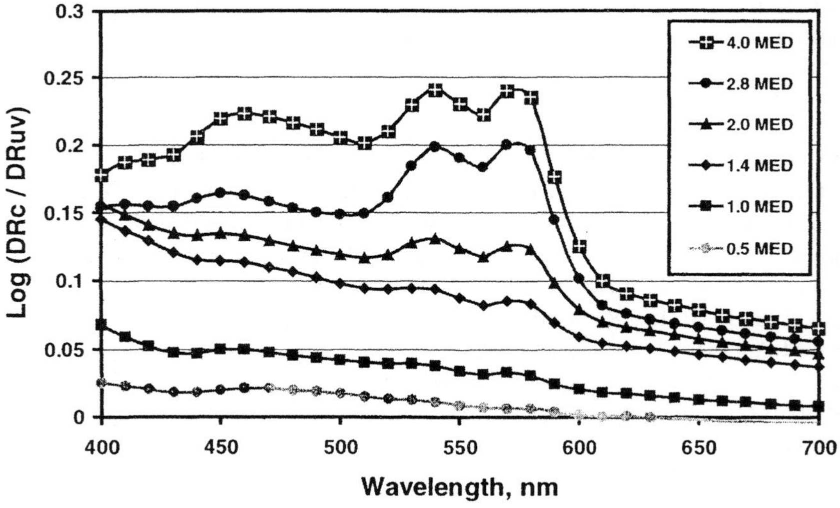

Figure 6. Diffuse reflectance spectrum of Subject 16 (Asian) at day 8.

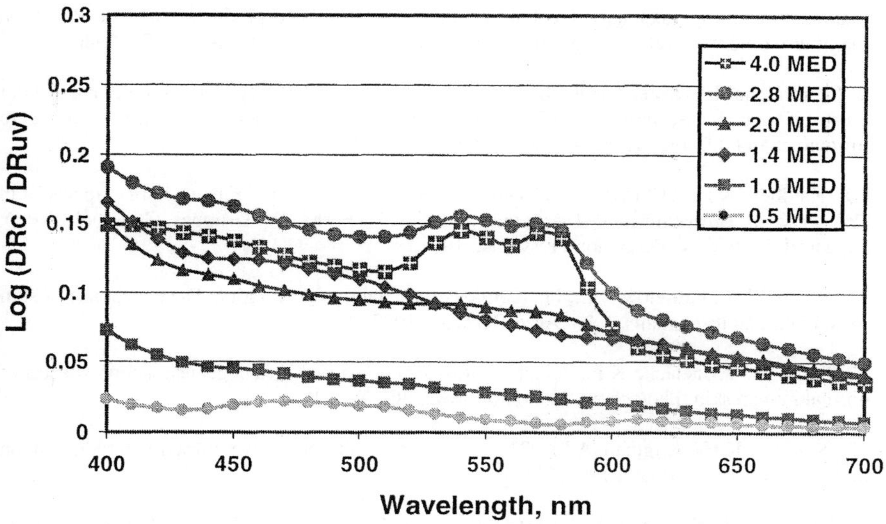

Figure 7. Diffuse reflectance spectrum of Subject 16 (Asian) at day 18.

Note: The mention of commercial products or their use in connection with material reported here is not to be construed as either an actual or implied endorsement of such products by the U.S. Food and Drug Administration.

REFERENCES

1. DeLeo V.A. Tanning salons. Proceedings of National Conference on Environmental Hazards to the Skin. Schaumberg, Ill: American Academy of Dermatology, 1994: 37-41.

2. Looking Fit Tanning Fact Book 2000/2001. Virgo Publishing, (March 15, 2000) Phoenix, Arizon, USA.

3. Diffey B.L. Use of UVA sunbeds for cosmetic tanning. Br J Dermatol 1986; 115: 67-76.

4. American Cancer Society, Statistics 2000, Skin Cancers (http://www3.cancer.org)

5. Leveque J.L., Porte G., de Rigal J., Corcuff P., Francois A.M., Saint Leger D: Influence of chronic exposure on some biophysical parameters of the human skin: An in vivo study. J Cutaneous Aging & Cosmetic Dermatology 1: 123-127, 1988/89.

6. Godar D.E., Swicord M.L., Kligman L.H."Photoaging." In *Environmental UV Radiation and Health Effects,* Schopka H-J, Steinmetz M., eds. Munich-Neuherberg, Germany: BfS-ISH-Berichte, 1995.

7. Department of Health and Human Services, Food and Drug Administration. 21 CFR Part 1040 (Docket No. 82N-0188). Sunlamp Products; Performance Standard - Final Rule. Federal Register, Vol. 50, No. 173, Friday, September 6, 1985.

8. Department of Health and Human Services. Food and Drug Administration. Letter from the Director, Office of Compliance, Center for Devices and Radiological Health, to all

manufacturers, importers, and potential manufactures of sunlamp products. Subject: Policy on maximum timer interval and exposure schedule for sunlamp products. August 21, 1986.

9. International Electrotechnical Commission - IEC 335-2-27 Safety of household and similar electrical appliances, Part 2: Particular requirements for appliances for skin exposure to ultraviolet and infrared radiation, Geneva, Switzerland (1995).

10. Kaidbey K.H. "Ultraviolet-induced melanogenesis in man: Quantitative aspects." In *Psoralens in Cosmetics and Dermatolog,y* Cahn J., Forlot P., Gruppes C., et al. (eds.): Elmsford, New York, Pergamon Press, Inc., 1981.

11. Pathak M.A., Fanselow D.L., Photobiology of melanin pigmentation: Dose/response of skin to sunlight and its contents. J Am Acad Dermatol 1983; 9: 724-733.

12. Parrish J.A., Jaenicke K.F., Anderson R.R. Erythema and melanogenesis action spectra of normal human skin. Photochem. Photobiol. 1982; 36: 187-191.

13. Kaidbey K.H., Kligman A.M., The acute effects of long-wave ultraviolet radiation on human skin. J Inv Dermatol 1978; 72:253-256.

14. Lavker R.M., Kaidbey K.H. Redistribution of melanosomal complexes within keratinocytes following UV-A irradiation: a possible mechanism for cutaneous darkening in man. Arch Dermatol Res 1982; 272: 215-228.

15. Chardon, A., Moyal D., Hourseau C. "Persistent Pigment-Darkening Response as a Method for Evaluation of Ultraviolet A Protection Assays." In *Sunscreens: Development, Evaluation, and Regulatory Aspects*, N. J. Lowe, N. A. Shaath, M. A. Pathak (eds.) New York, NY, Marcel Dekker, Inc., 1997.

16. CIE Research Note: A reference action spectrum for ultraviolet induced erythema in human skin. CIE J. 6: 17-22 (1987).

17. Scotto J, Fears TR and Fraumeni JF, Jr., Incidence of nonmelanoma skin cancer in the United States. NIH Publicaion no. 83-2433, 1983.

18. Godar D.E., Wengraitis S.P., Shreffler J., Sliney D.H., UV Doses of Americans. Photochem and Photobiol 2001;73: 621-629.

19. Caswell M., The kinetics of the tanning response to tanning bed exposures. Photodermatol Photoimmunol Photomed 2000; 16: 10-14.

20. Caswell M., Dowdy J.C., Sayre R.M., Maintenance of a tan induced by a tanning bed. 10th Annual Mtg of the Photomed Soc March 1,2001, Washington, DC.

21. Kollias N. and Baqer A., On the assessment of melanin in human skin in vivo. Photochem. Photobiol. 1986; 43: 49-54.

22. Bech-Thomsen N. A quantitative study of the melanogenic effect of multiple suberythemal doses of different ultraviolet radiation sources. Photodermatol Photoimmunol Photomed 1994; 10: 53-56.

23. Takiwaki H., Serup J. "Measurement of Erythema and Melanin Indices." In *Handbook of Non-Invasive Methods in the Skin*, J. Serup, G.B.E. Jemec, eds. Boca Raton, FL: CRC Press, 1995.

EFFECTIVENESS OF TANNING PROCEDURES

Jan C. van der Leun
(Ecofys, Utrecht, The Netherlands) and H. van Weelden (Dept. of Dermatology, University Medical Center, Utrecht, The Netherlands).

The widespread habit of tanning light-skinned people with ultraviolet radiation is controversial. The tanning industry could not exist without a broad demand from their customers. On the other hand, there are many experts who emphasize the damage done to the skins of these customers. These opposing forces have resulted in regulations for tanning equipment, which also set some restrictions on what can be done in tanning salons. The ultimate measure, to forbid this entire business, was not taken in any country, as far as I know.

What can we, as scientifically informed people, do in such a situation? Even for those among us who think that the risk is the all-important issue, there is little point in repeating this message over and over. That has not changed the development of the tanning business in the past, and is not likely to do that in the future, either. I think it is best that we recognize the existing situation and bring in our knowledge to promote a sensible dealing with the ultraviolet radiation from artificial sources. That is the same as we are doing with the ultraviolet radiation from our much more powerful source, the Sun, which we cannot forbid anyway. This attitude has prevailed in the Netherlands in recent years, and has led to a consensus process between the Dutch Cancer Society, the organized tanning business, the cosmetics trade and representatives of dermatology and photobiology. We succeeded in agreeing on a consensus document: "Indoors and Outdoors: Sun Wisely", Dutch Cancer Society, 1995. Tanning salons that cooperate have the "Ten golden Rules" from this document on their walls, and the leaflets of the Dutch Cancer Campaign on their tables. I think that this is a more positive influence than we could ever had by fighting each other.

The weighing of benefits and risks is not a customary scientific activity. But it would be a good step in the right direction if we could advise the tanners, how to achieve their tan with no more UV radiation than is necessary. In my first years as a photobiologist in an Institute of Dermatology, I met a young woman. Her main reason for coming to the clinic was, to be informed about UV effects. She had taken exposures from a UV-lamp; at that time this was a medium-pressure mercury arc. She had taken 10 minutes exposure every day, for three years. On my question why she did this, the answer was: "in order to have a tan". She was, however, about as white as a person can be. This is an extreme example of an inefficient way of doing. Apparently it is possible to take a massive accumulated dose of UV radiation without even achieving the tan. In less extreme cases it is also possible to achieve the tan, but with much more UV radiation than would have been needed.

What do we know about how to arrange an effective tanning procedure? One principle is clear: in a course of tanning exposures, the adaptation of the skin to the UV radiation has to be overcome. Especially with UV-B radiation, this adaptation is powerful. Typically, an exposure of the skin to 1 MED (minimal erythema dose) causes the sensitivity of the skin to decrease by about 20%, with the maximal effect about 10 days after the exposure. If this is not compensated by increasing the exposure times, the rest of the course will soon become ineffective. How great should the increments be? Too little will cause ineffectiveness, too much will lead to sunburn. The optimal will have to be found out; it is likely to be different for different combinations of UV-B and UV-A. It is well possible that the most effective course will be one that elicits erythema as well as pigmentation. Then comes again a weighing: some of the deleterious effects of UV radiation on the skin appear to depend not only on the total dose received by the skin, but also on sunburn episodes. This will also have to be taken into account. For doing this, the community will again need the best information scientists can give.

EFFECTS OF TOPICALLY APPLIED GLYCOLIC ACID (GA) ON THE SENSITIVITY OF HUMAN SKIN TO UV-INDUCED DAMAGE.

K. Kaidbey[1], B. M.. Sutherland[2], P.V. Bennett[2], D.A. Dennis[3], W.G. Wamer[3], C. Barton[3] and A. Kornhauser[3], [1]*Ivy Laboratories, Philadelphia, PA 19104;* [2]*Brookhaven National Lab, Upton, NY 11973;* [3]*Food and Drug Administration, Washington, DC 20204*

Abstract

Glycolic acid (GA), an alpha-hydroxy acid, (AHA) has been widely used in a large number of cosmetics products for daily use over long periods of time (Andersen, 1998). Sustained by the quest for a youthful appearance, the use of AHA's in skin care products has increased dramatically in the last decade. These products claim to reduce wrinkles, spots and other signs of aging (Van Scott, 1989). There is little information available to date about their long-term effects. One of the important question relates to the acute and chronic effects of AHA's on the sensitivity of human skin to sun exposure. This study was conducted to investigate the effects of GA, applied in a cosmetic formulation, in modifying UV-sensitivity, and to determine whether the effect was reversible after discontinuing GA applications.

Background

Hydroxyacids are organic carboxylic acids classified into alpha- and beta-types according to their molecular structure. Both alpha- and beta-hydroxy acids (HA) are used widely as active ingredients in a large number of cosmetic products. Many aspects concerning the mechanism(s) of action of HA still remain unknown.

Objectives

This study was designed to furnish information on whether daily treatment with a commonly used AHA could alter the sensitivity of human skin to UV-induced damage. A change, such as an increase in UV transmission can presumably result in greater damage to epidermal cells, which can be monitored by the formation of sunburn cells, by the formation of cyclobutyl-pyrimidine dimers (CPD's) and by a lowering of the Minimal Erythema Dose (MED). Another objective of this study was to determine if the sensitivity to UV persists after discontinuing the treatments with glycolic acid.

Methods

29 healthy subjects of both sexes (skin type I-III) were recruited. The subjects were divided into two groups: in group 1 (n=16), photosensitivity was evaluated by determination of sunburn cells (SBC's) and the minimal erythema dose (MED); in group 2, (n=13) (CPD's) were evaluated. The test material consisted of 10% GA, pH 3.5, in a cosmetic formulation (vehicle). The vehicle cream was formulated to resemble a standard cosmetic formulation for a skin care product, containing emulsifiers, surfactants, and preservatives. The test material and vehicle were applied (2 mg/cm^2) under supervision to assigned rectangular areas (5x10 cm) over the mid-back region daily (6x per week), for 4 consecutive weeks. An additional area served as an untreated control. Within each of the rectangular areas, one circular subsite (2 cm in diameter) was established for obtaining the biopsies needed for quantitation of SBC'c and CPD's. At the end of treatment, a single dose of 1.5 MED (determined for each subject on previously untreated skin) of UVB was delivered to treated and untreated sites on all subjects. The selection of the exposure doses was based on the skin type of the individual, determined by questionnaire. Shave biopsies (4 mm), from each subsite, were obtained 24 hrs after UV exposure in group 1 for enumeration of SBC's, and immediately after UV exposure in group 2 for determination of CPD's. At the end of an additional week (week 5), during which no topical treatments were administered, both the MED and the formation of UV-induced SBC's were redetermined.

UV Source and Exposure

Two UVR sources were used. A 150-watt xenon arc solar simulator equipped with a WG 320 (1mm) and a UG 5 (1mm) filters was employed for determining the MED in the treated and untreated skin sites at the end of the 4 week and 5 week periods. The output of the solar simulator was monitored with a calibrated thermopile. UVB and UVA at skin level were measured with an IL 443 radiometer (International Light, Inc., Newburyport, MA) using specific UVB and UVA sensors. UVB intensity was 6.0 mW/cm^2, and UVA intensity was 41.0 mW/cm^2. For the induction of SBC's, a bank of four 20-inch fluorescent FS20 bulbs was used. The bulbs were filtered with a 0.15mm thick sheet of cellulose acetate to remove UVC ($\lambda < 290$nm). The source-to-skin distance was 12 inches. UVR intensity at skin level was measured with an IL 443 radiometer (International Light, Inc., Newburyport, MA) using specific UVB and UVA sensors. The average UVB intensity at skin level was 0.14mW/cm^2, while the average UVA intensity was 0.08mW/cm^2.

Processing of Skin Speciments

For determination of SBC'c, skin speciments, obtained from shaved biopsies, were fixed in 10% formalin, embedded in paraffin, sectioned and stained with heamatoxylin-eosin. The number of SBC's was determined in approximately 16 sections at 50u intervals. 70 high power fields were counted from each biopsy and the average number was calculated. All specimens were counted in a blinded manner [Kaidbey, 1990]. For determination of CPD's, shave biopsies from the designated subsites were processed and analyzed by an alkaline agarose gel method, developed by Sutherland [Sutherland, 1980], for quantitating single strand breaks in nanogram quantities of DNA. After electrophoresis together with molecular lengths standards, the DNA was neutralized, stained with ethidium bromide, photographed, and the density profiles recorded with a computer controlled scanner. The single strand breaks were induced by pyrimidine dimer-specific endonuclease cleavage. The method has been used previously to determine CPD's from biopsies of human skin irradiated with UV [Bennett, 1996].

Statistical Evaluations All the data were analyzed by an analysis of variance (ANOVA) followed by a restricted t-test comparison.

Results

Minimal Erythema Dose (MED). The MED values in each test sites are shown in Fig. 1. The mean MED in the normal untreated skin was 88 mJ/cm^2, while the corresponding mean MED in the GA treated skin was 71 mJ/cm^2. The observed difference was statistically significant ($p<0.001$). The mean MED in the GA treated skin was significantly ($p<0.001$) lower than the mean MED in the vehicle treated site. There was no statistically significant difference between the mean of the MED for the untreated sites and sites receiving the vehicle. One week after discontinuing topical treatments (week 5, Fig. 1), the differences in the mean of the MED between all three sites were no longer statistically significant.

Sunburn Cells (SBC's). The mean number of SBC's in all test sites are shown in Fig. 2. The statistical analysis showed that at week four, the mean number of SBC's at the site treated with GA (2.97) was significantly higher than the mean number of SBC's at the site treated with vehicle (1.53) ($p<0.001$) or untreated sites ($p<0.001$). There was no statistically significant difference found between the mean number of SBC's at sites treated with the vehicle and untreated sites. At week five, the mean number of SBC's was not significantly different for any of the three sites.

Cyclobutyl-Pyrimidine Dimers (CPD's). Samples from one subject were inadequate for analysis. The number of CPD's from the remaining 12 subjects are shown in Fig. 3. Each number represents the average of three or four determinations of CPD's per million bases. The mean number of CPD's was highest in the glycolic acid treated skin (42.9) while the means in the vehicle treated skin and untreated but UV-irradiated skin were 39.7 and 40.1, respectively. The 8.1% increase in CPD's in the GA treated sites as compared to the sites treated with vehicle alone, was not statistically significant. Only the untreated, non irradiated site (mean number 3.5) was significantly different from the other three sites.

Discussion

Alpha hydroxy acids, particularly glycolic acid, have been used widely in cosmetic products for many years, particularly for improving the appearance of dry skin (Van Scott, 1974), photo-aged skin and various pigmentation and discoloration disorders (Van Scott,1989; Elson, 1992). In 1998 the Cosmetic Ingredient Review (CIR) Expert Panel published a safety assessment of alpha-hydroxy acids which, in summary, stated that:" Based on the available information included in this report, the CIR Expert panel concludes that glycolic and lactic acid, their common salts and simple esters, are safe in cosmetics products at concentrations less than 10%, at final formulations pH greater than 3.5, when formulated to avoid increasing sun sensitivity or when direction for use include the daily use of sun protection (Andersen, 1998). In addition, the Expert Panel stated that from the available data, the AHA's are not carcinogenic or mutagenic, are not reproductive or developmental toxins and are not sensitizers. The Panel concluded that three areas warranted further investigation including dermal irritation, enhanced skin penetration of other chemicals after topical application with AHA's, and sun sensitivity.

This study was designed to investigate the effects of GA, applied in a cosmetic formulation, in modifying UVB-sensitivity in human skin, and to determine whether the effect was reversible after discontinuing GA applications for a relatively short period of one week. In previous clinical studies it was determined that topical administration of GA increased the sensitivity of treated skin to UVB, as judged by the induction of SBC's. (Andersen, 1998).

In the present double blinded study a more comprehensive approach was undertaken to investigate the effect of GA on sensitivity to UVB by using three biomarkers as endpoints: MED, SBC's and CPD's. As seen from Fig. 1, and 2, a moderate dose of UVB resulted in a decrease of MED (18%) and an increase in the number of SBC's (90%) after four weeks of topical

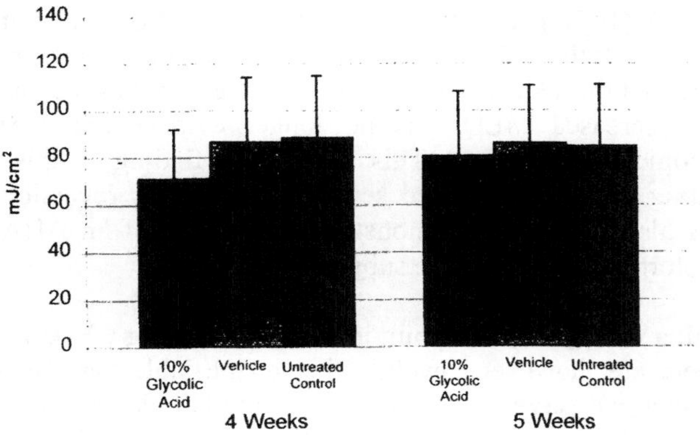

Figure 1. MED at Weeks 4 and 5

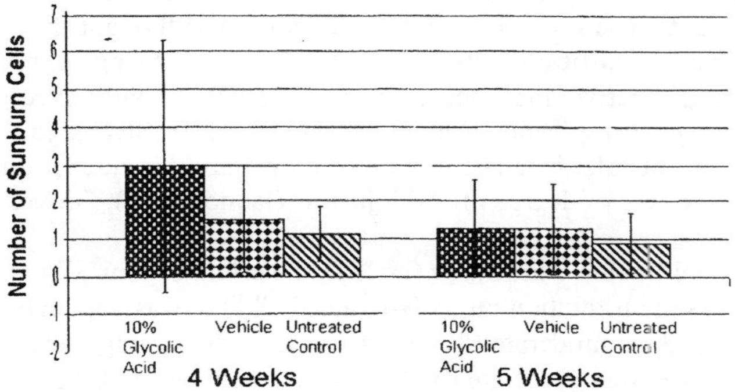

Figure 2. Average Number of UV-Induced SBC's/High Power Field

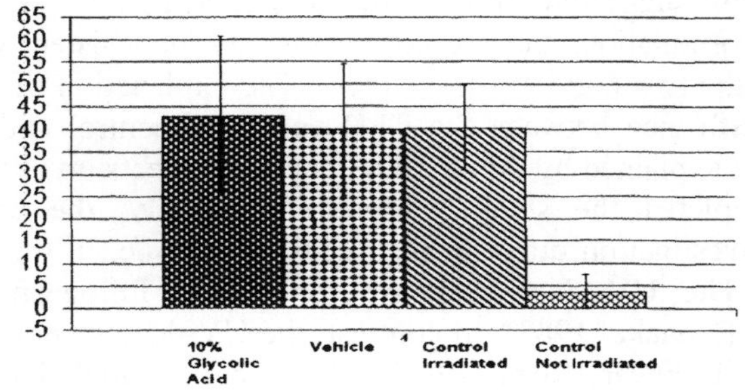

Figure 3. Average Number of Cyclobutyl Dimers/10^6 Bases

application of GA (10%; pH 3.5) in a cosmetic formulation, as compared to the vehicle cream. Both of those values were statistically significant. This study also found that that skin sensitivity to UVB, measured as increased SBC formation and decreased MED, is no longer evident one week after discontinuing topical applications. This is the first finding in this type of studies, indicating that AHA-induced sensitivity to UV is reversible. This investigation is also the first to demonstrate the effects of an AHA on the MED and SBC formation in the same subject.

Erythema is still a widely used endpoint in studies involving UV-induced skin sensitivity. There are, however, several other quantifiable parameters other than erythema that can serve as a measure of acute UV damage to the skin. One characteristic feature of a sunburn reaction is the formation within the epidermis of dyskeratotic cells, commonly referred to as sunburn cells (Daniels, 1961; Grove, 1980;Young, 1987). The SBC is derived from a keratinocyte showing by hematoxylin-eosin staining a shrunken chromatin, an eosinophilic cytoplasm, and a pyknotic nucleus. It is thought that such cells have sustained significant DNA damage and are being eliminated by apoptosis, a protective mechanism preventing DNA with excessive UV-induced damage to replicate (Danno, 1987). Because SBC's can easily be recognized by histologic characteristics, they may provide a quantitative index of assessment of degree of UV injuries (Danno, 1987); Grove, 1980).

In an additional double blind study (n=13), the effect of topically applied GA and vehicle on the formation of UVB-induced CPD's was conducted. As can be seen in Fig. 3, a moderate (8.1%) increase in the average level of CPD's induced by UVB was associated with topical treatment of GA, compared to the average level of CPD's at the site treated with the vehicle or with UVB only. This increase, however, was not statistically significant. The statistical analysis indicated that only the untreated, not irradiated site was significantly different from the other three sites. The results derived from the CPD measurements are still suggestive of an increased skin sensitivity induced by UVB and the GA formulation, and are not necessarily inconsistent with the data obtained with the two previous biomarkers. The explanation of the lack of a statistical significance between the CPD values of control and treated sites can be partially explained by the small number of subjects involved; only 12 volunteers completed the study. More significantly, the endpoints measured in this investigation differ in their sensitivities, i.e. in their UVB-dose responses. The UV dose response for SBC's follow nonlinear, exponential pattern (Nonaka, 1983). In contrast, the UV-dose response curve for CPD formation is linear (Roza, 1991). Therefore, a small increase in incident UV radiation results in a small increase in CPD formation. As a result, CPD formation is inherently a less sensitive quantitative measure to a change in UV radiation than SBC formation.

UV-induced CPD's represent an important endpoint for assessing UV-induced damage with a clear mechanistic and prognostic interpretation. An increase in the level of CPD's, associated with topical application of GA, would clearly indicate that repeated exposure of skin to GA increases the penetration of UV radiation into skin. Furthermore, the mutagenic potential of DNA photolesions, in particular CPD's, have been established in vitro systems, as well as in animal and human skin (Young, 1999). UV radiation-induced mutations are highly specific to dipyrimidine sites and are a direct consequence of photoproduct formation (Young, 1999). There is considerable evidence to date that UVR-induced mutation are important in development of skin cancer (Brash, 2001).

The mechanism underlying the effects of AHA's on increased sensitivity to UV radiation is still poorly understood. Increased penetration of UV radiation into the epidermis, resulting from long term topical application of formulations containing AHA's is a mechanism compatible with currently available data (Van Scott, 1974). Anti-inflammatory effects (Perricone, 1996) and antioxidant activities of AHA's in human skin homogenates (Morreale, 1997) were also reported as contributing mechanisms by which AHA's express their effects. A recent study reported data on modification of the expression of a number of cell-cycle regulatory proteins in UV plus GA-treated mouse skin, as compared to UV-irradiated skin (Hong, 2001). In an other recently published paper, the effect of alpha- and beta-hydroxy acids on cell proliferation in mouse skin was described (Sams, 2001). The authors demonstrated that continuous treatment of mouse skin with these hydroxy acids, applied topically in a cosmetic formulation, resulted in increased basal cell proliferation rates as well as in increased epidermal thickness. The authors state that "if the kinetics of human skin response to AHA are similar to those of SKH-1 mice, then an individual could be susceptible to DNA damage (e.g., caused by solar exposure) up to 12 hours after use of cosmetics containing alpha- or beta-hydroxy acids".

Conclusion

By measuring three commonly used biomarkers, it has been demonstrated that topically applied GA (10%; pH 3.5), in a standard cosmetic formulation, increases the sensitivity of human skin to UVB as compared to the vehicle alone. The effect is reversible after discontinuing the GA applications for one week.

References

Andersen, F. A., (1998). Final report on the safety assessment of glycolic acid, ammonium, calcium, potassium, and sodium glycolates, methyl, ethyl, propyl and butyl glycolates, and lactic acid, ammonium, calcium, potassium, sodium and TEA-lactates, methyl, ethyl, isopropyl, and butyl lactates, and lauryl, myristyl, and cetyl lactates. Int. J. Toxicol. 17(Suppl), 27.

Bennett, P. V., Gange, R. W., Hacham, H., Hejmadi, V. S., Moran, M., Ray, S. and Sutherland, B. M. (1996). Isolation of high-molecular-length DNA from human skin. Biotechniques, 21:458-463.

Brash, DE and Bale, AE.(2001) Molecular biology of skin cancer. In: DeVita, VT, Hellman, S, and Rosenberg, SA (Eds.) Cancer: Principles and Practice of Oncology, 6th ed., Philadelphia: Lippincott-Raven, pp. 1971-1975.

Danno, K., Horio, T., (1987). Sunburn cell: Factor involved in its formaton. Photochem. and Photobio. 45,: 683-690.

Daniels, F. Jr., Brophy, D., and Lobitz, W. C. Jr. (1961). Histochemical responses of human skin following ultraviolet irradiation. J.Invest. Dermatol. 37, 351-357.

Elson, M. L., (1992). The utilization of glycolic acid in photoaging. Cosmet. Dematol. 5, 36. 12-15.

Grove, G. L. and Kaidbey, K. H., (1980). Sunscreens prevent sunburn cell formation in human skin. J. Invest. Dermatol. 75: 363-364.

Hong, J. T., Kim, E. J., Ahn, K. S., Jung, M. J., Yun, Y. P., Park, Y.K. and Lee, S. H., (2001). Inhibitory effect of glycolic acid on skin tumorigenesis in SKH-1 hairless mice and its mechanism of action. Molecular Carcinogenesis , 31: 152-160.

Kaidbey, H. K., (1990). The photoprotective potential of the new superpotent sunscreens. J AM Acad Dermatol, 22: 449-452.

Morreale, M. and Livrea MA., (1997). Synergistic effect of glyconic acid on the antioxidant activity of alpha-hydroxy acids on photoaged skin: A pilot clinical, histologic, and ultrastructural study. Mol. Biol. Int., 42: 1093-1102.

Nonaka, N., Kaidbey, K. H. and Kligman, A. M., (1983). The influence of UVA and visible radiation on acute damage by short-wave UVR ($\lambda < 320$). J. Invest. Dermatol. 81: 524-527.

Perncone, NV. And Dinardo, JC., (1996). Photoprotective and anti-inflammatory effects of topical glycolic acid. Dermatol Surg. 22: 435-437.

Roza, L., De Gruijl, F. R., Bergen Henegouwen, J. B. A., Guikers, K., Van Weelden, H., Van DerSchans, G P., and Baan, R. A., (1991). Detection of photorepair of UV-induced Thymine dimers in human epidermis by immunofluorescnce microscopy. J Invest dermatol 96:903-907.

Sams, R. L., Couch, L. H., Miller, B. J., Okerberg, C. V., Warbritton, A., Wamer, W. G., Beer, J. Z., and Howard, P. C. (2001). Basal cell proliferation in female SKH-1 mice treated with α- and β-hydroxy acids. Toxicology and Applied Pharmacology 175, 76-82.

Sutherland, B. M., Harbor, L. C. and Kochevar, I. E., (1980). Pyrimidine dimer formation and repair in human skin. Cancer Res. 40: 3183-3185.

Van Scott, E. J., and Yu, R. J. (1974). Control of keratinization with alphahydroxy acids and related compounds. Topical treatment of ichthyotic disorders. Arch. Dermatol. 100, 586-590.

Van Scott, E. J., and Yu, R. J. (1989). Alpha hydroxy acids: Procedures for use in clinical practice. Cutis 43, 222-228.

Young, A. R. (1987). The sunburn cell. Photodermantology $: 127-134

Young, A.R., (1999) in Photodermatology, The Molecular and Genetic Effects of Ultraviolet Radiation Exposure on Skin Cells; Ed.: J.L.M. Hawk; Arnold, pp. 25-42

QUANTITATIVE EVALUATION OF PIGMENTATION FOLLOWING REPEATED UV IRRADIATION BY MICROFLASH SPECTROPHOTOMETER

Farhad Nasrollahi Chimeh, Michael F. Holick,
Section of Endocrinology, Nutrition and Diabetes, Department of Medicine, Boston University School of Medicine, Boston, MA

Abstract

The goal of this clinical research project was to determine quantitatively the change in pigmentation in adults following repeated UV irradiation. Healthy male and female adult volunteers (18 years and older) with skin types II-V participated in the study. The total duration of this study was 3 months. Total body irradiation was performed three times a week using a commercial tanning bed with the UVB (λ =290-320) output of about 5%. Irradiation time was determined based on the subject's skin type and the specifications provided by the tanning bed manufacturer and sanctioned by the FDA. To assess the degree of pigmentation induced by repeated UV irradiation we used Microflash spectrophotometer which is an instrument that measures color quantitatively in the range of 400-700nm. In this study we measured skin pigmentation before and immediately after each irradiation. We evaluated 16 subjects who participated in the study. The microflash spectrophotometer measurements showed the following results:

The mean increase in immediate pigmentation for skin type 2, 3, 4 and 5 was 2.9±0.26%, 4.59±0.38%, 4.55±0.46% and 1.8% respectively. The mean increase in delayed pigmentation for skin type 2, 3, 4 and 5 was 46.25±5.43%, 56.8±6.41%, 42.33±7.96% and 12.69% respectively.

Introduction

The most important determinant of normal human skin color is primarily the color, shape, size and distribution of melanosomes. These cytoplasmic organelles in which the melanin is synthesized and deposited are the product of melanocytes. Melanocytes are located in the basal layer of epidermis and transfer melanin containing melanosomes to the keratinocytes through arborizing nerve cell-like dendrites. At the dendrite-keratinocyte junction, the melanin is either actively phagocytized by keratinocytes or passively deposited in their cytoplasm after cell membrane fusion. Another important factor in pigmentation of skin melanogenesis) is the function of a copper-containing monooxygenase, tyrosinase. This is a bifunctional enzyme that converts tyrosine to dihydroxyphenylalanine (dopa) and then to dopaquinone which is finally cyclized and further oxidized to give rise to the polymer melanin. Increased pigmentation of skin (melanogenesis) is a recognized effect of UV irradiation. In human skin, the pigmentary response to UV is a biphasic process. The immediate skin pigmentation (immediate tanning, IT) is a transient light or dark brown coloration which is predominantly induced by UVA (320-400 m) and to a lesser extent by visible light (400-700nm). Immediate tanning occurs within minutes in individuals and fades within one or several hours depending on the UV dose and skin type. Since IT can be induced in a matter of a few minutes, it appears that it reflects photooxidation of melanin and a change in the distribution of the melanosomes from perinuclear location to the dendritic processes and then to keratinocytes. No change is observed in the number of melanosomes, melanocytes or in the tyrosinase activity during the immediate tanning phase.

Late onset pigmentation (delayed tanning, DT) can be visually recognized 48-72 hours after UV irradiation. Delayed pigmentation is the result of increased synthesis and number of melanosomes, increased arborization of dendrites, increase in the number of melanocytes and increased activity of tyrosinase. This occurs within days after exposure and lasts weeks to months. This phenomenon is predominantly induced by UVB (290-320nm) and to a lesser extent by UVA (320-400nm).

The goal of this clinical research project was to determine quantitatively the change in melanin pigmentation that occurs within 10 minutes after each irradiation session (IT) and also the change in pigmentation that occurs during the study i.e., the pigmentation in the baseline visit compared with the last visit after 3 months (DT) in adults with different skin types following repeated exposures to UV irradiation in a tanning bed similar to what an adult would receive in a tanning salon. These tanning beds emitted radiation in the ultraviolet region that mimics the spectral power distribution of natural sunlight.

Methods

Sixteen healthy male and female (9 male and 7 female) adult volunteers (18-68 years old) with skin types 2-5 participated in the study. The number of subjects with skin type 2,3,4 and 5 were five, seven, three and one, respectively. None of them were using artificial tanning devices. The study was performed from January to March. The total duration of this study was 3 months per subject. Total body irradiation was performed three times a week during the study using a commercial tanning bed with the UVB output of about 5%. Irradiation time was determined based on the subject's skin type and the specifications provided by the tanning bed manufacturer and sanctioned by the FDA. To assess the degree of pigmentation induced by repeated UV irradiation we used Microflash spectrophotometer (Datacolor International, Charlotte, NC).

Microflash spectrophotometer analyzed the light energy reflected or transmitted by a sample. This instrument measured the photometric characteristics of the material in the visible spectrum in the range of 400-700nm with the resolution of 5nm. The microflash spectrophotometer had a xenon lamp as its light source, an analyzer system combined with a detector with a computer based processing unit for the colorimetric calculations.

The L*a*b* system recommended by the (CIE) Commission International de l'Eclairage was used to assess the skin color. The L* index (luminance) stands for relative lightness ranging from total black to total white. The a* index represents the range between red and green and the b* index represents the range between yellow and blue.

In this study we measured immediate tanning (IT) by comparing the skin pigmentation before each irradiation with the pigmentation within 10 minutes after each irradiation session.

Also we assessed the delayed tanning (DT) by comparing the skin pigmentation in the baseline visit before irradiation with the pigmentation in the last visit (after 3 months) before irradiation.

Results

The mean increase in immediate pigmentation for skin type 2, 3, 4 and 5 was 2.9±0.26%, 4.59±0.38%, 4.55±0.46% and 1.8% respectively (Figure 1A).

The mean increase in delayed pigmentation for skin type 2, 3, 4 and 5 was 46.25±5.43%, 56.8±6.41%, 42.33±7.96% and 12.69, respectively (Figure 1B).

Conclusion

The skin type designation does not correspond well to the constitutive and facultative skin color.

The results of this study show that either self-reported burning/tanning histories are an unreliable means for skin typing or skin phototyping based on the Fitzpatrick's skin typing system can not be an adequate predictor of cutaneous pigmentation response to UV irradiation.

There is overlap in the degree of both immediate and delayed tanning for subjects with different skin colors and also skin types.

These variabilities in immediate and delayed pigmentation could be due to differences in genetic make-up, the role of melanin shield in prevention of further pigmentation, the difference in the activity of tyrosinase or synthesis of melanosomes in different skin types.

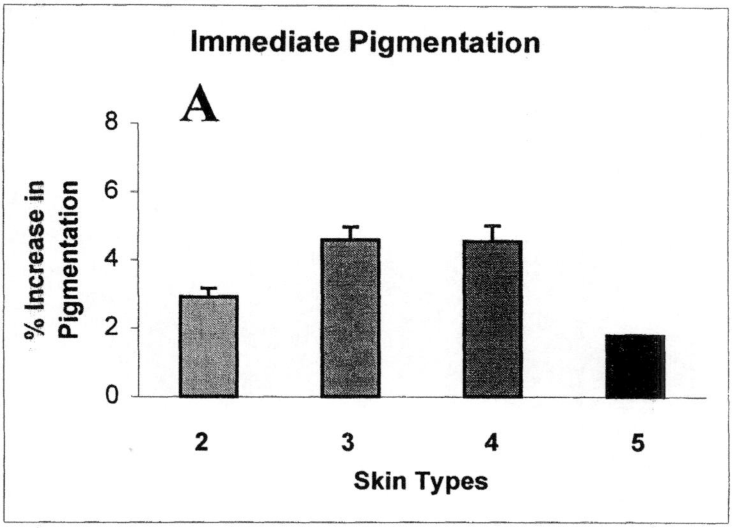
Immediate Pigmentation
A
% Increase in Pigmentation
8
6
4
2
0
2
3
4
5
Skin Types

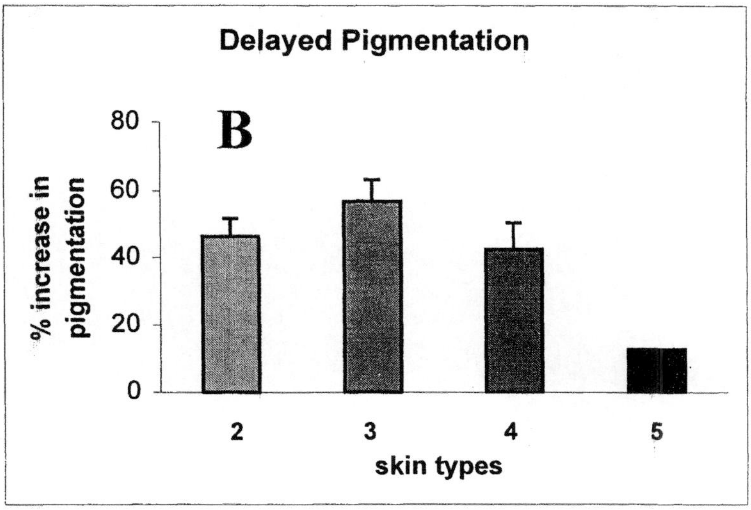
Delayed Pigmentation
B
% increase in pigmentation
80
60
40
20
0
2
3
4
5
skin types

VALIDATION OF A SUN-EXPOSURE QUESTIONNAIRE FOR ADOLESCENT GIRLS

Jennifer Cobb[a], Susan Sullivan[a], Clifford J. Rosen[ab], Michael F. Holick[c], Tai Chen[c], Michael Kimlin[d], Alfio Parisi[d]

[a]Department of Food Science and Human Nutrition, University of Maine, Orono, ME
[b]Maine Center for Osteoporosis Research and Education and St. Joseph's Hospital, Bangor, ME
[c]Vitamin D, Skin and Bone Research Laboratory, Boston University School of Medicine, Boston, MA
[d]Centre for Astronomy & Atmospheric Research, University of Southern Queensland, Queensland, Australia

Introduction

Sunlight is a major contributor to an individual's vitamin D status. Determinants of sun exposure have been used in conjunction with dietary data to study the effect of sunlight on bone health of elderly populations. In order to assess the impact of sun exposure on vitamin D status or bone mass of children, a sun-exposure questionnaire must be devised for children and tested. Because sun exposure is not equivalent over the course of a day, a factor that affects cutaneous previtamin D_3 synthesis, the questionnaire should account for the varying strengths of ultraviolet B (UVB) radiation throughout a day. This study attempts to improve upon previous methods of quantifying sun exposure and to further investigate *in vitro* synthesis of previtamin D_3 at varying UVB levels.

Materials and Methods

Thirty-five girls (ages 9-15) from the vicinity of Bangor, ME (44.5°N) wore a polysulphone (PS) film dosimeter and kept an activity log for one day on August 17, 2000. PS film is used to monitor broadband UVB exposure by measuring the change in optical absorbance (ΔA) using a spectrophotometer set at 330 nm. The dose of UVB radiation (kJ/m^2) associated with each dosimeter reading was calculated by calibrating the dosimeters to the source spectrum (1). The following day, the subjects referred to their activity logs to report time spent outdoors the previous day. Minutes spent outdoors during each hour of the day were adjusted for the strength of the UVB radiation during that hour. Total adjusted minutes were compared with PS dosimeter readings for each subject. A series of ampules containing 7-dehydrocholesterol (7-DHC) was exposed on an unshaded flat surface each hour of the study day in Orono, ME. The percent of previtamin D_3 detected in each ampule was determined by high performance liquid chromatography (HPLC) as described (2). Subtracting 0.8 from the percent previtamin D_3 detected in the 7-DHC ampules estimates cutaneous previtamin D_3 synthesis (3).

Broadband UVB data generated from a meter at the United States Department of Agriculture UVB Monitoring Site in Presque Isle, ME was used to determine the average UVB irradiance for each hour of the day on August 17, 2000. For each subject, reported minutes outdoors during each hour were multiplied by that hour's fractional equivalent of peak hour UVB to obtain total minutes spent outdoors adjusted to a standard of peak UVB. Table 1 indicates that adjusting the minutes spent outdoors for the strength of the sun can result in a substantially different number than total minutes spent outdoors.

Table 1. Example of Adjusting Subject Minutes Spent Outdoors

TIME OF DAY	MINUTES OUTDOORS REPORTED BY SUBJECT	ADJUSTMENT FACTOR*	ADJUSTED MINUTES OUTDOORS
6-7 AM		0.01	0
7-8 AM		0.03	0
8-9 AM	15	0.19	2.85
9-10 AM	23	0.38	8.74
10-11 AM	26	0.54	14.04
11 AM-12 PM		1	0
12-1 PM		0.98	0
1-2 PM		0.92	0
2-3 PM		0.46	0
3-4 PM	8	0.51	4.08
4-5 PM	13	0.3	3.9
5-6 PM	30	0.09	2.7
6-7 PM	19	0.04	0.76
TOTAL	**134**		**37.07**

* Determined from hourly UVB irradiance relative to peak

Results

Figure 1 illustrates that subject self-reported minutes outdoors adjusted for the time of day correlated with UVB exposure measured by the PS dosimeters (r=0.64). Adjusting minutes for the strength of the sun resulted in a stronger correlation with PS dosimeters than when total

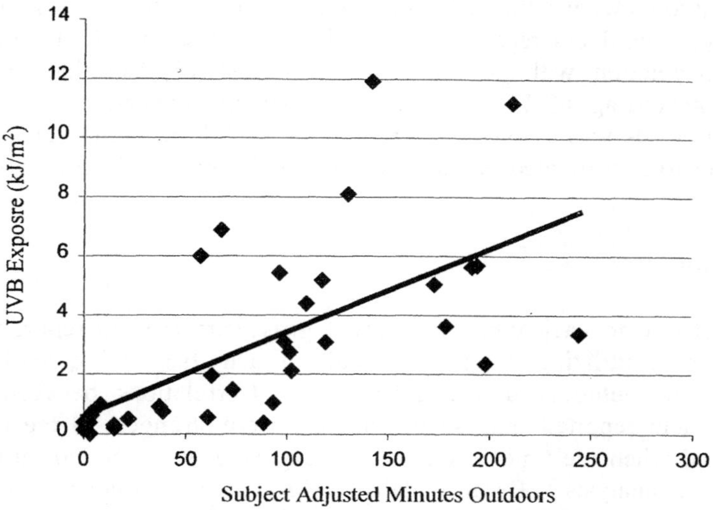

Figure 1. Adjusted Minutes Outdoors Reported by the Subjects vs. PS Dosimeter Reading

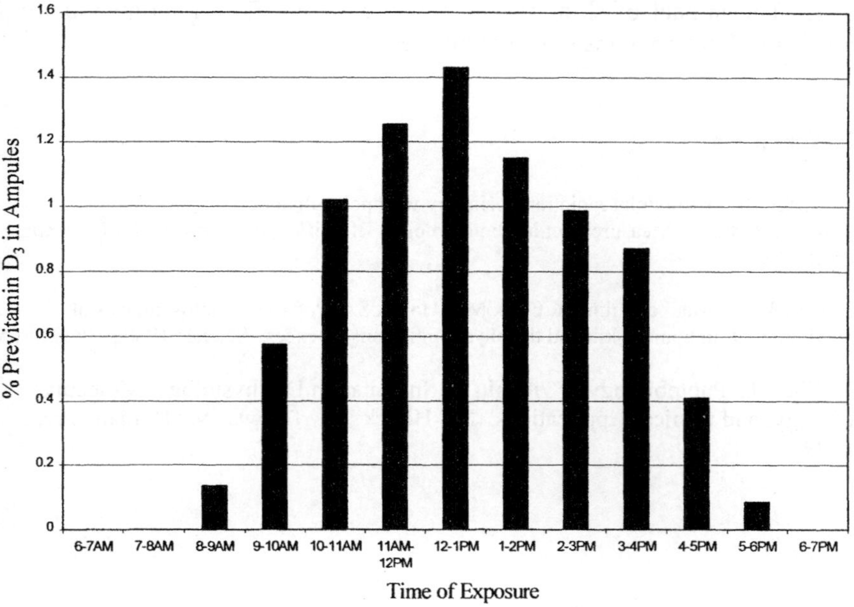

Figure 2. Hourly In Vitro Detection of Previtamin D3 in Orono, ME (August 17, 2000)

minutes outdoors were compared with PS dosimeters (r=0.57). The adjusted minutes spent outdoors reported by the older subjects, ages 12-15, had a stronger correlation with dosimeter readings (r=0.76) than that of the younger subjects, ages 9-11 (r=0.52). The maximum conversion of 7-DHC to previtamin D_3 was 1.43%, occurring from 12-1 PM (Figure 2). This corresponds to an estimated skin synthesis of 0.63% previtamin D_3.

Conclusions

Relative to each other, adolescent girls were able to report time spent outdoors utilizing a 24-hour recall of activities and quantifying minutes spent outdoors during each hour. Correlations between PS dosimeters and reported time spent outdoors were strongest in the older subjects and when the time of day of sun exposure was factored into the questionnaire analysis. Because daily UVB data is available in many locations in the United States, the adjustment factor for the time of day can be easily adapted for use in a variety of research settings. The ampule data reveal that on this heavily clouded August day in Maine, it is estimated that skin synthesis of previtamin D_3 would only have occurred between the hours of 10 AM and 4 PM, and the percent previtamin D_3 produced each hour was low. The impact of skin synthesis of previtamin D_3 on serum vitamin D levels in Maine remains to be determined.

References

1. Diffey BL. Ultraviolet radiation dosimetry with polysulphone film. In: Radiation Measurement in Photobiology. BL Diffey, ed. New York: Academic Press, 1989.

2. Holick MF, MacLaughlin JA, Clark MB, Holick SA, Potts JT. Photosynthesis of previtamin D_3 in human skin and the physiologic consequences. Science 1980;210:203-205.

3. Chen T. Photobiology of vitamin D. In: Vitamin D: Physiology, Molecular Biology, and Clinical Applications. MF Holick, ed. Totowa, NJ: Humana Press, 1999.

Vitamin D and Bone Health

Seasonal Changes in the IGF Regulatory System and The Relationship to Bone Loss in Elderly Women

Clifford J. Rosen[1], Subburaman Mohan[2], Michael F. Holick[3]
St. Joseph Hospital, Bangor, ME[1]; Jerry L. Pettis VA Medical Center, Loma Linda, CA[2]; Boston University School of Medicine and Boston Medical Center[3]

Introduction

We and others have previously reported that among healthy elderly women living in Maine with low calcium intake, seasonal changes in serum 25OH vitamin D led to secondary increases in parathyroid hormone (PTH), osteocalcin, and urinary N-telopeptide (N-Tx)(1). Indeed, such women were found to be at high risk for osteoporotic fractures based on significant bone loss (-2 to –3% per winter season) from the femoral neck over two consecutive winters(1). Moreover in one randomized trial, calcium supplementation prevented bone loss in these women, and blunted the seasonal changes in PTH, osteocalcin and NTx. Despite these data, and other studies confirming seasonal bone loss among women living in northern latitudes, the mechanism responsible for rapid bone loss remains unclear(2). We hypothesized that changes in the IGF regulatory system could lead to altered bone formation/ and or increased bone resorption, thereby resulting in rapid bone loss. Hence, we measured IGF-I and three IGF binding proteins in the serum of women who were randomized to placebo calcium during this two year trial(1).

Methods

Details of the cohort of women studied in this randomized trial are reported elsewhere(1). In this sub-study, we examined the cross-sectional and longitudinal changes in serum IGF-I and IGFBPs among the 17 women who completed the two-year trial with a placebo calcium. Table I provides the demographics and baseline biochemical data on these women. Serum IGF-I was measured by RIA, as described elsewhere. IGFBP-3 was determined by an IRMA assay (DSL, Webster, Tx). IGFBP-4 and IGFBP-5 were measured by RIA in the laboratory of Dr. Subburaman Mohan as described previously(3). PTH and osteocalcin were measured by IRMA, urinary N-Tx was determined by ELISA as reported previously(1).

Results

All the women provided informed consent that was approved by St. Joseph Hospital IRB. The demographic information about the placebo-controlled cohort are noted in Table I. These women were elderly but healthy. They did not suffer from chronic diseases, did not travel during the winter months, and they did not have a history of osteoporotic fractures. Mean BMD T- scores of the spine was +1.08 and of the femoral neck T= -0.78. Twenty women were initially randomized, 17 completed two years of the study. Baseline, 6 month and 24 month data were analyzed in these women. The six-month time point was in the month of February. The starting month of the study was October as was the 24-month completion time.

Serum IGFBP-4 correlated closely with serum PTH at baseline($r=0.41$, $p=0.003$). Change in IGFBP-4 correlated inversely with changes in total dietary calcium intake($r=-0.57$, $p=0.02$). By multivariate analysis the strongest predictor of winter time bone loss in all women, and in particular the placebo group noted above was dietary calcium intake.

Discussion

Seasonal bone loss in women with poor calcium intakes has been recognized for several years(1,2,4). The cause of this loss and the relative risk of subsequent fracture due to rapid bone turnover has not been elucidated, although clearly changes in 25OHD begin a cascade of events In this study, the women receiving placebo calcium were physically active, healthy and without osteoporosis, but they resided at latitude 45.5 degrees N. We previously reported that during the winter months, women leaving in this northern Maine community had a significant drop in 25OH D(4). Moreover, there was virtually no non- enzymatic conversion of pre-D to pro D in vitro during the month of December even on a cloudless day(4). So the fact that these women showed a decline in serum 25OHD is not totally surprising, nor is the loss of nearly 3% of their femoral bone mass during the winter months. However, it is important to note that even though summer 25OHD levels return to baseline, there is no recovery from winter changes. Indeed as seen in Table II, the overall changes in BMD, and biochemical markers were due almost exclusively to what was happening during the winter.

The purpose of this sub-study was to investigate potential mechanisms of bone loss during the winter in women not receiving calcium or vitamin D supplementation. We focused on the IGF regulatory system because of its importance in relation to bone remodeling and acquisition(5). Only one of the IGF or IGFBP markers changed significantly during the winter, and that was IGFBP-4. This binding protein has been shown to be an inhibitor of IGF-I action in vitro, and is induced by PTH(6,7). IGFBP-3 and IGFBP-5, in contrast, tend to be facilitators of IGF-I action on bone. However, in this study, the winter time changes in IGF-I, IGFBP-3 and IGFBP-5, were minor, and if anything, tended to decline over 24 months. Thus, there is an imbalance between changes in the stimulatory and antagonistic serum IGFBPs. Since PTH rose during the winter, and correlated with IGFBP-4, it seems likely that one major mechanism related to seasonal bone loss is a reduction in bone formation. On the other hand, there is no doubt that bone resorption increases during the winter, as witnessed by the significant increases in osteocalcin and NTx (see Tables I and II). Whether this is also a function of activation of resorption by PTH, or other factors, including cytokines remains uncertain.

Summary

Seasonal bone loss is a potentially major problem for postmenopausal women living in northern latitudes, even if they have few risk factors for osteoporosis. Major changes in PTH, osteocalcin, N telopeptide and IGFBP-4 suggest that one mechanism whereby these women lose bone is through a secondary increase in circulating parathyroid hormone, triggered by a decline in cutaneous synthesis of vitamin D. At a minimum, preventive efforts to block bone loss in elderly women must include the use of adequate calcium and vitamin D supplementation.

Table I.

Demographic Baseline Data on 17 women randomized to placebo calcium supplementation for two years (±SEM).

Age yrs	**71±1.0**
BMI-	**27.8±2.0**
Calcium Intake-mg/d	**699±64**
PTH-pg/ml	**29.5±2.4**
Osteocalcin-ng/ml	**7.7±1.0**
Ntx- BCE nmol/mmolCr	**44.2±3.2**
25OH D-ng/ml	**23.9±2.7**
IGF-I-ng/ml	**137±12**
IGFBP-3 ng/ml	**3156±190**
IGFBP-4 ng/ml	**515±38**
IGFBP-5ng/ml	**187±16.7**
FN-BMD- g/cm2	**0.80±0.02**
LS-BMD-g/cm2	**1.10±0.05**
Urinary Calcium Cr	**0.21±0.03**

Table II.

Percent Change in IGF- I and IGFBPs during the first winter and over 24 months of the study (±SEM)

Marker under study	% Change during winter	% change over 24 months
Femoral Bone Density	**-3.2±2.0***	**-3.0±1.8***
Dietary calcium	**-2.0±6**	**0.5±10**
25OH D	**-23±6***	**0.00±5**
PTH	**+22±5***	**49±7****
NTx	**+12±6**	**38±18***
Osteocalcin	**+25±5****	**61±10****
IGF-I	**+3.0±8**	**-7.6±3.6***
IGFBP-3	**-0.8+10**	**-4.0±6.0**
IGFBP-4	**+13±6***	**15.1±6.0***
IGFBP-5	**+0.3±10**	**-0.4±5.0**

*p<0.05; **p<0.001

References

1.Storm D, Smith-Porter E, Musgrave KO, Vereault D, Patton C, Kessenich CR, Eslin R, Mohan S, Chen T, Holick MF and Rosen CJ. Calcium supplementaiton preventsSeasonal bone loss and changes in biochemical markers of bone turnover in elderly New England women: A Randomized Placebo-Controlled Trial. J Clin Endocrinol Metab 83;3817-3826

2.Dawson-Hughes B, Harris S, Krall EA, Dallal GE. Rates of bone loss in postmenopausal women randomly assigned to one of two dosages of vitamin D. Am J Clin Nutr 1995; 61: 1140-1145.

3.Honda Y, Lansdale EE, Strong DD, Baylink DJ, Mohan S. Recombinant synthesis of IGFBP-4: development, validation and application of an RIA for IGFBP-4. J Clin Endocrinol Metab 1996; 81: 1389-1399.

4.Rosen CJ,Morrison T,Hunter SJ,Holick MF,Chen T,Zhou H,Wen-Wei and Storm D. Elderly Women in Northern New England Exhibit Seasonal Changes in Bone Mass and Calciotropic Hormones. Bone and Mineral 1994:25;83-92.

5.Rosen CJ. IGF-I in Osteoporosis. Clinics in Laboratory Medicine 2000;20:591-602

6.Scharla SH,Strong DD, Rosen, CJ et al. 1,25 Dihydroxyvitamin D increases expression of IGFBP-4 in human osteoblast like cells **in vitro** and elevates IGFBP-4 serum levels **in vivo.**J Clin Endocrinol Metab.1993;77:1190-1197.

7.Rosen C, Donahue LR, Hunter S, Holick M, Kavookjian H,Kirshenbaum A, Mohan S and Baylink DJ. The 24/25kD serum insulin-like growth factor binding protein is increased in elderly women with fractures. J Clin Endocrinol Metab. 1992;74:24-28.

VITAMIN D: IMPORTANCE FOR BONE HEALTH, CELLULAR HEALTH AND CANCER PREVENTION

Michael F. Holick, Ph.D., M.D.
Vitamin D, Skin and Bone Research Laboratory; Section of Endocrinology, Diabetes, and Nutrition: Departments of Medicine, Dermatology and Biophysics and Physiology; Boston University Medical Center, Boston, MA.

I. Introduction

Vitamin D is known as the sunshine vitamin(1,2). How important is sunlight in providing humans with their vitamin D requirement? Part of the problem is that vitamin D is rare in foods. Most vertebrates require a source of vitamin D (1). Rodents, dogs, cows, chickens all develop severe growth retardation and bony deformities that are consistent with rickets when they do not get either an adequate dietary source of vitamin D, or exposure to sunlight (Fig 1). Vitamin D deficiency is a major health concern at many zoos (3,4). A mated pair of Komodo dragons at Washington's National Zoo were finally successful in having a live hatch only after they received vitamin D in their diet. The hatchlings survived and thrived as a result of being exposed to an ultraviolet B radiation source (Fig 2). Primates that are kept in glass enclosures or kept indoors are at high risk of developing vitamin D deficiency and its consequences. Not only do they develop severe growth retardation and bony deformities, but also have poor dentition and poor tooth eruption. Even polar bears are at risk of developing vitamin D deficiency. At the Denver Zoo, it was observed that polar bear cubs were listless and unable to stand and walk. They were found to be severely vitamin D deficient (5). Addition of vitamin D to their diet rapidly cured the rickets, and they became very active and grew well.

II. Historical perspective on Vitamin D deficiency in humans

As the Industrial Revolution began to envelop northern Europe, it was observed that children who lived in crowded inner polluted cities such as Glasgow, London and Leiden were developing severe growth retardation and had skeletal deformities on their rib cage and bowing or concavity of

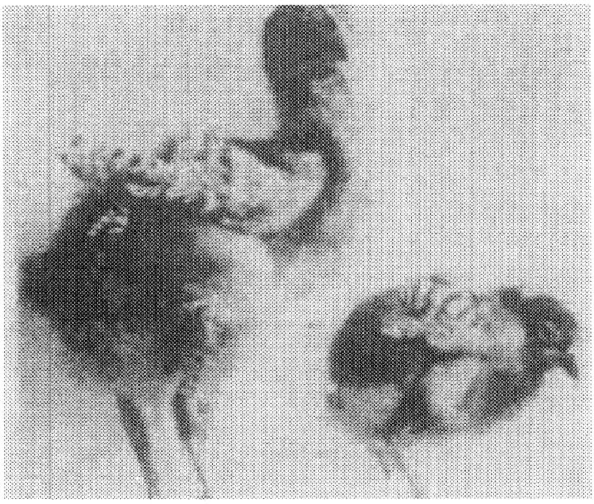

Figure 1. Chickens that were reared from the same clutch of eggs. The chicken on the left received a diet that was fortified with vitamin D. The chicken on the right received the same diet that was not fortified with vitamin D. Note: The severe stunting of growth due to vitamin D deficiency (56).

Figure 2. A komodo dragon baby that recently hatched from a fertile clutch of eggs from the breeding pair of komodo dragons at the National Zoo. The parents received vitamin D supplementation and the baby komodo dragons were housed in an enclosure that included a fluorescent light source that contained ultraviolet B radiation to produce vitamin D_3 in their skin.

their legs(1,2). This disease became the scourge of the industrial revolution. It had been estimated that upwards of 90% of children living in Leiden, the Netherlands had some manifestation of this insidious bone deforming disease (1,2,6).

Sniadecki reported in 1822 that strong and obvious is the influence of sunlight for the prevention and treatment of "English disease" which was the common term used for rickets (7). This insightful observation was based on his realization as a practicing physician in Warsaw that children he saw, in the inner city, had a high incidence of rickets while children that lived in the rural areas around Warsaw did not. He had encouraged sunbathing as a means of preventing and treating the disease (7). Palm also realized, based on his insightful observations, that children who lived in the poorest countries in India and the Orient had no significant rickets despite the fact that they had extremely poor diet and a poor hygienic environment (8). He reasoned that the cause for children developing rickets in London, Glasgow and in other industrialized cities throughout northern Europe was because they were not exposed to sunlight. He urged and encouraged sunbathing as a means for the prevention and cure of this devastating bone disease (8).

Young women who had rickets as children often had a flat and deformed pelvis making it difficult for childbirth leading to a high incidence of infant and maternal morbidity and mortality. This led Murdock Cameron in Glasgow, Scotland and other surgeons to use Cesarean sectioning as a means of delivering children from women who had had rickets (Fig.3). Indeed, Cesarean sectioning became popular because of rickets at the turn of the 20th Century.

Rickets was to children in the 17th 18th and 19th centuries as cancer is to adults today. As a result, many scientists and physicians initiated studies to better understand the cause and to find an effective treatment for the disease. In the early 1800's Bretonneau treated a young child who had acute rickets with cod liver oil and noted that the patient was cured of his disease (9,10). However, great skepticism was raised by the medical community for such a simple remedy to cure this bone disease. Koch (11) reported in 1911 that rickets was caused by an infectious disease. The breakthrough came in 1918 when Mellanby (12) reported that he could induce rickets in beagles by keeping them in a sunless environment and feeding a diet of oatmeal. He confirmed what had been a common folklore treatment for rickets that the treating of beagles with cod liver oil prevented and cured rickets.

Figure 3. These children are the first children to be delivered by Caesarian sectioning by Murdock Cameron in Glasgow from women that had flat pelvises due to rickets. Although the children were delivered without complication, note that because of vitamin D deficiency they suffered severe stunting in their growth (56).

In 1919 Huldschinsky (13) treated several children with rickets with radiation from a mercury arc lamp and demonstrated within four months radiologic healing of their rachitic bones. This prompted Steenbock (14) Hess and Weinstock (15) to independently suggest that exposure of foods to ultraviolet radiation would be an effective method for providing anti-rachitic properties to the food to prevent and cure rickets. This led Steenbock (14) to suggest that irradiation of a common food ingested by children i.e. milk would be ideal. Initially, the precursor of vitamin D was added to milk followed by irradiation. Once the structure of vitamin D was determined (16) and a simple method for making large quantities of vitamin D was accomplished, vitamin D was added to milk. As a result of the fortification of vitamin D to milk and other foods in the 1930's, rickets became a rare disease in the United States, Canada and Europe. However there has been a resurgence of rickets in Europe for the past 50 years, because vitamin D fortification of dairy products was halted because of an unfortunate outbreak of vitamin D intoxication, that was due to the injudicious fortification of milk and other foods with large amounts of vitamin D in the early 1950's (17). Even in the United States, rickets has reappeared especially in African American infants and young children due in part to exclusive breast feeding without any vitamin D supplementation (18).

III. Sources of Vitamin D

Cod liver oil was, and still is today, considered to be an excellent source of vitamin D_3. It was also recognized by Trousseau (10) that liver oils from a variety of fish including cod, and tuna as well as aquatic mammals including whale and seal were useful for the treatment of rickets. It is now recognized that fatty fish, such as salmon and mackerel contain vitamin D_3 within the fat interspersed in the flesh (19). As a result, it is a reasonable source of vitamin D_3 as long as it is ingested at least two to three times a week (2,19).

In Siberia, many of the schools irradiate their children with a mercury arc lamp during the winter in order to promote vitamin D_3 production in their skin to prevent them from developing rickets.

Sunlight is the most important source of vitamin D for humans. The skin has a large capacity to produce vitamin D_3 (1,2,20-22). Minimum exposure to sunlight a few times a week during seasons that promote vitamin D_3 production is usually more than adequate to satisfy the body's requirement (2). During the summer, exposure to sunlight is very efficient for producing vitamin D_3 in the skin at most latitudes around the globe (Fig.2)(2,23). However, as the zenith angle becomes more oblique with the change of season into fall and winter, very few ultraviolet B photons with energies between 290 and 315 nm (which are responsible for making previtamin D_3 in the skin) are able to reach the earth's surface (23,24). As the zenith angle increases the high energy ultraviolet B photons are efficiently absorbed by the ozone layer that envelops the earth resulting in very few of them reaching the earth surface at higher latitudes. This is the reason why during the late fall and winter at latitudes ~ 40° above and below the equator, very little if any vitamin D_3 can be produced in the skin (2,23,24,25). Below a latitude of about 35° there are enough ultraviolet B photons that reaches the earth's surface to produce vitamin D_3 in the skin all year (2,23,24).

IV. Mechanism for the Production of Vitamin D_3 in the Skin

The high-energy solar ultraviolet B photons with energies between 290 (atmospheric ozone absorbs solar radiation below 290 nm) and 315nm penetrate into the skin and are absorbed by 7-dehydrocholesterol that is in the plasma membrane of the keratinocytes in the epidermis and fibroblasts in the dermis (26,27). The net effect of this absorption results in an opening of the B ring of 7-dehydrocholesterol to form an unstable molecule known as previtamin D_3 (Fig.4). The double bonds in previtamin D_3 are thermodynamically unstable and undergo a rearrangement to form a more stable conjugated triene system that is the backbone for the vitamin

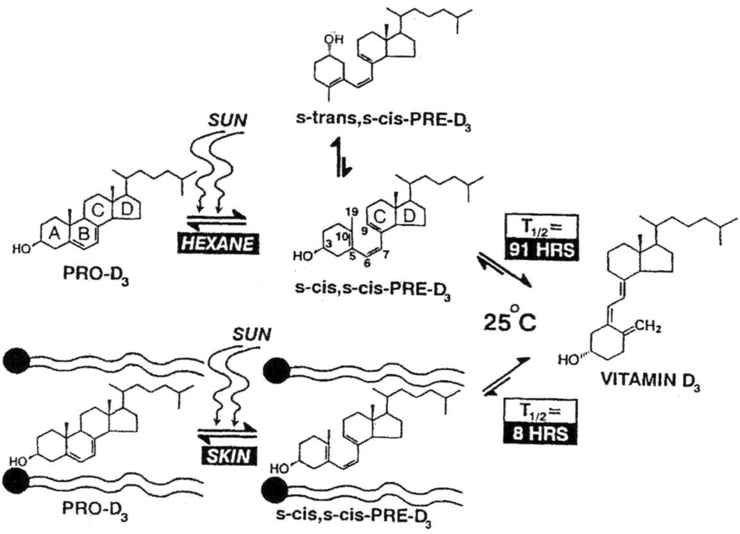

Figure 4. Photolysis of provitamin D_3 (pro-D_3) into previtamin D_3 (pre-D_3) and it thermal isomerization of vitamin D_3 in hexane and in lizard skin. In hexane pro-D_3 i photolyzed to s-cis,s-cis-pre-D_3. Once formed, this energetically unstabl conformation undergoes a conformational change to the s-trans,s-cis-pre-D_3. Onl the s-cis,s-cis-pre-D_3 can undergo thermal isomerization to vitamin D_3. The s-cis,s cis conformer of pre-D_3 is stabilized in the prospholipid bilayer by hydrophili interactions between the 3β-hydroxyl group and the polar head of the lipids, as we as by the van der Waals interactions between the steroid ring and side-chain structur and the hydrophobic tail of the lipids. These interactions significantly decrease th conversion of the s-cis,s-cis conformer to the s-trans,s-cis conformer, thereb facilitating the thermal isomerization of s-cis,s-cis-pre-D_3 to vitamin D_3. Reproduce with permission (28).

D_3 structure (Fig 4)(2,26,27). It is now recognized that it is important for this photochemical reaction to occur in the plasma membrane (27,28). Normally, when previtamin D_3 is incubated at the body temperature of 37° C in an inert organic solvent such as hexane, it takes more than 24 hours for it to convert to vitamin D_3. However, because 7-dehydrocholesterol (7-DHC) is trapped in the lipid bilayer of the plasma membrane its structure is less flexible. Because it is sandwiched in between the fatty acid side-chains when 7-DHC is exposed to ultraviolet B radiation, it can only form a very unstable conformer known as s-cis, s-cis-previtamin D_3 (27,28). This very unstable form of previtamin D_3 rapidly converts to vitamin D_3. The more stable conformer of previtamin D_3 is s-trans, s-cis-previtamin D_3. This thermodynamically stable form will not convert to vitamin D_3 and must change its conformation to the unstable s-cis,s-cis previtamin D_3 conformer. That is the reason why it takes much longer in an inert environment for previtamin D_3 to convert to vitamin D_3 compared to that in biologic membranes (27,28)(Fig 4).

As the previtamin D_3 undergoes a rearrangement of its triene system to form vitamin D_3 its elongated structure makes it unsuitable to remain in the plasma membrane thereby resulting in its ejection into the extracellular space. It ultimately diffuses into the dermal capillary bed where it is bound to the vitamin D binding protein and other lipoproteins in the circulation (29).

V. Factors that Affect Cutaneous Vitamin D Synthesis

Sunscreens have become extremely popular because of the wide publicity regarding the alarming increase in skin cancer, including the most aggressive and deadly form, melanoma (30). Thus it has been recommended that before going outdoors all Caucasians should put a sunscreen on with at least a sun protection factor of 15 in order to prevent any of the damaging consequences from exposure to sunlight. Sunscreens absorb ultraviolet B (UVB) photons, which are the most damaging to the skin. The absorption of UVB photons by a sunscreen on the surface of the skin results in a significant reduction in the cutaneous production of vitamin D_3. When used properly, an SPF of 8 and 15 will reduce the cutaneous production of vitamin D_3 by 97.5 and 99.5% respectively (2,31,32).

Melanin is a very effective natural sunscreen and therefore can reduce by as much as 100 fold the cutaneous production of vitamin D_3 when compared to a light skinned Caucasian of Celtic origin (33). This is part of the explanation for why African Americans in the United States have a high incidence of vitamin D deficiency (34,35).

VI. How Much Sun Exposure is Adequate to Provide Vitamin D?

Is it necessary to bake outside in order to produce enough vitamin D_3 in the skin to satisfy the body's requirement? We have conducted a study in young adults with skin types 2 and 3 and exposed them to 1 minimal erythemal dose of simulated solar ultraviolet B radiation in a tanning bed. Similar groups of healthy young adults were also given between 10,000 and 50,000 units of vitamin D_2 orally one time. An evaluation of the blood levels of vitamin D 24-hours after exposure to one MED revealed that it was similar to taking approximately 10,000 IU of vitamin D_3 a day (36). Now that the new adequate intake recommendation for vitamin D is 200, 400 and 600 IU for children and adults up to 50, 51-70 and 71+ years respectively (36,37) it has been estimated that if only 6% of the body receives one MED of solar UVB exposure, it is equivalent to taking approximately 800 to 1000 IU of vitamin D orally. This is the basis for the recommendation that sub-erythemal exposure equal to about 25 to 30% of an MED of hands, face and arms to sunlight in the spring, summer and fall in temperate latitudes is more than adequate to satisfy one's vitamin D requirement.

Although aging reduces the ability of the skin to produce 7-dehydrocholesterol by as much as 70% (age 80 yrs)(26,28), the skin has such a huge capacity to produce vitamin D_3, that even elderly exposed to 15 or 30 minutes of sunlight two to three time a week are able to substantially raise their blood levels of 25-hydroxyvitamin D [25(OH)D; a measure of vitamin D status](22,39). Chuck and Diffey recently reported that installation of a UVB radiation source installed in a nursing home led to the residents maintaining normal blood levels of 25(OH)D (40). They suggested that installation of a UVB radiation source to produce vitamin D_3 in the skin of nursing home residents should be the standard of care in Great Britain.

VII. Vitamin D Deficiency and its Consequences

Vitamin D deficiency in neonates and young children leads to growth retardation, skeletal abnormalities poor tooth eruption and poor dentition. It is relatively easy to recognize these signs and symptoms of vitamin D deficiency in children. Thus, vitamin D deficiency rickets is usually quickly diagnosed and rapidly treated without very few long-term consequences.

The detection of vitamin D deficiency in adults is much more difficult, since there are no observable consequences such as deformities of the skeleton. However vitamin D deficiency in adults can have consequences that can have a long ranging impact on the individual's health. Vitamin D deficiency cannot be detected by measuring a serum calcium. The reason for this is that the body maintains a normal serum calcium concentration at the expense of the skeleton. When a person becomes vitamin D deficient, there is an increase in the production of parathyroid hormone (PTH) in the parathyroid glands. This leads to elevated levels of PTH, which have effects on the kidney, including increasing tubular reabsorption of calcium to help maintain serum calcium levels (22,41)(Fig 5). However, PTH also causes phosphate wasting into the urine, resulting in a low serum phosphorus level. It is the low serum phosphorus level that leads to an inadequate calcium phosphate product resulting in a skeletal mineralization defect that causes rickets in children and osteomalacia (adult rickets) in adults. The excessive production of PTH also results in an increase in the mobilization of the mineral and matrix from the skeleton that can often precipitate or exacerbate osteoporosis and increased risk of fractures (22,41).

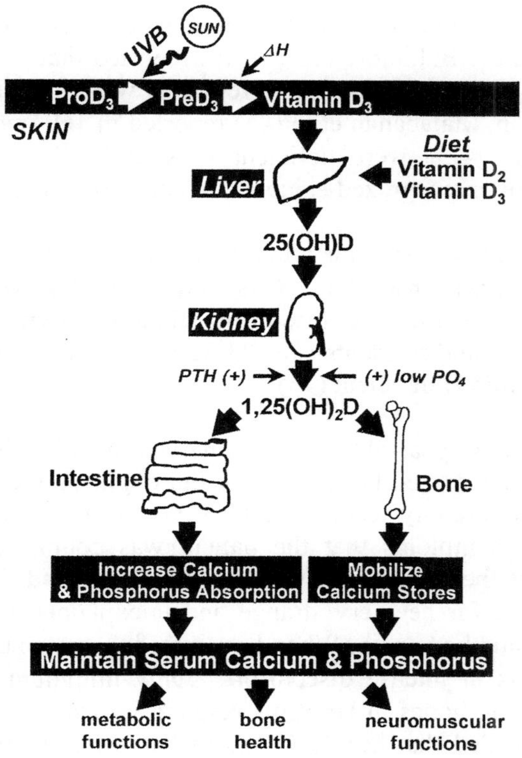

Figure 5. Schematic representation for the synthesis of vitamin D and its metabolism and biologic functions on calcium and phosphorus homeostasis. 7-dehydrocholesterol (provitamin D_3;proD_3 in the skin is photolyzed to previtamin D_3 (preD_3) during its exposure to ultraviolet B (UVB) radiation from sunlight. Once formed, previtamin D_3 undergoes a thermally-induced (ΔH) transformation to vitamin D_3. Vitamin D_3 from the skin and vitamin D_3 and vitamin D_2 from the diet enter into the blood and are metabolized to their 25-hydroxy counterparts, 25-hydroxyvitamin D (25(OH)D). Once formed, 25(OH)D is metabolized in the kidney to $1,25(OH)_2D$. Parathyroid hormone (PTH) and low serum phosphorus both enhance (+) the production of $1,25(OH)_2D$. Once formed, $1,25(OH)_2D$ regulates serum calcium, and phosphorus levels by increasing the efficiency of intestinal calcium and phosphorus absorption, and by mobilizing calcium stores from bone. The ultimate role of vitamin D is to maintain the serum calcium, and phosphorus levels within the normal range to sustain a wide variety of metabolic functions, physiologic functions as well as to optimize bone health.

In addition to its effect on the mineralization of the skeleton, vitamin D deficiency can cause severe bone discomfort as well as muscle weakness and muscle pain (42,43).

It is well recognized that adults are at risk of developing vitamin D deficiency. In the Baltimore area, it was reported that 54% of community dwellers and 38% of nursing home residents were deficient in vitamin D (44). In Boston, Malabanan et al (45) reported in 169 out patients ages 49 –83 years, that 41% were deficient in vitamin D. Thus, vitamin D deficiency is an unrecognized epidemic in adults over the age of 50 years.

Are younger adults at risk for vitamin D insufficiency? We conducted a study at our medical institution. It is well known that young adults often never see the light of day because of their indoor activities. We evaluated medical students and residents ages 23-32 years of age. We observed that 37% were vitamin D deficient (46).

I recently saw a 43-year-old African American female who presented with global muscle aches and bone pain. A thorough, very extensive work-up by her primary care physician did not reveal any abnormality that could be related to the symptoms that the patient was complaining of. It was concluded that the patient was seeking sympathy, and that there was no cause or reason for her very strange and unexplained muscle weakness, muscle aches and bone pain. When I evaluated the patient, it was clear that the patient was in intense discomfort. Only minimum touching of her sternum and long bones in her arms resulted in exquisite discomfort. An evaluation revealed that her 25(OH)D (a barometer of vitamin D status) was undetectable and that she was suffering severe vitamin D deficiency. This was in part caused by her increased skin pigmentation, poor dietary intake of vitamin D and a stomach surgery that may have caused fat malabsorption of vitamin D. Treatment with large doses of vitamin D_2 i.e. 50,000 IV vitamin D_2 once a week for 3 months resulted in complete resolution of her symptoms.

VIII. Vitamin D and the Cancer Connection

Garland et al. suggested in the late 1980's and early 1990's that individuals living at higher latitudes had a higher risk of dying of colon and breast cancer (47,48). They were the first to suggest that there may be a connection with inadequate production of vitamin D_3 in individuals who live at higher latitudes. It was known at that time that vitamin D had to be activated first in the liver and then in the kidney to 1,25-dihydroxyvitamin D [1,25$(OH)_2D_3$](Fig 5). It was also known that 1,25$(OH)_2D_3$ was effective in inhibiting proliferation and inducing terminal differentiation of cancer cells (41). However, it was also known that increased exposure to sunlight or ingestion of pharmacologic amounts of vitamin D did not result in an increase in blood levels of 1,25$(OH)_2$D. This was because the kidney tightly regulated its production of 1,25$(OH)_2D_3$. Thus, the conundrum was how was it possible that increased exposure to sunlight could decrease risk of some cancers since vitamin D's active form was not increased in the circulation? The mystery was finally solved when the gene for the 25-hydroxyvitamin D-1α-hydroxylase (1-OHase) was cloned. This resulted in the quick identification of the presence of this enzyme in normal and malignant prostate cells, colon and colon cancer cells as well as a wide variety of other tissues including the skin (49-52). It has been concluded that vitamin D may play a very important role in cellular health. Increasing the production of vitamin D in the skin results in higher circulating levels of 25(OH)D. 25(OH)D is the substrate for the 1-OHase that is present in prostate, colon, breast and other tissues. Thus, all of these tissues have the ability to make 1,25$(OH)_2$D which may in turn help regulate cellular growth, and possibly prevent them from becoming cancerous. (Fig 6)

IX. Effect of Skin Type on Circulating 25(OH)D Levels

We recruited healthy adults with skin types 2,3,4 and 5 and exposed them to tanning bed radiation along the guidelines recommended by the Tanning Bed Manufacturers, and sanctioned by the FDA. We measured circulating concentrations of 25(OH)D once a week for three months. It was observed that those with skin type 2 had on average 120% increase in 25(OH)D and those with skin types 3 and 4 about 80%. Those individuals with skin type 5 had about 50% increase in their circulating concentrations of 25(OH)D (Fig 7). It was concluded that circulating concentrations of 25(OH)D might be a more sensitive indicator of skin type than their historical recounting of their sensitivity to burning from sun exposure (53).

X. Vitamin D Metabolism of vitamin D in the Skin

The skin is the largest organ in the body and is responsible for making vitamin D_3 during exposure to sunlight. It has recently been suggested by Lehman et al (54). that the skin can make 1,25$(OH)_2D_3$ from vitamin D_3. We have used quantitative RT/PCR and have identified a vitamin D-25 hydroxylase and 25(OH)D-1α-OHase in human epidermis (55).

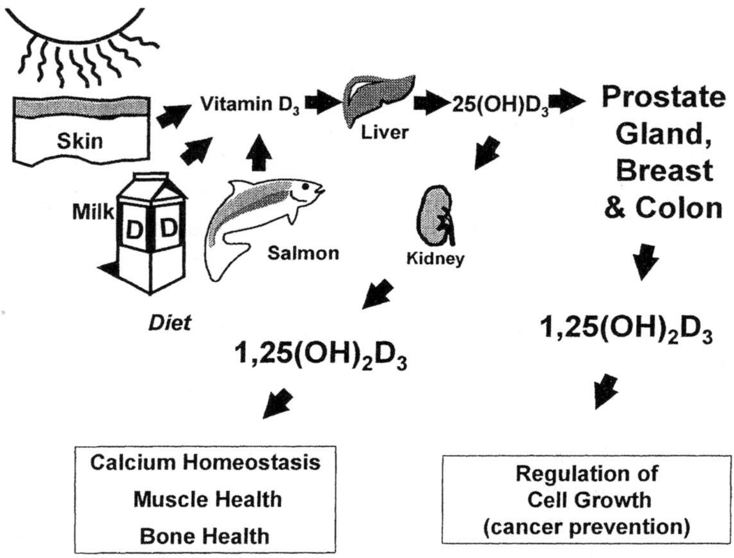

Figure 6. Two functions of vitamin D. The major function of vitamin D is t maintain serum calcium within a physiologically acceptable range for neuromuscula health, bone health and for a wide variety of metabolic processes. This i accomplished by producing 1,25-dihydroxyvitamin D_3 [1,25$(OH)_2D_3$] in the kidney 1,25$(OH)_2D_3$ enters the circulation and regulates intestinal calcium transport in th intestine and can mobilize calcium stores from bone when needed. Vitamin D ma also play an important role in cellular health. The prostate gland, breast, colon, ski among other organs have the enzymatic machinery to convert 25-hydroxyvitmai D[25$(OH)D_3$] to 1,25$(OH)_2D_3$. It is believed that once formed, it helps regulat cellular growth and therefore may in part be responsible for decreasing risk of man of the more common cancers including colon, prostate, breast and ovarian cancer. I is also thought that once formed, 1,25$(OH)_2D_3$ in these tissues carry out it physiologic function in regulating cell growth, and then is rapidly metabolized an destroyed. Thus, the extra renal tissues are incapable of acting as endocrine tissue for releasing 1,25$(OH)_2D_3$ into the circulation. This is probably the reason wh patients with renal failure have essentially undetectable levels of 1,25$(OH)_2D_3$ eve though there skin and a variety of other tissues are making 1,25$(OH)_2D_3$. (Reproduce with permission)(43).

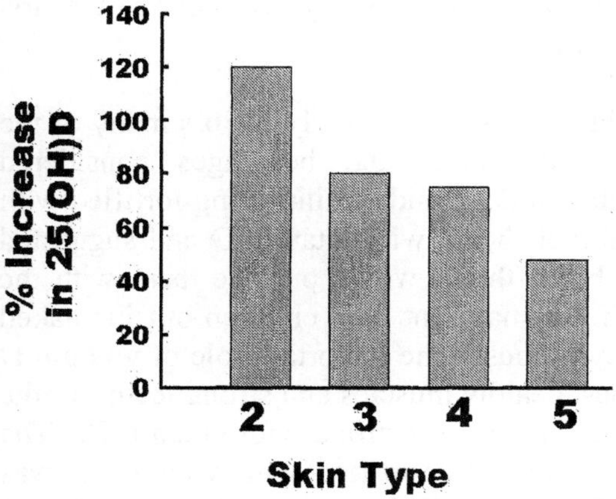

Figure 7. Percent increase in 25-hydroxyvitaminD [25(OH)D] in healthy adults with skin types 2, 3, 4 and 5 who were exposed to sun tanning bed radiation 3 times a week at the recommended levels by the tanning industry and sanctioned by Food and Drug Administration.

To better understand the role of the 1-OHase in human skin we made a plasmid construct with the 1-OHase gene and transfected normal human keratinocytes. We observed that the transfection of keratinocytes with the 1-OHase gene resulted in a 78% increase in the efficiency in conversion of $25(OH)D_3$ to $1,25(OH)_2D_3$. More importantly, when cultured keratinocytes that were transfected with the 1-OHase gene compared to keratinocytes transfected with an empty vector plasmid followed by incubation with $25(OH)D_3$ those cells that had increased 1-OHase activity had a more marked antiproliferative response to $25(OH)D_3$ compared to the control group (55).

Thus, it is possible that the presence of the vitamin D making machinery may be important for regulating skin cell growth, preventing some skin cancers such as squamous cell carcinoma, and may even be important in the prevention of melanoma.

Conclusion

I have hopefully convinced you that vitamin D is no longer considered to be a boring fat-soluble vitamin, but is a molecule that has multiple dynamic personalities. It is critically important for skeletal development, growth and maintenance from birth until death. It helps maintain bone health and prevents osteomalacia and osteoporosis in adults. It can maximize not only skeletal health, but also cellular health. There is the intriguing possibility that vitamin D plays an important role in maintaining cellular health and may help prevent many of the common cancers such as cancer of the colon, breast, prostate and ovaries. It may also be important in preventing these cancers from spreading aggressively. Vitamin D may act as one of the first sentinels to help regulate cell growth.

Vitamin D was discovered in the 1920's, and by early 1930's many of the companies who produced various foods and beverages considered fortifying their products with vitamin D. Besides milk, being fortified with vitamin D Bond Bread fortified their bread with vitamin D and suggested that by feeding children Bond bread that it would provide them with the same amount of vitamin D as if they had sent their children outside naked to play in sunlight (Fig 8). They espoused the important role of vitamin D in maintaining good strong bones, healthy muscles and strong teeth. In the 1940's in Chicago Richter's hot dogs were fortified with vitamin D. The most amazing product however that was fortified with vitamin D was introduced in the early 1930's. Schlitz fortified its beer with vitamin D and their advertisements suggested that by ingesting beer fortified with sunshine vitamin D it would provide extra energy, and during the winter provide both men and women with an adequate amount of vitamin D (Fig 9). However, once alcoholic beverages could no longer promote any health claims, vitamin D was removed from beer. Today very few foods other than milk and some breads and cereals are fortified with vitamin D. Because of the outbreak of vitamin D intoxication in the 1950's in Europe, vitamin D fortification of milk and other food products has been forbidden in most European countries. The exception is that some margarines and cereals are fortified with vitamin D in some European counties. There needs to be a reconsideration of whether additional foods should be fortified with vitamin D. Juice drinks would be an ideal vehicle for such a fortification program and would have a significant impact in vitamin D status for all age groups.

I hope that I have provided you a new appreciation for the multi-faceted personality of vitamin D. It is not only critically important for human skeletal and muscle health, but also may be equally important for cellular health and cancer prevention.

Figure 8. A Bond Bread advertisement espousing the importance of vitamin D for bone and muscle health in children. In the early 1930's Bond Bread fortified their product with vitamin D and used it as a marketing tool for their bread (57).

Figure 9. In 1932 through 1936 Schlitz fortified their beer with vitamin D to market it as a unique nutrient enriched product. However, in 1936 the FDA forbid any nutrient claims for alcoholic beverages and vitamin D was removed from beer (57).

Acknowledgement

This work was supported in part by a grant from the NIH, # M01RR-00533 and a grant from the Indoor Tanning Association

References

1. Holick, MF (1994) McCollum Award Lecture, Vitamin D: new horizons for the 21st century. Am J Clin Nutr 1994;60,619-630.

2. Holick, M.F. A perspective on the beneficial effects of moderate exposure to sunlight: bone health, cancer prevention, mental health and well being. In: Sun Protection in Man P.U. Giacomoni, (ed) Melville, NY: Elsevier Publishing, 2001.

3. Allen ME, Oftedal OT, Ullrey DE, Effect of dietary calcium concentration on mineral composition of fox, geckos(hemidactylus garnoti) and Cuban tree frogs (osteopilus septentrionalis). J Zoo Wildlife Med 1993;24:118-128.

4. Ullrey, DE, Bernard JB. Vitamin D: Metabolism, sources, unique problems in zoo animals, meeting needs. In Fowler ME, Miller RE editors. Zoo & Wild Animal Medicine. Philadelphia: W.B. Sanders Company,63-78 1999.

5. Kenny DE, Irlbeck NA, Eller JL. Rickets in two hand-reared polar bear (Ursus Maritimus) cubs. J Zoo Wildlife Med 1999;30:132-140.

6. Schmorl,G Die pathologische Anatomie der rachitischen Knochenerkrankung mit besonderer Berücksichtigung ihrer Histologie und Pathogenese. (The pathological anatomy of rachitic bone disease with a special focus on its histology and pathogenesis.) Ergeb.Inn. Med Kinderheilkd., 1909; p.403.

7. Sniadecki J, Sniadecki J. (1768-1838) on the cure for rickets (1840). Cited by Mozolowski W. Nature 1909;143:121.

8. Palm TA. The geographic distribution and etiology of rickets, Practitioner 1890;45:270-342.

9. Mayer J. Armand Trousseau and the arrow of time. Nutr Rev 1957;15:321-323.

10. Trousseau A. Rickets. In: M.M. Peter (ed) Clinique Medicale de L'Hotel-Dieu De Paris (4th edition). Librairie J.-B. Bailliere et Fils, Paris, pp. 490-532, 1873.

11. Koch J. Untersuchungen uber die localixation der bakterian das verhalten des knochenmarkes und die veranderungen der knochen, insbesondere der epiphysen, bei infektionskranheiten. Mit bemerkungen zur theorie der rachitis. Zeitschr F Hyg U Infetionskrankh 1911; IX:436.

12. Mellanby T. The part played by an 'accessory factor' in the production of experimental rickets. J Physiol 1918;52:11-14.

13. Huldschinsky K. Heilung von Rachitis durch Künstliche Höhensonne. (Curing rickets by artificial UV-radiation.) Dtsch. Med Wochenschr 1919;45:712-713.

14. Steenbock H. The induction of growth-promoting and calcifying properties in a ration exposed to light. Science 1924;50:224-225.

15. Hess AF, Weinstock M. Antirachitic properties imparted to inert fluids and green vegetables by ultraviolet irradiation. J Biol Chem 1924;62:301-313.

16. Havinga E. Vitamin D, example and challenge. Experientia 1973;29:1181-1193.

17. Jacobus CH, Holick MF, Shao Q, Chen TC, Holm IA, Kolodny JM, El-Hajj Fuleihan G, Seely E. Hypervitaminosis D associated with drinking milk. N Engl J Med 1992;326:1173-1177.

18. Welch T, Bergstrom W, Tsang R. Vitamin D deficient rickets: The reemergence of a once conquered disease. J Pediatr 2000;137:143-145.

19. Takeuchi A, Okano T, Tadada M, Kobayashi T. Possible origin of extremely high contents of vitamin D_3 in some kinds of fish liver. Comp Biochem Physiol 1991;100A:483-487.

20. MacLaughlin M.F.J. Aging decreases the capacity of human skin to produce vitamin D_3. J Clin Invest 1985;76:1536-1538.

21. Slovik DM, Adams JS, Neer RM, Holick MF, Potts JT Jr. Deficient production of 1,25-dihydroxyvitamin D in elderly osteoporotic patients. N Engl J Med. 1981;305:372-374.

22. Holick MF. Calcium/vitamin D diagnostics and therapeutics. In: C. Rosen (ed), Clinics in Laboratory Medicine. W.B. Saunders, Philadelphia, PA, pp.569-590, 2000.

23. Chen TC Functional metabolism and molecular biology of vitamin D action. In: M.F. Holick (ed), Vitamin D: Physiology, Molecular Biology, and Clinical Applications. Human Press, Totowa, NJ, pp. 30-56 1999.

24. Webb AR, Kline L, Holick MF. Influence of season and latitude on the cutaneous synthesis of vitamin D_3: exposure to winter sunlight in Boston and Edmonton will not promote vitamin D_3 synthesis in human skin. J Clin Endocrinol Metab 1998;67:373-378.

25. Lu Z, Chen, TC, Holick MF. Influence of season and latitude on the cutaneous synthesis of vitamin D_3. In: M.F. Holick, A. Kligman (eds), Proceedings, Biologic Effects of Light Symposium. De Gruyter, Berlin, pp. 53-56 1992.

26. Holick MF, MacLaughlin JA and Doppelt SH. Regulation of cutaneous previtamin D_3 photosynthesis in man: Skin pigment is not an essential regulator. Science 1981;211:590-593.

27. Tian XQ, Chen TC, Matsuoka LY, Wortsman J, Holick MF. Kinetic and thermodynamic studies of the conversion of previtamin D_3 to vitamin D_3 in human skin. J Biol Chem 1993;268:14888-14892.

28. Holick MF, Tian XQ, Allen M. Evolutionary importance for the membrane enhancement of the production of vitamin D_3 in the skin of poikilothermic animals. Proc Natl Acad Sci 1995;92:3124-3126.

29. Cooke NE, David EV. Serum vitamin D-binding protein is a third member of the albumin and alpha fetoprotein gene family. J Clin Invest 1985;76:2420-2424.

30. The Skin Cancer Foundation. Bad to worse: Melanoma increases again. Sun & Skin News 1998;15:1-2.

31. Matsuoka LY, Ide L, Wortsman J, MacLaughlin JA and Holick MF. Sunscreens suppress cutaneous vitamin D_3 synthesis. J Clin Endocrinol Metab 1987:64:1165-1168.

32. Matsuoka LY,Wortsman J, Hanifan N. Holick MF. Chronic sunscreen use decreases circulating concentrations of 25-hydroxyvitamin D: A preliminary study. Arch Dermatol 1988;124:1802-1804.

33. Clemens TL, Henderson SL, Adams JS, Holick MF. Increased skin pigment reduces the capacity of skin to synthesize vitamin D_3. Lancet 1982; i: 74-76.

34. Bell NH, Greene A, Epstein S, Oexmann MJ, Shaw S, Shary J. Evidence for alteration of the vitamin D-endocrine system in blacks. J Clin Invest 1985;76:470-473.

35. Krall EA, Sahyoun N, Tannenbaum S, Dallal GE, Dawson-Hughes B. Effect of vitamin D intake on seasonal variations in parathyroid hormone secretion in postmenopausal women. N Engl J Med 1989;321:1777-1783.

36. Holick MF. Vitamin D requirement for humans of all ages: New increased requirement from women and men 50 years and older. Osteoporosis Int.1998;8(suppl.) S24-29.

37. Holick MF. (Contributor, Panel on Calcium and Related Nutrients) Calcium. In: Dietary Reference Intakes for Calcium, Phosphorus, Magnesium, Vitamin D, and Fluoride. Institute of Medicine. National Academy Press 1999.

38. Holick MF, Matsuoka LY, Wortsman J. Age, vitamin D and solar ultraviolet. Lancet 1989; Nov.4:1104-1105.

39. Reid IR, Gallagher DJA, Bosworth J. Prophylaxis against vitamin D deficiency in the elderly by regular sunlight exposure. Age Ageing 1985;15:35-40.

40. Chuck A, Todd J, Diffey B. Subliminal ultraviolet-B irradiation for the prevention of vitamin D deficiency in the elderly: a feasibility study. Photoderm Photoimmun Photomed 2001;17:4,171-174.

41. Holick MF. Vitamin D: photobiology, metabolism, mechanism of action, and clinical applications. In: M.J. Favus (ed), Primer on the Metabolic Bone Diseases and Disorders of Mineral Metabolism (4th edition). Lippincott-Raven, Philadelphia, PA, pp.92-98 1999.

42. Glerup H, Mikkelsen K, Poulsen L, et al. Commonly recommended daily intake of vitamin D is not sufficient if sunlight exposure is limited. J Intern Med 2000;66:419-424.

43. Holick MF. The sunlight 'D'ilemma: risk of skin cancer or bone disease and muscle weakness. Lancet 2001:357:4-5.

44. Gloth FM, Tobin JD, Sherman SS, Hollis BW. Is the recommended daily allowance for vitamin D too low for the homebound elderly? J Am Geriatr Soc 1991;39:137-141.

45. Malabanan A, Veronikis IE, Holick MF. Redefining vitamin D insufficiency. Lancet 1998;351:805-806.

46. Tangpricha V, Flanagan JN, Pearce EN, Chen TC, Holick MF. Vitamin D deficiency among medical students and housestaff physicians. Proceedings of The Endocrine Society's 83rd Annual Meeting. June 20-23; Denver The Endocrine Society Press Publishers 2001.

47. Garland CF, Garland FC, Shaw EK, Comstock GW, Helsing KJ, Gorham ED. Serum 25-hydroxyvitamin D and colon cancer: Eight-year prospective study. Lancet 1989;18: November 1176-1178.

48. Garland FC, Garland CF, Gorham ED Young JF. Geographic variation in breast cancer mortality in the United States: A hypothesis involving exposure to solar radiation. Prev Med 1990;19:614-622.

49. Skowronski RJ, Peehl DM, Feldman D. Actions of vitamin D_3 analogs on human prostate cancer cell lines: Comparison with 1,25dihydroxyvitamin D_3. Endocrinology 1995;136:20-26.

50. Schwarz GG, Whitlatch LW, Chen TC, Lokeshwar BL, Holick, MF. Human prostate cells synthesize 1,25-dihydroxyvitamin D_3 from 25-hydroxyvitamin D_3. Cancer Epidemiol Biomark Prev 1998;7:391-395.

51. Tangpricha, V, Flanagan JN, Whitlatch LW, Tseng CC, Chen TC, Holt PR, Lipkin MS, Holick MF. 25-hydroxyvitamin D-1α-hydroxylase in normal and malignant colon tissue. Lancet 2001;357:1673-1674.

52. Cross HS, Bareis P, Hofer H, Bischof MG, Bajna E, Driwanek S, Bonner E, Peterlik M. 25-Hydroxyvitamin D_3-1α-hydroxylase and vitamin D receptor gene expression in human colonic mucosa is elevated during early cancerogenesis. Steroids 2001;66:287-292.

53. Nasrollahi F, Holick MF. Quantitative evaluation of pigmentation following repeated UV irradiation by Microflash spectrophotometer. In: M. F. Holick (ed.)Proceedings- Biologic Effects of Light. Boston: Kluwer Academic Publishers Proceedings 2001(In press).

54. Lehman, B. HaCaT cell line as a model system for vitamin D_3 metabolism in human skin. J Invest Dermatol 1997;108:78-82.

55. Flanagan JN, Whitlatch LW, Chen TC, Zhu XH, Holick MT, Kong X-F and Holick MF. Enhancing 1α-hydroxyloase activity with the 25-hydroxyvitamin D-1α-hydroxylase gene in cultured human keratinocytes and mouse skin. J Invest Dermatol 2001;116:910-914.

56. Photographs provided by Dr. Iain Boyle, University of Glasgow, Dept. of Medicine for medical education only. Original source unknown.

57. Photographs provided by Dr. Michael F. Holick, Boston University School of Medicine, Dept. of Medicine for medical education only. Original source unknown.

IMPACT OF VITAMIN D AND CALCIUM ON BONE AND MINERAL METABOLISM IN OLDER ADULTS

Bess Dawson-Hughes, M.D.
USDA Human Nutrition Research Center on Aging at Tufts University, Boston, MA 02111

INTRODUCTION

Osteoporosis has reached epidemic proportions in the United States and elsewhere. Many features of our lifestyle contribute to bone loss and bone fragility, including lack of physical exercise, smoking, excess alcohol consumption, and inadequate intakes of calcium and vitamin D. In this paper, I will consider how calcium and vitamin D intakes influence the skeleton, the impact of season on vitamin D levels, criteria for determining how much vitamin D is needed, and feasible approaches to meeting the vitamin D requirement.

CALCIUM AND VITAMIN D PHYSIOLOGY

An inadequate intake of calcium and an inadequate blood level of vitamin D, either alone or in combination, influences calcium-regulating hormone levels. As shown in Figure 1, a deficiency of either nutrient results in reduced calcium absorption and a lower circulating ionized calcium concentration. The latter stimulates the secretion of parathyroid hormone (PTH), a potent bone-resorbing agent. In most older adults, bone remodeling is "uncoupled" in favor of resorption, so that a modest deficit in net bone calcium content occurs with each remodeling event. Over time, this leads to measurable and significant bone loss and increased risk of fracture.

Intestinal calcium absorption occurs by active transport and by passive diffusion. In individuals with low to moderate calcium intakes, absorption occurs largely by active transport, a process mediated by 1,25-dihydroxyvitamin D [$1,25(OH)_2D$]. As calcium intake increases above approximately 500 mg per day, passive diffusion accounts for an increasing proportion of absorbed calcium (1). Estrogen enhances calcium absorption indirectly by enhancing $1,25(OH)_2D$ production in the kidney. It also has a direct effect on intestinal responsiveness to $1,25(OH)_2D$ (2). There is an age-related decline in calcium absorption efficiency in men and women (3). This may be related to loss of intestinal vitamin D receptors or resistance of these receptors to the action of $1,25(OH)_2D$ (4). In addition to calcium intake, vitamin D and estrogen levels, and aging, season and race also influence calcium absorption efficiency.

↓ Calcium
↓ Calcium intake → absorption ← ↓ Vitamin D
↓
↓Circulating [Ca^{++}]
↓
↑PTH
↓
↑Bone remodeling
↓ ↓
↑ Bone loss → ↑ Fracture risk

Figure 1. Consequences of calcium and/or vitamin D insufficiency.

Vitamin D is acquired from diet and from skin synthesis. Aging induces changes in vitamin D metabolism. As demonstrated by MacLaughlin and colleagues (5) there is a several-fold decline in the amount of 7-dehydrocholesterol, a precursor to vitamin D, in the epidermal layer of skin with aging. In addition to having a decreased capacity to photosynthesize vitamin D, elderly people often avoid sun exposure or use sun screens that block access of ultraviolet B rays to the skin. There is no consensus at this time on whether the intestinal absorption of vitamin D, which takes place largely in the distal ileum, changes with aging.

VITAMIN D AND SEASON

Season is a major determinant of vitamin D levels in people residing in the temperate zone. In 1988, Webb and colleagues (6) demonstrated that skin synthesis of vitamin D does not occur between October and March in Boston, latitude 42°N. This was postulated because the ultraviolet B rays that photo-convert 7-dehydrocholesterol to pre-vitamin D do not reach the earth's surface during these months. A clinical correlation of seasonal variation in ultraviolet B exposure has been demonstrated in a cohort of 333 postmenopausal women who were recruited over a 12-month period for a calcium intervention study (7). A plot of the baseline measurements according to the month of enrollment reveals a mean peak 25-hydroxyvitamin D (25OHD) concentration of 95 nmol/L in August and a mean nadir concentration of 63 nmol/L in March. Among the subset of women with dietary vitamin D intakes estimated at 220 IU and lower, the mean nadir

25OHD concentration was 55 nmol/L. The serum concentration of 25OHD is the best clinical indicator of vitamin D status because it reflects contributions from both dietary and skin sources of the vitamin and because its serum concentration fluctuates the least after an acute change in either sun exposure or dietary vitamin D intake.

In the Krall study cited above (7), serum PTH levels varied with time of year in a pattern reflecting the mirror image of the serum 25OHD levels. The wintertime rise in serum PTH level raised concern that bone loss may be increased during the wintertime. We have conducted two prospective studies to examine this. In the first study (8), 246 healthy, postmenopausal women were recruited and all were placed on supplemental calcium. Half of the members, randomly selected, received 400 IU/d of vitamin D and the other half received a placebo. We examined rates of bone loss over the 6-month winter/spring interval when PTH levels were highest and separately over the summer/fall 6-month interval when serum PTH levels were lowest. This study revealed that the vitamin D supplemented group had reduced wintertime bone loss from the spine and total body and had a net benefit to bone mineral density at the end of this 1-year study. To confirm this finding and to test other vitamin D doses, a second study involving 247 healthy postmenopausal women was conducted over a 2-year period (9). All volunteers were again placed on calcium supplements and randomized to treatment with either 700 IU/d of vitamin D or 100 IU/d. At the spine, change in bone mineral density was similar in the two groups. At the femoral neck, however, loss was greater in the group receiving 100 IU/d of vitamin D than in the group receiving 700 IU/d. Seventy % of the benefit from the higher dosage of vitamin D accrued during the winter/spring months. From these two studies, we conclude that wintertime declines in serum 25OHD and the accompanying increases in PTH result in accelerated wintertime and net bone loss in healthy older individuals.

VITAMIN D, CALCIUM, AND FRACTURES

Vitamin D supplementation has been shown to reduce bone loss in several randomized controlled studies (8, 10) , but the effects of vitamin D on fracture incidence have been variable. Heikinheimo, et al. (11) reported that an annual intra-muscular injection of vitamin D reduced fracture rates in older Finnish men and women. In contrast, another large prospective controlled trial found that supplementation with 400 IU/d of vitamin D did not alter hip fracture rates in elderly Dutch men and women (12). The reasons for the discrepant findings in these two studies (11,12) are not entirely clear but the difference may have been related to different degrees of vitamin D deficiency in the two populations at baseline. Inadequate information is available to ascertain this for sure.

Combined vitamin D and calcium supplementation has been studied in a cohort of more than 3,000 very elderly French nursing home residents who were supplemented with 1,200 mg of calcium as the tri-phosphate together with 800 IU/d of vitamin D or double placebo (13). The supplements significantly reduced the incidence of hip fractures and of other non-vertebral fractures within 18-months, and the reduction was sustained throughout the 3-year study. The nursing home residents were at high risk for fracture on the basis of their mean age of 84 years, their low self-selected calcium intakes (approximately 500 mg/d), and their low initial mean 25OHD level of 24 nmol/L (measured in a subset). A second study of combined calcium and vitamin D supplementation was conducted in a lower risk group - ambulatory healthy men and women age 65 and older (mean age 71 years) who resided in their homes in the Boston area (14). In this randomized trial, half of the subjects received 500 mg/d of calcium as the highly absorbable citrate malate plus 700 IU/d of vitamin D or double placebo. Over the 3-year intervention period, the subjects taking calcium and vitamin D not only had reduced rates of bone loss from the spine, hip, and total body but they also had a 50% lower incidence of clinical fractures. This study demonstrated that increasing intakes of calcium and vitamin D is beneficial to men and women consuming the typical American diets containing an average of 700 mg/d of calcium and 200 IU/d of vitamin D. The Boston cohort was followed for an additional 2 years after they discontinued the study supplements (15). During this interval they were free to take any supplements that they chose and were reminded of the recommendations to consume 1,200 mg/d of calcium and 600 IU/d of vitamin D. The subjects returned for bone mineral density measurements at annual intervals. Over the 2-year follow-up period, all gains in bone mineral density at the spine and hip were lost so that the two treatment arms were indistinguishable in bone density and in change in bone density at the end of the 5-year period. Thus, in order to sustain the benefits offered by increased calcium and vitamin D intakes, the higher intakes need to be maintained.

DEFINITION OF VITAMIN D ADEQUACY

Even subtle declines in serum 25OHD are associated with rises in circulating PTH concentrations (Figure 1). Vitamin D adequacy is often defined as the level of 25OHD needed to maximally suppress serum PTH levels. Several approaches have been taken to identify the optimal level of 25OHD. Malabanan et al. (16) assessed PTH responses to treatment with vitamin D according the initial 25OHD levels of the subjects. In this study, 35 subjects were treated with 50,000 IU/wk for 8 weeks. Subjects with 25OHD levels of the subjects below 50 nmol/L had significant declines in serum PTH with treatment but those with initial 25OHD concentrations above this level did not. Other investigators have documented an inverse association between serum 25OHD and PTH levels but noted that the association is hyperbolic, not linear (7, 17-20) . Analyses were performed in these five cross-sectional data sets to identify the inflection point, or the 25OHD level above which serum PTH no longer decreases significantly. One estimate of the inflection point was 30 nmol/L, three estimates were within the fairly narrow range of 75 to 90 nmol/L, and one estimate was 110 nmol/L. These findings are summarized in Table 1. Some of the factors that may have influenced the

Table 1. Estimates of serum 25OHD levels needed to maximally suppress serum PTH

Author	25OHD, nmol/L
Lips (17)	30
Malabanan (16)	50
Peacock (19)	75
Chapuy (20)	78
Krall (7)	90
Dawson-Hughes (18)	110

estimates in Table 1 include subject age, ambient calcium intake, and 25OHD assay differences. The latter however does not appear to be fully explanatory (21). It should be noted that the variability around any of these estimates is quite large. Until new evidence emerges indicating a better goal, it is probably wise to maintain 25OHD levels at 80 nmol/L or above - that is, in the upper half of the normal range. This of course implies that many subjects in the reference population for the assay had some degree of vitamin D insufficiency, an assumption supported by the prevalence of low basal 25OHD levels of the subjects in many of the studies discussed above.

By any of the cut points in Table 1, vitamin D deficiency is commonplace in several segments of the population in Boston. Fifty % of patients admitted to the medical service of a local hospital had 25OHD levels below 37 nmol/L (22), 50% of patients admitted to a different local hospital with acute hip fractures had 25OHD levels below 30 nmol/L (23), and 80% of residents of a local nursing home had 25OHD levels below 37 nmol/L (24). In contrast, only 14% of patients seen in a local outpatient thyroid clinic had 25OHD levels below 40 nmol/L (25).

VITAMIN D INTAKE AND 25OHD CONCENTRATION

It is useful to know what vitamin D intake is needed to raise the serum 25OHD to a specific desired level. We examined this in men and women who participated in a calcium and vitamin D intervention study (14). The men and women in this study consumed an average of 200 IU/d of vitamin D in their diets and were supplemented with an additional 700 IU/d of vitamin D or with placebo. Serum 25OHD levels were measured throughout the year. After one year of supplementation, nearly 90% of the men and 87% of the women taking 700 IU/d of vitamin D had 25OHD levels of 80 nmol/L and above. In contrast, only about half of the men and one third of the women who were taking placebo had levels of 80 nmol/L and above (Table 2). The results for the placebo group indicate that current dietary intakes of vitamin D fall short in much of this elderly population.

Table 2. Percentages of elderly men and women whose 25OHD concentrations reached selected levels after one year of supplementation with 700 IU/d of vitamin D and 500 mg/d calcium or placebo (N=376)

Group	25OHD, nmol/L	Men, %	Women, %
Vitamin D + calcium	20	100.0	100.0
	40	100.0	99.0
	60	97.7	96.9
	80	89.8	86.6
	100	75.0	71.1
Placebo	20	100.0	95.1
	40	96.6	84.5
	60	77.3	58.3
	80	56.8	28.2
	100	35.2	15.5

Another way to examine the association of vitamin D intake with serum 25OHD is to determine the intake needed to bring the 25OHD concentration to a desired level in the wintertime when the skin synthesis is making little to no contribution to vitamin D. We recently completed a survey in Boston in which we assessed the vitamin D supplement use and serum 25OHD levels in 272 thyroid clinic outpatients during the months of January, February and March (25). Significant determinants of 25OHD levels in the group as a whole were supplement use, race, weight, milk intake, and recent southern travel. In the largest relatively homogenous subset, Caucasian women who had not recently traveled south, 65% of the subjects who took 400 IU/d of supplemental vitamin D had 25OHD levels equal to or greater than 80 nmol/L whereas only 30% of un-supplemented women reached this level. The percentages of supplemented and un-supplemented women reaching other levels of 25OHD are shown in Table 3.

Table 3. Percentages of women with selected 25OHD levels among 54 women taking no supplements and 46 women taking 400 IU/d

Plasma 25OHD, nmol/L	No supplements %	400 IU/d %
≥ 20	96.3	100.0
≥ 40	72.2	97.8
≥ 60	44.4	89.1
≥ 80	29.6	65.2

We conclude that current dietary intakes of vitamin D are not sufficient to bring serum 25OHD concentrations of much of the elderly population to the level needed to prevent wintertime rises in PTH and the accompanying bone loss. Similarly, dietary vitamin D intakes are not sufficient to prevent bone loss throughout the year in institutionalized subjects and others who have limited sun exposure.

VITAMIN D INTAKE RECOMMENDATIONS

In 1997, the National Academy of Sciences released updated recommendations for intakes of calcium, vitamin D and other nutrients (26). The vitamin D recommendations were based on 1) achieving the serum 25OHD level associated with no seasonal variation (e.g., wintertime rise) in serum PTH concentration and 2) reduced rates of bone loss. The vitamin D intake recommendations are 400 IU/d for men and women age 51-70 years and 600 IU/d for men and women age 71 and older. For the first time, the recommendations also included a safe upper limit (defined as the maximal intake that can safely be consumed by normal subjects) of 2,000 IU/d. National surveys do not provide information about dietary vitamin D intakes. Based on intake estimates from several convenience samples, however, dietary intakes of vitamin D by the elderly are consistently below the intakes now recommended. Appropriate increases in vitamin D intake should have a significant favorable impact on bone loss and fracture risk in the elderly.

References

1. Ireland P, Fordtran JS. Effect of dietary calcium and age on jejunal calcium absorption in humans studied by intestinal perfusion. J Clin Invest 1973; 52:2672-2681.

2. Gennari C, Agnusdei D, Nardi P, Civitelli R. Estrogen preserves a normal intestinal responsiveness to 1,25-dihydroxyvitamin D_3 in oophorectomized women. J Clin Endocrinol Metab 1990; 71:1288-1293.

3. Bullamore JR, Wilkinson R, Gallagher JC, Nordin BE, Marshall DH. Effect of age on calcium absorption. Lancet 1970; 2:535-537.

4. Ebeling PR, Sandgren ME, DiMagno EP, Lane AW, DeLuca HF, Riggs BL. Evidence of an age-related decrease in intestinal responsiveness to vitamin D. relationship between serum 1,25-dihydroxyvitamin D_3 and intestinal vitamin D receptor concentrations in normal women. J Clin Endocrinol Metab 1992; 75:176-182.

5. MacLaughlin J, Holick MF. Aging decreases the capacity of human skin to produce vitamin D_3. J Clin Invest 1985; 76:1536-1538.

6. Webb AR, Kline L, Holick MF. Influence of season and latitude on the cutaneous synthesis of vitamin D_3: exposure to winter sunlight in Boston and Edmonton will not promote vitamin D_3 synthesis in human skin. J Clin Endocrinol Metab 1988; 67:373-378.

7. Krall EA, Sahyoun N, Tannenbaum S, Dallal GE, Dawson-Hughes B. Effect of vitamin D intake on seasonal variations in parathyroid hormone secretion in postmenopausal women. N Eng J Med 1989; 321:1777-1783.

8. Dawson-Hughes B, Dallal GE, Krall EA, Harris S, Sokoll LJ, Falconer G. Effect of vitamin D supplementation on wintertime and overall bone loss in healthy postmenopausal women. Ann Intern Med 1991; 115:505-512.

9. Dawson-Hughes B, Harris SS, Krall EA, Dallal GE, Falconer G, Green CL. Rates of bone loss in postmenopausal women randomly assigned to one of two dosages of vitamin D. Am J Clin Nutr 1995; 61:1140-1145.

10. Ooms ME, Roos JC, Bezemer PD, van der Vijgh WJ, Bouter LM, Lips P. Prevention of bone loss by vitamin D supplementation in elderly women: a randomized double-blind trial. J Clin Endocrinol Metab 1995; 80:1052-1058.

11. Heikinheimo RJ, Inkovaara JA, Harju EJ, Haavisto MV, Kaarela RH, Kataja JM et al. Annual injection of vitamin D and fractures of aged bones. Calcif Tissue Int 1992; 51:105-110.

12. Lips P, Graafmans WC, Ooms ME, Bezemer PD, Bouter LM. Vitamin D supplementation and fracture incidence in elderly persons. A randomized, placebo-controlled clinical trial. Ann Intern Med 1996; 124:400-406.

13. Chapuy MC, Arlot ME, Duboeuf F, Brun J, Crouzet B, Arnaud S et al. Vitamin D_3 and calcium to prevent hip fractures in the elderly women. N Eng J Med 1992; 327: 1637-1642.

14. Dawson-Hughes B, Harris SS, Krall EA, Dallal GE. Effect of calcium and vitamin D supplementation on bone density in men and women 65 years of age or older. N Eng J Med 1997; 337: 670-676.

15. Dawson-Hughes B, Harris SS, Krall EA, Dallal, GE. Effect of withdrawal of calcium and vitamin D supplements on bone mass in elderly men and women. Am J Clin Nutr 2000; 72:745-50.

16. Malabanan A, Veronikis IE, Holick MF. Redefining vitamin D insufficiency. Lancet 1998; 351:805-806.

17. Lips P, Wiersinga A, van Ginkel FC, Jongen MJ, Netelenbos JC, Hackeng WH et al. The effect of vitamin D supplementation on vitamin D status and parathyroid function in elderly subjects. J Clin Endocrinol Metab 1988; 67:644-650.

18. Dawson-Hughes B, Harris SS, Dallal GE. Plasma calcidiol, season, and serum parathyroid hormone concentrations in healthy elderly men and women. Am J Clin Nutr 1997; 65:67-71.

19. Peacock M. Effects of calcium and vitamin D insufficiency on the skeleton. Osteoporos Int 1998; 8: S45-51.

20. Chapuy MC, Preziosi P, Maamer M, Arnaud S, Galan P, Hercberg S et al. Prevalence of vitamin D insufficiency in an adult normal population. Osteoporos Int 1997; 7:439-443.

21. Lips P, Chapuy MC, Dawson-Hughes B, Pols HA, Holick MF. An international comparison of serum 25-hydroxyvitamin D measurements. Osteoporos Int 1999; 9:394-397.

22. Thomas MK, Lloyd-Jones DM, Thadhani RI, Shaw AC, Deraska DJ, Kitch BT et al. Hypovitaminosis D in medical inpatients. N Eng J Med 1998; 338:777-783.

23. LeBoff MS, Kohlmeier L, Hurwitz S, Franklin J, Wright J, Glowacki J. Occult vitamin D deficiency in postmenopausal US women with acute hip fracture. JAMA 1999; 281:1505-1511.

24. Webb AR, Pilbeam C, Hanafin N, Holick MF. An evaluation of the relative contributions of exposure to sunlight and of diet to the circulating concentrations of 25-hydroxyvitamin D in an elderly nursing home population in Boston. Am J Clin Nutr 1990; 51:1075-1081.

25. Margiloff L, Harris SS, Lee S, Lechan R, Dawson-Hughes B. Vitamin D insufficiency in both supplemented and unsupplemented ambulatory care patients. Calcif Tissue Int 2001: In press.

26. Standing Committee on the Scientific Evaluation of Dietary Reference Intakes. Dietary reference intakes: calcium, phosphorus, magnesium, vitamin D, and fluoride. Institute of Medicine. Washington, DC: National Academy Press, 1997.

HYPOVITAMINOSIS D MYOPATHY.

Henning Glerup and Erik F. Eriksen
Dept of Endocrinology & Metabolism C, AAS, Aarhus University Hospital, DK-8000 Aarhus C, Denmark

INTRODUCTION

For many years myopathy has been known to be part of the osteomalacic symptom complex [1-7]. Generally, hypovitaminosis D myopathy (HDM) has been considered a symptom occasionally seen in patients suffering from osteomalacic bone disease, rather than an early presenting symptom of vitamin D deficiency. However, already in 1975 in a group of 30 patients with biopsy proven osteomalacia, Skaria et al [5] found that myopathy was the presenting symptom in 30%, but was clinically detectable in 96.7% of patients. Thus, myopathy is a prominent symptom in osteomalacia. Recently it was shown by Glerup et al [8] that HDM precedes bone involvement in osteomalacia caused by vitamin D deficiency.

Hypovitaminosis D myopathy is often misdiagnosed [9], due to several reasons. First of all is it important to understand that the decrease in muscle strength is a continuous and gradual process, whereas the loss of functional ability is quantal. This means, that the patients can have a considerable loss of muscle strength, before they complain of muscle weakness. Their main complaint would rather be fatigue. They won't complain over muscle weakness, until they are unable to walk or rise from a squatting position unaided. Another reason for missing the diagnosis is, that many of the symptoms are rather unspecific (diffuse muscle pain, deep bone pain, arthralgias, paresthesia) [4,6,9], which often leads to alternative diagnoses like rheumatic disease, polymyalgia, psychoneurotic disorders, fibromyalgia, malignant diseases etc.

EFFECTS OF VITAMIN D ON STRIATED MUSCLE

The vitamin D receptor (VDR) has been identified in striated muscle [10-13]. The presence of a nuclear VDR in muscle enables vitamin D to mediate muscular effects through genome mediated de novo protein synthesis. In agreement with this notion, vitamin D has been shown to stimulate synthesis of a number of important muscle proteins (troponin C, actin, proteins in the sarcoplasmatic reticulum and the inner membrane of mitochondria) [14-16].

In addition to the classical genomic effects of vitamin D a number of non-genomic effects of vitamin D has been identified [17-19]. The non-genomic effects are mediated through binding of vitamin D to membrane receptors resulting in formation of intracellular second messengers, which can cause rapid regulation of the activity of the intracellular enzyme and ion-pump systems.

Calcium homeostasis in muscle cells

The Ca^{++} concentration in the cytosol of the resting muscle is very low (10^{-7} – 10^{-8} mol/l), 10.000 times lower than extracellular levels (2 x 10^{-3} mol/l). This gradient can only be maintained by an energy demanding pumping activity. Ca-ATPase is localised both in the outer cell membrane (sarcolemma) pumping Ca^{++} out of the cell, as well as in the sarcoplasmatic membrane pumping Ca^{++} into the sarcoplasmatic reticulum (SR). Quantitatively the content of Ca-ATPase of the SR are much higher than in sarcolemma, which makes the SR the most important organelle for regulation of intracellular calcium homeostasis.

During muscle contraction Ca^{++} are liberated from the SR. Relaxation of the muscle occurs when Ca^{++} has been pumped back into the SR. Thus Ca-ATPase plays a pivotal role in the regulation of calcium homeostasis and function of muscle cells.

In 1974 Curry et al [20] demonstrated that the ATP-dependent Ca^{++} uptake of isolated vesicles of the SR was reduced in muscles from vitamin D depleted rabbits compared to vitamin D repleted animals. If the vitamin D depleted animals were given vitamin D a few before the experiment Ca^{++} uptake was normalised. In 1978 Rodman and Baker [21] published a study with an *in vivo* model of vitamin D depleted and repleted rats for electrophysiological investigations of the muscle kinetics of the soleus muscle. Vitamin D depleted animals exhibited a prolonged time-to-peak, contractile tension and prolonged relaxation half-life. Again this could be normalised by a few days vitamin D treatment prior to the experiment. The result of their study is best explained by decreased activity of the Ca-ATPase, which is normalised during vitamin D treatment.

Toury et al [22] investigated the calcium content in subcellular fractions of vitamin D depleted and repleted rats. In all subcellular fractions except the cytosol they found lower Ca^{++}-content in vitamin D depleted compared to vitamin D repleted animals. In the cytosol the calcium content was higher in vitamin D depleted animals compared to vitamin D repleted, indicating lower activity of the Ca-ATPase. In EM of muscle biopsies exchangeable calcium was visualised with pyroantimonate-osmium. The vitamin D repleted animals revealed much more intensive staining of the areas around the terminal cisternae of the SR, the T-tubules and at the active sites of the myofibrils. This means that the net effect of vitamin D on intracellular calcium homeostasis is an augmentation of intracellular flux and active transportation of calcium and an increased calcium pool available for the muscle contraction.

Are effects of vitamin D on muscle Mediated via 25-hydroxy-vitamin D_3 or 1,25-dihydroxyvitamin D_3?

Knowing that the VDR receptor is present in muscle, it would be expected that 1,25-dihydroxyvitamin D_3 was the vitamin D metabolite, primary responsible for the muscular effects of vitamin D. However, in clinical studies [8] correlation analyses have shown, that muscle power correlated positively to s-25-OHD ($r = 0.42$, $p < 0.001$), inversely to s-PTH ($r = -0.36$, $p < 0.001$) but was not correlated to s-1,25-OH_2D. In multivariate regression analysis only 25-OHD came out as a significant determinant. Our results therefore strongly suggest, that normal serum levels of 25-OHD are essential for maintaining a normal muscle function; whereas we could demonstrate only minor importance of PTH and no effect at al of 1,25-OH_2D. Other researchers have previously suggested a direct effect of 25-OHD on striated muscle. Birge and Haddad [23] found that 25-OHD stimulated protein synthesis and increased intracellular ATP content in striated muscle more efficiently than did 1,25-OH_2D. Further more Pointon et al [14] and de Boland et al [15] demonstrated a stimulation of the synthesis of troponin-C, actin and proteins of the sarcoplasmatic reticulum, that seemed most likely to be an action of 25-OHD rather than 1,25-OH_2D. On the other hand muscular effects of 1,25-OH_2D are well documented in several in vitro studies [17], but we believe, that our study strongly suggests, that in vivo 25-OHD is by far the most important effector in striated muscle. Additional research is needed to clarify how teh muscular effects of 25-OHD are mediated.

Our conclusion, that 25-OHD is the most important in vivo effector in striated muscle, is supported by the findings of Grady et al [24]. In an in vivo study they failed to demonstrate any muscular effects of treatment with 0.5 μg calcitriol daily. Serum levels of 25-OHD among the participants were, however, above > 60 nmol/l. If 25-OHD is the primary in vivo effector, no hypovitaminosis D myopathy would be expected among these participants, and consequently no effect of treatment with calcitriol should be expected.

CLINICAL STUDIES

In a prospective study Young et al [7] followed the progress in muscle function in 12 osteomalacic patients during vitamin D treatment. They used an isometric dynamometer model for measurement of the Quadriceps muscle strength in combination with muscle biopsies from the lateral Vastus muscle. In the biopsies prior to vitamin D treatment the most prominent finding was atrophy of the faster and more powerful type II fibres. A significant improvement in both muscle strength and muscle biopsies was demonstrated after 3 months of treatment, but full restitution of the muscles would need treatment for 6 to 12 months.

In our own study [8] we followed 55 vitamin D deficient participants during vitamin D treatment. 22 healthy persons with normal vitamin D status acted as controls. Our most important results were:

1. *Involved muscle groups.* In a screening of several muscle groups the weight bearing antigravity muscles were found to be most effected by vitamin D deficiency.
2. *Relation to bone involvement.* Myopathy was equally decreased in vitamin D deficient individuals (s-25-OHD < 20 nmol/l) with and without bone involvement indicated by increased serum levels of alkaline phosphatase. Thus it seems that hypovitaminosis D myopathy precedes hypovitaminosis D osteopathy.
3. *Effect of vitamin D treatment on muscle power and kinetics.* Maximal knee voluntary extension power improved significantly during 6 months of vitamin D treatment (259.4 ± 11.0 N to 320.7 ± 14.3 N, $p < 0.001$). In electrically stimulated muscle power measurements, vitamin D depleted individuals were found to have weaker and slower muscles (decreased maximal production rate (MPR) and maximal relaxation rate (MRR)) Muscle kinetics were completely normalised during vitamin D treatment (**Fig. 1**).

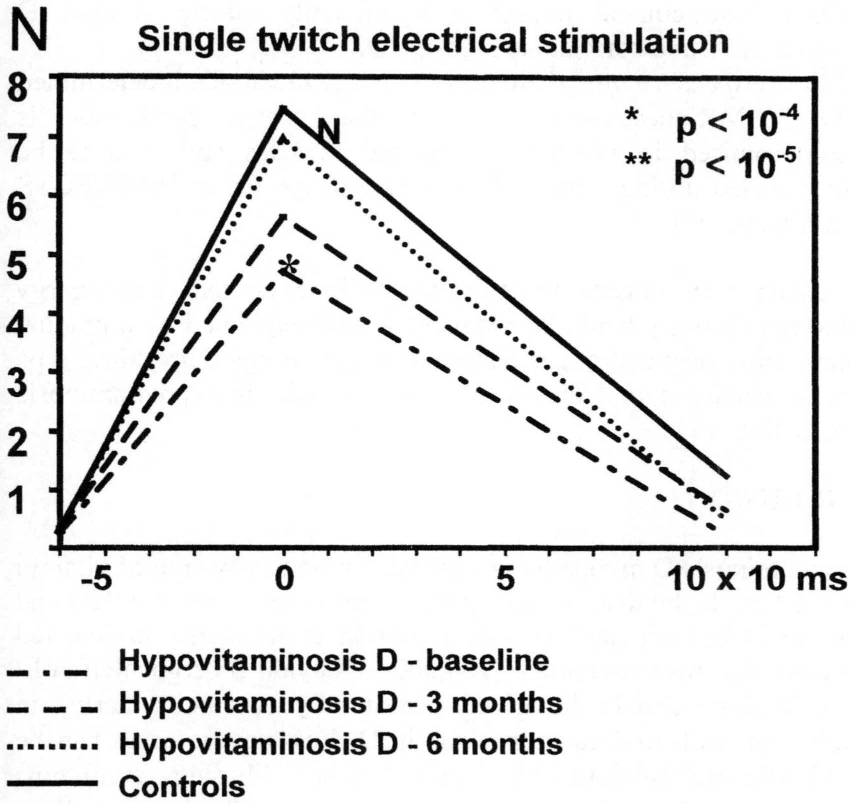

Figure 1. Electrical stimulation of muscle contraction and subsequent relaxation in vitamin D deficiencmy, controls and after 3 and 6 months of vitamin D + calcium treatment. The ascending slope expresses the maximal production rate(MPR) and the descending slope the maximal relaxation rate (MRR).

4. *Ca*-ATPase content improved significantly during vitamin D treatment [25] and correlate to s-25-OHD.
5. *P^{31}-MR-spectroscopy.* Estimates of energy metabolism determined by P^{31}-MR-spectroscopy indicate that energy production is compromised in vitamin D depleted muscles and tend to be normalised during vitamin D treatment (reported at IBMS/ECTS in Madrid 2001).

The changes in muscle kinetics, Ca-ATPase content and energy production during vitamin D treatment is best explained by a gradual normalisation of the stronger faster and more energy demanding type II fibres, which are well known to undergo atrophy in hypovitaminosis D myopathy.

CONCLUSION

Hypovitaminosis D myopathy is a prominent and early sign of vitamin D deficiency. If the disease is suspected, serum levels of 25-OHD and PTH should be estimated. If severe vitamin D deficiency is detected high dose vitamin D treatment (100,000 – 300,000 IU ergocalciferol / cholecalciferol) should be given parenterally (intramuscularly) in combination with oral calcium+vitamin D. Less severe cases can be treated with oral calcium and vitamin D alone. Muscular symptoms will improve within a few weeks to months and usually be normalised within six months.

Serum levels of 25-OHD seems to be the most important determinant of hypovitaminosis D myopathy. It is, however, still unknown, whether 25-OHD acts directly on muscle tissue, or possible through a local synthesis of 1,25-OH_2D within the muscle cell [26]. With the recent demonstration of intracellular 1α-hydroxylases in other tissues, the latter hypothesis seems to be the most plausible.

References

1. Smith R, Stern G. Muscular weakness in osteomalacia and hyperparathyroidism. J. Neurol. Sci. 1969; 8:511-520.

2. Smith R, Stern G. Myopathy, osteomalacia and hyperparathyroidism. Brain 1967; 90:593-602.

3. Chalmers J, Conacher WDH, Gardner DL, Scott PJ. Osteomalacia - A common disease in elderly women. J.Bone Joint Surg 1967; 49B:403-423.

4. Mallette LE, Patten BM, Engel WK. Neuromuscular disease in secondary hyperparathyroidism. Ann.Intern.Med. 1975; 82:474-483.

5. Skaria J, Katiyar BC, Srivastava TP, Dube B. Myopathy and neuropathy associated with osteomalacia. Acta Neurol.Scand. 1975; 51:37-58.

6. Schott GD, Wills MR. Muscle weakness in osteomalacia. Lancet 1976; 1:626-629.

7. Young A, Edwards RHT, Jones DA, Brenton DP. Quadriceps muscle strength and fibre size during the treatment of osteomalacia. In: Stokes IAF, editor. Mechanical factors and the skeleton. 1981:137-145.

8. Glerup H, Mikkelsen K, Poulsen L, Hass E, Overbeck S, Andersen H, et al. Hypovitaminosis D myopathy without biochemical signs of osteomalacic bone involvement. Calcif.Tissue Int. 2000; 66 (6):419-424.

9. Glerup H, Eriksen EF. Osteomalacia and servere vitamin D deficiency - A review of the clinical and paraclinical findings and guidelines for the treatment with vitamin D. Ugeskr.Laeger 1999; 161:2515-2521.

10. Costa EM, Blau HM, Feldman D. 1,25-dihydroxyvitamin D3 receptors and hormonal responses in cloned human skeletal muscle cells. Endocrinology 1986; 119:2214-2220.

11. Bischoff HA, Borchers M, Gudat F, Duermueller U, Theiler R, Stahelin HB, et al. In situ detection of 1,25-dihydroxyvitamin D-3 receptor in human skeletal muscle tissue. Histochem J 2001; 33:19-24.

12. Boland R, Norman A, Ritz E, Hasselbach W. Presence of a 1,25-dihydroxy-vitamin D3 receptor in chick skeletal muscle myoblasts. Biochem.Biophys.Res.Commun. 1985; 128:305-311.

13. Simpson RU, Thomas GA, Arnold AJ. Identification of 1,25-dihydroxyvitamin D3 receptors and activities in muscle. J.Biol.Chem. 1985; 260:8882-8891.

14. Pointon JJ, Francis MJ, Smith R. Effect of vitamin D deficiency on sarcoplasmic reticulum function and troponin C concentration of rabbit skeletal muscle. Clin.Sci. 1979; 57:257-263.

15. de Boland AR, Albornoz LE, Boland R. The effect of cholecalciferol in vivo on proteins and lipids of skeletal muscle from rachitic chicks. Calcif.Tissue Int. 1983; 35:798-805.

16. de Boland AR, Boland RL. Effects of vitamin D3 on in vivo labelling of chick skeletal muscle proteins with [3H]leucine. Z.Naturforsch.C. 1984; 39:1015-1016.

17. Boland R. Role of vitamin D in skeletal muscle function. Endocr.Rev. 1986; 7:434-448.

18. de Boland AR, Boland RL. Rapid changes in skeletal muscle calcium uptake induced in vitro by 1,25-dihydroxyvitamin D3 are suppressed by calcium channel blockers. Endocrinology 1987; 120:1858-1864.

19. de Boland AR, Boland RL. Non-genomic signal transduction pathway of vitamin D in muscle. Cell Signal. 1994; 6:717-724.

20. Curry OB, Basten JF, Francis MJ, Smith R. Calcium uptake by sarcoplasmic reticulum of muscle from vitamin D-deficient rabbits. Nature 1974; 249:83-84.

21. Rodman JS, Baker T. Changes in the kinetics of muscle contraction in vitamin D- depleted rats. Kidney Int 1978; 13:189-193.

22. Toury R, Stelly N, Boisonneau E, Convert M, Dupuis Y. Relationship between vitamin D status and deposition of bound calcium in skeletal muscle of the rat. Biol.Cell 1990; 69:179-189.

23. Birge SJ, Haddad JG. 25-hydroxycholecalciferol stimulation of muscle metabolism. J.Clin.Invest. 1975; 56:1100-1107.

24. Grady D, Halloran B, Cummings S, Leveille S, Wells L, Black D, et al. 1,25-Dihydroxyvitamin D3 and muscle strength in the elderly: a randomized controlled trial. J.Clin.Endocrinol.Metab. 1991; 73:1111-1117.

25. Glerup H, Andersen H, Clausen T, Charles P, Eriksen EF. Vitamin-D regulates Ca-ATPase in striated muscle: Studies on osteomalacic myopathy. J.Bone Miner.Res. 1997; 12:46

26. Schwartz GG, Whitlatch LW, Chen TC, Lokeshwar BL, Holick MF. Human prostate cells synthesize 1,25-dihydroxyvitamin D3 from 25-hydroxyvitamin D3. Cancer Epidemiol.Biomarkers.Prev. 1998; 7:391-395.

EFFECTS OF DIETARY VITAMIN D AND UVB IRRADIANCE ON VOLUNTARY EXPOSURE TO ULTRAVIOLET LIGHT, GROWTH AND SURVIVAL OF THE PANTHER CHAMELEON *Furcifer pardalis*

Gary W. Ferguson, William H. Gehrmann, Stephen H. Hammack, Tai C. Chen, Michael F. Holick

Texas Christian University, Fort Worth, TX 76129 (GWF, WHG), Fort Worth Zoo, Fort Worth, TX 76110 (SHH), Boston Medical Center, Boston, MA 02118 (TCC, MFH)

INTRODUCTION

Vertebrates require vitamin D for several bodily processes including the regulation of calcium and phosphorus levels[1]. Vitamin D is obtained in two ways. First, vitamin D can be obtained from the diet[2]. Second, vitamin D can be produced in the skin upon exposure to ultraviolet B irradiation (UVB, 280-315nm)[2].

The endogenous production of the hormonally active form of vitamin D (1,25-dihydroxyvitamin D_3 or calcitriol) involves a complex sequence of events[2]. It begins in the skin with the photolysis of provitamin D_3 (7-dehydrocholesterol) to previtamin D_3 upon exposure to UVB. Previtamin D_3 is then thermally isomerized to vitamin D_3 (cholecalciferol). Vitamin D_3 is then hydroxylated in the liver to 25-hydroxyvitamin D_3 (calcifidiol); and again in the kidney to become calcitriol. Regulation of internal levels of vitamin D may occur at any of several points along the sequence[3].

Ectothermic vertebrates utilize basking behavior to regulate their body temperature[4]. Through behavior alone some species are able to maintain constant, elevated body temperatures for extended periods in a fashion similar to most endotherms. Most biologists consider ectothermic basking behavior to function solely for thermoregulation. However, recently some turtles have been shown to bask in a manner that has little influence on body temperature[5]. Furthermore, some lizards and crocodiles have been shown to undergo seasonal shifts in thermal preference for unknown reasons[6,7]. Recently, we showed for the panther chameleon that dietary vitamin D may influence voluntary exposure to ultraviolet irradiation in a fashion that suggests vitamin D regulation[8]. Thus, lower dietary vitamin D_3 was associated with higher UVB exposure and vice versa. We report the results of a repetition of the experiment[8] with modifications and reemphasize that the regulation of vitamin D may involve basking behavior, in addition to the currently recognized physiological mechanisms.

METHODS

From October 1999 through January 2001 fifteen neonate female panther chameleons *Furcifer pardalis*, hatched in our laboratory at TCU, were maintained in isolation in terraria in the laboratory. The husbandry and terrarium set up has been described previously[8,9]. Each subject was included in one of four experimental treatments as part of a 2-factor cross-classified design. Two levels of UVB (high and low) were cross-classified with two levels of dietary vitamin D (high and low). Administration of dietary vitamin D was exactly as described in the previous experiment[8]. Crickets were fed commercially prepared experimental grain diets (Zeigler Bros. Corp) containing either 100 IU/g vitamin D or none. The measurement of UV and visible light was with Spectronics (model DM-365N, DM-300N) and GE (Type 214) meters previously described[9].

Administration of light treatments differed from the previous experiment[8] in some respects (Table 1). As in the previous experiment, an FS-40 sunlamp illuminated for one hour provided the high UVB treatment (maximum 102 $\mu W/ cm^2$). A board placed on top of one end of the cage provided a gradient down to 0 $\mu W/ cm^2$. Chameleons could freely move through the gradient. In contrast to the previous experiment[8] only three vs five treatments/week were provided. Five treatments per week had produced skin pathology (tumors) in some non experimental animals, so fewer treatments were considered safer for the animals. Also, the low UVB treatment was changed as follows. In the previous experiment[8], the low UVB treatment was provided by illumination with a blacklight (GE BL-40), which incidentally provided a much higher UVA environment than the sunlamp (Table 1). In the current experiment illumination during the low UVB treatment was provided by illumination with a sunlamp (FS-40) blocked by two panes of window glass (4mm thick). This reduced both the UVB and the UVA irradiance compared to that of the high UVB treatment. UVB irradiance ranged from 0 to 1 µW/ cm^2 in the gradient of the low UVB treatment. Daily visible illuminance (12h) in the current study was provided by GE cool white fluorescent tubes. Whereas Vita-Lite tubes (Duro Test) were used in the previous experiment[8]. The cool white tubes provided about 50% of the illuminance of the previous experiment (Table 1) but, as in the previous experiment, illuminance was

Table 1. Comparison of light environments (maximum available in the gradient) of current experiment (A) with that of previous experiment[8] (B).

	UVB µW/ cm^2	UVA µW/ cm^2	visible lux	cage temp EC
low UVB (A)	1.0	4.4	507	26.2
low UVB (B)	2.1	150	1,184	31
high UVB (A)	70	8.8	538	25.8
high UVB (B)	89.0	12	1,184	31

similar for both the high and low UVB treatments.

During a UVB treatment the location of each chameleon was recorded every 10 minutes (6 locations). Using a map of zones of irradiance within a terrarium (Table 2) and a curve (Fig. 1) relating UVB irradiance to potential provitamin D conversion in in-vitro models[10], each location of the lizard was scored regarding effective UVB exposure as follows: 1-Maximum Effective Exposure Score, 2- Intermediate Effective Exposure Score, or 3- Minimum Effective Exposure Score. For each treatment the percent of the 6 locations that were in a maximum effective exposure zone (score=1) was calculated. The mean percent for a given lizard was calculated for each four-month period of the experiment. Those values (means) were analyzed with ANOVA for treatment effects. There were 178 treatment periods for each lizard surviving the entire experiment. While lizards could freely move throughout the terrarium during the light treatment, they concentrated activity in either the strongest (score=1) or weakest (score=3) zones of UVB irradiance (Table 2).

Table 2. Use of UVB gradient by panther chameleons in 38x23x28 cm terraria and effective exposure score. Observational frequency behavior (FQ) data were generated for 14 weeks and included 42 observational periods (6 readings for each period) on the 8 juvenile lizards in the High UVB light treatment. UVB Irradiance was measured with a Spectroline UVB meter (DM-300N). Percent conversion is provitamin D_3 to photoproducts in in-vitro models[10]. See text for explanation of exposure score.

FQ	ZONES	UVB $\mu W/CM^2$	% CONVERSION	EFFECTIVE EXPOSURE SCORE 1=MAX 3=MIN
0.08	1	102	22.7	1
0.23	2	70	21.5	1
0.01	3	36	17.2	1
0.01	4	29	10.7	2
0.01	5	4	6.2	3
0.64	6	0	0	3

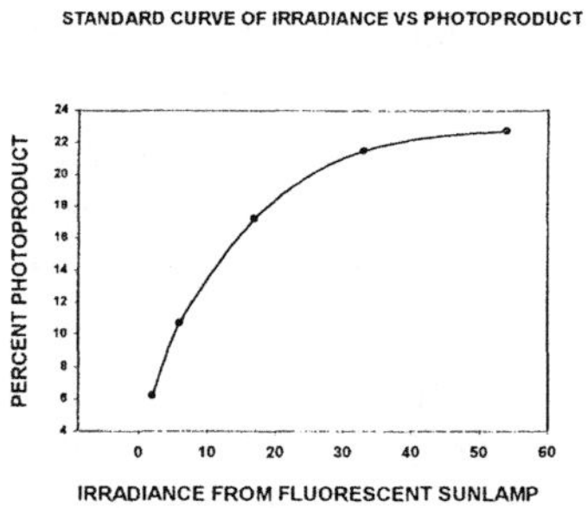

Figure 1. Relationship of irradiance ($\mu W/cm^2$) of sunlamp to photoproduct conversion of provitamin D_3 in in-vitro models[10].

Surviving lizards were raised to maturity and bred to non-experimental captive raised males. Lizard mass was measured biweekly throughout the study. Mass of the crickets fed to the lizards was monitored throughout the study. Dietary mass averaged 75% of lizard body mass every 2 weeks during the first 4 months of the study, 50% during the second 4 months, and 30% thereafter. Lizards in all treatment groups were fed the same amount. Blood was drawn from the survivors at the end of the experiment to measure treatment effects on circulating calcifidiol (25-OH vitamin D). Cricket and feed samples were frozen for analysis of vitamin D_3 at the beginning and end of the experiment to monitor possible vitamin D decline as the feed aged. One egg from each clutch of each female that successfully reproduced (8 of the 15 females) was frozen for analysis of treatment effects on egg vitamin D_3 content. At least one female representing 3 of the 4 treatment groups successfully reproduced. The results of biochemical analysis of calcifidiol and vitamin D_3 are pending. Available data were compiled in LOTUS 1-2-3 (release 5) files and analyzed using SYSTAT 6.0, SIGMA PLOT 4.0, and SIGMASTAT 2.0.

RESULTS

There were statistically significant effects (null probability <0.05) of both dietary vitamin D and UVB irradiance on percent maximum exposure during the first two 4-month age periods of the experiment (Figs. 2&3, Tables 3). Age period 1 was the juvenile growth period. Age period 2 was the adolescent growth period. Age period 3 was the first adult reproduction period. Percent maximum exposure was greater for individuals fed low dietary vitamin D. Percent maximum exposure was greater for individuals exposed to the high UVB treatment. Percent maximum exposure increased with age regardless of treatment. The treatment effects disappeared upon maturity. Thus, a repeated-measures ANOVA, including all three periods, showed a significant time effect but no significant UVB or dietary vitamin D treatment effects or interactions (results not shown).

There was a statistically significant dietary vitamin D effect on growth (Fig. 4, Table 4). Growth was slower for individuals fed the high dietary vitamin D.

There was significantly lower survival during the experiment for individuals fed high dietary vitamin D (Table 5).

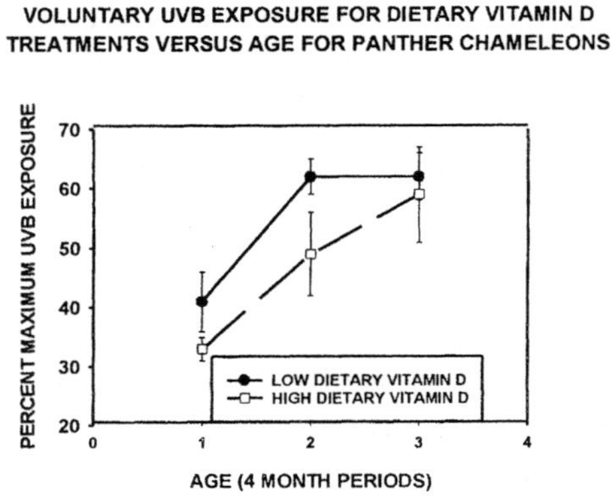

Figure 2. Voluntary UVB exposure of panther chameleons fed high versus low dietary vitamin D. Vertical lines represent one standard error

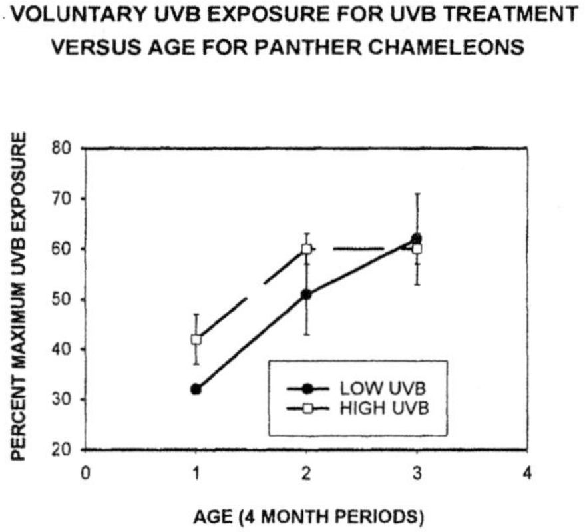

Figure 3. Voluntary UVB exposure of panther chameleons exposed to low versus high irradiances of UVB during treatments. Vertical lines represent one standard error.

Table 3. Repeated measures ANOVA of UVB irradiance and dietary vitamin D effects on percent maximum voluntary exposure to artificial UVB source of juvenile and adolescent panther chameleons (age periods 1 &2) in the laboratory.

Between Subjects

Source	D F	Mean Square	F ratio	Null Probability
UVB high or low	1	0.056	5.43	0.048
Dietary Vitamin D high or low	1	0.074	7.17	0.028
UVB x Vitamin D	1	0.002	0.218	0.653
Error	8	0.01		

Within Subjects

Source	D F	Mean Square	F ratio	Null Probability
Age	1	0.184	19.71	0.002
Error	8	0.009		

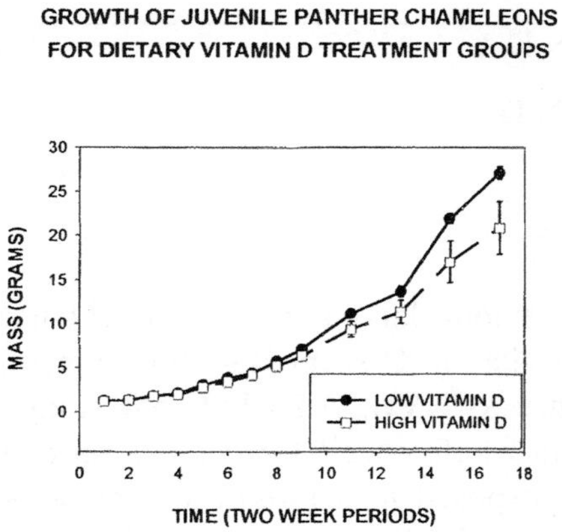

Figure 4. Growth of panther chameleons fed low versus high dietary vitamin D. Vertical lines represent standard errors.

Table 4. Repeated measures ANOVA of Dietary Vitamin D effects on growth (mass versus time) of juvenile and adolescent panther chameleons.

Between Subjects

Source	D F	Mean Square	F ratio	Null Probability
Dietary Vitamin D high or low	1	88.11	5.36	0.043
Error	10	16.45		

Within Subjects

Source	D F	Mean Square	F ratio	Null Probability
Time	8	761.3	249.8	<0.001
Time x Vitamin D	8	15.14	4.97	<0.001
Error	80	3.05		

Table 5. Survival of panther chameleons for one year on low versus high vitamin D diets. Fisher Exact probability = 0.026

	SURVIVED	DIED
LOW VITAMIN D	7	0
HIGH VITAMIN D	3	5

DISCUSSION

As in the previous laboratory experiment using artificial UVB sources[8] low dietary vitamin D seemed to elicit a compensatory increase in voluntary exposure to UVB. An additional study was conducted in outdoor enclosures (see reference[11] for description of study design) with panther chameleons exposed to a mosaic of natural sunlight. A similar compensatory increase in UVB exposure for individuals fed a low vitamin D diet was observed (Karsten, unpublished Master's thesis). Thus, the effect is repeatable with either artificial or natural UVB sources. An extension of the research is planned to determine the ability of female panther chameleons to regulate their internal vitamin D (calcifidiol) levels using behavior. Can relatively constant levels of calcifidiol be maintained with varied dietary intake only if the animals are allowed to regulate their UVB exposure? If so, basking behavior will have been shown to be an important component in the regulation of vitamin D in some vertebrates in addition to the previously documented physiological mechanisms.

There were some interesting differences between the results of the current experiment and those of the previous laboratory experiment[8]. Animals in the current study showed an ontogenetic increase in voluntary exposure that was not observed in the previous experiment[8]. Voluntary exposure by juveniles in the current study was lower than in the previous study[8]. In the previous study the diet effect remained strong in adults but disappeared at maturity in the current study. While the causes of the differences in the results of the two studies remain unclear, voluntary exposure to UVB has been shown to be a labile behavior. All factors that might contribute to basking behavior must be closely monitored in future studies.

The high dietary vitamin D in the current study had deleterious effects on growth and survival. Future studies should include reduced high dietary levels that do not have pathological effects.

ACKNOWLEDGEMENTS

Specimens used in this study were provided by TRUE CHAMELEONS David A Roberts, proprietor. Funding for the study was in part from a grant from the Cleveland Metroparks Zoo. Partial funding for biochemical analyses was from NIH grant (AR 36963) to MFH.

REFERENCES

1. Holick MF. Vitamin D: physiology, molecular biology, and clinical applications. Totowa, NJ: Humana; 1999.

2. Holick MF. Evolution, biologic functions, and recommended dietary allowances for vitamin D. In: Holick MF, ed. Vitamin D: physiology, molecular biology, and clinical applications, Totowa, NJ: Humana, 1999;1-16.

3. Holick MF. Vitamin D In: Shils ME, Olson, JA Shike M, Ross AC, eds. Modern nutrition in health and disease. Baltimore, MD: Williams and Wilkins, 1999;329-45.

4. Huey RB. Temperature, physiology, and the ecology of reptiles" In: Gans C, Pough FA, eds. Biology of the Reptilia, Vol.12, New York, NY: Academic Press, 1982;25-91.

5. Manning B, Grigg GC. Basking is not of thermoregulatory significance in the "basking" freshwater turtle *Emydura signata*. Copeia 1997;1979:579-84.

6. Christian KA, Bedford GS. Seasonal changes in thermoregulation by the frillneck lizard *Chlamydosaurus kingii*. Ecology 1995;76:124-32.

7. Seebacher F, Grigg GC. Patterns of body temperature in freshwater crocodiles *Crocodylus johnstoni*: thermoregulation versus thermal conformity, seasonal acclimation and the effect of social interactions. Copeia 1997;1997:549-57.

8. Jones JR, Ferguson GW, Gehrmann WH, Holick MF, Chen TC, Lu Z. Vitamin D nutritional status influences voluntary behavioral photoregulation in a lizard" In: Holick MF, Jung EG, eds. Biologic effects of light 1995, M.F. Berlin: Walter de Gruyter, 1996;49-55.

9. Ferguson GW, Jones JR, Gehrmann WH, et al. Indoor husbandry of the panther chameleon *Chamaeleo [Furcifer] pardalis*: effects of dietary vitamins A and D and ultraviolet irradiation on pathology and life-history traits. Zoo Biol 1996;15:279-99.

10. Lu Z, Chen TC, Holick MF Influence of season and time of day on the synthesis of vitamin D_3 In: Holick MF, Kligman AM, eds. Biologic effects of light, New York, NY: Walter de Gruyter, 1992;53-6.

11. Ferguson GW, Carman EN, Gehrmann WH, Holick MF, Chen T, Photobiosynthesis in lizards: opportunity, ability and behavioral regulation In: M.F. Holick, E.G. Jung, eds. Biologic Effects of Light 1998, Boston, MA: Kluwer, 1999;103-10.

PRODUCTION OF PREVITAMIN D_3 BY A MERCURY ARC LAMP AND A HYBRID INCANDESCENT/MERCURY ARC LAMP

E.A. Holick, Z. Lu, M. T. Holick, T.C Chen, J. Sheperd and M.F. Holick
Vitamin D, Skin and Bone research Laboratory, Department of Biophysics and Physiology and the Department of Medicine, Boston University Medical Center, Boston, MA 02118 and KBD Inc. Erlanger, Kentucky

Introduction

Most vertebrates including reptiles and amphibians need a source of ultraviolet B radiation in order to promote an adequate production of vitamin D to satisfy their body's requirement (1). Reptiles routinely sun themselves, not only to warm their bodies, but also to produce vitamin D_3 in their skin. Amphibians exposed to sunlight also have the ability to produce vitamin D_3 in their skin (1). It is estimated that there are over ten million households in the United States that have a reptile or amphibian as a pet. Often these animals are housed in a glass enclosure, and are exposed to incandescent lighting. Since incandescent lighting does not emit any ultraviolet B radiation, these animals depend solely on their diet for their vitamin D requirement. Frequently, diets including vegetable matter or live or dead animals do not contain an adequate amount of vitamin D to satisfy their requirements resulting in these animals developing severe vitamin D deficiency. Vitamin D deficiency causes rickets and osteomalacia that can lead to fractures, muscle weakness and ultimately, death. Many zoos also house reptiles and amphibians indoors, in glass enclosures and often experience vitamin D deficiency in these prized animals. The appreciation that these animals require a source of ultraviolet B radiation has prompted the lamp manufactures and distributors to produce florescent lamps that emit ultraviolet B radiation that is similar to lamps used in tanning salons. These lamp systems have produced mixed results in preventing vitamin D deficiency bone disease in captive reptiles and amphibians. The goal of this research program was to evaluate a new hybrid incandescent lamp that contains a mercury arc system that emits radiation with energies above 290 nm. This system was compared to a mercury arc Sperti lamp system, and natural sunlight for its capacity to produce previtamin D in an ampoule system (2).

Methods

Crystalline 7-dehydrocholesterol was dissolved in methanol at a concentration of 50 μg/ ml. Borasilicate ampoules were filled with 1 cc of the 7-DHC solution. The ampoules were exposed to a Sperti lamp (Fig 1) (KBD Inc. Erlanger, KY) for 10 minutes at varying distances. Ampoules were placed 6 inches from the Westron incandescent lamps (Fig 2) and samples were obtained at various times for up to one hour. Ampoules were also placed on the roof of a ten-story building within the Boston Medical Center campus in the middle of June on a cloudless day at noontime.

After the samples were collected, they were immediately stored at -20°C. The ampoules were opened and the contents removed. An aliquot of each sample was taken to dryness and chromatographed on a straight phase high performance liquid chromatography system as previously described (2).

Results

Exposure of ampoules to the Sperti lamp system revealed that at a distance of 1 foot, approximately 8% of 7-DHC was converted to previtamin D_3 in 10 minutes (Fig 3). There was almost a 4-fold reduction in the percent production of previtamin D_3 in ampoules that were exposed to the Sperti lamp for 10 minutes at a distance of 2 feet. The amount of previtamin D_3 continued to decrease as the distance from the Sperti lamp increased. Essentially no previtamin D_3 was detected in samples that were exposed to the Sperti lamp system at a distance of 4 feet from it (Fig 3).

Regulating the lamp power with a transformer had a significant impact on previtamin D formation in the ampoules as shown in Figure 4.

A comparison of previtamin D_3 formation between the Sperti sunlamp system and natural sunlight in Boston in June is shown in Figure 5. Percent conversion of 7-DHC to previtamin D_3 in June in Boston at noontime was approximately 1.0, 2.5 and 3.5 % at 10, 20 and 30 minutes respectively. In comparison at one minute, the Sperti lamp system at a distance of 12 inches converted 1% of 7-DHC to previtamin D_3 and after 10 minutes, 10% of 7-DHC was converted to previtamin D_3. (Fig 5).

An evaluation of the effectiveness of the Westron lamp to produce previtamin D_3 in the ampoule system was performed. Ampoules were exposed to 6 different Westron Lamps at a distance of 6 inches for up to 1 hour. As can be seen in Figure 6 between 1 and 2% of 7-DHC was converted to previtamin D_3 after 15 minutes and after one hour of exposure, between 3 and 5% of 7-DHC was converted to previtamin D_3.

Conclusion

It is recognized that ultraviolet radiation with energies between 255 and 315 nm are effective in photolyzing 7-DHC to previtamin D_3 (Fig 7). An evaluation of the spectral output for the Sperti mercury arc lamp system shows significant emission of ultraviolet radiation above 280 nm (Fig 8). The spectral output of the Westron lamp revealed emission of ultraviolet radiation with energies above 290 nm (Fig 9). Thus, both the Sperti sunlamp system and the Westron sunlamp systems emit radiation that are effective in converting 7-DHC to previtamin D_3. Our results conclusively demonstrate that both the Sperti lamp system and Westron lamp systems have the capacity to produce previtamin D_3. Thus, both systems should be of value both for inducing vitamin D_3 in reptiles and amphibians that are housed indoors as pets or in zoos. In addition, they may be of value in preventing vitamin D deficiency in individuals who either can not absorb vitamin D from dietary sources due to fat malabsorption syndrome, or because they are infirmed and are unable to be exposed to sunlight.

We have used the Westron lamp as a means of preventing vitamin D deficiency in a mated pair of Euromastyx mali lizards as demonstrated in Figure 10.

In conclusion, it is reasonable that vertebrates who normally obtain their vitamin D requirement from exposure to sunlight, and are kept as pets or as specimens in zoos should be exposed to a source of ultraviolet B radiation to satisfy their vitamin D requirement if they are not getting an adequate amount from their diet. Therefore commercial fluorescent lamps and magnified incandescent lamps that have a spectral outputs in the ultraviolet B and have adequate power levels such as the Sperti sunlamp system, Westron sunlamp system are good sources for this vital vitamin D photobiologic process.

References:

1. Holick,MF. Phylogenetic and evolutionary aspects of vitamin D from phytoplankton to humans. In: P.K.T. Pang and M.P. Schreibman, (eds), Vertebrate Endocrinology: Fundamentals and Biomedical Implications, Vol. 3. Academic Press, Inc. (Harcourt Brace Jovanovich) Orlando, pp. 7-34, 1989.

2. Tian, X.Q., and Holick, M.F. A liposomal model that mimics the cutaneous production of vitamin D_3. J Biol Chem 1999; 274:4174-79.

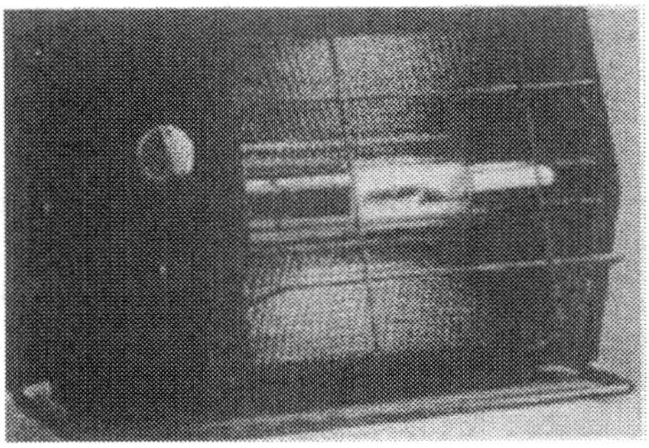

Figure 1. The Sperti sunlamp that was used in the described experiments.

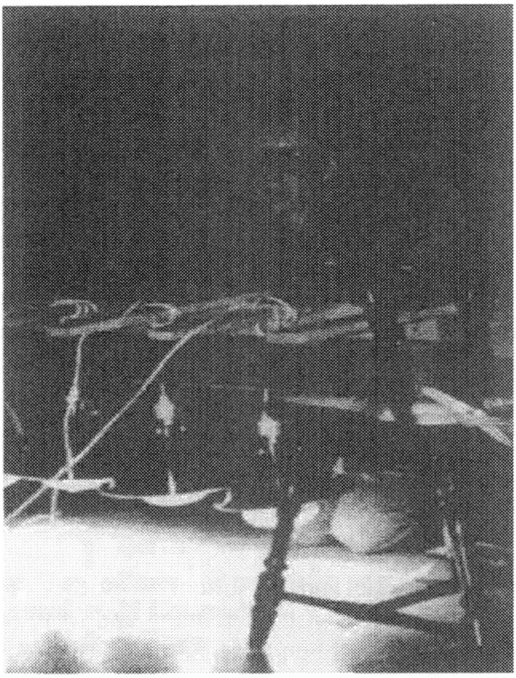

Figure 2. Experimental setup for evaluating the effectiveness of the Westron lamp to produce previtamin D_3 in borasilicate ampoules containing 7-dehydrocholesterol in methanol.

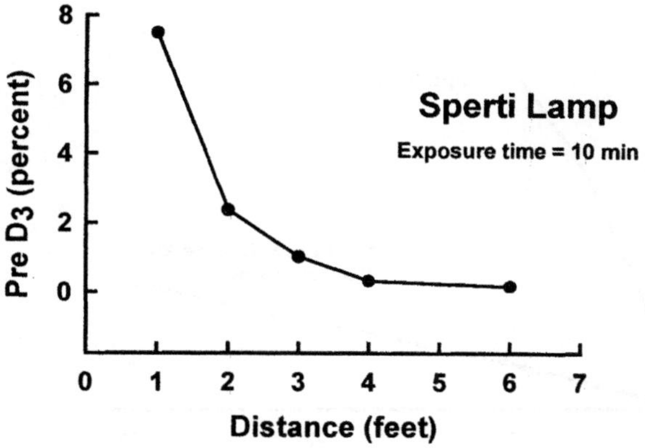

Figure 3. Percent previtamin D_3 produced from 7-dehydrocholesterol in methanol ampoules that were exposed to the Sperti lamp for 10 minutes at varying distances.

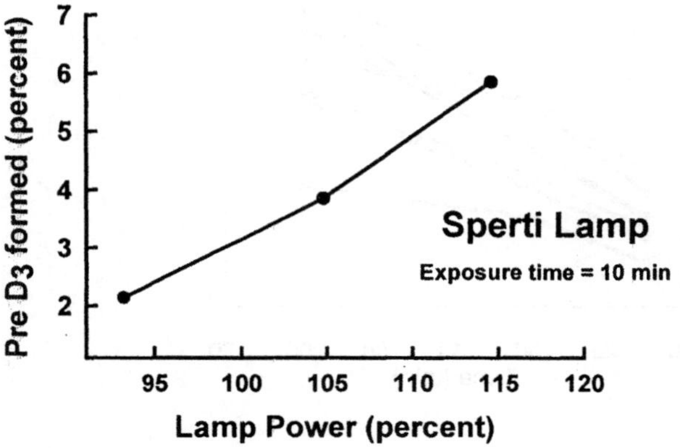

Figure 4. The effect of lamp power of the Sperti lamp on previtamin D_3 formation from 7-dehydrocholesterol after 10 minutes of exposure.

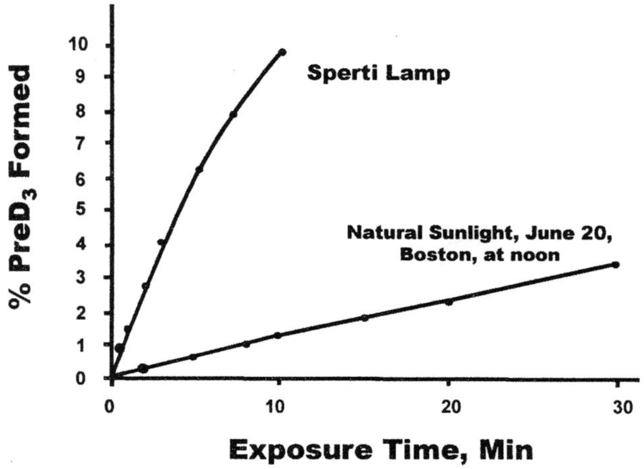

Figure 5. Ampoules of 7-dehydrocholestol were exposed to a Sperti lamp or to sunlight in June in Boston at noontime on a cloudless day. The amount of previtamin D_3 formed was determined at various times of exposure.

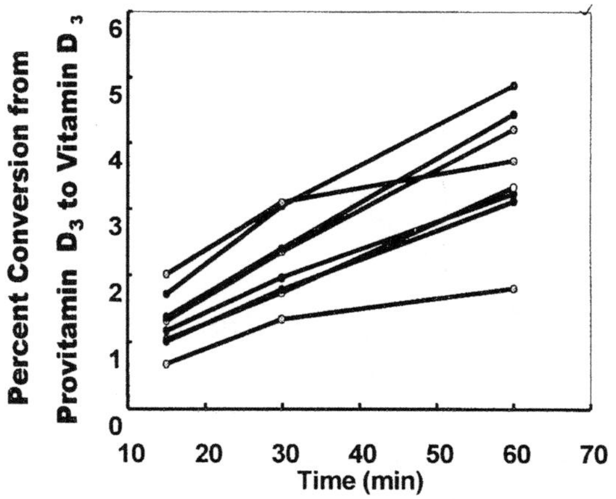

Figure 6. Six Westron lamps were evaluated for their ability to convert 7-dehydrocholesterol (provitamin D_3) to previtamin D_3 at various times.

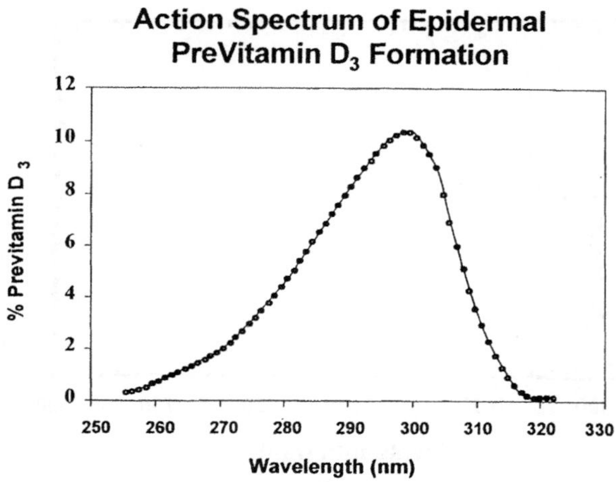

Figure 7. Action spectrum for the production of previtamin D_3 in human skin. The percent conversion of 7-dehydrocholesteol to previtamin D_3 was determined at various wavelength +5nm.

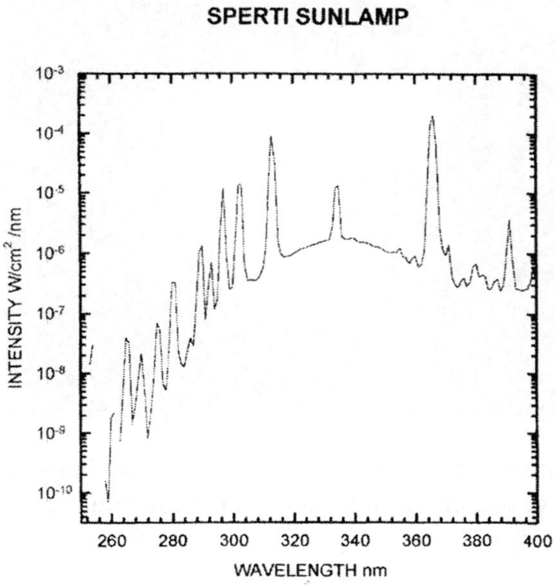

Figure 8. Spectral output for the Sperti sunlamp.

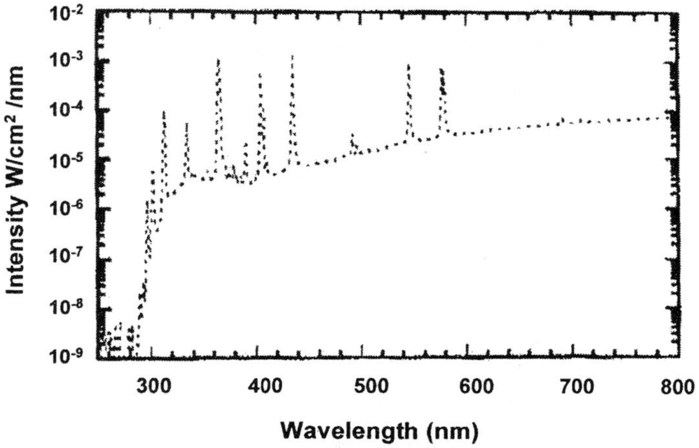

Figure 9. Spectral output for the Westron Lamp

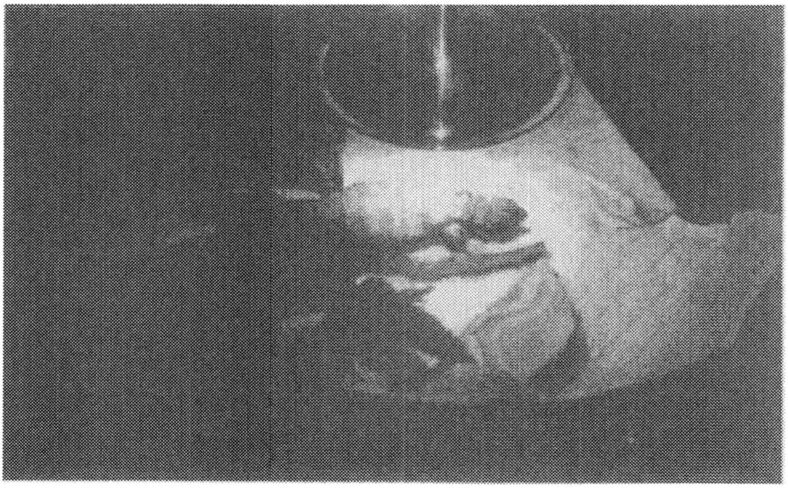

Figure 10. Exposure of Euromastyx mali to a Westron lamp which provided an adequate source of vitamin D_3 for these relatively rare reptiles.

VITAMIN D INSUFFICIENCY IS COMMON AND UNDERDIAGNOSED IN AFRICAN AMERICAN FEMALE PATIENTS

S. Shewakramani, V. Tangpricha, D. Rakita, M.F. Holick
Vitamin D, Skin & Bone Research Laboratory
Boston University School of Medicine

INTRODUCTION

Traditionally African American women have been considered to be at a low risk for developing osteoporosis, as they tend to attain higher peak bone mass and have slower bone turnover when compared to Caucasian women. As a result, at the age of 50, Caucasian women greatly outnumber African American women in the osteoporotic population (1). African American women, due to the increased amount of melanin in their skin, convert less 7-dehydrocholesterol to previtamin D_3 in the epidermis when exposed to the same amount of solar radiation (2,3) leading to lower 25-hydroxyvitamin D_3 [25(OH)D] levels in their circulation when compared to Caucasian women (4). 1,25-dihydroxyvitamin D_3 [1,25$(OH)_2$D], the active form of vitamin D which is formed in the kidneys from 25(OH)D, is essential for efficient calcium uptake from the intestine. Lower 25(OH)D levels and the generally low calcium intake in this population places African American women at a high risk for developing osteoporosis. In fact, at the age of 80, the number of African Americans with osteoporosis begins to approach the number of Caucasian women with osteoporosis (5).

The lower 25(OH)D levels in African American women is physiologically significant, meaning that supplementing the diet of African American patients with vitamin D will decrease the rate of bone turnover (4). Therefore, one can conclude that low levels of 25(OH)D in African American women are a major risk factor for developing osteoporosis and osteomalacia, and supple-menting African American women's diets with this essential vitamin can help prevent the progression of these highly prevalent and debilitating conditions.

In this report, we evaluated the prevalence of vitamin D insufficiency in a post-menopausal African American population, and determined the frequency by which primary care physicians order a serum

25(OH)D to determine the vitamin D status of their African-American patients.

METHODS

We obtained approval from the Institutional Review Board to review the medical records of 116 inner-city post-menopausal African American women who had undergone a bone mineral density (BMD) evaluation within the last ten years. This population consisted of patients with osteoporosis, osteopenia, and normal BMD values. We recorded information on their evaluation for osteoporosis including whether a 25(OH)D level was determined. We also recorded which physician ordered each test.

RESULTS

According to WHO definitions (osteopenia BMD T-score: -1 to -2.5 and osteoporosis -2.5 and lower), 28.4% and 39.7% of patients in this study were osteoporotic and osteopenic, respectively, by BMD. Patients with normal BMD values comprised the remaining 31.9% of the population. Osteoporotic and osteopenic patients were found to have 25(OH)D levels of 15.3 ± 10.5 ng/mL (mean ± SEM) and 15.7 ± 10.6 ng/mL respectively, while patients with normal BMD averaged 19.1 ± 8.1 ng/mL (normal > 20 ng/mL). Vitamin D deficiency (25(OH)D < 20 ng/mL) was present in 75.0% of osteoporotic patients, 68.8% of osteopenic patients, and 63.2% of patients with normal BMD values. Primary care physicians ordered 29.2% of 25(OH)D testing in osteoporotic patients, 40.6% of 25(OH)D testing in osteopenic patients, and 52.6% of 25(OH)D testing in patients with normal BMD values. Specialists, including endocrinologists and rheumatologists ordered the rest of the 25(OH) testing.

	Osteoporotic patients	Osteopenic patients	Normal BMD patients
n	33	46	37
25(OH)D levels (ng/mL) (Normal > 20 ng/mL)	15.3 ± 10.5	15.7 ± 10.6	19.1 ± 8.1
% Deficient	75.0%	68.8%	63.2%
% Tested by Primary Care	29.2%	40.6%	52.6%

DISCUSSION

Cumulatively, 69.9% of African American women in our sample were vitamin D deficient (25(OH)D$\leq$20 ng/ml). Osteoporotic and osteopenic patients tended to have lower 25(OH)D_3 levels than normal patients, but this difference was not significant ($p = 0.14$). A follow-up study with a larger patient sample will be needed to further investigate this trend. Nonetheless, the high frequency of vitamin D deficiency in this sample show the need for increased awareness among physicians for 25(OH)D_3 testing in African American patients.

In this study, primary care physicians ordered only 38.6% of 25(OH)D_3 tests, with endocrinologists and rheumatologists ordering the rest. A similar study should be conducted among Caucasian patients to compare the attitudes of primary care physicians when dealing with osteoporosis and vitamin D deficiency in post-menopausal women of different races.

REFERENCES

1. Aloia J.F., Vaswani A., Yeh J.K., Flaster E. Risk for osteoporosis in black women. Calcified Tissue International 1996; 59(6):415-23.

2. Clemens T.L., Adams J.S., Henderson S.L., and Holick, M.F. Increased skin pigment reduces the capacity of the skin to synthesize vitamin D. Lancet 1982;1:74-76.

3. Holick M.F. McCollum Award Lecture, 1994: Vitamin D: new horizons for the 21st century. American Journal Clinical Nutrition 1994;60:619-630.

4. Harris S.S., Soteriades E., Coolidge J.A., Mudgal S., Dawson-Hughes B. Vitamin D insufficiency and hyperparathyroidism in a low income, multiracial, elderly population. Journal of Clinical Endocrinology & Metabolism 2000; 85(11):4125-30.

5. Kyriakidou-Himonas M., Aloia J.F., Yeh J.K. Vitamin D supplementation in post menopausal black women. Journal of Clinical Endocrinology and Metabolism 1999; 84(11):3988-23.

Non-Calcemic Effects of Vitamin D

THE ROLE OF ULTRAVIOLET RADIATION ON CARDIOCIRCULATORY REGULATION AND ON CARDIOVASCULAR RISK

Rolfdieter Krause[1], Jeanette Dobberke[1], Malte Buehring[1], Werner Hopfenmueller[2], Heinrich Kaase[3], Tai C. Chen[4], Michael F. Holick[4]
1: Department of Natural Medicine & 2: Institute of Medical Statistics, Benjamin Franklin Medical Center, Free University of Berlin, and 3: Institute of Lighting Technics, Technical University of Berlin, Germany; 4: Vitamin D, Skin, and Bone Research Laboratory, Boston University Medical Center, Boston, MA, USA.

I. Historical aspects of sunlight and cardiocirculatory regulation

The benefit of the exposure to sunlight as a treatment for cardiocirculatory and cardiovascular disorders has been reported for more than 6,000 years, since the times of the Egyptian emperors Ramses and Nofretete. The medical use of natural heliotherapy has also been reported from Hippokrates ("Corpus Hippocraticum"), from health care habits of ancient Rome and from the physicians of the Islamic culture in ancient Iran [29].
A renaissance of heliotherapy took place in the last decades of the nineteenth and in the beginning of the twentieth century. All over Europe, especially in rural areas and at middle and high altitudes (500 – 2000 m above sea level) the exposure to natural sunlight for preventing or treating resistant infectious diseases (e.g. tuberculosis), bone health as well as treating cardiocirculatory disorders was developed [14].

In Germany, natural heliotherapy became an established measure of curative treatment against cardiocirculatory disorders, arrhythmias, and hypertension at the beginning of the twentieth century [9]. Moreover, between 1940 and 1950 reports were published from Sweden about children and adolescents who were artificially UV-irradiated in wintertime to prevent recurrent viral infections [36] and from Germany about miners who were irradiated with artificial ultraviolet light for the compensation of the lack of sunlight while working in the mines and for rehabilitating their physical condition [28]. Through these experiences it became evident for the first time that serial irradiation with (artificial) sunlight had cardiocirculatory effects similar to those after endurance training.

II. UVB and cardiocirculatory and blood pressure regulation

Bühring [7] conducted serial whole body irradiations of healthy young people and demonstrated that only a UV-irradiation lamp with a sun simulating UV-output (that contained UVB radiation; λ=290-320nm) showed beneficial effects on the cardiovascular system, whereas in the UVA-irradiated (320-400nm) or non-irradiated controls showed no effect. Schuh et al. [40] demonstrated the same effects after full-daytime exposure at a high altitude (2000 m above sea level); both groups showed that there was a reduction of heart rate at rest and during ergometric exercise and also a reduction of the accumulation of lactic acid at submaximal and maximal ergometric work load. These empirical observations were supported by epidemiological observations over the last four decades that there was a seasonal variation in blood pressure in normotensive as well as in untreated and treated hypertensive persons, with a decrease of blood pressure during summer and an increase during winter [3, 37]. Rostand [38] reported, using a meta-analysis of the data of the INTERSALT-study, that there was a lower level of blood pressure in normal people and a lower prevalence of hypertension the nearer people lived to the equator. The prevalence of hypertension rose with the distance from the equator both to northern and to southern latitudes. Data from the WHO-MONICA-project showed that the incidence of myocardial infarction and the death rate from myocardial infarction as well as sudden death is statistically significantly higher in countries of northern latitudes all over the world [48].

A relationship between latitude and cardiovascular disease mortality was discussed by Gardner [11] from data of England and Wales, from other parts of Europe by Smith [46], and by Fabsitz [10] from the United States. Scragg et al. [42, 43, 44] first focussed on a possible relationship between ultraviolet irradiation and the seasonality of cardiovascular disease mortality. In a community-based study he found that the incidence of myocardial infarction is inversely associated with the serum level of 25-hydroxyvitamin D_3.

III. Vitamin D and cardiocirculatory system and blood pressure regulation

An association between the elevation of blood pressure and the serum concentration of 25-hydroxyvitamin D_3 was observed in patients with essential hypertension [15] and in animals (Dahl salt-sensitive rats) [47]. Moreover, Kristal-Boneh et al. [25] found a correlation between 1,25-dihydroxyvitamin D_3 and the height of blood pressure.

Therefore the question was: is there evidence that sufficient exposure to sunlight itself or the activation of the vitamin D metabolism can provide benefits on cardiocirculatory regulation, can it influence the vascular tone, can it lower (elevated) blood pressure, and can it reduce cardiovascular risk?

Experimental data of Merke et al. [31, 32] demonstrated that vascular smooth muscle cells and endothelial cells of the (bovine) aorta possess the vitamin D receptor (VDR), and that these cells and tissues were able to synthesize 1,25-dihydroxyvitamin D_3 extrarenally when a sufficient amount of 25-hydroxyvitamin D_3 is added to the cell cultures.

On the other hand, vitamin D is the natural regulator of calcium absorption from the intestine and of calcium homeostasis. Animal experiments by Weishaar [50] have shown that the contractility of myocardiac muscle cells is dependent on the content of vitamin D_3. Also, experiments from Shan et al. [45] have shown that 1,25-dihydroxyvitamin D_3 is involved in the regulation of cytosolic free calcium and therefore in the regulation of the calcium channels. The calcium balance (extra- and intracellular) might be influenced by the serum and/or tissue levels of 25-hydroxyvitamin D_3 and of 1,25-dihydroxyvitamin D_3. The activity of the calcium channels is important for blood pressure control. This is the reason calcium channel antagonists are used in the pharmacotherapy of essential hypertension. In addition, Watson et al. [49] have shown that the intensity of coronary calcification is inversely correlated to the concentration of 1,25-dihydroxyvitamin D_3.

IV. Vitamin D and parathyroid hormone/PTHrP

Vitamin D is also the physiological regulator in the feedback system for parathyroid hormone (PTH). As known from clinical observations in patients with primary and secondary hyperparathyroidism, PTH plays an independent role in the regulation of blood pressure and hypertension. There are some reports [5, 13, 30, 51] to suggest that the serum levels of PTH are higher in hypertensive patients than in normotensive people. After surgery of the adenomatous parathyroid gland in patients with primary hyperparathyroidism as well as with secondary hyperparathyroidism elevated blood pressure decreased immediately [12].

In our studies on patients with mild to moderate essential hypertension the serum levels of PTH were in the upper normal range during wintertime in Berlin/Germany, corresponding with levels of 25-hydroxyvitamin D_3 that were in the lower normal or below normal range. After serial irradiation with solar-simulating UV-radiation 25-hydroxyvitamin D_3 [25(OH)D] significantly increased 2- to 2.5 fold and PTH decreased slightly by about 20 %. Parallel with the changes in 25(OH)D and PTH levels blood pressure decreased [18]. Another regulator of the vascular tone, PTHrP, is expressed in different cells of the cardiovascular system, in arterial cardiomyocytes, and in smooth muscle cells. From in vitro experiments by Schlüter and Piper [39] we know that PTHrP has vasorelaxing effects.

Thus, vitamin D appears to play a key role in mediating effects on the cardiovascular system. These effects may be triggered by the complexity of the extracellular and intracellular calcium balance. Vitamin D may also be responsible for some of the vasoactive actions of the parathyroid hormone and/or the parathyroid hormone related peptide.

V. Extrarenal synthesis of 1,25-dihydroxyvitamin D_3

With a growing sensitivity of the laboratory assays of vitamin D metabolites in the early 1980s, it was recognized, that anephric patients had in their circulation not only the precursor 25-hydroxyvitamin D_3 but also 1,25-dihydroxyvitamin D_3 [26]. Therefore it was concluded that cells and tissues other than renal cells or tissues must have the capacity to synthesize 1,25-dihydroxyvitamin D_3. It was also observed that in some forms of granulomatous diseases and some lymphomas high levels of 1,25-dihydroxyvitamin D_3 were present [4, 34]. From work done by Bikle [2], Merke [31] and Schwartz [41] we know that many other cells and tissues are able to synthesize extrarenally 1,25-dihydroxyvitamin D_3.

VI. Vitamin D receptor (VDR) and VDR gene polymorphism

It is now established that vitamin D is not a "vitamin" like other vitamins but a hormone. The final metabolite 1,25-dihydroxyvitamin D_3 (calcitriol) is a hormonal active substance, that belongs to the family of the steroid hormones. Similar to other hormones, the transmission of $1,25(OH)_2D$ to tissues and cells is mediated by a specific receptor, the vitamin D receptor (VDR). For their most prominent physiological role these receptors were found in the small intestine, bone and kidney as well as the skin. But the VDR is also found in the heart, in smooth muscle cells, vascular endothelial cells, the parathyroid cells, and in many other organs.

About 10 years ago a vitamin D receptor gene polymorphism and its potential association with osteoporosis was first described by Morrison et al. [33]. Some epidemiological observations [8, 27] allowed to postulate a possible correlation between the loss of bone density on one hand and elevated blood pressure on the other hand in association with the genotypes of the VDR-gene polymorphism.

VII. Effect of UVA and UVB irradiation on blood pressure

Thus, there are many suggestions that there might be an association between exposure to sunlight, vitamin D metabolism, and the regulation of the cardiocirculatory system in healthy people and in patients with cardiocirculatory dysregulation and/or cardiovascular complications.

Within the last 10 years our group has evaluated the effect of UV radiation from different lamps on the cardiovascular system to dissect out the role of UV radiation and vitamin D.

A. Patients with end-stage renal disease on maintenance hemodialysis

Serial irradiation with UVB from sun-simulating lamps was used for whole- or partial-body (legs only) [16, 24] exposure of patients with endstage renal disease (ESRD) on maintenance dialysis (n= 19)(22). We observed that by both procedures increased circulating levels of vitamin D_3, 25(OH)D, and 1,25$(OH)_2$D. The time course of increase was different for each metabolite: vitamin D_3 was the first to increase to a maximum of 7-fold. 25-Hydroxyvitamin D_3 increased in a continuous manner to a maximum of 3-fold. 1,25-dihydroxyvitamin D_3 showed a delayed rise up to a 2-fold from the baseline.

In some of the ESRD patients we saw that the vitamin D metabolism was normalized [22, 23]. A small subgroup (n=9) was observed over a period of 15 months (6 months irradiation and 9 months follow-up). The 25-hydroxyvitamin D_3 levels remained at significantly higher levels at about 2-fold of the initial values while vitamin D_3 and 1,25$(OH)_2$D decreased to baseline values. PTH decreased slightly with the known delayed time course. A significant decrease of blood pressure at rest and during ergometric exercise was also observed in these patients (table 1).

	syst. BP (rest)	syst. BP (max)	diast. BP (rest)	diast. BP (max)
%change	-17.6 %	-20.4 %	-7.6 %	+2.6 %

Table 1: Percentage change to baseline of systolic (syst.) and diastolic (diast.) blood pressure (BP) at rest and at maximum ergometric work load (max).

As a result of the UV irradiations some patients were able to reduce their antihypertensive pharmacotherapy medications. Moreover, in a subgroup of patients the heart rate variability (HRV) was monitored before and after serial UV-irradiation. After UV irradiation an improvement of some parameters of HRV was observed [1].

B. Coronary patients

In a group of patients with coronary heart disease and/or after coronary bypass surgery [17] we confirmed the earlier observations of Bühring et al. [6, 7], who showed that serial irradiations with a sun simulating UV-spectrum (three times a week for one month) was able to increase the general physical capacity, to decrease heart rate at submaximal and maximal work loads, and also to diminish the accumulation of lactic acid. These adaptations were the same as after endurance training for the therapeutic endpoint in sports medicine using physical activity in patients with coronary heart disease: lowering the heart rate at rest and during exercise prolonging the period of the diastole and lowering the work effort and therefore the myocardial need of oxygen and cardiac risk. Additionally, we found changes in the antioxidative redox system after irradiation [35]. It is known that this may also reduce coronary risk via cytokines and/or the NO system.

Therefore, there is good evidence that a combination of physical training and serial heliotherapy (especially during wintertime in northern latitudes) will have a cumulative effect in reducing cardiovascular risk in normal adults and in people in cardiac rehabilitation rehabilitation at high risk.

C. Patients with essential hypertension

In patients with mild to moderate (still untreated) essential hypertension (pilot study design, n= 18) it was confirmed that UVB-irradiation was able to lower blood pressure [18]. By irradiating this group three times a week for a period of six weeks and after six weeks of follow up we observed that only an UVB containing sun simulating spectrum caused a significant decrease on systolic (-6 mmHg) and diastolic (-6 mmHg) blood pressure assessed by 24-hour ambulatory blood pressure monitoring (24-h-ABPM). In the same group, a lowering of diastolic and systolic blood pressure, as well as a decrease of the accumulation of lactic acid after exhaustive bicycle ergometry was also achieved [21]. After six weeks of irradiation the serum levels of 25-hydroxyvitamin D_3 showed a rise of 162%, whereas 1,25-dihydroxyvitamin D_3 was unaffected. PTH decreased by 15%. No comparable effects on blood pressure or on circulating concentrations of vitamin D_3 and its metabolites were seen after irradiation with UVA.

D. Effect of Vitamin D_3 on blood pressure

Do these blood pressure lowering effects only depend on the increase of the circulating levels of vitamin D and/or on vitamin D-PTH-feedback or are there other effects of sunlight which cannot be mimicked by vitamin D replacement therapy

To answer this question a group of patients with mild essential hypertension were randomized to artificial irradiation twice weekly, over 10 weeks with either an UVB containing sun simulating spectrum or with UVA. Additionally, the UVA-group received an oral administration of vitamin D_3, in total a cumulative dose of 200,000 IU D_3 [20]. Application of treatment was single-blinded. Again, we found a significant decrease both of systolic and diastolic blood pressure measured by 24-h-ABPM and by selfmeasurement only in the UVB-irradiated group. The number of responders (decrease of diastolic BP > 5 mmHg, and/or of systolic BP > 10 mmHg) was higher in the UVB-irradiated group than in the UVA-plus-vitamin D-group although 25-hydroxyvitamin D_3 increased similarly and significantly in both groups. Moreover, the blood pressure lowering effect on systolic and diastolic blood pressure in the UVB-group was sustained for 5 and 9 months, respectively [19]. Like in all other trials (including the ESRD patients) no significant changes in (total) calcium, phosphate, sodium, catecholamines or endothelin (ET1) were observed.

VIII. Model of the role of ultraviolet irradiation on cardiocirculatory regulation and adaptation of heart rate and elevated blood pressure

In an approach to summarize the data discussed in the literature and the results of our own studies we constructed a model. This model shows different branches by which irradiation with ultraviolet radiation influences cardiocirculatory regulation, either by the mediation via the vitamin D system, or independent of vitamin D.

A. The activation of the vitamin D system addresses different pathways:

1. The "classical" pathway includes the increase of circulating vitamin D_3, 25-hydroxyvitamin D_3, and 1,25-dihydroxyvitamin D_3 which enhances intestinal calcium uptake and may also influence the calcium channel system, comparable to the effects of calcium antagonistic pharmacotherapy.

2. The feedback regulation of $1,25(OH)_2D$ on with the VDR of the parathyroid gland, leading to a decrease of circulating levels of parathyroid hormone (even within the normal range) and a suppression of the blood pressure elevating capacity of PTH.
3. The extrarenal synthesis of 1,25- dihydroxyvitamin D_3 and 24,25-dihydroxyvitamin D_3 directly in the skin and their paracrine action via the vitamin D receptor in the smooth muscle cells of the resistance vessels in the subcutaneous tissue that may lead to vasodilatation.

B. The non-vitamin D dependent action of UVB radiation may involve PTHrP and the antioxidative redox-system:

1. UVB irradiation may activate the cutaneous synthesis of PTHrP which is known to cause vasodilation.
2. The activation of the antioxidative redox-system, either in the extracellular or in the intracellular compartment may influence the balance of some cytokines directly as well as indirectly, e.g. nitric oxide (NO-system).

The decrease of heart rate might be a result of direct effects of 1,25-dihydroxyvitamin D_3 on the VDR of the heart, which are located in the right atrium, anatomically near the rhythm controlling cells of the heart. Additionally, there may be non-vitamin D-mediated pathways via the autonomic nervous system regulation, similar to the effects known after physical endurance training.

Conclusions

For reducing cardiovascular risk the lowering of elevated blood pressure is as important as the reduction of heart rate at rest and during work load. Each factor reduces the work load of the heart myocardial oxygen consumption. Maintenance of normal sinus cardiac rhythm also has a The increase in heart rate variability (and rhythmicity) is an additional positive influence on cardiocirculatory regulation and corresponds with a decrease of risk of sudden cardiac death. There is good evidence that regular and "physiologic", i.e. suberythemal exposure to sunlight has a capacity to reduce elevated cardiocirculatory and cardiovascular risk. Therefore, "heliotherapy" can be a useful addition to the non-pharmacological treatment of hypertension and coronary heart disease.

References

(1) Abel H.-H., Krause R., Klein B., Klamroth R., Bennhold I., Buehring M.: Does UV(B) irradiation affect autonomic neuropathy in hemodialysis patients? Nephrol. Dial. Tranplant. 1996; 11: 1473

(2) Bikle D.D., Nemanic M.D., Whitney J.O., Elias P.W.: Neonatal human

foreskin keratinocytes produce 1,25-dihydroxyvitamin D_3. Biochem. 1986; 25: 1545 – 1548

(3) Brennan P.J., Greenberg G., Miall W.E., Tompson S.G.: Seasonal variation in arterial blood pressure. Br. Med. J. 1982; 285: 919 – 923

(4) Breslau N.A., McGuire J.L., Zerwekh J.E.: Hypercalcemia associated with increased serum calcitriol levels in three patients with lymphoma. Ann. Intern. Med. 1984; 100: 1 – 7

(5) Brickman A.S., Nyby M.D., von Hungen K., Eggena P., Tuck M.L.: Calcitropic hormones, platelet calcium, and blood pressure in essential hypertension. Hypertension 1990; 16(5): 515 – 522

(6) Bühring M., Britzke K., Krause R., Boldt F., Klamroth R., Bocionek P., Kühn G.: Serielle UV-Exposition mit einem natürlichen Strahlenspektrum (UVA und UVB) verbessert die Kreislaufregulation und die aerobe Kapazität (Laktatstoffwechsel) bei Patienten mit koronarer Herzerkrankung. Phys. Rehab. Kur. Med. 1996; 6:16 – 18

(7) Bühring M.: Kreislauf- und metabolische Effekte serieller UV-Expositionen. Z. Phys. Med. Baln. Med. Klim. 1986; 15: 170 – 172

(8) Cappuccio F., Meilahn E., Zmuda J.M., Cauley J.A.: High blood pressure and bone-mineral loss in elderly white women: a prospective study. Lancet 1999; 354: 971 – 975

(9) Chlebarov S., Pratzel H.G.: Phototherapie. In: Lehrbuch der Physikalischen Medizin und Rehabilitation. Eds.: Schmidt K.L., Drexel H., Jochheim K.-A.; Gustav-Fischer-Verlag, Stuttgart, Jena, New York 1995; pp. 198 – 206

(10) Fabsitz R., Feinleib M.: Geographic patterns in county mortality rates from cardiovascular diseases. Am. J. Epidemiol. 1980; 111: 315 – 328

(11) Gardner M.J., Crawford M.D., Norris N.J.: Patterns of mortality in middle and early old age in the county boroughs of England and Wales. Br. J. Prev. Soc. Med. 1969; 23: 133 – 140

(12) Goldsmith D.J.A., Covic A.A., Venning M.C., Ackrill P.: Blood pressure reduction after parathyroidectomy for secondary hyperparathyroidism: further evidence implicating calcium homeostasis in blood pressure regulation. Am. J. Kid. Dis. 1996; 6: 819 – 825

(13) Grobbee D.E., Hackeng W.H.L., Birkenhäger J.C., Hofman A.: Raised plasma intact parathyroid hormone concentrations in young people with mildly raised blood pressure. Br. Med. J. 1988; 296: 814 – 816

(14) Holick M.F.: Biologic effects of light: historical and new perspectives. In: Biologic Effects of Light 1998. Eds.: Holick M.F., Jung E.G.; Kluwer Academic Publishers, Boston, London, Dordrecht 1999; pp.11 – 32

(15) Kokot F., Pietrek J., Srokowska S.: 25-hydroxyvitamin D in patients with essential hypertension. Clin. Nephrol. 1981; 16: 188 – 192

(16) Krause R., Matulla B., Chen T.C., Holick M.F., Schmidt-Gayk H., Roth H.-J., Bennhold I., Buehring M.: Regular UV(B)-irradiation: an effective way to normalize calcitriol deficiency in hemodialysis patients. Clin. Lab. 1996; 42: 305 – 307

(17) Krause R., Bennhold I., Britzke K., Klamroth R., Buehring M., Chen T.C., Holick M.F., Schmidt-Gayk H., Roth H.-J., Boldt F., Kirchner M.: Reduction of cardiac risk factors in coronary and hemodialysis patients after UV(B) therapy. In: Biologic Effects of Light 1995. Eds.: Holick M.F., Jung E.G.; de Gruyter, Berlin, New York 1996; pp.70 – 72

(18) Krause R., Buehring M., Hopfenmueller W., Holick M.F., Sharma A.M.: Ultraviolet B and blood pressure. Lancet 1998; 352: 709

(19) Krause R., Dobberke J., Hopfenmueller W., Chen T.C., Holick M.F., Buehring M.: UV-irradiation during wintertime prevents onset of mild hypertension around the year. Dtsch. med. Wschr. 2000; 125 (Suppl. 3): S44

(20) Krause R., Dobberke J., Hopfenmueller W., Sharma A.M., Hocher B.,

Neumayer H.-H., Chen T.C., Holick M.F., Buehring M.: Is the blood pressure lowering of UV-B independent from vitamin D? J. Hypert. 2000; 18(Suppl. 4): S7

(21) Krause R., Dobberke J., Schnaufer H., Kratz B., Hopfenmueller W., Roeckl T., Huettler M., Beneke R., Boening D., Sharma A.M., Popow I.N., Lewin G., Chen T.C., Holick M.F., Buehring M.: Serielle Bestrahlung mit sonnenaehnlichem UV-Spektrum senkt Ruhe- und Belastungsblutdruck und verbessert die Nierenfunktion. Nieren- und Hochdruckkrankheiten 2000; 29(12): 628 – 630

(22) Krause R., Klamroth R., Bennhold I, Holick MF, Chen TC, Molzahn M: UV(B) irradiation reduces cardiac risk in hemodialysis (HD) patients. J. Am. Soc. Nephrol 1996; 7: 1452 – 1453

(23) Krause R., Matulla-Nolte B., Albrecht C., Schmidt-Gayk H., Chen T.C., Holick M.F., Buehring M., Introd. by Neumayer H.-H.: UVB can normalize vitamin D, 25(OH)D_3, and calcitriol in ESRD patients without side effects: a result of extrarenal synthesis of 1,25$(OH)_2D_3$? J. Am. Soc. Nephrol. 1999; 10: 621A

(24) Krause R., Winter P., Mienert K., Dobberke J., Neumayer H.-H., Roth H.-J., Schmidt-Gayk H., Bühring M.: UV(B) irradiation during hemodialysis: new treatment of vitamin D deficiency. Nephrol. Dial. Transplant. 2001; 16: A88

(25) Kristal-Boneh, E., Froom, P., Harari, G., Ribak, J.: Association of calcitriol and blood pressure in normotensive men. Hypertension 1997; 30: 1289 – 1294

(26) Lambert P.W., Stern P.H., Avioli R.C., Brackett N.C., Turner R.T., Greene A., Fu I.Y., Bell N.H.: Evidence for extrarenal production of 1alpha,25-dihydroxyvitamin D in man. J. Clin. Invest. 1982; 69: 722 – 725

(27) Lee B.K.W. et al.: Association of blood pressure and hypertension with lead measures and polymorphisms in the vitamin D receptor. Environ. Health Perspect. 2001; 109: 383 – 389

(28) Lehmann G., Szakall A.: Versuche über die Hebung der Leistungsfähigkeit durch Ultraviolettbestrahlung. Arbeitsphysiologie 1944; 13: 101 – 113

(29) Lentner A.: Geschichte der Lichttherapie. Eigenverlag RWTH Aachen 1992

(30) McCarron D., Pingree P.A., Rubin R.J., Gaucher S.M., Molitch M., Krutzik S.: Enhanced parathyroid function in essential hypertension: a homeostatic response to a urinary calcium leak. Hypertension 1979; 2: 162 – 168

(31) Merke J., Hofmann W., Goldschmidt D., Ritz E.: Demonstration of 1,25$(OH)_2$ vitamin D_3 receptors and actions in vascular smooth muscle cells in vitro. Calcif. Tissue Int. 1987; 41: 112 – 114

(32) Merke J., Milde P., Lewicka S., Hügel U., Klaus G., Mangelsdorf D.J., Haussler M.R., Rauterberg E.W., Ritz E.: Identification and regulation of 1,25-dihydroxyvitamin D_3 receptor activity and biosynthesis of 1,25-dihydroxyvitamin D_3. J. Clin. Invest. 1989; 83: 1903 – 1915

(33) Morrison N.A., Cheng Qi J., Tokita A., Kelly P.J., Crofts L., Nguyen T.V., Sambrook P.N., Eisman J.A.: Prediction of bone density from vitamin D receptor alleles. Nature 1994; 367: 284 – 287

(34) Papapoulos S.E., Clemens T.L., Fraher L.J.: 1,25-dihydroxycholecalciferol in the pathogenesis of the hypercalcemia of sarcoid. Lancet 1979; 1: 627 – 630

(35) Roeckl T., Huetler M., Beneke R., Boening D., Schnaufer H., Kratz B., Krause R., Buehring M.: UV-Irradiation affects antioxidants in the intravascular compartment. In: Biologic Effects of Light 1998; Eds.: Holick M.F., Jung, E.G; Kluwer Academic Press, Boston, London, Dordrecht 1999; pp. 53 – 55

(36) Ronge H.E.: Ultraviolet irradiation with artificial illumination. Acta physiol. scand. 1948 (Suppl. 49); 15

(37) Rose G.: Seasonal variation in blood pressure in man. Nature 1961; 4760: 235

(38) Rostand S.G.: Ultraviolet light may contribute to geographic and racial blood pressure differences. Hypertension 1997; 30(2): 150 – 155

(39) Schlüter K.-D., Piper H. M.: Cardiovascular actions of parathyroid hormone and parathyroid hormone-related peptide. Cardiovasc. Res. 1998; 37: 34 – 41

(40) Schuh A., Kneist W., Schmidt H.J.: Steigerung der Ausdauerleistungsfähigkeit von durchschnitlich trainierten Personen durch natürliche Sonnenstrahlung (Heliotherapie). Phys. Rehab. Kur. Med. 1993; 3:95 – 99

(41) Schwartz G.G., Whitlatch L.W., Chen T.C., Lokeshwar B.L., Holick M.F.: Human prostate cells synthesize 1,25-dihydroxyvitamin D_3 from 25-hydroxyvitamin D_3. Cancer Epidemiol. Biomark. Prev. 1998; 7: 391 – 395

(42) Scragg R., Holdaway I., Jackson R.: Plasma 25-hydroxyvitamin D_3 and its relation to physical activity and other heart disease risk factors in the general population. Ann. Epidemiol. 1992; 2: 697 – 703

(43) Scragg R., Jackson R., Holdaway I.M., Lim T., Beaglehole R.: Myocardial infarction is inversely associated with plasma 25-hydroxyvitamin D_3 levels: a community-based study. Int. J. Epidemiol. 1990; 19(3): 559 – 563

(44) Scragg R.: Seasonality of cardiovascular disease mortality and the possible protective effect of ultraviolet radiation. Int. J. Epidemiol. 1981; 10(4): 337 – 341

(45) Shan J., Resnick L.M., Lewanczuk R.Z., Karpinski E., Li, B., Pang P.K.T.: 1,25-dihydroxyvitamin D as a cardiovascular hormone. Effects on calcium current and cytosolic free calcium in vascular smooth muscle cells. Am. J. Hypertens. 1993; 6: 983 – 988

(46) Smith W.C., Tunstall-Pedoe H.: European regional variation in cardiovascular mortality. Br. Med. Bull. 1984; 40: 374 – 379

(47) Thierry-Palmer M., Carlyle K.S., Williams M.D., Tewolde T., Caines-McKenzie S., Bayorh M.A., Emmett N.L., Harris-Hooker S.A., Sanford G.L., Williams E.F.: Plasma 25-hydroxyvitamin D concentrations are inversely associated with blood pressure of Dahl salt-sensitive rats. J. Steroid Biochem. Molec. Biol. 1998; 66(4): 255 – 261

(48) Tunstall-Pedoe H., Kuulasmaa K., Mähönen M., Tolonen H., Ruokokoski E., Amouyel P.: Contribution of trends in survival and coronary-event rates to changes in coronary heart disease mortality: 10-year results from 37 WHO MONICA project populations. Lancet 1999; 253: 1547 – 1557

(49) Watson K.E., Abrolat M.L., Malone L.L., Hoeg J.M., Doherty R., Demer L.L.: Active serum vitamin D levels are inverely correlated with coronary calcification. Circulation 1997; 96(6): 1755 – 1760

(50) Weishaar R.E, Kim S.N, Saunders D.E, Simpson R.U.: Involvement of vitamin D_3 with cardiovascular function III. Effects on physical and morphological properties. Am. J. Physiol. 1990; 258: E134 – E142

(51) Young E.W., Morris C.D., Holcomb S., McMillan G., McCarron D.A.: Regulation of parathyroid hormone and vitamin D in essential hypertension. Am. J. Hypertens. 1995; 8: 957 – 964

VITAMIN D AUTOCRINE SYSTEM AND CANCER

Tai C. Chen and Michael F. Holick
Vitamin D, Skin and Bone Research Laboratory, Section of Endocrinology, Department of Medicine, Boston University Medical Center, Boston, MA 02118 USA

Abstract

Epidemiological and biochemical evidence strongly implicates a role for the active form of vitamin D, 1α,25-dihydroxyvitamin D [1α,25$(OH)_2$D], in the growth and differentiation of human breast, colon and prostate cells. Recently, our laboratory demonstrated the presence of mRNA and enzyme activity for 25-hydroxyvitamin D-1α-hydroxylase (1α-OHase), the enzyme responsible for the synthesis of 1α,25$(OH)_2$D, in human cancer cells derived from those tissues. We further demonstrated a marked decrease in 1α-OHase activity in the primary cultures and cell lines of prostate cancer cells, compared to those found in the primary cultures of normal prostate cells. Transient or stable transfection of LNCaP prostate cancer cells with 1α-OHase cDNA greatly increased the 1α-OHase enzyme activity, and thereby their cellular response to the antiproliferative activity of 25(OH)D. These findings suggest that, by synthesizing 1α,25$(OH)_2$D within the cell, the enzyme may play an autocrine role in modulating the proliferation and differentiation of breast, colon and prostate cells. Thus, higher circulating concentrations of 25(OH)D may be necessary for the extra-renal 1α-OHase to work at its maximum capacity to produce an adequate amount of 1α,25$(OH)_2$D in tissues such as the breast, colon and prostate to regulate cell growth and protect them from becoming malignant. This may explain why people who live closer to the equator are less likely to die from cancers of the breast, colon and prostate.

Introduction

Breast, colon and prostate cancer are among the most frequently diagnosed cancers in the US and Western European countries. Although the causes for these cancers are poorly understood, they all arise from the transformation of epithelial cells. Vitamin D insufficiency has been implicated as a potential risk factor in the development of breast, colon and prostate cancer.

Vitamin D_3 is formed in the skin after exposure to ultraviolet B portion (UVB) of the solar spectrum (290-315 nm). The production of vitamin D_3 is inversely related to latitude, degrees of skin pigmentation and age (1). To become biologically active, vitamin D must be hydroxylated first in the liver to form 25(OH)D, the major circulating metabolite of vitamin D, and then in the kidney at the 1α-position catalyzed by 25 hydroxyvitamin D-1α-hydroxylase (1α-OHase), a mitochondrial cytochrome P_{450} enzyme, to form 1α,25$(OH)_2D_3$. Another cytochrome P-450 enzyme that plays a critical role in vitamin D metabolism is 25 hydroxyvitamin D-24R-hydroxylase (24R-OHase). The substrates for 24R-OHase include 25(OH)D and 1α,25$(OH)_2$D. 24R-OHase is likely to be responsible for the clearance of excess 25(OH)D and 1α,25$(OH)_2$D in circulation (2).

In addition to its role in calcium and phosphorus homeostasis (1), 1α,25$(OH)_2$D and its analogues have been shown to regulate the proliferation and differentiation of a variety of normal and cancerous cells, which possess vitamin D receptor or VDR, including breast, colon and prostate cells. Thus, a potential role for 1α,25$(OH)_2D_3$ and its analogues in breast, colon and prostate cancer prevention and therapy have been suggested.

Vitamin D and Cancer

Mason et al. reported in 1975 that mortality and incidence rates of breast cancer in North America increased with increasing latitude (3). Subsequent research by Garland et al. (4) and Schwartz et al. (5) also demonstrated a positive association of breast cancer and prostate cancer mortality, respectively, with low sunlight levels. Because vitamin D synthesis depends on the availability of ultraviolet B (295-310 nm) radiation, the north-south gradient in cancer mortality is reminiscent of vitamin D deficiency syndrome. Along this line, evidence began to accumulate linking serum 25(OH)D with colon and prostate cancer (6-10). Overall results suggest that one cause of cancer for breast, colon and prostate may be due to vitamin D insufficiency. On the other hand, a mechanism involving the endogenous production of vitamin D and/or its metabolites may have a protective role for these three cancers.

The critical biochemical evidence to support a role for vitamin D in cancers was the demonstration of specific vitamin D receptor (VDR) for 1α,25$(OH)_2$D in breast, colon, and prostate cells and the demonstration that 1α,25$(OH)_2$D could inhibit the proliferation and promote the differentiation of these cancer cells in vitro and in vivo (Figure 1) (11-13). Furthermore, there is a correlation between the effects of 1α,25$(OH)_2$D on cancer growth and the expression of VDR in vitro and in vivo (14-18). In a study when rats were fed high fat, low calcium and low vitamin D diets, the animals had a significant increase in mammary tumors, whereas increased vitamin D supplementation inhibited tumorigenesis (19).

Collectively, these findings have led to the use of 1α,25$(OH)_2D$ in clinical trials for breast, colon and prostate cancer (11,12,20,21). However, it also caused hypercalcemia and calciuric side effects. Studies are underway using non-calcemic analogs of 1α,25$(OH)_2D_3$.

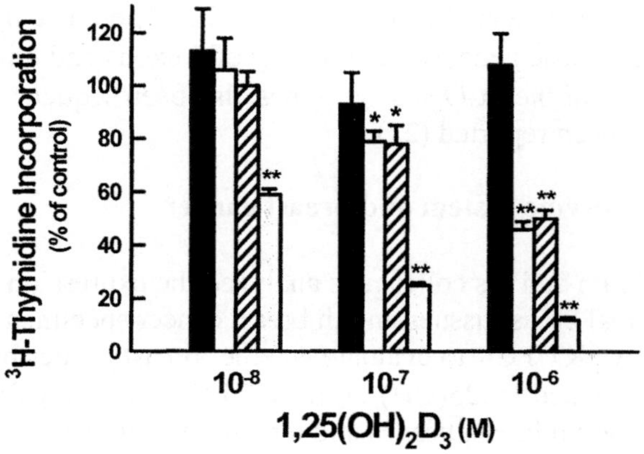

Figure 1. Effect of 1α,25$(OH)_2D_3$ on the ^{3}H-thymidine incorporation in cultured cell lines, LNCaP [▭] and DU145 [■] PC-3 [▨] and primary culture [▭] of prostate from a cancer patient. Bars represent the mean ± SD of 5-8 determinations. *$p<0.05$, **$p<0.01$ vs. controls.

Autocrine System of Vitamin D

There are two published reports indicating a significant inverse relationship between lower serum levels of 1α,25-dihydroxyvitamin D_3 [1α,25$(OH)_2D_3$] and higher prostate cancer risk (22,23). However, the inverse relationship has not been observed by all investigators (24,25). This discrepancy in published studies point to the importance of intra-prostate levels of 1α,25$(OH)_2D_3$ rather than serum levels as the risk factor for prostate cancer. In addition, under normal physiological conditions, the serum 1α,25$(OH)_2D$ is mainly derived from the action of renal 1α-OHase which is tightly regulated by calcium, phosphorus and parathyroid hormone (PTH) (2), and the 1α,25$(OH)_2D$ level does not fluctuate significantly with changing levels of serum 25(OH)D. Therefore, it was difficult to understand why vitamin D deficiency with low circulating levels of 25(OH)D was associated with breast, colon and prostate cancer mortality rates, when the serum levels of 1α,25$(OH)_2D$ may not be related to higher cancer risk (24,25). One explanation for this dilemma is that those cells may have a 1α-OHase that can convert 25(OH)D to 1α,25$(OH)_2D$. Thus, the 1α,25$(OH)_2D$ level within the cells could be influenced by the serum level of 25(OH)D, the substrate of 1α-OHase.

Extra-renal 1α-OHase is a type I (mitochondrial) cytochrome p450 enzyme located at the inner membrane of mitochondria (26). The enzyme functions as an oxidase by transferring electrons from NADPH to a molecular oxygen. 1α-OHases from mouse, rat and human have been cloned and their sequences were shown to be identical among different species. The gene is located on chromosome 12. There is only one copy of the human 1α-OHase gene, which consists of 9 exons and 8 introns. The promoter region of the1α-OHase gene has also been sequenced and its regulation has been reported (27).

Vitamin D autocrine system and breast cancer

Reichrath and his colleagues analyzed the expression for 1α-OHase in normal breast tissues and in breast cancer specimen by semiquantitative RT/PCR to evaluate whether breast tissues had the capability to produce 1α,25$(OH)_2D_3$ from 25$(OH)D_3$. They found that five out of seven normal breast tissues and ten out of fourteen breast cancer samples analyzed expressed mRNA for 1α-OHase (personal communication). mRNA and protein for 1α-OHase were also detected in MCF-7 cells by RT/PCR and western analysis, respectively (Figure 2).

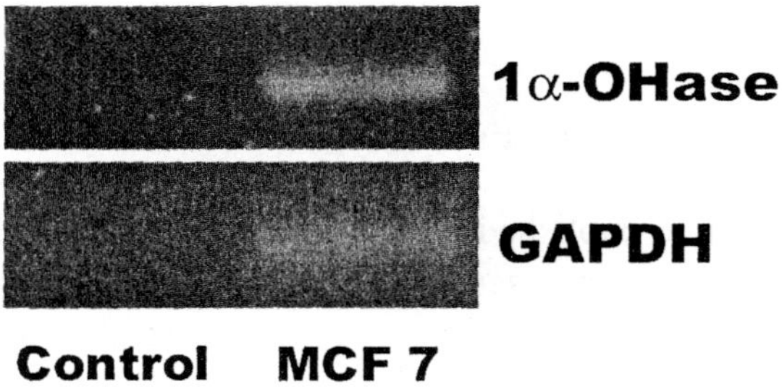

Figure 2. Western blot analysis of human 1α-OHase protein in MCF-7 breast cancer cells.

Vitamin D autocrine system and colon cancer

Experiments using human colon cancer cell lines, including Caco-2 cells, have demonstrated that 1α,25$(OH)_2D_3$ and its synthetic analogs are potent anti-proliferative and pro-differentiation agents (12). Since the major circulating form of vitamin D is 25(OH)D, and serum 1α,25$(OH)_2$D levels are tightly regulated in the circulation regardless of 25(OH)D levels, it was unclear why higher 25(OH)D levels or more sunlight exposure would protect colon from developing cancer.

Cross *et al* demonstrated that Caco-2 cells could produce $[^3H]1\alpha,25(OH)_2D_3$ from $[^3H]25(OH)D_3$ in vitro (28). Later, the same group showed that the expression of 1α-OHase gene as well as VDR gene increased in parallel during the early phase of carcinogenesis, whereas at the late stage of carcinomas only low levels of the mRNAs were detected. Our laboratory has confirmed the presence of the 1α-OHase message by real time quantitative reversed-transcription (RT)/ polymerase chain reaction (PCR) and protein by western blot in human colon tissue (Figure 3) (29). We further used sensitive real-time RT/PCR to measure the expression levels of 1α-OHase in malignant and adjacent normal colon tissues from three patients and normal tissues from five people. We found that 1α-OHase gene was expressed in all samples from normal and malignant colon. The most differentiated colon cancer sample that we had analyzed had about the same amount of 1α-OHase mRNA as the normal adjacent tissue. The less differentiated cancer tissues however had 20 to 30-fold more expression than adjacent normal tissues. The findings agree with those by Cross et al (30). This increase in expression could be because of resistance to $1\alpha,25(OH)_2D$, increased degradation of $1\alpha,25(OH)_2D$, or ineffective production of $1\alpha,25(OH)_2D$. Thus, 1α-OHase in the colon cells may produce the growth-regulating hormone $1\alpha,25(OH)_2D$ within the cells. Furthermore, therapies that increase levels of the 1α-OHase activity in malignant tissue could increase the local production of $1\alpha,25(OH)_2D$ and thereby reduce tumor proliferation and promote tumor cell differentiation without inducing systemic hypercalcemia.

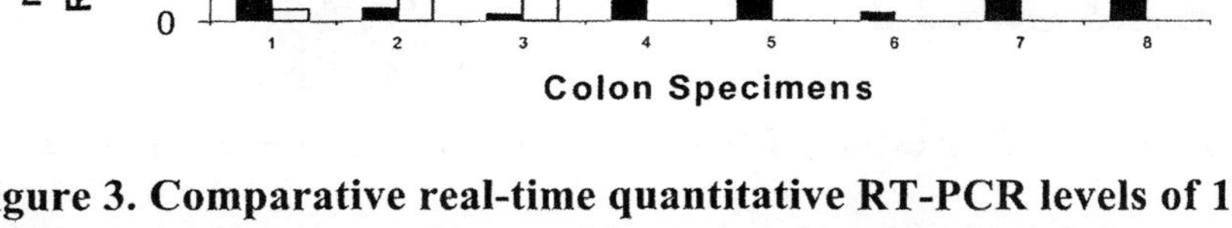

Figure 3. Comparative real-time quantitative RT-PCR levels of 1α-OHase in malignant and normal human colon tissue in eight surgical specimens. Specimens 1-3 had both the malignant and adjacent normal colon tissues for comparison. The Y-axis represents the relative quantity of the 1α-OHase expression relative to the expression of the constitutive cellular marker GAPDH. Dark bars represent normal colon tissue. Open bars represent the adjacent malignant colon tissue. (Tangpricha *et al.* The Lancet 2001)

Vitamin D autocrine system and prostate cancer

Epidemiological and biochemical data support a role for vitamin D in the growth and differentiation of human prostatic cells. These findings prompted us to ask whether prostatic cells from human prostate cancer cell lines and primary cultures of prostate cells could convert $25(OH)D_3$ to $1\alpha,25(OH)_2D_3$. We found that DU145 and PC-3 cancer cell lines produced 0.31 ± 0.06 and 0.07 ± 0.01 pmol of $1\alpha,25(OH)_2D_3$/mg protein/h, respectively. No measurable $1\alpha,25(OH)_2D_3$ was detected in LNCaP cells. The normal and benign prostatic hyperplasia primary cultures produced 3.08 ± 1.56 and 1.05 ± 0.31 pmol of $1\alpha,25(OH)_2D_3$/mg protein/h, respectively. Furthermore, the production of $1\alpha,25(OH)_2D_3$ in the prostate cells was completely inhibited in the presence of clotrimazole, a cytochrome P_{450} inhibitor (31). A comparison of 4 different primary cultures of prostate cancer cells, two BPH primary cultures and three prostate cancer cell lines (LNCaP, DU145 and PC-3) with two primary cultures from normal prostate cells illustrated that the normal prostates had higher 1α-OHase activity than cells from BPH (40% of the normal activity), which in turn had higher 1α-OHase activity than prostate cancer cells (17% of the normal activity). No activity could be detected in LNCaP cells.

The expression of 1α-OHase gene was also studied using northern blot analysis, and semiquantitative RT/PCR. It demonstrated the presence of mRNA for 1α-OHase in PC-3 and DU145 and in primary cultures of normal prostate epithelial cells (Figures 4 & 5).

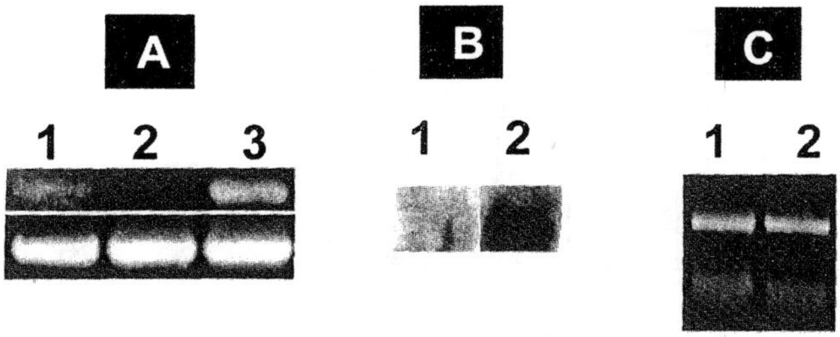

Figure 4. Expression of 1α-OHase in prostate cancer cell lines

Figure 4A shows PCR products for 1α-OHase (upper panel) and GAPDH (lower panel) from DU145 (lane 1), LNCaP (lane 2), and keratinocytes (lane 3). The RT/PCR data for LNCaP in Panel A were confirmed by the northern blot analysis shown in Panel B of Figure 4. PC-3 cells also had comparable expression of 1α-OHase by RT/PCR (Figure 4C lane 2, upper band) as DU145 (Panel C lane 1, upper band). The lower bands illustrate the PCR product for GAPDH. Keratinocyte

RNA was also reversed transcribed, amplified by PCR and served as a positive control.

The primary culture of cells derived from a normal prostate (Figure 5, lane 2, upper band) expressed 1α-OHase message as much as keratinocytes (Figure 5, lane 1, upper band,). The lower bands in figure 5 illustrate the PCR product for GAPDH. This high level of 1α-OHase gene expression may explain why 25(OH)D_3 was effective in inhibiting the proliferation of those cells (Figures 6 & 7).

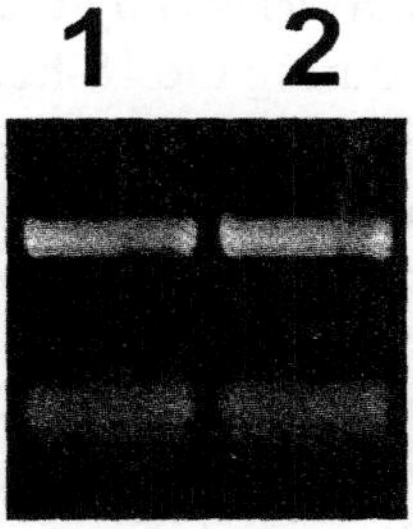

Figure 5. Expression of 1α-OHase (upper bands) in keranotinocytes (lane 1) and in the primary culture derived from normal prostate epithelial cells (lane 2). Lower bands indicate the PCR product for GAPDH.

When primary cultures of prostate cells were treated with 25(OH)D_3 or 1α,25$(OH)_2D_3$ for 7 days, followed by counting the number of cells, the two compounds caused a similar inhibition in cell proliferation (Fig. 6). When ^{3}H-thymidine incorporation assay was used for cell proliferation assay, we demonstrated that 25(OH)D_3 was also highly active in inhibiting primary cultured normal prostate cell proliferation cells after 18 hours of treatment (Fig. 7).

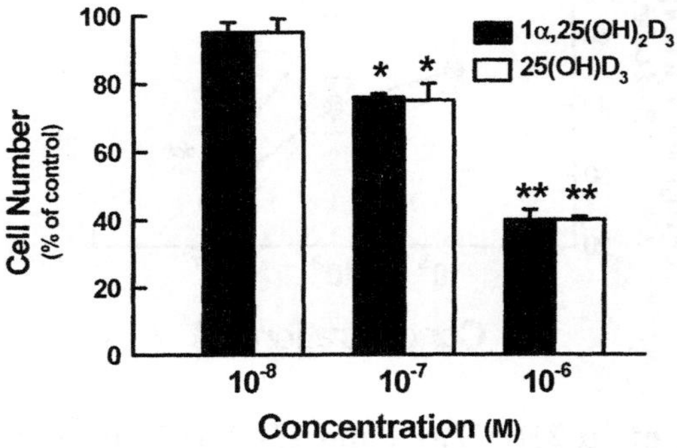

Figure 6. Effect of 1α,25$(OH)_2D_3$ and 25(OH)D_3 on cell proliferation in the primary cultures of normal prostate cells. Results are presented as the means ± standard deviation of 9 determinations. *$p<0.05$, **$p<0.001$ vs. controls. (Chen et al. Clin Cancer Res 6:901-908, 2000)

25(OH)D_3 caused 33±7%, 46±4%, and 70±3% inhibition at 10^{-9}, 10^{-8} and 10^{-7} M, respectively, as compared to 55±4%, 60±4% and 73±4% inhibition by 1α,25$(OH)_2D_3$. In both assays, 25(OH)D_3 and 1α,25$(OH)_2D_3$ were equipotent in inhibiting the proliferation of normal prostate cells at 10^{-7} M. Because 25(OH)D_3 has only 1/500th to 1/1000th of binding affinity for VDR as compared to 1α,25$(OH)_2D_3$, 25(OH)D_3 usually has little biological activity on its own, such as in LNCaP cells which has little or no 1α-OHase activity (31). The most likely explanation of the results in Fig. 7 is that portion of 25(OH)D_3 was converted to 1α,25$(OH)_2D_3$ by the high levels of 1α-OHase present in normal prostate cells during the 18-hour incubation. The production of 1α,25$(OH)_2D_3$ in these cells suggests an autocrine/paracrine role for 1α,25$(OH)_2D_3$ in which it locally modulates cell proliferation and differentiation (32).

Because LNCaP cells contain VDRs and their growth can be inhibited by 1α,25$(OH)_2D_3$, but the cells do not have 1α-OHase activity and were not responsive to the addition of 25(OH)D_3 in inhibiting cell proliferation, we wonder whether the lack of response to 25(OH)D_3 in LNCaP cells could be restored in cells either transiently or stably transfected with 1α-OHase cDNA plasmid.

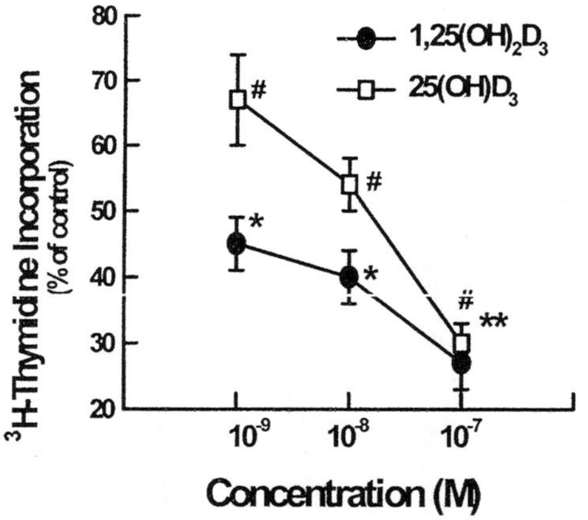

Figure 7. Effect of 1α,25$(OH)_2D_3$ and 25(OH)D_3 on ^{3}H-thymidine incorporation in primary cultures of normal prostate cells. Results are presented as the means ± standard deviation of 5-8 determinations. #p<0.05 vs. controls. *p<0.05, **p>0.1. Between the two compounds (Chen et al. Clin Cancer Res 6:901-908, 2000)

High performance liquid chromatography analysis revealed that 36±5% of ^{3}H-25(OH)D_3 was converted to ^{3}H-1α,25$(OH)_2D_3$ in LNCaP cells transiently transfected with 1α-OHase-cDNA plasmid (right panel) compared to less than 1% conversion by LNCaP cells mock- transfected with pCR 3.1 vector alone (left panel).

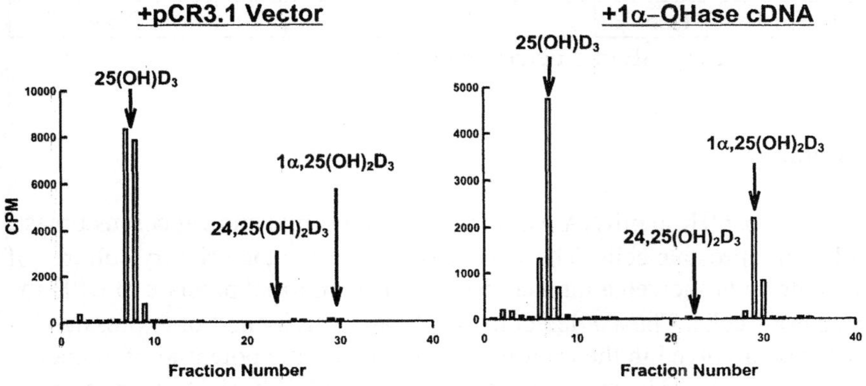

Figure 8. HPLC chromatograms of the lipid extracts from LNCaP cells transfected with pCR 3.1 Vector control (left panel) or with 1α-OHase cDNA plasmid (right panel).

When it was established that LNCaP cells transfected with 1α-OHase had enhanced the 1α-OHase activity, we then evaluated whether the increased conversion of 25(OH)D_3 to 1α,25$(OH)_2D_3$ would result in an increased sensitivity to the antiproliferative effects of 10^{-8} M 25(OH)D_3. LNCaP has little or no 1α-OHase (31) and 10^{-8} M 25(OH)D_3 had no antiproliferative activity in LNCaP cells. Table 1 shows that transfection with sense 1α-OHase cDNA plasmid into LNCaP cells caused a significant decrease in ^{3}H-thymidine incorporation (35±6% inhibition), compared to no inhibition in LNCaP cells transfected with pCR3.1 vector alone (control). The increase in cellular sensitivity to the antiproliferative activity of the pro-hormone, 25(OH)D_3, is also evident in cells stably transfected with 1α-OHase-cDNA.

Table 1. Effect of transfection of 1α-OHase cDNA into LNCaP cells on the ^{3}H-thymidine incorporation into DNA in the presence of 10^{-8} M 25(OH)D_3.

Treatment	Transient Transfection (% of Control)*	Stable Transfection (% of Control)*
pCR 3.1 Vector	100 ± 4 %	91 ± 3 %
Sense 1α-OHase	63 ± 6 %	44 ± 3 %

* Means±SE of 8 determinations

Summary

1α-OHase mRNA was expressed in normal and cancerous breast, colon and prostate cells. The 1α-OHase activity in the primary cultures of prostate cells showed a marked decrease from normal prostate to BPH to cancerous cells. Thus, a defect in 1α-OHase activity and/or expression might be involved in the progression and metastatic potential of human prostate cancer. The antiproliferative activity of 25(OH)D_3 in primary cultured prostate epithelial cells is likely due to its conversion to 1α,25$(OH)_2D_3$ by the actions of 1α-OHase present in the cells. Transfection of 1α-OHase cDNA into cells lacking 1α-OHase, such as LNCaP cells, enhanced the conversion of 25(OH)D_3 to 1α,25$(OH)_2D_3$, and therefore increased the antiproliferative activity of 25(OH)D_3.

Future Prospects

There are two types of 1α-OHase: one found in renal proximal tubules and the other in non-renal tissues including prostate. The renal 1α-OHase is responsible for providing circulating levels of 1α,25$(OH)_2$D for the maintenance of calcium and phosphorus homeostasis. The renal enzyme is tightly regulated by PTH, calcium and phosphorus. The function of 1α,25$(OH)_2$D produced within the cells of non-renal tissues, such as breast, colon and prostate, is mainly to regulate the proliferation and differentiation of those cells. How the extra-renal 1α-hydroxylase is regulated is an important question in the autocrine regulation of cellular proliferation and differentiation. Preliminary data obtained from our laboratory suggest that the regulation of 1α-hydroxylase enzyme in prostate tissue may be different from that of renal tissue.

Current treatment strategies for prostate cancer, principally radical prostatectomy or radiation therapy, are curative only in men with localized disease. For men with locally advanced or metastatic prostate cancer, the main therapeutic option, androgen ablation, is only palliative. New potentially curative approaches are urgently needed. One promising approach is gene therapy. Based on the epidemiological and biochemical data generated during the past decade, a target-specific gene therapeutic strategy that combines 1α-OHase gene delivery to cancer cells with the systemic administration of 25(OH)D could be developed. Because increases in systemic levels of 25(OH)D could result in increased local production of $1\alpha,25(OH)_2D$ in cancer cells without inducing systemic hypercalcemia, the introduction of the 1α-OHase gene into cancer cells, together with the systemic administration of 25(OH)D, could offer a novel and less toxic approach for treating breast, colon and prostate cancer than treatment with $1\alpha,25(OH)_2D$.

References

1. Holick MF Vitamin D: photobiology, metabolism, and clinical applications. In: DeGroot L, Besser H, Burger HG, e al (eds) Endocrinology, 3rd Ed. W.B. Saundrs, Philadelphia, PA, U.S.A. pp.990-1013, 1995.
2. Jones, G. Metabolism and catabolism of vitamin D, its metabolites, and clinically relevant analogs. In Vitamin D: Molecular biology, physiology, and clinical applications. Pages 57-84, Holick MF, ed. Humana Press, Totowa, New Jersey, 1999).
3. Mason, TJ, McKay FW, Hoover R, et al. Atlas of Cancer Mortality for U.S. Countries: 1950-69. U.S. Government Printing Office, Washington D.C. (DHEW Publication No. NIH 75-780, 1975.
4. Garland FC, Garland CF, Gorham MPH, Young BA. Geographic variation in breast cancer mortality in the United States: a hypothesis involving exposure to solar radiation. Prev Med 19: 616-622, 1990.
5. Hanchette CL and Schwartz GG. Geographic patterns of prostate cancer mortality. Evidence for a protective effect of ultraviolet radiation. Cancer 70:2861-2869, 1992.
6. Garland CF, Garland FC. Do sunlight and vitamin D reduce the likelihood of colon cancer? Inter J Epidemiol 9:227-231, 1980.
7. Kampman E. Giovannucci E. van't Veer P. Rimm E. Stampfer MJ. Colditz GA. Kok FJ. Willett WC. Calcium, vitamin D, dairy foods, and the occurrence of colorectal adenomas among men and women in two prospective studies. Am J Epidemiol 1994; 139: 16-29.
8. Kearney J, Giovannucci E, Rimm EB, et al. Calcium, vitamin D, and dairy foods and the occurrence of colon cancer in men. *Am J Epidemiol* 1996;143(9):907-17.
9. Garland CF, Comstock GW, Garland FC, Helsing KJ, Shaw EK, Gorham ED. Serum 25-hydroxyvitamin D and colon cancer: eight-year prospective study [see comments]. *Lancet* 1989;2(8673):1176-8.
10. Ahonen MH, Tenkanen L, Teppo L, Hakama M, Tuohimaa P. Prostate

cancer risk and prediagnostic serum 25-hydroxyvitamin D levels (Finland). Cancer Causes and Control 11:847-852, 2000.
11. Colston K. Vitamin D and breast cancer: Therapeutic potential of new vitamin D analogs. In: Feldman D, Glorieux F, and Pike W (eds) Vitamin D. Academic Press, pp. 1107-1123, 1997.
12. Brasitus TA, Sitrin MD. Chemoprevention of colon cancer by vitamin D_3 and its metabolites/analogs. In: Feldman D, Glorieux F, and Pike W (eds) Vitamin D. Academic Press, pp. 1141-1154, 1997.
13. Miller GJ. Vitamin D and prostate cancer: biologic interactions and clinical potentials. Cancer and Metastasis Reviews 17:353-360, 1999.
14. Lointier P, Meggouh F, Dechelotte P, Pezet D, Ferrier C, Chipponi J, Saez S. 1,25-Dihydroxyvitamin D_3 receptors and human colon adenocarcinoma. Br J Surg 78:435-439, 1991.
15. Meggouh F, Lointier P, Pezet D, Saez S. Evidence of 1,25-Dihydroxyvitamin D_3 receptors in human digestive mucosa and carcinoma tissue biopsies taken at different levels of the digestive tract in 152 patients. J steroid Biochem 36:143-147, 1990.
16. Shabahang M, Buras RR, Davoodi F, Schumaker LM, Nauta RJ, Evans SR. Cancer Res 53:3712-18, 1993.
17. Zhuang SH, Schwartz GG, Cameron D, Burnstein KL. Vitamin D receptor content and transcriptional activity do not fully predict antiproliferative effects of vitamin D in human prostate cancer cell lines. Mol Cellular Endocrinology 126:83-90, 1997.
18. Colston KW, Berger U, Coombes RC. Possible role for vitamin D in controlling breast cancer cell proliferation. Lancet 1: 185-191, 1989.
19. Carroll KK, Jacobson EA, Eckel LA, Newmark HL. Calcium and carcinogenesis of the mammary gland. Am J Clin Nutr. 54, 206S-208S, 1991.
20. Osborn JL, Schwartz GG, Smith DC, Bahnson R, Day R, Trump DL. Phase II trial of oral 1,25-dihydroxyvitamin D (calcitriol) in hormone refractory prostate cancer. Urol Oncol 1:195-198, 1995.
21. Gross C, Stamey T, Hancock S, Feldman D. Treatment of early recurrent prostate cancer with 1,25-dihydroxyvitamin D_3 (calcitriol). J Urology 159:2035-2040, 1998.
22. Schwartz GG. Ph.D. Dissertation, Dept. Epidemiology, Univ of North Carolina at Chapel Hill, 1993.
23. Corder EH, Guess HA, Hulka BS, Friedman GD, Sadler M, Vollmer RT, Lobaugh B, Drezner MK, Vogelman JH, Orentreich N. Vitamin D and prostate cancer: A prediagnostic study with stored sera. Cancer Epidemiol Biomarkers & Prev. 2, 467-472, 1993.
24. Nomura AMY, Stemmermann GN, Lee J, Kolonel LN, Chen TC, Turner A, Holick MF. Serum vitamin D metabolite levels and the subsequent development of prostate cancer (Hawaii, United States). Cancer Causes and Control 9:425-432, 1998.

25. Gann PH, Ma J, Hennekens CH, Hollis BW, Haddad JG, Stampfer MJ. Circulating vitamin D metabolites in relation to subsequent development of prostate cancer. Cancer Epidemiol Biomark Prev 5:121-126. 1996.
26. Miller WL, Portale AA. Vitamin D 1α-hydroxylase. TEM 11:315-319, 2000
27. Kong XF, Zhu XH, Pei YL, Jackson DM, Holick MF. Molecular cloning, characterization, and promoter analysis of the human 25-hydroxhvitamin D_3-1α-hydroxylase gene. Proc Natl Acad Sci 96:6988-6993, 1999.
28. Cross HS, Peterlik M, Reddy GS, Schuster I. Vitamin D Metabolism in Human Colon Adenocarcinoma-derived Caco-2 Cells. J Steroid Biochem Mol Biol. (62)1: 21-28, 1997.
29. Tangpricha V, Flanagan JN, Whitlatch LW, Tseng CC, Chen TC, Holt PR, Lipkin MS, Holick MF. 25-hydroxyvitamin D_3 in normal and malignant colon tissue. The Lancet. 2001;367(9269) 1673-4.
30. Cross HS, Bareis P, Hofer H, Bischof MG, Bajna E, Kriwanek S, Bonner E, Peterlik M. Steroids 66:287-292, 2001.
31. Schwartz GG, Whitlatch LW, and Chen TC, Lokeshwar BL, Holick MF. Human prostate cells synthesize 1,25-dihydroxyvitamin D_3 from 25-hydroxyvitamin D_3. Cancer Epidemiol, Biomark & Prev 7:391-395, 1998.
32. Chen TC, Schwartz GG, Burnstein KL, Lokeshwar BL, Holick MF. The in vitro evaluation of 25-hydroxyvitamin D_3 and 19-nor-1α,25-dihydroxyvitamin D_2 for prostate cancer therapy. Clin Cancer Res 6: 901-908, 2000.

Acknowledgments

This work was supported in part by grants MO1RR00533 from NIH, 4118PP1017 and 41211159016 from The Commonwealth of Massachusetts, and PCRP-PC001404 from Department of Defense.

THE EPIDERMAL VITAMIN D SYSTEM

Siegfried Segaert[1,2], Petra De Haes[1,2] and Roger Bouillon[1]
[1]Laboratory for Experimental Medicine and Endocrinology and [2]Department of Dermatology, University of Leuven, Belgium.

The epidermis as a source of vitamin D and active vitamin D metabolites

The cutaneous photosynthesis of vitamin D_3 represents the main source of vitamin D in humans. It is formed from 7-dehydrocholesterol (7DHC or provitamin D_3), which is present in large amounts in cell membranes of keratinocytes of the basal and spinous epidermal layers (1). By the action of ultraviolet B light (UVB light with a wavelength of 290-315 nm) the B ring of 7DHC can be broken to form previtamin D_3. Previtamin D_3 has very low or no affinity for vitamin D binding protein, precluding its entrance into the circulation. In the lipid bilayer of membranes, the unstable previtamin D_3 is further isomerized to vitamin D_3 by thermal energy (1). The conformational change due to this isomerization can project vitamin D_3 into the circulation, where it is caught by vitamin D binding protein and transported to the liver and kidney for further metabolization to 1α,25-dihydroxyvitamin D_3 [1,25$(OH)_2D_3$] (1).

The production of previtamin D_3 is a non-enzymatic photochemical reaction which is not subject to regulation other than substrate (7DHC) availability or intensity of UVB irradiation. 7DHC is the last precursor in the de novo biosynthesis of cholesterol. The enzyme 7DHC Δ^7-reductase (or sterol Δ^7-reductase) catalyzes the production of cholesterol from 7DHC. Inactivating mutations of the recently cloned 7DHC Δ^7-reductase gene (2) are the hallmark of the autosomal recessive Smith-Lemli-Opitz syndrome, characterized by high tissue and serum 7DHC levels and multiple anomalies including craniofacial dysmorphism and mental retardation, due to the lack of cholesterol synthesis (3). These patients exhibit increased vitamin D_3 and 25-hydroxyvitamin D_3 (25OHD_3) serum concentrations (4). Likewise, animals which are pre-treated with a specific sterol Δ^7-reductase inhibitor, also exhibit an augmented vitamin D_3 synthesis following UVB irradiation (5). In a similar fashion, the high incidence of Smith-Lemli-Opitz syndrome may be explained by the heterozygote advantage, provided by increased production of vitamin D_3 in carriers of a dysfunctional 7DHC Δ^7-reductase gene (2). To keep its provitamin D_3 levels elevated, skin in vivo exhibits low sterol Δ^7-reductase activity, probably by the presence of an endogenous inhibitor of the enzyme (6).

With increasing age however, the cutaneous stores of provitamin D_3 decrease together with photoproduction of vitamin D_3 (1). In some animals like cats, the high cutaneous sterol Δ^7-reductase activity hampers photoproduction of vitamin D, making it a true vitamin in these animals (7). Taken together, the activity of 7DHC Δ^7-reductase is highly correlated with dependence on nutritional sources of vitamin D. Therefore the regulation of its activity may represent a future pharmacological target for influencing the magnitude of the cutaneous 7DHC stores and vitamin D_3 production in response to sunlight.

Apart from substrate (7DHC) availability, the photochemical synthesis of vitamin D_3 in the skin largely depends on the amount of UVB photons that strike the basal epidermal layers. Glass, sunscreens, clothes and skin pigment absorb UVB and blunt vitamin D_3 synthesis (1). Latitude, time of day and season are factors that influence the intensity of solar radiation and the cutaneous production of vitamin D_3. Therefore there is a risk for shortage of vitamin D supplies during and after winter (1).

Nature has built in a few feedback mechanisms to minimize the risk that prolonged sun exposure would cause vitamin D intoxication. Cutaneous previtamin D_3 and vitamin D_3 are photosensitive and will be photodegraded to inactive sterols, when they are not translocated to the circulation (1). This makes that only a maximum of 10 to 15% of the provitamin D_3 will be converted to vitamin D_3. Sunlight induced melanin synthesis, acting as a natural sunscreen, provides an additional negative feedback (1). In addition, $1,25(OH)_2D_3$ downregulates its own biosynthesis and induces its own catabolism by regulating 1α- and 24-hydroxylase activity (8). Finally, UVB suppresses the expression of the vitamin D receptor (VDR) in a dose-dependent fashion (9), maybe further adding to this feedback.

Epidermal keratinocytes not only produce vitamin D_3, they also possess vitamin D receptors, rendering them responsive to $1,25(OH)_2D_3$ treatment (8). Furthermore they express CYP1α very abundantly, enabling them to convert $25OHD_3$ to $1,25(OH)_2D_3$ (10). Cutaneous 1α-hydroxylase activity is under tight feedback control by $1,25(OH)_2D_3$ (8). The mitogenic growth factors EGF and TGFα stimulate production of $1,25(OH)_2D_3$ (11) whereas the triggering of keratinocyte differentiation by density arrest or calcium exerts the opposite effect (8). The generation of calcitriol thus appears to be restricted to actively cycling cells, suggesting its involvement in epidermal growth and differentiation. The regulation of keratinocyte 1α-hydroxylase by the cytokines TNFα and IFNγ (8), which may be released by infiltrating inflammatory cells, maybe refers to a function for vitamin D in local immunomodulation in the skin.

In addition to 1α-hydroxylase, recent evidence indicates the cutaneous presence of CYP27 (12,13) as well as 25-hydroxylase activity in epidermal cells (14,15). The combined presence of vitamin D production, 25-hydroxylase, 1α-hydroxylase and VDR in the epidermis suggests the existence of a unique epidermal photo-endocrine vitamin D system in which keratinocytes may supply their own needs for $1,25(OH)_2D_3$. In support of this hypothesis, Lehmann demonstrated production of calcitriol following UVB irradiation of 7DHC supplemented keratinocytes (16). In addition, our unpublished data indicate that UVB-irradiated keratinocytes, when pretreated with a 7DHC Δ^7-reductase inhibitor, secrete $1,25(OH)_2D_3$ and exhibit a marked upregulation of the expression of 24-hydroxylase (the most sensitive vitamin D-responsive gene).

The epidermal keratinocyte as a target cell for $1,25(OH)_2D_3$

Since the discovery of vitamin D receptors in the skin, numerous research groups have focused on investigating $1,25(OH)_2D_3$ effects on keratinocyte function. Calcitriol at pharmacological concentrations is a potent inducer of keratinocyte growth arrest and differentiation (8). These properties were the basis for the application of vitamin D analogues in the treatment of psoriasis (17). Although various vitamin D regulated genes have been identified in keratinocytes (Table), the primary mechanism of action for vitamin D induced cell cycle arrest remains elusive. Calcitriol treated keratinocytes fail to progress from the G_1 to the S phase of the cell cycle, leaving retinoblastoma protein in its dephosphorylated growth suppressive state (18). Keratinocyte transformation by the adenovirus E1A gene product confers resistance to the growth inhibitory effects of $1,25(OH)_2D_3$, further suggesting that E1A binding proteins such as retinoblastoma family members (and/or the transcriptional co-integrator p300/CBP) are required for the vitamin D action (19). The suppression of phosphorylation of the retinoblastoma protein by vitamin D is preceded by induction of the cyclin-dependent kinase inhibitors $p21^{WAF1}$ and $p27^{KIP1}$ (18), which prevent the activity of G_1 cyclin-dependent kinases on retinoblastoma. $p21^{WAF1}$ is a transcriptional target for $1,25(OH)_2D_3$ (20) but it is presently unknown to which extent this direct vitamin D effect contributes to the growth suppression. Indeed, the effects of calcitriol on $p21^{WAF1}$ accumulation and on cell cycle kinetics take 24h to several days (18) and are preceded by changed expression of immediate early genes such as c-myc and c-fos (21,22) and by upregulation of phospholipase C with increased phosphoinositide turnover (23). Indirect effects through autocrine or paracrine growth factors such as TGF β1 and β2 (18,24), TNFα (25) and PTHrP (26) are also likely to contribute to the antimitotic vitamin D effect together with induction of 17β-hydroxysteroid dehydrogenase causing changes in local oestrogen metabolism (27). Increased differentiation is illustrated by induction of type I transglutaminase and cornified envelope precursors such as involucrin (8). A stimulatory effect on stratification by translocation of E-cadherin to assembling adherens junctions (28) will enhance the prodifferentiative actions of calcitriol, which further require a functional protein kinase C (28,29). The increased secretion of extracellular matrix molecules such as fibronectin and osteopontin in vitamin D treated keratinocytes (24,30) may prove beneficial for wound healing but also enhances anchorage-independent growth and possibly invasiveness of transformed keratinocytes (30).

Most of the studies concerning growth and differentiation were carried out with rather high pharmacological concentrations of $1,25(OH)_2D_3$ and may not add to knowledge about the physiological role of vitamin D in the epidermis. Indeed, some reports indicated that (low concentrations) of calcitriol might exert mitogenic activity rather than a growth inhibiting effect on keratinocytes (31,32). These proproliferative effects were most obvious in cells that were committed to differentiate (31) and correlate with activation of the mitogen activated protein kinase pathway (33). Increased proliferation was also observed following application of vitamin D compounds to normal mouse skin (34). In contrast, hyperproliferative epidermis (in psoriasis or induced by application of hyperplasiogens) responds to vitamin D derivatives with growth arrest (17,35). With respect to the prodifferentiating actions of vitamin D, paradoxical negative effects have also been reported (36). These observations indicate that the activation state of the keratinocyte (non cycling versus actively cycling) appears to determine the vitamin D effects with respect to proliferation and differentiation. This may be explained by the regulation of VDR expression by environmental factors that determine the proliferation and differentiation state of the cells such as interaction with the extracellular matrix (37), cell density, extracellular calcium concentration (38), growth factors (39), and ultraviolet B irradiation (9).

Keratinocytes that are activated by pro-inflammatory cytokines such as IL-1, TNFα and IFNγ, are known to participate in the cutaneous inflammation process by the production of cytokines, chemokines and cell adhesion molecules that allow communication with infiltrating immune cells. Many of these functions are also targeted by vitamin D: the release of the chemokines RANTES and IL-8 is inhibited by tacalcitol (40) and the pro-inflammatory cytokine IL-6 is also negatively regulated by active vitamin D (own unpublished data). The receptor for the immunosuppressive cytokine IL-10 is upregulated (41) and IL-10 release was also stimulated in calcipotriol treated psoriatic skin (42), although its cellular source was not investigated.

Like in other target tissues, autoregulation of the vitamin D system also takes place in epidermal keratinocytes. $1,25(OH)_2D_3$ exerts a strong negative feedback on its own production (8), in spite of the maintenance of CYP1α mRNA levels (Bikle, unpublished; Segaert, unpublished): (post-)translational regulation or targeting of cytochrome p450 cofactors (ferridoxin or ferridoxinreductase) conceivably might explain this discrepancy. In contrast, 24-hydroxylase activity is strongly induced by $1,25(OH)_2D_3$ (8) via a very potent upregulation of CYP24 mRNA levels (37,38). Finally, keratinocyte VDR protein is stabilized by its ligand in the absence of overt regulation of its mRNA (18).

Physiological function of the epidermal vitamin D system

Despite the numerous studies that describe the effects of vitamin D derivatives on cultured skin cells or on normal or diseased skin, the exact physiological function of $1,25(OH)_2D_3$ and VDR in the epidermis remains elusive. The postnatal alopecia in VDR deficient mice or patients (43-45), which cannot be cured by normalization of mineral ion homeostasis (46), firmly indicates that VDR is directly implicated in hair follicle function but not morphogenesis. However, the underlying mechanisms as well as the target genes in the hair follicle remain to be identified. Recent elegant studies refer to a ligand-independent role for VDR in anagen initiation in the epithelial (keratinocyte) component of the hair follicle apparatus (47,48). Retinoid X receptor (RXR)α, VDR's heterodimeric partner receptor in the epidermis, appears to be involved as well. This is clearly shown by the (postnatal) alopecia due to failure of anagen initiation following selective RXRα ablation in mouse keratinocytes during adult life (49) or during skin morphogenesis (50).

The early appearance of cutaneous VDR in the phylogenetic tree (with high levels observed in frog skin) may refer to an initial role for VDR in cutaneous calcium absorption in aquatic animals (51). Although the developmental role of VDR (if any) remains to be delineated (as demonstrated by normal development in VDR knockout mice), VDR is also expressed early during ontogeny, with the skin as one of the first organs for VDR to appear (52). Moreover, during the neonatal period, when high proliferative activity in epidermis and hair follicles is manifest, there is a sharp increase in cutaneous VDR levels, before a similar rise is seen in the gut (53). An association between VDR expression and proliferation was also noticed in cultured keratinocytes (37-39). Together with the mitogenic effects of $1,25(OH)_2D_3$ on normal skin (34) these data raise the hypothesis that $1,25(OH)_2D_3$ and VDR may participate in the maintenance of keratinocyte proliferation in the epidermis. The paradoxical antiproliferative and pro-differentiative pharmacological effects of vitamin D analogues in hyperproliferative conditions such as psoriasis or cultured keratinocytes (8,17) may be explained by increased VDR expression under these circumstances (37-39).

Finally, in view of the close interplay between UVB and the generation of vitamin D (and calcitriol), active vitamin D compounds may serve as an endogenous defense mechanism against harmful events caused by UVB. Indeed, we (unpublished) and others (54-56) recently identified a strong photoprotective effect of calcitriol against UVB mediated events in cultured keratinocytes, that may be related to its capability to induce metallothionein, a protein with antioxidant properties (54,55).

References

1. Holick M.F. McCollum award lecture, 1994: vitamin D – new horizons for the 21st century. Am J Clin Nutr 1994; 60:619-630
2. Kelley R.I. RSH/Smith-Lemli-Opitz syndrome: mutations and metabolic morphogenesis. Am J Hum Genet 1998; 63:322-326
3. Cunniff C., Kratz L.E., Moser A., Natowicz M.R., Kelley R.I. Clinical and biochemical spectrum of patients with RSH/Smith-Lemli-Opitz syndrome and abnormal cholesterol metabolism. Am J Med Genet 1997; 68:263-269
4. Chen T.C., Lu Z., Shao Q., Tint G.S., Matsuoka L., Wortsman J., Holick M.F. Vitamin D metabolism in patients with Smith-Lemli-Opitz syndrome (abstract). Photodermatol Photoimmunol Photomed 1995; 11:63
5. Bonjour J.-P., Trechsel U., Granzer E., Klöpffer G., Müller K., Scholler D. The increase in skin 7-dehydrocholesterol induced by an hypocholesterolemic agent is associated with elevated 25-hydroxyvitamin D_3 plasma level. Pflügers Arch 1987; 410:165-168
6. Pillai S., Bikle D.D., Elias P.M. Vitamin D and epidermal differentiation: evidence for a role of endogenously produced vitamin D metabolites in keratinocyte differentiation. Skin Pharmacol 1988; 1:149-160
7. Morris, J.G. Ineffective vitamin D synthesis in cats is reversed by an inhibitor of 7-dehydrocholesterol-delta7-reductase. J Nutr 1999; 129:903-908
8. Bikle D.D., Pillai S. Vitamin D, calcium and epidermal differentiation. Endocr Rev 1993; 14:3-19
9. Courtois S.J., Segaert S., Degreef H., Bouillon R., Garmyn M. Ultraviolet B suppresses vitamin D receptor gene expression in keratinocytes. Biochem Biophys Res Commun 1998; 246:64-69
10. Fu G.K., Lin D., Zhang M.Y.H., Bikle D.D., Shackleton C.H.L., Miller W.L., Portale A.A. Cloning of 25-hydroxyvitamin D-1α-hydroxylase and mutations causing vitamin D-dependent rickets type 1. Mol Endocrinol 1997; 11:1961-1970
11. Lehmann B. HaCaT cell line as a model system for vitamin D_3 metabolism in human skin. J Invest Dermatol 1997; 108:78-82
12. Ichikawa F., Sato K., Nanjo M., Nishii Y., Shinki T., Takahashi N., Suda T. Mouse primary osteoblasts express vitamin D_3 25-hydroxylase mRNA and convert 1α-hydroxyvitamin D_3 into 1α,25-dihydroxyvitamin D_3. Bone 1995; 16:129-135
13. Lehmann, B., Tiebel, O. and Meurer, M. (1999) Arch Dermatol Res Expression of vitamin D_3 25-hydroxylase (CYP27) mRNA after induction by vitamin D_3 or UVB

radiation in keratinocytes of human skin equivalents – a preliminary study. 1999; 291:507-510

14. Lehmann B., Pietzsch J., Kampf A., Meurer M. Human keratinocyte line HaCaT metabolizes 1α-hydroxyvitamin D_3 to 1α,25-dihydroxyvitamin D_3 (calcitriol). J Dermatol Sci 1998; 18:118-127
15. Lehmann B., Rudolph T., Pietzsch J., Meurer M. Conversion of vitamin D_3 to 1α,25-dihydroxyvitamin D_3 in human skin equivalents. Exp Dermatol 2000; 9:97-103
16. Lehmann B., Knuschke P., Meurer M. UVB-induced conversion of 7-dehydrocholesterol to 1α,25-dihydroxyvitamin D_3 (calcitriol) in the human keratinocyte line HaCaT. Photochem Photobiol 2000; 72:203-209
17. Fogh K., Kragballe K. Vitamin D_3 analogues. Clin Dermatol 1997; 15:705-713
18. Segaert S., Garmyn M., Degreef H., Bouillon R. Retinoic acid modulates the anti-proliferative effect of 1,25-dihydroxyvitamin D_3 in cultured human epidermal keratinocytes. J Invest Dermatol 1997; 109:46-54
19. Park K., Bae H., Heydemann A., Roberts A.B., Dotto P., Sporn M.B., Kim S.-J. The E1A oncogene induces resistance to the effects of 1,25-dihydroxyvitamin D_3 on inhibition of growth of mouse keratinocytes. Cancer Res 1994; 54:6087-6089
20. Liu M., Lee M.-H., Cohen M., Bommakanti M., Freedman L.P. Transcriptional activation of the Cdk inhibitor p21 by vitamin D_3 leads to the induced differentiation of the myelomonocytic cell line U937. Genes Dev 1996; 10:142-153
21. Matsumoto K., Hashimoto K., Nishida Y., Hashiro M., Yoshikawa K. Growth-inhibitory effects of 1,25-dihydroxyvitamin D_3 on normal human keratinocytes cultured in serum-free medium. Biochem Biophys Res Commun 1990; 166:916-923
22. Sebag M., Gulliver W., Kremer W. Effects of 1,25-dihydroxyvitamin D_3 and calcium on growth and differentiation and on *c-fos* and p53 gene expression in normal human keratinocytes. J Invest Dermatol 1994; 103, 323-329.
23. Xie Z., Bikle D.D. Cloning of the human phospholipase C-γ1 promoter and identification of a DR6-type vitamin D-responsive element. J Biol Chem 1997; 272:6573-6577
24. Kim H.-J., Abdelkader N., Katz M., Mc Lane J.A. 1,25-dihydroxyvitamin D_3 enhances antiproliferative effect and transcription of TGF-β1 on human keratinocytes in culture. J Cell Physiol 1992; 151:579-587
25. Geilen C.C., Bektas M., Wieder T., Kodelja V., Goerdt S., Orfanos C.E. 1,25-dihydroxyvitamin D_3 induces sphingomyelin hydrolysis in HaCaT cells via tumor necrosis factor α.J Biol Chem 1997; 272:8997-9001
26. Kremer R., Karaplis A.C., Henderson J., Gulliver W., Banville D., Hendy G.N., Goltzman D. Regulation of parathyroid hormone-like peptide in cultured normal human keratinocytes. J Clin Invest 1991; 87:884-893
27. Hughes S.V., Robinson E., Bland R., Lewis H.M., Stewart P.M., Hewison M. 1,25-dihydroxyvitamin D_3 regulates estrogen metabolism in cultured keratinocytes. Endocrinology 1997; 138:3711-3718
28. Gniadecki, R., Gajkawska, B. and Hansen, M. 1,25-dihydroxyvitamin D_3 stimulates the assembly of adherens junctions in keratinocytes: involvement of protein kinase C. Endocrinology 1997; 138:2241-2248
29. Ohba M., Ishino K., Kashiwagi M., Kawabe S., Chida K., Huh N.-H., Kuroki T. Induction of differentiation in normal human keratinocytes by adenovirus-mediated introduction of the η and δ isoforms of protein kinase C. Mol Cell Biol 1998; 18:5199-5207
30. Chang P.-L., Prince C.W. 1α,25-dihydroxyvitamin D_3 enhances 12-O-tetradecanoylphorbol-13-acetate-induced tumorigenic transformation and

osteopontin expression in mouse JB6 epidermal cells. Cancer Res 1993; 53:2217-2220

31. Gniadecki, R. Stimulation versus inhibition of keratinocyte growth by 1,25-dihydroxyvitamin D_3: dependence on cell culture conditions. J Invest Dermatol 1996; 106:510-516
32. Garach-Jehoshua O., Ravid A., Liberman U.A., Koren R. 1,25-dihydroxyvitamin D_3 increases the growth-promoting activity of autocrine epidermal growth-promoting activity of autocrine epidermal growth factor receptor ligands in keratinocytes. Endocrinology 1999; 140:713-721
33. Gniadecki R. Activation of Raf-mitogen-activated protein kinase signaling pathway by 1,25-dihydroxyvitamin D_3 in normal human keratinocytes. J Invest Dermatol 1996; 106:1212-1217
34. Gniadecki R., Serup J. Stimulation of epidermal proliferation in mice with 1α,25-dihydroxyvitamin D_3 and receptor-active analogues of 1α,25-dihydroxyvitamin D_3. Biochem Pharmacol 1995; 49:621-624
35. Sato H., Sugimoto I., Matsunaga T., Tsuchimoto M., Ohta T., Uno H., Kiyoki M. Tacalcitol (1,24$(OH)_2D_3$, TV-02) inhibits phorbol ester-induced epidermal proliferation and cutaneous inflammation and induces epidermal differentiation in mice. Arch Dermatol Res 1996; 288:656-663
36. Lu B., Rothnagel J.A., Longley M.A., Tsai S.Y., Roop D.R. Differentiation-specific expression of human keratin 1 is mediated by a composite AP-1/steroid hormone element. J Biol Chem 1994; 269:7443-7449
37. Segaert S., Garmyn M., Degreef H., Bouillon R. Anchorage-dependent expression of the vitamin D receptor in normal human keratinocytes. J Invest Dermatol 1998; 111:551-558
38. Segaert S., Garmyn M., Degreef H., Bouillon R. Suppression of vitamin D receptor and upregulation of retinoid X receptor α expression during squamous differentiation of cultured keratinocytes. J Invest Dermatol 2000; 114:494-501
39. Segaert S., Degreef H., Bouillon R. Vitamin D receptor expression is linked to cell cycle control in normal human keratinocytes. Biochem Biophys Res Commun 2000; 279:89-94
40. Fukuoka M., Ogino Y., Sato H., Ohta T., Komoriya K., Nishioka K., Katayama I. RANTES expression in psoriatic skin, and regulation of RANTES and IL-8 production in cultured keratinocytes by active vitamin D_3 (tacalcitol). Br J Dermatol 1998; 138:63-70
41. Michel G., Gaillis A., Jarzebska-Deussen B., Müschen A., Mirmohammadsadegh A., Ruzicka T. 1,25-$(OH)_2$-vitamin D_3 and calcipotriol induce IL-10 receptor gene expression in human epidermal cells. Inflamm Res 1997; 46:32-34
42. Kang S., Yi S., Griffiths C.E.M., Fancher L., Hamilton T.A., Choi J.H. Calcipotriene-induced improvement in psoriasis is associated with reduced interleukin-8 and increased interleukin-10 levels within lesions. Br J Dermatol 1998; 138:77-83
43. Yoshizawa T, Handa Y., Uematsu Y., Takeda S., Sekine K., Yoshihara Y., Kawakami T., Arioka K., Sato H., Uchiyama Y., Masushige S., Fukamizu A., Matsumoto T., Kato S. Mice lacking the vitamin D receptor exhibit impaired bone formation, uterine hypoplasia and growth retardation after weaning. Nat Genet 1997; 16:391-396
44. Li, Y.C., Pirro, A.E., Amling, M., Delling, G., Baron. R., Bronson, R. and Demay, M.B. Targeted ablation of the vitamin D receptor: An animal model of vitamin D-dependent rickets type II with alopecia. Proc Natl Acad Sci USA 1997; 94:9831-9835
45. Malloy P.J., Pike J.W., Feldman D. The vitamin D receptor and the syndrome of hereditary 1,25-dihydroxyvitamin D-resistant rickets. Endocr Rev 1999; 20:156-188

46. Li Y.C., Amling M., Pirro A.E., Priemel M., Meuse M., Baron R., Delling G., Demay M.B. Normalization of mineral ion homeostasis by dietary means prevents hyperparathyroidism, rickets and osteomalacia, but not alopecia in vitamin D receptor-ablated mice. Endocrinology 1998; 139:4391-4396
47. Sakai Y., Demay M. Evaluation of keratinocyte proliferation and differentiation in vitamin D receptor knockout mice. Endocrinology 2000; 141:2043-2049
48. Sakai Y., Kishimoto J., Demay M. Metabolic and cellular analysis of alopecia in vitamin D receptor knockout mice. J Clin Invest 2001; 107:961-966
49. Li M., Indra A.K., Warot X., Brocard J., Messaddeq N., Kato S., Metzger D., Chambon P. Skin abnormalities generated by temporally controlled RXRα mutations in mouse epidermis. Nature 2000; 407:633-636
50. Li M., Chiba H., Warot X., Messaddeq N., Gérard C., Chambon P., Metzger D. RXRα ablation in skin keratinocytes results in alopecia and epidermal alterations. Development 2001; 128:675-688
51. Li Y.C., Bergwitz C., Jüppner H., Demay M.B. Cloning and characterization of the vitamin D receptor from *Xenopus Laevis*. Endocrinology 1997; 138:2347-2353
52. Johnson J.A., Grande J.P., Roche P.C., Kumar R. Ontogeny of the 1,25-dihydroxyvitamin D_3 receptor in fetal rat bone. J Bone Miner Res 1996; 11:56-61.
53. Horiuchi N., Clemens T.L., Schiller A.L., Holick M.F. Detection and developmental changes of the 1,25-$(OH)_2$-D_3 receptor concentration in mouse skin and in intestine. J Invest Dermatol 1985; 84:461-464
54. Hanada K., Sawamura D., Nakano H., Hashimoto I. Possible role of 1,25-dihydroxyvitamin D_3-induced metallothionein in photoprotection against UVB injury in mouse skin and cultured keratinocytes. J Dermatol Sci 1995; 9:203-208
55. Lee J.-H., Youn J.I. The photoprotective effect of 1,25-dihydroxyvitamin D_3 on ultraviolet light B-induced damage in keratinocytes and mechanism of action. J Dermatol Sci 1998; 18:11-18
56. Masson R.S., Holliday C.J. "1,25-dihydroxyvitamin D contributes to photoprotection in skin cells." In *Vitamin D endocrine system Structural, biological, genetic and clinical aspects.* Norman A.W., Bouillon R., Thomasset M., eds. Riverside, CA: University of Riverside Printing and Reprographics, 2000 pp. 605-608

VITAMIN D AND THE BIG THREE: CANCERS OF THE COLON, BREAST, AND PROSTATE

Gary G. Schwartz
Wake Forest University School of Medicine, Winston-Salem, NC 27157

ABSTRACT

Geographic data showing an apparent protective effect of ultraviolet radiation on mortality from non-skin cancers have been published for more than 50 years. These data suggest that vitamin D may prevent some cancers, especially the common "endocrine" tumors of the colon, breast, and prostate, much as it prevents the bone disease, rickets. Until recently, many scientists have regarded this idea with skepticism. This skepticism is the result of two factors: the "ecologic fallacy" and the "endocrinologic fallacy". The ecologic fallacy concerns the potential of drawing inaccurate conclusions about individuals based on observations at the level of the group. The endocrinologic fallacy is the misconception that the kidney is the only important source of $1{,}25(OH)_2D$. Epidemiologic data indicating that low levels of ultraviolet radiation/vitamin D are associated with an increased risk of cancer in individuals, and experimental data indicating that the synthesis of $1{,}25(OH)_2D$ occurs at extra-renal sites, e.g. the colon and prostate, effectively removes these objections. The evolution of our understanding of the role of vitamin D in cancer parallels our understanding of the role of vitamin D in rickets. In both diseases, ecologic observations about solar radiation preceded experimental observations and were subsequently validated by them.

INTRODUCTION

The idea that sunlight and vitamin D inhibit the growth of human cancers is not new. In 1941, Frank Apperly, a pathologist, published geographic data demonstrating an inverse correlation between levels of ultraviolet radiation in North America and mortality rates from non-skin cancers (1). Nor were Apperly's observations the first. Based on an apparent deficit of non-skin cancer among U.S. Navy personnel, who experienced an excess of skin cancer, in 1936, Peller proposed that skin cancers induced a relative immunity to other cancers. He advocated the deliberate induction of skin cancers, which were easily detected and treated, as a form of vaccination against more occult and less treatable cancers (2). Conversely, Apperly reasoned that "the presence of skin cancer is really only an occasional accompaniment of *a relative cancer immunity* in some way related to exposure to ultraviolet radiation" (p. 191) . "A closer study of the action of solar radiation on the body", he concluded, "might well reveal the nature of cancer immunity" (p. 195).

Apperly's insightful observations on ultraviolet radiation and cancer, like those of Theobold Palm on the protective effects of ultraviolet radiation on rickets a half century earlier (3), passed virtually unnoticed by his contemporaries, only to be re-discovered by epidemiologists decades later.

Synthesis of 1,25$(OH)_2$D

The field of vitamin D and cancer has changed considerably since Apperly's time. We now know that the biologically active form of vitamin D is not vitamin D itself, but the dihydroxylated metabolite, 1,25$(OH)_2$D. The synthesis of 1,25$(OH)_2$D begins with the cutaneous production of vitamin D after exposure to sunlight or after the intestinal absorption of vitamin D_2 or vitamin D_3 obtained from the diet. To become biologically active, vitamin D must undergo two hydroxylations. The first hydroxylation occurs in the liver at the 25th carbon position, forming 25-OH-D, the prohormone and major circulating metabolite of vitamin D. The second hydroxylation occurs in the kidney at the 1 α position, forming 1,25$(OH)_2$D, the hormonally active metabolite (4). Although the kidney is the major source of 1,25$(OH)_2$D, the enzyme that converts 25-OH-D to 1,25$(OH)_2$D, 1α-hydroxylase, is also present in several types of non-renal cells, e.g., activated macrophages, keratinocytes, and colon and prostate cells (5) (6). The production of 1,25$(OH)_2$D in these cells suggests an autocrine/paracrine role for 1,25$(OH)_2$D in which it locally modulates cell proliferation and differentiation (7) (8). Figure 1 shows the "classical" synthesis of 1,25$(OH)_2$D.

Apart from its role in mineral metabolism, 1,25$(OH)_2$D plays important roles in cell growth and differentiation. Many human cells, e.g., those of the colon, breast, and prostate, contain specific intracellular receptors (commonly called vitamin D receptors or VDRs) for 1,25$(OH)_2$D. Numerous studies have shown that in response to 1,25$(OH)_2$D, these cells show an increase in differentiation and a decrease in proliferation, invasiveness, and metastasis (see Figure 2). For a comprehensive review of laboratory studies of vitamin D and cancer, the reader may consult the recent work of Hansen et al (9).

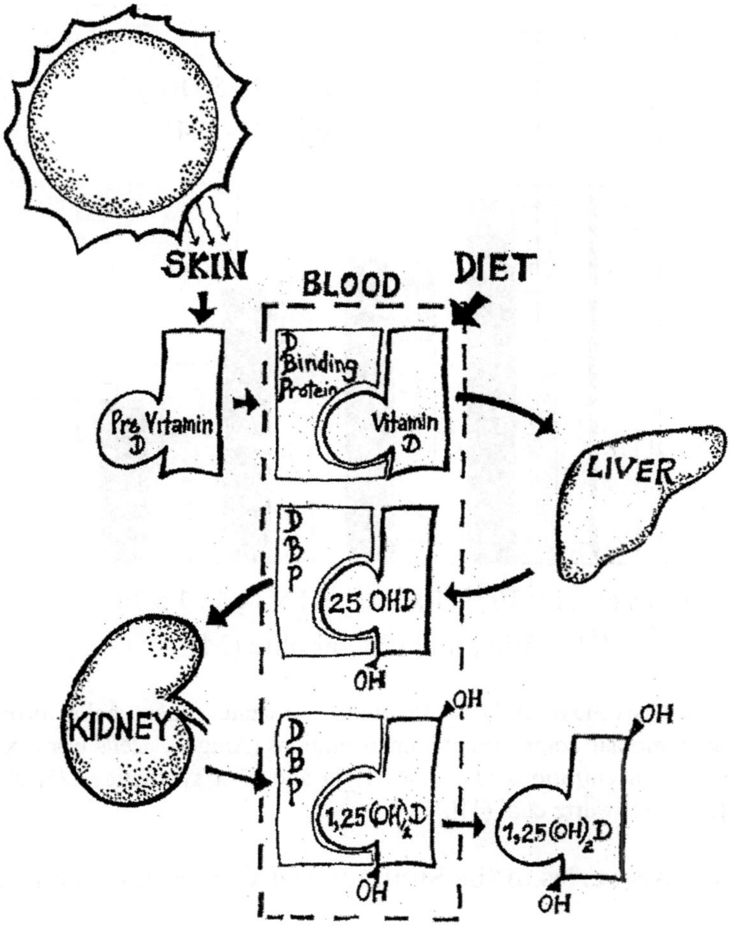

Figure 1. The "classic" view of 1,25$(OH)_2$D synthesis. Vitamin D metabolites in serum are bound to Vitamin D-Binding Protein (DBP), a serum transport molecule. We now know that, in addition to the kidney, colon and prostate cells also synthesize 1,25$(OH)_2$D from 25-OH-D.

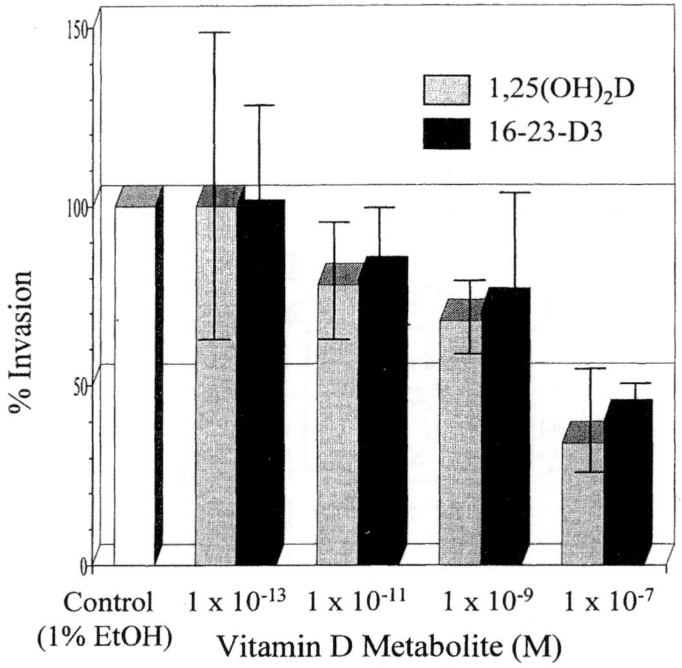

Figure 2. Invasiveness of DU 145 human prostate cancer cells through a basement membrane composed of human amnions (Amgel). Cells were exposed to different concentrations of $1{,}25(OH)_2D$ and of a synthetic $1{,}25(OH)_2D$-analog [from Schwartz et al (1)].

The Re-Discoveries of the Sunlight and Cancer Correlation

Several scientists unknowing rediscovered the inverse relationship between sunlight and cancer. In 1980, Garland and Garland recognized an inverse relationship between solar radiation and mortality from colon cancer, and in 1990, a similar relationship was described for cancer of the breast (1). In 1990, Schwartz and Hulka noted that the major risk factors for prostate cancer, older age, Black race, and residence at northern latitudes, all were associated with vitamin D deficiency (1). We proposed that vitamin D acts to maintain the normal phenotype of prostatic cells and that low levels of vitamin D permit the development of clinical prostate cancer from its subclinical precursors.

Because the major source of vitamin D is casual exposure to ultraviolet radiation, we studied the geographic distributions of ultraviolet radiation and of age-adjusted prostate cancer mortality rates among Whites in all 3,073 counties of the contiguous U.S. using trend surface analysis (13). We demonstrated that ultraviolet radiation and prostate cancer mortality are inversely correlated (P<0.0001), and exhibit opposite geographic trends (Figures 3 and 4). Exposure to ultraviolet radiation is a well-known determinant of vitamin D in normal populations. Thus, these data suggest that vitamin D may protect against clinical prostate cancer at the population level. However, these results are not proof that vitamin D inhibits prostate cancer in individuals because of the problem known as the "ecologic fallacy".

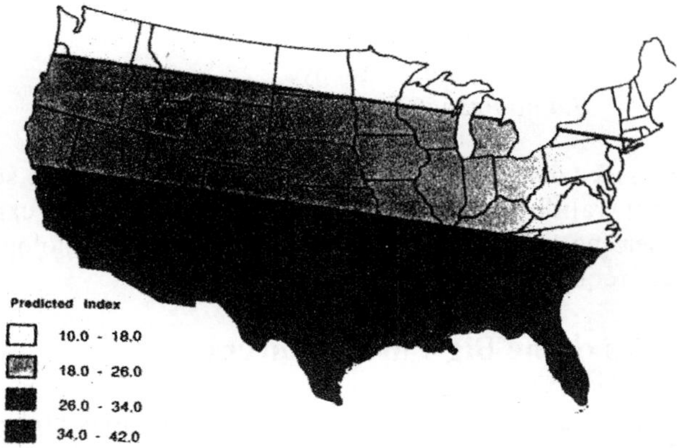

Figure 3. Linear trend surface map of ultraviolet radiation. Reprinted with permission from CANCER, 70, No. 12, 1992, 2861-2869.

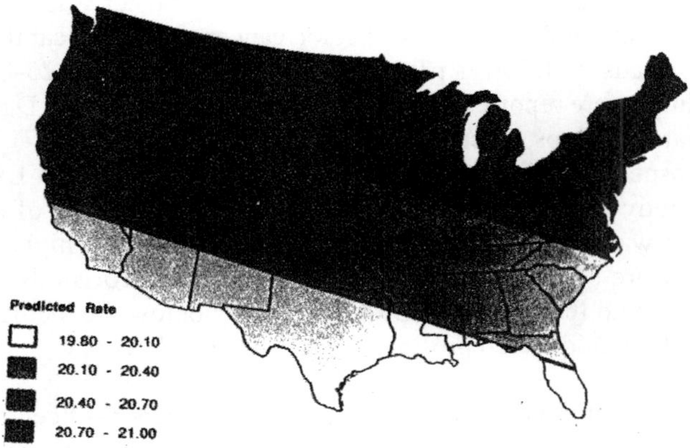

Figure 4. Linear trend surface map of (age-adjusted) prostate cancer mortality among White men. Reprinted from CANCER, 70, No. 12, 1992, 2861-2869.

The Ecologic Fallacy

The ecologic fallacy is the name given by epidemiologists to errors in reasoning that result from a confusion between levels of analysis. In short, correlations observed at the level of the group may be misleading when applied to individuals (14). The problem is not unique to epidemiology, it is a classical error in logic and was recognized at least since the time of Aristotle. The cartoonist Matt Groening gave this fallacy memorable expression in "The Cinematographer's Paradox":

The French are funny;
Comedies are funny;
Sex is funny;
But no French comedy about sex is funny!

Thus, although cancer mortality rates are lower in *regions* where sunlight is scarce, we cannot validly conclude that *individuals* with low exposure to sunlight will experience lower rates of cancer. Analytic epidemiologic studies on individuals are required in order to test that hypothesis.

Analytic Studies of the Big Three Cancers

The hypothesis that vitamin D and/or ultraviolet radiation may prevent cancer has been the subject of many epidemiologic studies. These studies vary greatly in their design and quality. Although there is some inconsistency among the results [e.g., (15)], the overall picture that emerges suggests that exposure to vitamin D and/or ultraviolet radiation may indeed protect against cancers of the big three.

The first prospective study of vitamin D and colon cancer was conducted by Garland et al who found that the risk of colon cancer was significantly increased among individuals with lower pre-diagnostic levels of serum 25-OH-D (16). Similar findings were reported in a recent study of serum vitamin D metabolites and adenomatous polyps, precursor lesions of colorectal cancer (17). Using data collected prospectively from the National Health and Nutrition Examination Follow-up Study (NHANES I), John et al found that the risk of subsequent breast cancer was significantly lower among women with higher levels of sunlight exposure (18). Similarly, in a nested case-control study in Finland, Ahonen et al found that men with 25-OH-D levels below the median were 1.7 times more likely to develop prostate cancer 11 years later (19).

It is noteworthy that the risk for all three of these cancers has been found to be associated with genotypic differences in the vitamin D receptor, e.g. (20)(21)(22). This suggests that at least some of the inconsistencies in epidemiologic studies of serum levels of vitamin D metabolites may be due to genetic variation in this receptor (23). For example, in the Physicians' Health Study, Ma et al observed a significant reduction in the risk of prostate cancer for men with the BB vs. the bb genotype, but only in men with 25-OH-D levels below the median (i.e, in men with vitamin D insufficiency) (24). Table 1 summarizes the current status of epidemiologic studies of vitamin D and the big three cancers.

Cancer Site	N/S Gradient	Low Vitamin D UV or 25OHD	Low $1,25(OH)_2D$	VDR Variants	$1\alpha OH$ expression
Breast	+	+	+ –	+ –	+
Colon	+	+ + –	+ –	+ + –	+ +
Prostate	+	+ + –	+ –	+ + –	+ +

Evidence

+ Positive — **Negative**

Table 1. Summary of current status of epidemiologic studies of vitamin D and cancers of the colon, breast, and prostate.

The "Endocrinologic Fallacy"

A second reason underlying the skepticism of some scientists about the causal nature of the inverse relationship between vitamin D and cancer mortality, even at the level of the individual, is a misconception about vitamin D. We define this as "**the endocrinologic fallacy**": the erroneous belief that the kidney is the only important source of $1,25(OH)_2D$. If this were true, then a causal relationship between ultraviolet radiation and cancer mortality would indeed be difficult to understand. The conceptual problem is this: Systemic levels of 25-OH-D are known to be dependent on exposure to ultraviolet radiation and vitamin D. However, 25-OH-D has little biologic activity. In normal individuals, systemic levels of the active metabolite, $1,25(OH)_2D$, are very tightly regulated and are not correlated with systemic levels of 25-OH-D (25,26). Thus, a mechanism by which exposure to ultraviolet radiation and vitamin D could result in higher levels of $1,25(OH)_2D$ in target tissues was unclear.

This problem was resolved by the demonstration that extra-renal sites, such as the keratinocyte, macrophage, and cells of the colon and prostate, synthesize their own $1,25(OH)_2D$ from 25-OH-D (27,28). The proof of this for the colon was provided in 1997 by Cross et al (29), and by Schwartz et al in 1998 for the prostate (30). Not only do colon and prostate cells synthesize 1,25$(OH)_2$D from 25-OH-D, but this ability effectively confers antiproliferative activity to 25-OH-D in these cells (31). Preliminary data indicate that at least some breast cells may also have this ability (32).

The existence of extra-renal sources of 1,25$(OH)_2$D in sites such as the colon, breast and prostate strongly supports the hypothesis that vitamin D may play an important role in the chemoprevention of cancers at these sites (33,34).

CONCLUSIONS

A protective role for exposure to vitamin D and the "big three" cancers is supported by epidemiologic, laboratory, and, increasingly, therapeutic clinical trials (35,36). Our understanding of the role of vitamin D in these cancers has progressed greatly beyond ecologic (correlation) studies. Cells of the colon, breast and prostate possess specific receptors for 1,25$(OH)_2$D, and in response to this hormone, modulate their growth, invasion, and metastasis. Moreover, 1,25$(OH)_2$D is now recognized to be an autocrine hormone in the colon and prostate. The evolution of our understanding of the role of vitamin D in cancer parallels our understanding of the role of vitamin D in rickets. In both instances, epidemiologic observations about solar radiation significantly preceded laboratory observations and were subsequently validated by them.

Acknowledgments

This research was supported by NCI R03CA 48440 and NCI R01CA68565.

References

1. Apperly FL. The relation of solar radiation to cancer mortality in North America. Cancer Res 1941;1:191-195.

2. Peller S. Carcinogenesis as a means of reducing cancer mortality. Lancet 1936;2:552-556.

3. Palm TA. The geographical distribution and etiology of rickets. Practitioner 1890;45:270-279, 321-342.

4. Holick MF. Vitamin D: Biosynthesis, metabolism, and mode of action. In DeGroot LJ (Ed.), Endocrinology (3rd ed.), Philadelphia, W.B. Saunders, 1995, pp. 990-1014.

5. Barbour GL, Coburn JW, Slatopolsky E, et al. Hypercalcemia in an anephric patient with sarcoidosis: Evidence for extra-renal generation of 1,25-dihydroxyvitamin D. N Engl J Med 1981;305:440-443.

6. Bikle, DD, Nemanic MF, Gee E, Elias P. $1,25(OH)_2D_3$ production by human keratinocytes. Kinetics and regulation. J Clin Invest 1986;78:557-566.

7. Dusso A, Brown A, Slatopolsky E. Extrarenal production of calcitriol. Sem Nephrol 1994; 14:144-155.

8. Bouillon R, Garmyn M, Verstuyf A, et al. Paracrine role for calcitriol in the immune system and skin creates new therapeutic possibilities for vitamin D analogs. Eur J Endocrinol 1995;133:7-16.

9. Hansen CM, Binderip L, Hamberg KJ, Carlberg C. Vitamin D and cancer: Effects of $1,25(OH)_2D_3$ and its analogs on growth control and tumorigenesis. Front Biosc 2001; 6:D820-D848.

10. Schwartz GG, Wang M-H, Zhang M, et al. 1α,25-Dihydroxyvitamin D (calcitriol) inhibits the invasiveness of human prostate cancer cells. Cancer Epidemiol Biomark Prev 1997;6:727-732.

11. Garland CF, Garland FC. Do sunlight and vitamin D reduce the likelihood of colon cancer? Int J Epid 1980;9:227-231.

12. Schwartz GG, Hulka BS. Is vitamin D deficiency a risk factor for prostate cancer [Hypothesis]. Anticancer Res 1990;10:1307-1311.

13. Hanchette CL, Schwartz GG. Geographic patterns of prostate cancer mortality: Evidence for a protective effect of ultraviolet radiation. Cancer 1992;70: 2861-2869.

14. Last JM (ed.) A Dictionary of Epidemiology. Oxford University Press, New York, 1983.

15. Braun MM, Helzlsouer KJ, Hollis BW, Constock GW. Colon cancer and serum vitamin D metabolite levels 10-17 years prior to diagnosis. Am J Epidemiol 1995;142:608-611.

16. Garland CF, Comstock GW, Garland FC et al. Serum 25-Hydroxy vitamin D and colon cancer: Eight-year prospective study. Lancet 1989;2:1176-1178.

17. Platz EA, Hankinson SE, Hollis BW, et al. Plasma 1,25-Dihydroxy- and 25-Hydroxyvitamin D and adenomatous polyps of the distal colorectum. Cancer Epidemiol Biomark Prev 2000;9:1059-1065.

18. John EM, Schwartz GG, Dreon DM, Koo J. Vitamin D and breast cancer risk: the NHANES I Epidemiologic follow-up study, 1971-1975 to 1992. Cancer Epidemiol Biomark Prev 1999;8:399-406.

19. Ahonen MH, Tenkanen L, Teppo L, et al. Prostate cancer risk and prediagnostic serum 25-Hydroxyvitamin D levels (Finland). Cancer Causes Control 2000;11:847-852.

20. Slatter ML, Yakumo K, Hoffman M, Neuhausen S. Variants of the VDR gene and risk of colon cancer (United States). Cancer Causes Control 2001;12:359-364.

21. Correa-Cerro L, Berthon P, Haussler J et al. Vitamin D receptor polymorphisms as markers in prostate cancer. Hum Gen 1999;105:281-287.

22. Bretherton-Watt D, Given-Wilson R, Mansi JL, et al. Vitamin D receptor gene polymorphisms are associated with breast cancer risk in a UK Caucasian population. Br J Cancer 2001;85:171-175.

23. Kim HS, Newcomb PA, Ulrich CM et al. Vitamin D receptor polymorphisms and the risk of colorectal ademomas: Evidence of interaction with dietary vitamin D and calcium. Cancer Epidemiol Biomark Prev 2001;10:869-874.

24. Ma J, Stampfer MJ, Gann PH, et al. Vitamin D receptor polymorphisms, circulating vitamin D metabolites, and risk of prostate cancer in United States physicians. Cancer Epidemiol Biomark Prev 1998;7:383-390.

25. Chesney RW, Rosen JF, Hanstra AJ et al. Absence of seasonal variation in serum concentrations of 1,25-dihydroxyvitamin D despite a rise in 25-Hydroxy vitamin D in summer. J Clin Endocrin Metab 1981; 53: 139-142.

26. Dusso A, Finch J, Delmez J et al. Extrarenal production of calcitriol. Kidney Internat 1990; 38(Suppl. 29): S36-S40.

27. Bikle DD. 1,25$(OH)_2D_3$-regulated human keratinocyte proliferation and differentiation: Basic studies and their clinical application. J Nutr 1995;125:1709S-1714S.

28. Bell NH. Renal and nonrenal 25-Hydroxy vitamin D-1-α-Hydroxylases and their clinical significance. J Bone Min Res 1998;13:350-353.

29. Cross HS, Peterlink M, Reddy GS, Schuster I. Vitamin D metabolism in human colon adenocarcinoma-derived Caco-2 cells: Expression of 25-Hydroxy vitamin D3-1alpha-hydroxylase activity and regulation of side-chain metabolism. J Steroid Biochem Mol Biol 1997;62:21-28.

30. Schwartz GG, Whitlatch LW, Chen TC, et al. Human prostate cells synthesize 1,25-Dihydroxyvitamin D_3 from 25-Hydroxyvitamin D_3. Cancer Epidemiol Biomark Prev 1998;7:391-395.

31. Chen TC, Schwartz GG, Burnstein KL, et al. The *in vitro* evaluation of 25-Hydroxyvitamin D_3 and 19-nor-1α,25-Dihydroxyvitamin D_2 as therapeutic agents for prostate cancer. Clin Cancer Res 2000;6:901-908.

32. Friedrich M, Reichrath J, Chen TC et al. Expression of 25-Hydroxy vitamin D_3-1α-Hydroxylase in breast tissue. In AW Norman, R Bouillon, M Thomasset (eds.), Vitamin D Endocrine System. Proceedings of the Eleventh Workshop on Vitamin D. Riverside, CA, University of California, 2000, pp. 189-191.

33. Studzinski GP, Moore DC. Sunlight–can it prevent as well as cause cancer? Cancer Res 1995;55:4014-4022.

34. Guyton KZ, Kensler TW, Posner GH. Cancer chemoprevention using natural vitamin D and synthetic analogs. Ann Rev Pharm Toxicol 2001;41:421-442.

35. Osborne JL, Schwartz GG, Smith DC, et al. A phase II trial of oral 1,25-Dihydroxyvitamin D (calcitriol) in hormone refractory prostate cancer. Urol Oncol 1995;1:195-198.

36. Guilliford T, English J, Colston KW et al. A phase I study of the vitamin D analogue EB 1089 in patients with advanced breast and colorectal cancer. Br J Cancer 1998;78:6-13.

An ecologic study of the role of solar UV-B radiation in reducing the risk of cancer using cancer mortality data, dietary supply data, and latitude for European countries

William B. Grant, Ph.D.
12 Sir Francis Wyatt Place, Newport News, VA 23606-3660, U.S.A., wbgrant@infi.net

ABSTRACT

The ecologic approach is applied to determining for which cancers solar UV-B radiation reduces the risk. Cancer mortality rates from 16 western European countries for 1989-91 are used along with dietary supply factors for 1979-81, and latitude is used as an index of solar UV-B radiation. This study is a follow on to a recent study using cancer mortality rates in 500 state economic areas in the U.S.A. along with DNA-weighted UV-B radiation for July 1992 from the Total Ozone Mapping Spectrometer (TOMS) in which a protective role for UV-B was found for 12 types of cancer (Grant, submitted). Increasing latitude is found to be a risk factor for 7 types of cancer identified in Grant (submitted), cancers of the breast, colon, esophagus, kidney, ovary and prostate, and non-Hodgkin's lymphoma (NHL), as well as 2 additional cancers, cancer of the pancreas and multiple myeloma. Those cancers for which the U.S.A. results could not be confirmed tend to be the minor cancers. These results should form the basis for further studies on the protective role of solar UV-B radiation and/or vitamin D against cancer.

INTRODUCTION

The hypothesis that solar UV-B radiation (280-320 nm) exposure mediated through production of vitamin D explains a large portion of the observed geographical variations in mortality rates for several types of cancer in the U.S., Canada, and the former Soviet Union has been in the literature for two decades. There have been reports of inverse correlations between solar radiation and cancer for cancer of the breast (Gorham et al., 1989, 1990; Garland et al., 1990; John et al., 1999), colon (Garland and Garland, 1980), ovary (Lefkowitz and Garland, 1994), and prostate (Schwartz and Hulka, 1990; Hanchette and Schwartz, 1992). In addition, vitamin D or its analogues have been shown to play a number of roles in cell biology that reduce the risk of cancer, such as prostate cancer (Feldman et al., 2000)

A study based on cancer mortality in the U.S. from the *Atlas of Cancer Mortality* (Devesa et al., 1999), along with DNA-weighted solar UV-B radiation for July 1992 (Grant, submitted) confirms previous results that solar UV-B radiation is associated with reduced risk for four types of cancer (breast, colon, ovary, and prostate). In addition, it is found that solar UV-B is inversely correlated with mortality rates for four types of cancer for which vitamin D has been found to reduce the risk (bladder, kidney, non-Hodgkin's lymphoma (NHL), and rectum). It also finds similar results to four additional cancers not yet linked to vitamin D (esophagus, gallbladder, stomach and corpus uteri). The strength of the inverse association between solar UV-B radiation and these cancers is given in Table 1. For this U.S. study, it was assumed that most factors other than solar UV-B radiation were essentially the same throughout the U.S. This is generally true for dietary factors (U.S. Dept. of Agriculture, 1985), although not for others, such as the distribution of smoking and alcohol consumption (Williams et al., 1996). However, the variations in other factors contribute to the uncertainty of each data point, rather than to the trends. Several factors that were considered and accounted for were a large fraction of Hispanic people and food near the U.S-Mexican border, large immigration to a few states, and a longer life expectancy in the north central U.S. (important for prostate cancer).

Method

In performing the original, unpublished work presented here, statistical analyses were performed for 14 types of cancer. The ecologic approach was used with mortality rates averaged for 1989-1991, age adjusted using world-age-standardized rate based on world population from the World Health Organization (WHO) Databank (Table 2). In addition, dietary supply data are used since most cancers have a number of risk and risk reduction factors related to diet (see Table 1). Dietary data were not used in the study involving the Atlas of Cancer Mortality in the U.S. since it appeared that the variations in diet were not very large across the U.S. The dietary data are for the period 1979-81 unless 1975 is given with the results, in which case data for 1974-76 were used. The latitude is used as the index of solar UV-B radiation. The regression analyses were run for three groupings of countries: all 19; those less the 3 former socialist economies (FSE); those less the 3 Scandinavian countries.

Table 1. Strength of UV-B link to various cancers, as well as dietary links (Potter et al., 1997 and other sources).

Cancer	UV-B Correlation	Risk (Reduction) Factors
Bladder	Med.	Stimulants, fish, (fruits), (vegetables)
Breast	High	Animal (vegetable) products, alcohol, (fish)
Colon	High	Animal (vegetable) products, sugar, (nonfat milk), (fish)
Esophagus	Low	Wine, hot drinks, (fruits), (vegetables), (fish)
Gallbladder	High	Total energy
Kidney	Low	Total energy, animal products
Multiple myeloma	---	
NHL	Low	Sugars, (cereals), (fruit), (fish)
Ovary	Medium	Animal fats, (fruits), (vegetables)
Pancreas	---	Energy, meat, (fruits), (vegetables), (fish)
Prostate	Low	Calcium/nonfat milk, (lycopene/tomatoes)
Rectum	High	Animal (vegetable) products, sugar, (nonfat milk), (fish)
Stomach	Med.	Salt, hot peppers, (fruits), (vegetables), (fish)
Uteri, Corpus	Med.	Total energy, animal fat, (fruits), (vegetables), (fish)

Results

There were few data on cancer for the FSE countries, and when the data were included in the analysis, the results were not as good as when these countries were omitted. Strong positive correlations with latitude were found for 2 of the cancers, multiple myeloma and NHL when all countries were included, and 5 of the 14 cancers investigated when the FSE countries were excluded: cancers of the breast, kidney, ovary, pancreas and prostate. The results are given in Table 3. Note that dietary factors played very important roles as risk or risk reduction factors in addition to latitude. When the three Scandinavian countries were also omitted, cancer of the colon and esophagus were also found to have latitude as a significant risk factor (Table 4).

Table 2. Mortality rates for cancers of the breast, kidney, pancreas, and non-Hodgkin's lymphoma for 1989-1991, age adjusted using world-age-standardized rate based on world population from the World Health Organization (WHO) Databank.

Country	Breast	Kidney		Pancreas		NHL	
	F	M	F	M	F	M	F
Austria	22.00	5.33	2.47	8.70	5.83	3.63	2.23
Belgium	26.27			7.33	4.37		
Bulgaria	15.43					2.17	1.27
Denmark	27.47	4.50	2.77	8.47	6.87	4.63	3.07
Finland	16.67			9.37	6.80	5.20	3.70
France	19.67	4.10	1.73	7.30	3.80	4.17	2.53
Hungary	22.87					3.66	2.13
Ireland	27.83	3.10	1.77	8.43	5.70	4.93	3.97
Italy	20.80	4.03	1.57	7.57	4.50	4.33	2.60
Netherlands	26.87	4.47	2.13	8.00	5.60	5.07	3.20
Norway	18.63	3.23	1.93	7.70	5.27	5.13	3.13
Portugal	17.93	1.73	0.90	5.67	3.20	3.23	1.86
Romania	14.77						
Sweden	17.60	4.90	2.93	7.93	6.67	4.93	2.90
Switzerland	23.97	4.23	1.87	7.90	5.37	4.67	2.87
UK	28.30		3.83	1.80	7.30	5.07	

Discussion

The results presented here appear to confirm and extend the results found for the U.S. (Grant, submitted). Latitude is found to be a risk factor for 7 of the 12 cancers for which DNA-weighted UV-B was found to be a risk in the U.S. There is also support in the literature for kidney cancer (Nagakura et al., 1986; Fujioka et al., 2000) and NHL (Hartge et al., 1996; Freedman et al., 1997). In addition, there are two novel findings: cancer of the pancreas and multiple myeloma. However, it has been reported that vitamin D analogues can inhibit pancreatic cancer cell growth (Zugmaier et al., 1996; Colston et al., 1997; Kawa et al., 1997; Pettersson et al., 2000). These two results are further supported in work in progress using ground-based UV-B data from the U.S. Department of Agriculture's UV-B Monitoring Program (http://uvb.nrel.colostate.edu/) along with cancer mortality data from Devesa et al. (1999) (Grant et al., in preparation).

Table 3. Ecologic study regression results for cancers for which latitude is a significant risk reduction factor. The mortality data are given in Table 2. The dietary supply data (Food and Agriculture Organization, 1991, 1996) are for 1979-81, or 1974-6 if (1975) listed with the factors. The latitude is used as the index of solar UV-B radiation.

Cancer	M/F	Risk (Reduction) Factors	N*, r, F**, p
Breast (p: 0.006, 0.033)	F	(fish), latitude	15, 0.714, 6.2, 0.014
Kidney (p: <0.001, 0.047)	M	latitude, (fish) (1975)	13, 0.781, 7.8, 0.010
Kidney (p: 0.004, 0.070)	F	latitude, (fish)	13, 0.873, 16.0, <0.001
Multiple myeloma	M	latitude	12, -0.889, 39, <0.001
Multiple myeloma	F	latitude	12, -0.899, 42, <0.001
NHL	M	latitude	16, 0.812, 27.0, <0.001
NHL (p: 0.035, 0.195)	M	latitude, alcohol	16, 0.838, 15.3, <0.001
NHL (p: 0.081, 0.091, 0.137)	M	latitude, alcohol, nonfat milk	16, 0.868, 12,2, <0.001
NHL	F	latitude	16, 0.804, 25.7, <0.001
Ovary (p: <0.001, 0.023, 0.058)	F	sugar, alcohol, latitude	11, 0.989, 105, <0.001
Pancreas (p: <0.001, <0.001, 0.001, 0.003, 0.009, 0.031)	M	latitude, fruit, (fish), alcohol, cereals, (%urban)***	15, 0.985, 42,5, <0.001
Pancreas (p: <0.001, <0.001, 0.003, 0.007, 0.009)	M	latitude, fruit, cereals, alcohol, (fish)	15, 0.971, 30.0, <0.001
Pancreas (p: <0.001, 0.020)	F	latitude, (fish)	15, 0.896, 24.3, <0.001

Table 3, continued.

Cancer	M/F	Risk (Reduction) Factors	N*, r, F**, p
Pancreas (p: 0.002, 0.071, 0.128)	F	latitude, (fish), sugar	15, 0.917, 19.4, <0.001
Prostate (p: 0.038, 0.055, 0.214)	M	latitude, alcohol, nonfat milk	15, 0.820, 7.5, 0.005

* N is the number of countries; ** F is t^2 (Student t-test); *** Farmaghetti (1994)

Table 4. Regression results for selected cancers when FSE and Scandinavian countries are omitted.

Cancer	M/F	Risk (Reduction) Factors	N*, r, F**, p
Breast	F	latitude	12, 0.934, 69, <0.001
Colon (p: 0.013, 0.070, 0.189, 0.191, 0.213)	M	alcohol, nonfat milk, sugar, animal (vegetable) energy, latitude	10, 0.976, 16, 0.009
Colon (p: 0.020, 0.215, 0.246, 0.249)	F	latitude, (vegetable) animal energy, alcohol, sugar	10, 0.971, 21, 0.003
Esophagus (p: 0.002, 0.002, 0.033, 0.036)	F	nonfat milk, (animal energy), (vegetable energy), latitude	12, 0.934, 12, 0.003
Pancreas (p: 0.005, 0.005, 0.009, 0.021, 0.029)	M	cereals, alcohol, latitude, (fish), fruit	12, 0.983, 35.0, <0.001
Prostate (p: 0.009, 0.114)	M	latitude, alcohol	12, 0.794, 7.7, 0.011
Prostate (p: 0.049, 0.127, 0.178)	M	alcohol, latitude, nonfat milk	12, 0.842, 6.5, 0.015

The dietary findings reported here are in general accord with prior findings (e.g., Potter et al., 1997; Grant, submitted). It is interesting that fish supply is generally found to have a strong inverse correlation with cancer mortality for three organ cancers (breast, kidney, and pancreas) and often second to latitude. Fish has two compounds that reduce the risk of cancer – n-3 polyunsaturated fatty acids and vitamin D (Fernandez et al., 1999). The association of the non-fat portion of milk as a risk factor for NHL was reported in Grant (2000), although in that study, leukemia, Hodgkin's lymphoma, and NHL were lumped together in the analysis. Note that gender differences in dietary links can be due to either hormonal differences or different dietary preferences. For example, Type A men prefer fat, protein and total energy, while Type A women prefer sweets (Barker et al., 1996). Also, men are more likely to smoke and consume larger quantities of alcoholic beverages than women. However, some of the dietary links, such as fruit as a risk factor for pancreatic cancer for males, should be checked using other approaches since it is counterintuitive.

It is worthwhile to explore why the findings for the remaining 5 cancers in Grant (submitted) were not confirmed in this study: cancer of the bladder, gallbladder, rectum, stomach, and corpus uteri. Bladder has a number of risk factors, including drinking habits (water and other beverages) that are hard to model. For esophagus cancer, diet and drinking and smoking habits are very important and hard to model. Rectal cancer can be transmitted by personal contact involving viruses (Sobhani et al., 2001). Stomach cancer is strongly dependent on minor dietary factors, such as salt, preservatives, and hot peppers, all of which are hard to model (Palli et al., 2001). In addition, average stomach cancer mortality rates in Europe are twice those in the U.S. Average mortality rates for corpus uteri cancer in Europe is half that in the U.S., and reproductive history has been found to play a role, with more births associated with lower risk (Parazzini et al., 1998). In general, the cancers for which a protective role for latitude could not be found tend to be the less common cancers (bladder, gallbladder, rectum, and corpus uteri) or one for which minor dietary factors play an important role.

While solar UV-B radiation appears to be a risk-reduction factor for many types of cancer, it is not clear that people should be encouraged to seek more sun exposure without taking measures to reduce the risk of melanoma and other skin cancer. Instead, it might be safer to take more vitamin D supplements. The fact that wintertime serum vitamin D (25-hydroxyvitamin D) levels are higher in northern Europe than southern

Europe (van der Wielen et al., 1995) testifies to the efficacy of vitamin D supplementation and/or the ability to obtain it from dietary fish. Also, Vieth (2001) suggests that 4,000 I.U. per day of vitamin D is safe for most people. However, it would be wise for anyone supplementing with vitamin D to have their serum vitamin D levels monitored since vitamin D can interfere with calcium metabolism (K. Sullivan, Woodacre, Calif,. personal communication).

There are a number of questions to address regarding solar UV-B radiation and vitamin D, including:

1 – what are the appropriate serum levels of vitamin D metabolites to reduce the risk of cancer?

2 – is summertime sun exposure sufficient to reduce the risk of cancer yearlong?

3 – can the ecologic study results be confirmed through case-control and cohort studies?

4 – what is the relative benefit of vitamin D in reducing the risk of cancer compared to antioxidants such as vitamins C and E and selenium?

5 – is vitamin D supplementation fully equivalent to producing vitamin D through exposure to solar UV-B radiation?

6 – at what stage of tumor development is vitamin D effective in reducing the risk of further development?

Conclusion

Strong evidence has been found for risk reduction for about a dozen cancers through solar UV-B radiation (Grant, submitted) and through latitudinal gradients for 9 cancers in Europe. Since these are ecologic studies, it should be noted that factors other than solar UV-B might explain some of the geographical dependences. For example, in Europe, milk consumption increases with latitude, while wine consumption decreases with latitude. Other factors, such as exercise, pesticides, and viruses, also play roles in the etiology of the various cancers but are difficult to model in an ecologic study, yet may have latitudinal dependences. Thus, as for any ecologic study, the results should be taken as the basis of a hypothesis that should then be checked using other epidemiologic and clinical approaches.

References

Barker, ME, Thompson, KA, McClean, SI. Do Type As eat differently? A comparison of men and women. *Appetite* 1996;26:277-86.

Colston KW, James SY, Ofori-Kuragu EA, Binderup L, Grant AG. Vitamin D receptors and anti-proliferative effects of vitamin D derivatives in human pancreatic carcinoma cells in vivo and in vitro. Br J Cancer. 1997;76:1017-20.

Devesa, SS, Grauman, DJ, Blot, WJ, Pennello, GA, Hoover, RN, Fraumeni, JF Jr, *Atlas of Cancer Mortality in the United States, 1950-1994*. NIH Publication No. 99-4564, 1999. http://cancer.gov/atlasplus/new.html

Farmaghetti, R., (ed.). *The World Almanac and Book of Facts 1995*, Funk and Waggnalls Corp., Mahwah, NJ, 1994:744-830.

Feldman D, Zhao XY, Krishnan AV. Editorial/mini-review: Vitamin D and prostate cancer. Endocrinology 2000;141:5-9.

Fernandez E, Chatenoud L, La Vecchia C, Negri E, Franceschi S. Fish consumption and cancer risk. Am J Clin Nutr. 1999;70:85-90.

Food and Agriculture Organization (FAO) of the United Nations. *Food Balance Sheets*, Rome, Italy, 1991, 1996.

Freedman DM, Zahm SH, Dosemeci M. Residential and occupational exposure to sunlight and mortality from non-Hodgkin's lymphoma:composite (threefold) case-control study. BMJ 1997;314:1451-5.

Fujioka T, Suzuki Y, Okamoto T, Mastushita N, Hasegawa M, Omori S. Prevention of renal cell carcinoma by active vitamin D3. World J Surg. 2000;24:1205-10.

Garland FC, Garland CF, Gorham ED, Young JF. Geographic variation in breast cancer mortality in the United States: A hypothesis involving exposure to solar radiation. Prev Med 1990;19:614-22.

Gorham ED, Garland CF, Garland FC. Acid haze air pollution and breast and colon cancer mortality in 20 Canadian cities. Can J Publ Health 1989;80:96-100.

Gorham ED, Garland FC, Garland CF. Sunlight and breast cancer incidence in the USSR. Int J Epidemiol 1990;19:820-4.

Garland CF, Garland FC. Do sunlight and vitamin D reduce the likelihood of colon cancer? Int J Epidemiol 1980;9:227-31.

Grant, WB. An ecologic study of dietary and smoking links to lymphoma, *Altern Med Rev*, 2000;5:563-572.

Hanchette CL, Schwartz GG. Geographic patterns of prostate cancer mortality. Cancer 1992;70:2861-9.

Hartge P, Devessa SS, Graumen D, Fears TR, Fraumeni JF. Non-Hodgkin's lymphoma and sunlight. J Natl Cancer Inst 1996;88:298-300.

John EM, Schwartz GG, Dreon DM, Koo J. Vitamin D and breast cancer risk: the NHANES I epidemiologic follow-up study, 1971-1975 to 1992. National Health and Nutrition Examination Survey. Cancer Epidemiol Biomarkers Prev 1999;8:399-406.

Kawa S, Nikaido T, Aoki Y, Zhai Y, Kumagai T, Furihata K, Fujii S, Kiyosawa K. Vitamin D analogues up-regulate p21 and p27 during growth inhibition of pancreatic cancer cell lines. Br J Cancer. 1997;76:884-9.

Lefkowitz ES, Garland CF. Sunlight, vitamin D, and ovarian cancer mortality rates in U.S. women. Int J Epidemiol 1994;23:1133-6.

Nagakura K, Abe E, Suda T, Hayakawa M, Nakamura H, Tazaki H. Inhibitory effect of 1 alpha,25-dihydroxyvitamin D3 on the growth of the renal carcinoma cell line. Kidney Int. 1986;29:834-40.

Palli D, Russo A, Decarli A. Dietary patterns, nutrient intake and gastric cancer in a high-risk area of Italy. Cancer Causes Control. 2001;12:163-72.

Parazzini F, Negri E, La Vecchia C, Benzi G, Chiaffarino F, Polatti A, Francheschi S. Role of reproductive factors on the risk of endometrial cancer. Int J Cancer. 1998;76:784-6.

Pettersson F, Colston KW, Dalgleish AG. Differential and antagonistic effects of 9-cis-retinoic acid and vitamin D analogues on pancreatic cancer cells in vitro. Br J Cancer. 2000;83:239-45.

Potter, JD. *Food, Nutrition and the Prevention of Cancer: A Global Perspective.* World Cancer Research Fund, 1997.

Schwartz GG, Hulka BS. Is vitamin D deficiency a risk factor for prostate cancer? (hypothesis). Anticancer Res 1990;10:1307-12.

Sobhani I, Vuagnat A, Walker F, Vissuzaine C, Mirin B, Hervatin F, Marmuse JP, Cremieux AC, Carbon C, Henin D, Lehy T, Mignon M. Prevalence of high-grade dysplasia and cancer in the anal canal in human papillomavirus-infected individuals. Gastroenterology. 2001;120:857-66.

Total Ozone Mapping Spectrometer - data: http://jwocky.gsfc.nasa.gov/

U.S. Dept. of Agriculture, Human Nutrition Information Service, Food and Nutrient Intakes: Individuals in Four Regions, 1977-78, NFCS 1977-78, Report No. I-3, 1985.

Van der Wielen RPJ, Lowik MRH, van den Berg H, de Groot LCPGM, Haller J, Moreiras O, van Stavereen WA. Serum vitamin D concentrations among elderly people in Europe. Lancet 1995;346:207-10.

Vieth R, Chan PC, MacFarlane GD. Efficacy and safety of vitamin D(3) intake exceeding the lowest observed adverse effect level. Am J Clin Nutr 2001;73:288-294.

Williams, GD, Stinson, FS, Lane JD, Tunson, SL, Dufour, MC. Apparent alcohol consumption by state: National, state, and regional trends, 1977-94. National Institute on Alcohol Abuse and Alcoholism, Division of Biometry and Epidemiology, Surveillance Report #39. CSR, Inc., Washington, DC, 1996, Table 2.

World Health Organization, Databank. http://www-dep.iarc.fr/cgi-bin/cgisql/who_country.idc

Zugmaier G, Jager R, Grage B, Gottardis MM, Havemann K, Knabbe C. Growth-inhibitory effects of vitamin D analogues and retinoids on human pancreatic cancer cells. Br J Cancer. 1996;73:1341-6.

A REDUCED 25-HYDROXYVITAMIN D-1α-HYDROXYLASE ACTIVITY IN HUMAN PROSTATE CANCER CELLS CAN BE RESTORED BY GENE TRANSFER

Michael V. Young[a], Lyman W. Whitlatch[a], Gary G. Schwartz[b], John N. Flanagan[a], Kerry L. Burnstein[c], Bal L. Lokeshwar[c] Michael F. Holick[a] and Tai C. Chen[a]

[a]Vitamin D, Skin and Bone Research Laboratory and Endocrine Section, Department of Medicine, Boston University Medical Center, Boston, Massachusetts 02118, [b]Comprehensive Cancer Center of Wake Forest University, Winston-Salem, North Carolina 27157, [c]University of Miami School of Medicine, Miami, Florida 33101.

INTRODUCTION

Either absorbed in the intestine from the diet or synthesized in the skin through exposure to ultraviolet (UV) radiation, vitamin D undergoes two sequential enzymatic hydroxylations, first in the liver at the 25th carbon position, producing 25-hydroxyvitamin D_3 ($25(OH)D_3$), and then in the kidney, at the 1α position, yielding the active hormone, 1α,25-dihydroxyvitamin D_3 ($1\alpha,25(OH)_2D_3$)[1]. Since the discovery of extra-renal 25(OH)D-1α-hydroxylase (1α-OHase) in keratinocytes[2], this enzyme, responsible for activating vitamin D, has been demonstrated in various tissues, including several prostate cancer cell lines and primary cultures of prostate cells as shown by our group in 1998[3]. In that report, we observed that a primary culture of prostate cells from a patient with benign prostatic hypertrophy (BPH) had a lower 1α-OHase activity level than that obtained from a normal donor, and that prostate cancer cell lines had an even lower 1α-OHase activity level than that found in BPH cells. In fact, no detectable 1α-OHase activity was found in the LNCaP cancer cell line.

LNCaP is an androgen-dependent prostate cancer cell line that retains many of the characteristics of prostate epithelial cells and that is widely employed as an *in vitro* model of human prostate cancer[4]. LNCaP cells contain the vitamin D receptor (VDR) which mediates the antiproliferative action of $1\alpha,25(OH)_2D_3$ in prostate cells. However, LNCaP cells do not have 1α-OHase activity[3] and, thus, are not responsive to the addition of $25(OH)D_3$ which requires activation by this enzyme in order to effect cell inhibition[5]. These observations presented a potential role for the transfection of 1α-OHase cDNA plasmid into prostate cancer cells, as represented by the LNCaP cell line, which when introduced could possibly restore not only 1α-OHase activity but also the antiproliferative response to $25(OH)D_3$.

METHODS

Transient Transfection: 1α-OHase or 1α-OHase-GFP (green Fluorescence Protein) fusion plasmid was obtained as described[6]. Transfection was carried out using a reagent kit (Lipofect AMINE PLUS, Life Technologies, Gaithersburg, MD) as per the protocol supplied by the manufacturer. Twenty-four hours after transfection, cells were either analyzed for 1α-OHase activity or dosed with $25(OH)D_3$ (10^{-8}, 10^{-7}, 10^{-6} M) for the cell proliferation assay. The transfection efficiency for 1α-OHase-GFP cDNA ranged from 15 to 30% as visualized by green fluorescence under a fluorescent microscope.

Enzyme Activity and Proliferation assay: 1α-OHase enzyme activity was determined by reverse phase high performance liquid chromatography (HPLC) from treated cell extracts as previously described[3]. ^3H-thymidine incorporation into DNA was used as an index of cell proliferation and was performed as described previously[6].

RESULTS and Discussion

Table 1 demonstrates that normal prostate cell cultures had an average activity of 3.0±0.36 pmol/mg protein/hr, whereas BPH and prostate cancer cultures had an average activity of 1.2±0.28 and 0.5±0.15 pmol/mg protein/hr, respectively ($p < 0.01$ between normal and BPH, and between BPH and prostate cancer cultures). Thus, compared to the normal prostate primary cultures, enzyme activity was 60 and 85% lower in the BPH and prostate cancer cell primary cultures, respectively. DU145 and PC-3 had 0.2±0.1 and 0.06±0.01 pmol/mg protein/hr, respectively. No activity was detected in LNCaP cells.

To determine the functional activity of 1α-OHase or 1α-OHase-GFP in transfected LNCaP, the efficiency of ^3H-$25(OH)D_3$ to ^3H-$1\alpha,25(OH)_2D_3$ conversion in LNCaP cells transfected with 1α-OHase-GFP was assayed. Generating 4.95±0.69 pmol/mg protein/hr of $1\alpha,25(OH)_2D_3$, transfected LNCaP cells demonstrate a 1α-OHase activity level markedly greater than that of mock transfected cells which are capable of producing only 0.07±0.07 pmol/mg protein/hr.

Next, the biological significance of transfecting the 1α-OHase gene in LNCaP cells was evaluated by assessing the antiproliferative effect of $25(OH)_2D_3$ on these cells. As in figure 1, transfected LNCaP cells show a 37±6% reduction in ^3H-thymidine incorporation of as compared to mock transfected cells, indicating a reduction in cell proliferation upon treatment with $25(OH)_2D_3$.

Table 1. Synthesis of $1\alpha,25(OH)_2D_3$ by primary cultures of human prostate cells and prostate cancer cell lines

	$1\alpha,25(OH)_2D_3$ produced (pmol/mg protein/hour)
Primary Culture	
Normal Prostate Cells	3.0 ± 0.36* (n=3)
BPH Prostate Cells	1.2 ± 0.28 (n=2)
Cancerous Prostate Cells	0.5 ± 0.15 (n=4)
Prostate Cancer Cell Line	
DU145	0.2 ± 0.1 (n=3)
PC-3	0.06 ± 0.01 (n=3)
LNCaP	Undetectable (n=3)

* Means ± SE; n: number of replicates

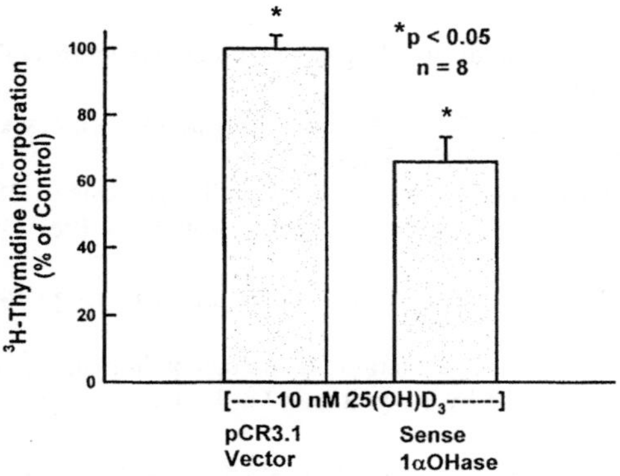

Figure 1. Effect of $25(OH)D_3$ on ^{3}H-thymidine incorporation into DNA in LNCaP cells transfected with sense 1α-OHase-cDNA or vector only.

While vitamin D, in its active form, 1α,25-dihydroxyvitamin D_3 ($1\alpha,25\ (OH)_2D_3$), most notably serves to maintain calcium and phosphorus homeostasis, it has been observed that this steroid hormone can also act as a potent inhibitory growth factor, regulating cellular proliferation and differentiation in numerous tissues. Interestingly, epidemiology demonstrates that several major risk factors for prostate cancer, old age, African American race and decreased exposure to uv radiation in northern latitudes, correlate well with low circulating concentrations of vitamin D[7]. Coupled with these epidemiological findings, biochemical and molecular studies demonstrating the presence of the vitamin D receptor (VDR) in prostate cells and the ability of the active hormone, $1\alpha,25(OH)_2D_3$, to decrease proliferation and to stimulate differentiation in primary prostate culture cells and various prostate cell lines[8,9], suggest the possibility that this hormone may be important not only to the regulation of normal prostate cell growth, but also, through either dysfunction, dysregulation, and/or decreased production may contribute to the increased and abnormal proliferation of prostate cells in the cancerous state.

Our *in vitro* results indicate that prostate cancer cells have a lower 1α-OHase activity than normal prostate primary cultures and that the introduction of the 1α-OHase gene to prostate cancer LNCaP cells enhances 1α-OHase conversion activity, thereby allowing $25(OH)D_3$ to exert an antiproliferative effect. Together, these observations not only further support an important role for vitamin D in prostate cancer development and/or progression, but also suggest a potential localized therapeutic strategy involving gene transfection. The combination of 1α-OHase gene delivery to prostatic cells with the systemic administration of 25(OH)D could potentially raise local levels of the active, antiproliferative hormone without inducing the hypercalcemic effect seen with the systemic administration of $1\alpha,25(OH)_2D_3$, and thus, could be postulated as a safe, effective method of slowing growth and causing the terminal differentiation of cancerous prostate cells.

REFERENCES

1. Jones G, Strugnell SA, DeLuca HF: Current understanding of the molecular actions of vitamin D. Physiol Rev 78:1193-1231, 1998.
2. Bikle DD, Nemanic MK, and Whitney JO, *et al*: Neonatal human foreskin keratinocytes produce 1,25-dihydroxyvitamin D_3. Biochemistry 25:1545-1548, 1986.
3. Schwartz GG, Whitlatch LW, and Chen TC, et al: Human prostate cells synthesize 1,25-dihydroxyvitamin D_3 from 25-hydroxyvitamin D_3. Cancer Epidemiol, Biomark & Prev 7:391-395, 1998.
4. Feldman D, Zhao XY, Krishnan AV: Vitamin D and prostate cancer. Endocrinology 141: 5-9, 2000.
5. Skowronski RJ, Peehl DM, Feldman D: Actions of vitamin D_3 analogs on human prostate cancer cell lines: Comparison with 1,25-dihydroxyvitamin D_3. Endocrinology 136:20-26, 1995.
6. Flanagan JN, Whitlacth LW, and Chen TC, *et al*: Enhancing 1α-hydroxylase activity with the 25-hydroxyvitamin D-1α-hydroxylase gene in cultured human keratinocytes and mouse skin. J Invest Dermatol 116:910-914, 2001.
7. Schwartz GG, and Hulka BS: Is vitamin D deficiency a risk factor for prostate cancer? (Hypothesis). Anticancer Res 10:1307-1311, 1990.
8. Schwartz GG, Oeler TA, and Uskokovic MR, *et al*: Human prostate cancer cells: Inhibition of proliferation by vitamin D analogs. Anticancer Res 14:1077-1081, 1994.
9. Chen TC, Schwartz GG, and Burnstein KL, *et al*: The in vitro evaluation of 25-hydroxyvitamin D_3 and 19-nor-1α,25-dihydroxyvitamin D_2 for prostate cancer therapy. Clin Cancer Res 6: 901-908, 2000
10. This work was supported in part by grants MO1RR00533 and RO1CA68565 from NIH, 4118PP1017 and 41211159016 from The Commonwealth of Massachusetts, and US Army PCRP-PC001404.

VITAMIN D, SUNLIGHT AND COLON CANCER: THE IMPLICATIONS FOR THE PRESENCE OF THE 1α-HYDROXYLASE IN NORMAL AND MALIGNANT COLON CANCER TISSUE

Vin Tangpricha[a], John N. Flanagan[a], Robert M. Beazley[b], Peter R. Holt[c], Mark S. Lipkin[d], Chi C. Tseng[e], Tai C. Chen[a], Michael F. Holick[a]

[a]*Vitamin D, Skin and Bone Research Laboratory. Section of Endocrinology, Diabetes & Nutrition Department of Medicine. Boston University School of Medicine. Boston, Massachusetts.*
[b]*Department of Surgery. Boston University School of Medicine. Boston, Massachusetts.*
[d]*Strong Cancer Prevention Center. The New York Hospital-Cornell Medical Center. New York, NY.*
[c]*Division of Gastoenterology. Department of Medicine. Columbia University College of Physicians and Surgeons. St. Luke's Roosevelt Hospital Center. New York, NY.*
[e]*Section of Gastroenterology. Department of Medicine. Boston University School of Medicine. Boston, Massachusetts.*

INTRODUCTION

Vitamin D deficiency has been implicated as a possible factor in the development of many cancers including cancer of the prostate, breast and colon[1,2]. Garland & Garland reported that mortality rates due to colorectal cancer in the United States are highest in areas with the least amount of solar radiation[3]. Countries with the lowest prevalence of colon cancer are located within 20 degrees of the equator[4]. Increasing latitude increases the risk of developing colon cancer with the exception of Japan where the diet is high in vitamin D from ingestion of fatty fish[5]. A prospective cohort study of 1954 men determined that men who ingested more than 150 IU of vitamin D had a 50% lower incidence of colon cancer than those who consumed less than 150 IU of vitamin D[6]. A twelve-year prospective study of over 89,000 female nurses found that the highest quintile of vitamin D intake was associated with a relative risk of 0.42 (95% CI 0.19 – 0.91) of developing colon cancer when compared to the lowest quintile of vitamin D intake[7]. A 6 year prospective study of almost 48,000 men revealed that men in the highest quintile of vitamin D intake had half the incidence rate of colon cancer than men in the lowest quintile of vitamin D intake[8]. A case-control study of 25,620 adults revealed that individuals with 25(OH)D levels greater than 20 ng/ml had one-third the risk of colon cancer compared to subjects who were vitamin D insufficient[9].

The majority of vitamin D is obtained from synthesis in the skin. The other source of vitamin D is from the diet. Vitamin D must undergo sequential hydroxylations to become active. The liver is responsible for hydroxylation in the 25 position to form 25-hydroxyvitamin D (25(OH)D), which is the major circulating form of vitamin D. The kidney hydroxylates 25(OH)D in the 1 position to form 1,25-dihydroxyvitamin D_3 (1,25$(OH)_2D_3$), the active form of vitamin D.

In vitro experiments with colon cancer cell lines have demonstrated that the addition of 1,25$(OH)_2D_3$ resulted in a dose dependant inhibition of growth and increased cellular differentiation[10,11]. 1,25$(OH)_2$D production is in the kidney is tightly regulated and thus the circulating levels of 1,25$(OH)_2$D do not significantly change even when serum 25(OH)D levels are markedly increased. Thus, it remained unclear why humans with higher 25(OH)D levels or more exposure to sunlight would be protected from developing colon cancer[3].

Cross *et al* demonstrated that Caco-2 (a human colon cancer cell line) could produce [H^3]1,25$(OH)_2D_3$ from [H^3]25$(OH)D_3$ in vitro[12]. The presence of the 1α-OHase message has been reported in human colon tissue[13,14]. We report additional evidence for the presence of the 1α-OHase message in malignant colon cancer tissue and discuss the possible roles for the presence of this enzyme in colon tissue as it relates to colon cancer.

METHODS

We obtained colon specimens from 5 anonymous donors. All five patients had colon cancer. Tissue was collected from the site of the tumor and the adjacent normal mucosa. An additional two colon cancer cell lines, SW48 and SW480 were evaluated. Samples were flash frozen and stored at –80 C until processing. RNA was extracted from the colon specimens using a SV Total RNA isolation kit (Promega, Madison, WI).

To investigate the expression levels of 1α-OHase in the colon specimens, we employed a highly sensitive real-time quantitative reverse transcriptase polymerase chain reaction (RT-PCR) method by using a 7700 sequence detector (ABI Prism systems, PE Biosystems Inc, Foster City, CA, USA). Single stranded cDNA was synthesized from the colon specimens using a Single Stranded cDNA Kit (Applied Biosystems, Foster City, CA, USA). For each sample, two hundred nanograms of cDNA was tested twice for the expression of 1α-OHase and cytokeratin-8 (CK-8), an endogenous reference gene to normalize between different samples. A set of sequence-specific primers and probe were designed and used for both 1α-OHase and CK-8 according to ABI Prism systems standard Cybr Green™ technology using the software Primer Express[15]. Using the ABI Prism 7700-sequence detector, we were able to detect the amount of fluorescent signals during each cycle of PCR. The number of signals determined the amount of message of cDNA that was bound to the fluorescent probe[15].

RESULTS

We observed the expression of 1α-OHase in the five normal colon specimens and adjacent normal mucosa obtained from 5 patients (figure 1). In three out of five specimens there was a greater than 20 fold induction of mRNA 1α-OHase expression in the cancer tissue compared with the normal adjacent mucosa. These three specimens included a tubular adenoma with severe dysplasia, an adenocarcinoma metastatic to lymph nodes and another adenocarcinoma with unknown lymph node status. The remaining two specimens, both adenocarcinomas with no lymph node invasion, showed no significant increase in mRNA expression in the colon cancer tissue. Both the human colon cancer cell lines, SW 48 and SW 480 expressed the mRNA for the 1α-OHase.

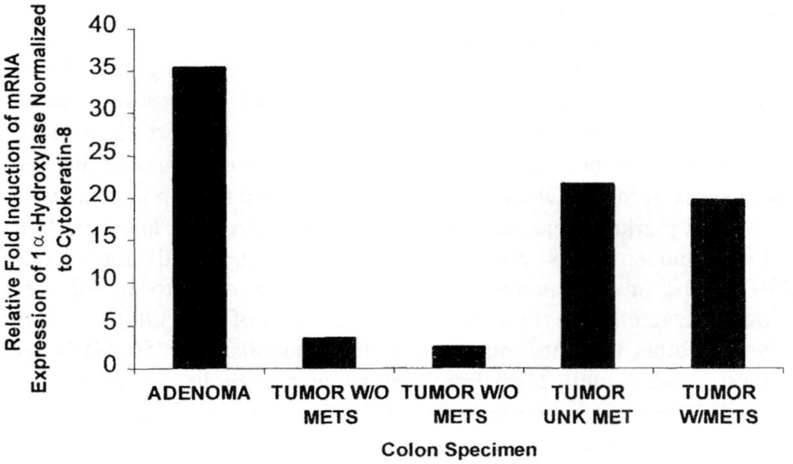

FIGURE 1

Real time RT-PCR results comparing the mRNA expression level of the 1α-OHase in malignant colon cancer tissue to normal adjacent mucosa obtained from five patients normalized to cytokeratin-8. The first bar (adenoma) represents a tubular adenoma with severe dysplasia. The second and third bars (tumor w/o mets) represent adenocarcinomas with no lymph no invasion. The fourth bar (tumor unk met) represents an adenocarcinoma with unknown lymph node status. The fifth bar (tumor w/mets) represents an adencarcinoma with lymph node invasion.

DISCUSSION

The discovery of the expression of the1α-OHase gene in the human colon provides a possible explanation to why vitamin D and ultraviolet radiation could protect humans from developing colon cancer[13]. The 1α-hydroxylase provides the colon cell the machinery to produce the antiproliferative hormone, $1\alpha,25(OH)_2D_3$ locally which in turn could act as a inhibitory autocrine/paracrine factor to maintain normal proliferation. Higher circulating levels of $25(OH)D_3$ may be important for the 1α-OHase in the colon to maximize the local production of this growth regulating hormone. Patients with lower circulating levels of 25(OH)D could have higher risk for colon cancer due to lack of adequate substrate to produce $1\alpha,25(OH)_2D_3$ to inhibit cell growth. Studies have suggested that colonic cell proliferation in patients at risk for developing colon neoplasia can be reduced with higher $25(OH)D_3$ levels[16]. It has been observed that the risk of colon cancer was decreased three fold in people who had 25(OH)D levels of greater than 20 ng/ml[16].

There are differences in the expression level of the 1α-OHase. We found that in an adenoma with severe dysplasia and an adenocarcinoma with lymph node metastasis had significantly elevated expression of 1α-OHase mRNA whereas adenocarcinoma with only local invasion showed nearly the same expression level compared with normal adjacent mucosa (figure 1). The reasons for this difference in 1α-OHase mRNA expression is unknown. The ability of the extra-renal 1α-OHase to convert $25(OH)D_3$ to $1\alpha,25(OH)_2D_3$ has been established in colon cancer cell lines[3]; however, the effectiveness of the1α-OHase enzyme in colon tissue in vivo is unknown.

The marked increase in the expression of the1α-OHase could be due to a resistance to its product $1,25(OH)_2D_3$, enhanced degradation of $1,25(OH)_2D_3$, or to a upregulation of the mechanism to inhibit cellular proliferative activity i.e. to increase its production of $1,25(OH)_2D_3$. Further analysis of other enzymes important to the metabolism of $25(OH)D_3$ need to be investigated, namely 24-hydroxylase (a major catabolic enzyme) .

In conclusion, we report that the presence of 1α-OHase in malignant and normal human colon tissue and in two human colon cancer cell lines. This finding provides an explanation for the protective mechanism of sunlight. Ultraviolet radiation increases circulating 25(OH)D levels which results in an increased substrate available for the local production of $1\alpha,25(OH)D_3$ in the colon by the 1α-OHase. The colon 1α-OHase may produce an important hormone, $1\alpha,25(OH)_2D$, that is necessary for the maintenance of optimal colon cellular health. Increases in the 1α-OHase expression could represent an adaptive response to colon cancer regulation. The explanation for the differences in 1α-OHase expression among colon cancer tissue and normal colon mucosa requires further investigation.

Acknowledgement

This work was supported in part by a grant from the Indoor Tanning Association

1 Gorham ED, Garland CF, Garland FC. Acid haze air pollution and breast and colon cancer mortality in 20 Canadian cities. *Can J Public Health* 1989;**80**(2):96-100.

[2] Schwartz GG, Hulka BS. Is vitamin D deficiency a risk factor for prostate cancer? *Anticancer Res* 1990;**10**(5A):1307-11.

[3] Garland CF, Garland FC. Do sunlight and vitamin D reduce the likelihood of colon cancer? *Int J Epidemiol* 1980;9(3):227-31.

[4] Muir C. Waterhouse J. Mack T. Cancer Incidence in Five Continents, vol 5. IARC Scientific Publication No. 88. Lyon: International Agency for Research on Cancer, 1987.

[5] Gorham ED. Int Journ Epidemiology. 1989.

[6] Kampman E. Giovannucci E. van't Veer P. Rimm E. Stampfer MJ. Colditz GA. Kok FJ. Willett WC. Calcium, vitamin D, dairy foods, and the occurrence of colorectal adenomas among men and women in two prospective studies. Am J Epidemiol 1994; 139: 16-29.

[7] Martinez ME, Giovannucci EL, Colditz GA, et al. Calcium, vitamin D, and the occurrence of colorectal cancer among women. *J Natl Cancer Inst* 1996;**88**(19):1375-82.

[8] Kearney J, Giovannucci E, Rimm EB, et al. Calcium, vitamin D, and dairy foods and the occurrence of colon cancer in men. *Am J Epidemiol* 1996;**143**(9):907-17.

[9] Garland CF, Comstock GW, Garland FC, Helsing KJ, Shaw EK, Gorham ED. Serum 25-hydroxyvitamin D and colon cancer: eight-year prospective study [see comments]. *Lancet* 1989;2(8673):1176-8.

[10] Cross HS. Farsoudi KH. Peterlik M. Growth inhibition of human colon adenocarcinoma-derived Caco-2 cells by 1,25-dihydroxyvitamin D3 and two synthetic analogs. Nauyn-Schmiedebergs Archives of Pharmacology. 347(1):105-110, 1993 Jan.

[11] Zhao X. Feldman D. Regulation of vitamin D receptor abundance and responsiveness during differentiation of HT-29 human colon cancer cells. Endocrinology. 132(4):1808-14, 1993 Apr.

[12] Cross HS, Bareis P, Hofer H, Bischof MG, Bajna E, Kriwanek S, Bonner E, Peterlik M. 25-Hydroxyvitamin D_3-1-alpha-hydroxylase and vitamin D receptor gene expression in human colonic mucosa is elevated during early cancerogenesis. Steroids 2001 Mar-May;66(3-5):287-92.

[13] Tangpricha V. Flanagan JN. Whitlatch LW. Tseng CC. Chen TC. Holt PR. Lipkin MS. Holick MF. 25-hydroxyvitamin D_3 in normal and malignant colon tissue. The Lancet. 2001;367(9269) 1673-4.

[14] Bareis P, Bises G, Bischof MG, Cross HS, Peterlik M. 25-hydroxy-vitamin d metabolism in human colon cancer cells during tumor progression. *Biochem Biophys Res Commun* 2001;**285**(4):1012-7.

[15] Brink N. Szamel M. Young AR. Wittern KP. Bergemann J. Comparative quantification of IL-1beta, IL-10, IL-10r, TNFalpha and IL-7 mRNA levels in UV-irradiated human skin in vivo. *Inflammation Research* 2000;49(6):290-6.

[16] Holt PR. Arber N. O'Connor J. Halmos B. McGlynn C. Moss SF. Pou R. Yang K. Fan K. Newmark H. Lipkin M. Serum 25 hydroxyvitamin D_3 inhibits proliferation of colonic epithelial cells in subjects at high risk for colon neoplasia. *Gastroenterology* 2000;118:A276.

Photoimmunology

ULTRAVIOLET A RADIATION-INDUCED SIGNAL TRANSDUCTION IN HUMAN KERATINOCYTES.

Jean Krutmann and Susanne Grether-Beck

Clinical and Experimental Photodermatology, Department of Dermatology, Heinrich-Heine-University, D-40225 Düsseldorf, Germany

INTRODUCTION

The ultraviolet (UV) radiation solar spectrum that reaches the earth consists of short wavelength UVC (< 280 nm), intermediate wavelength UVB (280-320 nm) and long wavelength UVA (320-400 nm) radiation. Under physiological conditions, human cells are exclusively exposed to UVB and UVA radiation, because ozone in the stratosphere completely absorbs wavelengths < 290 nm. The amount of UVA radiation reaching the earth's surface is approximately 20 times greater than that of UVB radiation, and solar UVA radiation contributes to photoageing and photocarcinogenesis of human skin and is responsible for triggering the most frequent photodermatoses, that is polymorphous light eruption (Urbach, 1992).

Human skin is additionally exposed to UVA radiation from artificial sources including high-intensity UVA irradiation devices, which in former years have mainly been used for cosmetical purposes, but are currently employed at an increasing rate for the treatment of inflammatory skin diseases such as atopic dermatitis, urticaria pigmentosa and connective tissue diseases (Krutmann et al., 1992; Stege et al., 1996; Stege et al., 1997). Analysis of the photobiological and molecular effects which are exerted in human skin cells by UVA radiation is thus of relevance for both, beneficial and detrimental effects associated with UVA irradiation.

Studies addressing the effects of UV radiation on mammalian cells have primarily focused on UVC, and to some extent, UVB irradiation. The majority of these studies used low pressure mercury germicidal lamps with an emission spectrum that peaks at 254 nm (Karin and Herrlich, 1989). These studies did not assess epidermal keratinocytes, which are the primary cellular target for UV radiation, and at the doses employed (> 20 J/m^2) most of the irradiated cells were metabolically active, but reproductively dead. It is without question that these studies have greatly helped to advance our knowledge about the molecular consequences which are induced in mammalian cells upon environmental stress. Analysis of this so-called UV response has shown that UVC and UVB irradiation of mammalian cells result in increased gene expression, and the cis- and transacting genetic elements responsible for gene expression by UVB or UVC radiation have been well characterized (reviewed in Bender et al., 1997). In order to fully understand the impact of solar radiation on human skin, it is also necessary to analyze the molecular mechanisms by which physiologically relevant doses of UVA radiation induce gene expression in human keratinocytes. These studies are of particular interest, because biological effects induced by short wavelength and long wavelength UV radiation involve different chromophores leading to different photobiological effects (Kochevar, 1995). It is therefore likely that the mechanisms by which UVA radiation induces transcriptional activation of human genes differ from those induced by UVC or UVB radiation.

The studies summarized in this review support this concept and indicate the existence of a unique and previously unrecognized signal transduction pathway, by which UVA radiation induces gene expression in human keratinocytes. This pathway appears to be mediated through the generation of singlet oxygen and to involve ceramides, mitochondria and Activator Protein-2 (AP-2).

Photobiological aspects of UVA radiation-induced gene expression

It is now well established that UVA radiation is capable of inducing gene expression in mammalian cells including epidermal (e.g. keratinocytes) and dermal (e.g. fibroblasts, skin infiltrating T-cells) cells. These genes include the heme-oxygenase-1 gene (HO-1), various cytokines, the adhesion molecule Intercellular Adhesion Molecule-1 (ICAM-1), matrix metalloproteinases, the FAS-Ligand molecule (CD95-L), and CL100, a non-receptor-type protein-tyrosine phosphatase. With the exception of HO-1 and CL-100, these genes are not specifically induced by UVA radiation, but may be upregulated to similar extents by UVB or UVC radiation as well. A growing number of recent studies indicates, however, that UVA radiation-induced expression of these genes involves photobiological mechanisms that differ from those used by UVB or UVC radiation.

Ultraviolet A radiation-induced expression of the HO-1, MMP-1, ICAM-1, FAS-L, and possibly also selected cytokine genes involve the generation of reactive oxygen species (Tyrrell, 1996; Krutmann and Grewe, 1995). In transformed human keratinocytes, UVA radiation-induced ICAM-1 expression was only observed in cells that had been pretreated with an inhibitor of de novo glutathione synthesis, indicating that UVA radiation-induced ICAM-1 expression occurred through a mechanism which depended on the thiol-status of the irradiated cells (Krutmann and Grewe, 1995; Morita et al., 1997). This effect was specific for UVA radiation-induced ICAM-1 expression, because UVB radiation increased ICAM-1 gene expression regardless of whether these cells had been depleted of endogenous glutathione or not. A similar induction pattern was also observed for selected cytokine genes including Interleukin (IL)-1α and IL-6. It should be noted that UVA radiation-induced expression of IL-8 and TNFα expression did not require lowering of endogenous glutathione levels (Morita et al., 1996).

The observation that UVA radiation-induced ICAM-1 expression in human keratinocytes is modulated through cellular glutathione levels points to a prominent role of reactive oxygen species (ROIs) in this system. Reactive oxygen species include hydrogen peroxide, lipid peroxide, singlet oxygen, hypochlorous acid, and N-chloramine compounds (Punchard and Kelly, 1996). Cardinal features of ROIs are their high reactivity, short half-lives and short migration distances. Methods to measure the generation of ROIs such as chemiluminescence or electron spin resonance have certain experimental restrictions, and their applicability to complex biological systems is therefore very limited. Experimental evidence concerning the role of ROIs in a specific biological process is thus circumstantial and based on a combination of strategies directed at enhancing or quenching ROI-mediated effects. A number of recent studies employing these methods has provided convincing evidence that UVA radiation-induced gene expression in mammalian cells is mediated through the generation of singlet oxygen. Induction of the HO-1 gene in UVA-irradiated human dermal fibroblasts was efficiently blocked by singlet oxygen quenchers such as sodium azide and L-histidine, it was enhanced if the cells were irradiated in the presence of deuterium oxide in which singlet oxygen has a longer halflife, and it was mimicked upon stimulation with Rose Bengal plus light, a known singlet oxygen generating system (Basu-Modak and Tyrrell, 1993). These data indicated that singlet oxygen generation is important for UVA radiation-induced gene expression. This conclusion was corroborated and extended in subsequent studies in which singlet oxygen was found to mediate UVA radiation-induced MMP-1 expression in human dermal fibroblasts (Wlaschek et al., 1995), UVA radiation-induced ICAM-1 expression in human epidermal keratinocytes (Grether-Beck et al., 1996), and UVA radiation-induced FAS-L upregulation in skin-infiltrating human T-helper cells (Morita et al., 1997). Singlet oxygen thus seems to be the primary effector in UVA radiation-induced gene expression. Analysis of the molecular mechanisms involved in UVA-induced gene expression have therefore also provided insight into the mechanisms by which singlet oxygen activates transcription factors and induces promotor activation.

Molecular aspects of UVA radiation-induced gene expression

Ultraviolet A radiation and singlet oxygen-induced promoter activation have recently been studied by using promoter constructs based on the human ICAM-1 promoter (Grether-Beck et al., 1996). Both stimuli were found to activate transcription factor AP-2, and deletion of the putative AP-2 binding site abrogated ICAM-1 promoter activation in UVA-irradiated or singlet oxygen-stimulated human keratinocytes. It has therefore been concluded that the AP-2 site served as both the UVA-responsive and the singlet oxygen-responsive element of the human ICAM-1 gene. Deletion of the AP-2 site did not abrogate UVB radiation-induced ICAM-1 promoter activation, indicating that UVB radiation and UVA radiation-induced gene expression do not only differ with regard to the photobiological mechanisms involved, but also depend on different molecular mechanisms (Krutmann and Grewe, 1995; Grether-Beck et al., 1996).

The critical role of AP-2 in UVA radiation-induced gene expression is further emphasized by very recent studies, in which the functional role of AP-2 and its alternative splice product AP2-B was studied for UVA radiation-induced gene expression. AP-2B results from alternative usage of exon 5 and differs from AP-2 in the C-terminus (Bauer et al., 1994). It lacks the dimerization domain required for DNA binding, but contains the transactivation domain and part of the DNA-binding domain (Buettner et al, 1993). Studies addressing the functional relevance of AP-2B have revealed that it serves as a dominant negative regulator of AP2 transactivation (Buettner et al., 1993; Duan and Clemmons, 1995). This concept is in line with the recent observation that in UVA-irradiated human epidermal keratinocytes, increased ICAM-1 expression was always observed at time points when the ratio of AP2/AP2-B was shifted towards AP-2 (Grether-Beck et al., 1997). Similarly, cells that had been stably transfected to overexpress AP-2 (Buettner et al., 1993) showed a 1000-fold increased ICAM-1 expression, as compared to the parental clone, whereas AP-2B overexpression was associated with a failure of cells to upregulate ICAM-1 expression above baseline levels following UVA irradiation (Grether-Beck et al., 1997). Additional data imply that this pattern of UVA radiation-induced gene expression in AP2 versus AP-2B overexpressing cells is not specific for ICAM-1, but may be observed for other UVA radiation- and possibly also singlet oxygen-inducible genes as well. The promoter regions of almost all of these genes contain one or several putative AP-2 binding sites, which in case of the HO-1 gene share a strong homology to the AP-2 site present in the human ICAM-1 promoter. It is thus likely that activation of transcription factor AP-2 may constitutes a general mechanism by which UVA radiation and/or singlet oxygen induce gene expression in human keratinocytes.

The role of ceramides and mitochondria in UVA radiation-induced signal transduction

Ceramide is a key component of intracellular stress responses. We have recently demonstrated that ceramide formation mediates UVA radiation-induced ICAM-1 and AP-2 activation (Grether-Beck et al., 2000). Similar to UVA radiation, ceramide stimulation of human keratinocytes induced ICAM-1 mRNA expression and activated the ICAM-1 promoter through transcription factor AP-2. Accordingly, ceramide activated AP-2 and ceramide-induced ICAM-1 reporter gene activation was abrogated through deletion of the AP-2 binding site. UVA radiation increased the level of ceramide in keratinocytes and inhibition of sphingomyelin synthesis in these cells prevented UVA radiation-induced ICAM-1 expression. Hitherto, two pathways have been identified for ceramide accumulation: hydrolysis from sphingomyelin through neutral and acid sphingomyelinases and de novo synthesis by ceramide synthase. UVA radiation did not activate any of these enzymes. Ceramide generation in UVA-irradiated cells, however, was inhibited by singlet oxygen quenchers and mimicked in unirradiated cells by a singlet oxygen generating system. In addition, UVA radiation and singlet oxygen both generated ceramide in protein-free, sphingomyelin-containing liposomes. This study indicates that singlet oxygen triggers a third, non-enzymatic mechanism of ceramide formation. In very recent studies we have also been able to define the mechanism by which UVA radiation-induced second messenger ceramides are capable of activating transcription factor AP-2 (Felsner et al. 2001). In initial studies we have observed that ceramide-induced AP-2 activation and ICAM-1 mRNA expression could be inhibited upon treatment of KC with inhibitors of the mitochondrial (mt) electron transport chain, indicating that mitochondria may serve a role in UVA radiation-induced signal transduction in human keratinocytes. Ceramide-induced apoptosis is known to involve the release of cytochrome C from the mitochondria through their voltage dependent ion channel (vdac; = permeability transition pore) into the cytoplasm. In our studies, stimulation of KC with C_2- or C_6-ceramide at concentrations which did not cause apoptosis but AP-2 activation and ICAM-1 induction, also lead to the release of mt cytochrome C into the cytoplasm of KC (Western blot analysis). Even more important, treatment of ceramide-stimulated KC with substances, which prevent opening of the mt vdac (e.g. Carboxyatractyloside and Bongkretic acid) inhibited (i) the release of mt cytchrome C into the cytoplasm, (ii) AP-2 activation (GEMSA) and (iii) increased ICAM-1 transcription (RT-PCR). This effect was specific because under identical experimental conditions IFN-g-induced ICAM-1 mRNA expression was not prevented. Mitochondrial cytochrome C, after its release into the cytoplasm of KC, might thus have the capacity to activate transcription factor AP-2 and thereby induce gene transcription. In order to test this possibility rh AP-2 was coincubated in-vitro with cytochrome C. Photometric analysis revealed that this interaction was associated with a redox change, i.e. cytochrome C was reduced, whereas AP-2 was oxidized. The latter effect was of functional relevance, because oxidized AP-2 exhibited a greatly increased binding capacity for AP-2 consensus sites (GEMSA). These studies indicate a previously unrecognized role of mitochondria in transcription factor activation and gene regulation in mammalian cells.

REFERENCES

Basu-Modak, S. and Tyrrell, R.M. (1993) Singlet oxygen: A primary effector in the ultraviolet A/ near-visible light induction of the human heme oxygenase gene. Cancer Res. 53, 4505-4519.

Bauer, R., Imhof, A., Pscherer, A., Kopp, H., Moser, M., Seegers, S., Kerscher, M., Tainky, M.A., Hofstaedter, F., Buettner, R. (1994). The genomic structure of the human AP-2 transcription factor. Nucl. Acids. Res. 22, 1413-1420.

Bender, K., Blattner, C., Knebel, A., Iordanov, M., Herrlich, P., Rahmsdorf, H.J. (1997). UV-induced signal transduction. Photochem. Photobiol B 37, 1-17.

Buettner, R., Kannan, P., Imhof, A., Bauer, R., Yim, S.O.,Glockshuber, R., van Dyke, M.W., Tainsky, M.A. (1993). An alternatively spliced mRNA from the AP-2 gene enodes a negative regulator of transcriptional activation by AP-2. Mol. Cell. Biol.13, 4174-4185.

Duan, C. and Clemmons, D.R. (1995). Transcription factor AP-2 regulates human insulin-like growth factor binding protein-5 gene expression. J. Biol. Chem. 270, 24844-24851.

Grether-Beck, S., Olaizola-Horn, S., Schmitt, H., Grewe, M., Jahnke, A., Johnson, J.P., Briviba, K., Sies, H., Krutmann, J. (1996). Activation of transcription factor AP-2 mediates UVA radiation-and singlet oxygen-induced expression of the human intercellular adhesion molecule 1 gene. Proc. Natl. Acad. Sci. USA 93:14586-14591.

Felsner, I., Grether-Beck, S., Schmitt-Brenden, H., Krutmann, J. (2001) Mitochondrial cytochrome C mediates ceramide-induced gene expression in human keratinocytes through oxidation of transcription factor AP-2.

Abstract. Arch. Dermatol. Res. 403, 30.

Grether-Beck, S., Schmitt, H., Grewe, M., Buettner, R., Krutmann, J. (1997) The balance between expression of transcription factor AP2 and its alternative splice product AP2B controls ultraviolet A radiation (UVAR)-induced gene expression in human cells. J. Invest. Dermatol. 108, 594.

Grether-Beck, S., Bonizzi, G., Schmitt-Brenden, H., Felsner, I., Timmer, A., Sies, H., Johnson, J.P., Piette, J., Krutmann, J. (2000) Non-enzymatic triggering of the ceramide signalling cascade by solar UVA radiation. EMBO J. 19, 5793-5800.

Karin, M., and Herrlich, P. (1989) Cis-and trans-acting elements responsible for induction of specific genes by tumor promotors, serum factors, and stress. In: Genes and Siganl Transduction in Multistage Carcinogenesis, Colburn, N.H., ed., Marcel Dekker, New York, pp. 415-440.

Kochevar, I.E. (1995). Primary processes in photobiology and photosensitation. In: Photoimmunology. Krutmann, J., and Elmets, C.A., eds. (Blackwell Science Ltd.) pp. 19-33.

Krutmann, J., Czech W, Diepgen T, Niedner R, Kapp A, Schöpf, E. (1992). High-dose UVA1 therapy in the treatment of patients with atopic dermatitis. J Am Acad Dermatol 26, 225-230.

Krutmann, J., and Grewe, M. (1995) Involvement of cytokines, DNA damage, and reactive oxygen intermediates in ultraviolet radiation-induced modulation of intercellular adhesion molecule-1 (ICAM-1) expression. J. Invest. Dermatol. 105, 67S-70S.

Morita, A., Grewe, M., Grether-Beck, S., Olaizola-Horn, S., Krutmann, J. (1997). Induction of proinflammatory cytokines in human epidermoid carcinoma cells by in vitro ultraviolet A1 irradiation. Photochem. Photobiol. 65, 630-635.

Morita, A., Werfel, T., Stege, H., Ahrens, C., Karmann, K., Grewe, M., Grether-Beck, S., Ruzicka, T., Kapp, A., Klotz, L.O., Sies, H., Krutmann, J. (1997) Evidence that singlet oxygen-induced human T-helper cell apoptosis is the basic mechanism of ultraviolet-A radiation phototherapy. J. Exp. Med. 186, 1763-1768.

Punchard, N.A., and Kelly, F.J. (1996). Free radicals. In: Free Radicals-A practical approach eds. N.A. Punchard and F.J. Kelly. (IRL press, Oxford, New York, Tokyo) pp.1-8.

Stege, H., Humke, S., Berneburg, M. et al., 1996. High-dose UVA1 therapy in the treatment of patients with localized scleroderma (morphea). J. Am. Acad. Dermatol. 36, 938-944.

Stege H., Schöpf, E., Ruzicka, T., Krutmann, J. (1996). High-dose UVA1 for urticaria pigmentosa. Lancet 347, 64.

Tyrrell, R.M. (1996). Activation of mammalian gene expression by the UV component of sunlight -from models to reality. Bioassays 18, 139-148.

Urbach, F., ed. (1992) Biological responses to ultraviolet A radiation. Valdenmar, Overland Park, Kansas.

Wlaschek, M. Briviba, K., Stricklin, G.P., Sies, H., and Scharfetter-Kochanek, K. (1995). Singlet oxygen may mediate the ultraviolet A-induced synthesis of interstitial collagenase. J. Invest. Dermatol. 104, 194-198.

UVB AND UVA-1 EFFECTS ON HIV ACTIVATION

Ponciano D. Cruz, Jr., M.D.
The University of Texas Southwestern Medical Center, Dallas, Texas

INTRODUCTION

Concern about safety of phototherapy for patients infected with the human immunodeficiency virus (HIV) exists because spread of the infection remains unabated, because HIV infection is associated with skin diseases responsive to phototherapy (including psoriasis, eosinophilic folliculitis, and pruritus), and because ultraviolet (UV) radiation has been shown to activate HIV and promote its proliferation.

EVIDENCE FOR UV-INDUCED HIV ACTIVATION

In vitro studies provide the clearest illustration of how UV radiation can promote HIV infection (1-5). In studies using HeLa cells in which the long terminal repeat promoter of HIV was fused to the reporter chloramphenicol acetyl transferase (CAT) gene, CAT activity following exposure to different UV wavelengths correlated with an upregulatory effect on HIV activation (5). UVC (<290 nm) stimulated CAT activity by 50- to 150-fold (2;5). PUVA, UVB+UVA2 (280-340nm) and sunlight also enhanced CAT activity, although by lesser magnitudes. By contrast, neither γ-radiation (<100nm) nor UVA1 (340-400nm) promoted CAT activity (5). Moreover, Stanley and colleagues (3) showed UVC, UVB or sunlight to activate HIV itself (not just a HIV LTR construct) in latently infected cultured cells. Finally, 3 groups of investigators utilized different transgenes in mice to test the UV effect on HIV (6-8). Cavard and associates (6) used bacterial β-galactosidase linked to the HIV-LTR; Morrey and colleagues (7) employed firefly luciferase linked to the HIV-LTR; and Vogel and associates (8) utilized the HIV LTR and *tat* genes directly. All three groups showed UV exposure to induce cutaneous expression of the protein encoded by the respective transgenes. Furthermore, their results using various UV wavelengths echoed the findings from *in vitro* studies (2-5) showing UVC, UVB, sunlight, and particularly PUVA (but not UVA) to be effective activators of HIV (6-8).

In my laboratory, we have developed an assay for measuring HIV-1 gag gene in skin of HIV-positive patients. This assay was adapted from widely available commercial plasma HIV assays. Our studies show that treatment of human skin with a single suberythmogenic dose of UVB leads to a 6-10 fold increase in the number of HIV copies (manuscript submitted for publication). This upregulated expression of HIV in human skin can be inhibited by known blockers of NFκB activation, including oligonucleotide decoys and pharmacologic agents like Na salicylate. By contrast, exposure of skin of HIV-positive patients to high-dose UVA-1 has no effect on cutaneous viral load.

POSTULATED MECHANISMS OF UV-INDUCED HIV ACTIVATION

The responsible mechanisms have not yet been precisely identified, although signal transduction pathways involving NFκB as cited above and p38 MAP kinase (9-11) have been implicated. In addition, direct effects of UV radiation on DNA may also contribute. UV-induced DNA damage leads to repair, which then activates an array of genes including viral genes integrated in the host genome (12). Circumstantial evidence for this scenario is the capacity of genotoxic agents such as mitomycin C, cis-platinum and anthracycline (as well as UVC, UVB and PUVA to activate HIV (2;13;14); whereas UVA, which is not genotoxic, does not activate the virus (3).

IS PHOTOTHERAPY SAFE FOR HIV-POSITIVE PATIENTS?

Most published clinical studies (15-22) on serial HIV-positive patients treated with UVB or PUVA suggest that, despite the foregoing laboratory evidence, phototherapy remains a safe modality for seropositive individuals. One explanation to reconcile the discrepancy maybe that UVB- or PUVA-induced increments in skin HIV need not lead to correspondingly sustained increments in plasma HIV, either because of low skin viral loads to begin or counteracting effects of UVB or PUVA's propensity to induce selective apoptosis of HIV-bearing lymphocytes that circulate through skin during phototherapy.

We are currently completing a randomized and controlled study evaluating the effect of UVB and PUVA therapy on both skin and plasma HIV burdens. Until more definitive information regarding the safety of phototherapy in HIV-infected individuals is clarified by studies like this, UVB and PUVA therapy should continue to be a viable therapeutic option in HIV-positive patients with UV-responsive skin diseases. However, the following questions should be raised at the time of decision making for treatment:

1. *Is the skin disease UV-responsive?* If yes, consider

phototherapy.

2. *Do alternative therapies offer less risk to the patient?* If yes, it may be judicious to first try alternative treatments.
3. *Is anticipated improvement in morbidity following phototherapy enough to justify potential risks?* If yes, proceed with phototherapy.
4. *Is the patient sufficiently reliable to show up for treatment visits?* If yes, proceed with phototherapy. If no, consider other treatments.
5. *Are there other contraindications to phototherapy (e.g., medication that confers photosensitivity)?* If yes, weigh the risk/benefit ratio.

UVA-1 phototherapy is definitely safe with respect to the risk of HIV activation, however, the light source is not widely available in the U.S.

REFERENCES

(1) Folks TM, Justement J, Kinter A, Dinarello CA, Fauci AS. Cytokine-induced expression of HIV-1 in a chronically infected promonocyte cell line. Science 1995; 228:800-802.

(2) Valerie K, Delers A, Bruck C, Thiriart C, Rosenberg H, Debouck C et al. Activation of human immunodeficiency virus type 1 by DNA damage in human cells. Nature 1988; 333:78-81.

(3) Stanley SK, Folks TM, Fauci AS. Induction of expression of human immunodeficiency virus in a chronically infected promonocytic cell line by ultraviolet irradiation. AIDS Res Hum Retroviruses 1989; 5:375-384.

(4) Zmudzka BZ, Strickland AG, Miller SA, Valerie K, Dall'Acqua F, Beer JZ. Activation of the human immunodeficiency virus promoter by UVA radiation in combination with psoralens or angelicins. Photochem Photobiol 1993; 58:226-232.

(5) Beer JZ, Olvey KM, Lee W, Zmudzka BZ. Reassessment of the differential effects of ultraviolet and ionizing radiation on HIV promoter: the use of cell survival as the basis for comparisons. Photochem Photobiol 1994; 59(6):643-649.

(6) Cavard C, Zider A, Vernet M, Bennoun M, Saragosti S, Grimber G et al. *In vivo* activation by ultraviolet rays of the human immunodeficiency virus type 1 long terminal repeat. J Clin Invest 1990; 86:1369-1374.

(7) Morrey JD, Bourn SM, Bunch TD, Jackson MK, Sidwell RW, Barrows LR et al. *In vivo* activation of human immunodeficiency virus type 1 long terminal repeat by UV

type A (UV-A) light plus psoralen and UV-B light in the skin of transgenic mice. J Virol 1991; 65(9):5045-5051.

(8) Vogel J, Cepeda M, Tschachler E, Napolitano LA, Jay G. UV activation of human immunodeficiency virus gene expression in transgenic mice. J Virol 1992; 66(1):1-5.

(9) Taher MM, Baumgardner T, Dent P, Valerie K. Genetic evidence that stress-activated p38 MAP kinase is necessary but not sufficient for UV activation of HIV gene expression. Biochem 1999; 38:13055-13062.

(10) Taher MM, Hershey CM, Oakley JD, Valerie K. Role of the p38 and MEK-1/2/p42/44 MAP kinase pathways in the differential activation of human immunodeficiency virus gene expression by ultraviolet and ionizing radiation. Photochem Photobiol 2000; 71(4):455-459.

(11) Taher MM, Oakley JD, Hershey CM, Valerie K. Activation of NF-κB and p38 MAP kinase is not sufficient for triggering efficient HIV gene expression in response to stress. Biochem 2000; 39:1709-1715.

(12) Stein B, Rahmsdorf HJ, Steffen A, Litfin M, Herrlich P. UV-induced DNA damage is an intermediate step in UV-induced expression of human immunodeficiency virus Type 1, collagenase, c-fos, and metallothionein. Molecular & Cellular Biology 1989; 9:5169-5181.

(13) Morison WL. Phototherapy and Photochemotherapy. New York: Raven Press, 1991.

(14) Lim HW, Soter NA. Clinical Photomedicine. New York: Marcel Dekker, 1993.

(15) Buchness MR, Lim HW, Hatcher VA, Sanchez M, Soter NA. Eosinophilic pustular folliculitis in the acquired immunodeficiency syndrome. N Eng J Med 1988; 318(18):1183-1186.

(16) Ranki A, Puska P, Mattinen S, Lagerstedt A, Krohn K. Effect of PUVA on immunologic and virologic findings in HIV-infected patients. J Am Acad Dermatol 24, 404. 1991.
Ref Type: Journal (Full)

(17) Pardo RJ, Bogaert MA, Penneys NS, Byrne GE, Ruiz P. UVB phototherapy of the pruritic papular eruption of the acquired immunodeficiency syndrome. J Am Acad Dermatol 1992; 26:423-428.

(18) Meola T, Soter NA, Ostreicher R, Sanchez M, Moy JA. The safety of UVB phototherapy in patients with HIV infection. J Am Acad Dermatol 1993; 29:216-220.

(19) Horn TD, Morison WL, Farzadegan H, Zmudzka BZ, Beer JZ. Effects of psoralen plus UVA radiation (PUVA) on HIV-1 in human beings: a pilot study. J Am Acad Dermatol 1994; 31:735-740.

(20) Fotiades J, Lim HW, Jiang SB, Soter NA, Sanchez M, Moy J. Efficacy of ultraviolet B phototherapy for psoriasis in patients infected with human immunodeficiency virus. Photodermatol Photoimmunol Photomed 1995; 11:107-111.

(21) Fotiades J, Soter NA, Sanchez MR, Moy JA. A three-year follow-up evaluation on 28 HIV-positive patients treated with ultraviolet B (UVB) phototherapy. J Invest Dermatol 1995; 104:660a.

(22) Breuer-McHam J, Marshall G, Adu-oppong A, Goller M, Mays S, Berger T et al. Alterations in HIV expression in AIDS patients with psoriasis or pruritus treated with phototherapy. J Am Acad Dermatol 1999; 40(1):48-60.

CLINICAL RELEVANCE OF UVA1-INDUCED T-CELL APOPTOSIS

Akimichi Morita, and *Jean Krutmann
*Department of Dermatology, Nagoya City University Medical School, and *Clinical and Experimental Photodermatology, Department of Dermatology, Heinrich-Heine-University of Düsseldorf, Germany*

Keywords: Phototherapy - Ultraviolet A radiation - Apoptosis - Singlet Oxygen - FAS - Atopic Dermatitis - Cutaneous T cell lymphoma

1. ATOPIC DERMATITIS

Introduction

Ultraviolet (UV) radiation has been used for decades in the treatment of patients with a variety of T-cell-mediated skin diseases such as atopic dermatitis. Among the different phototherapeutic modalities that are currently available recent interest has focused on the use of irradiation device that allow exposure of human skin to wavelengths in the longer UVA range, that is UVA-1 (340-400 nm) [1]. The almost complete absence of erythemogenic wavelengths shorter than 340 nm allow exposure of human skin to single doses of up to 130 J/m^2. These clinical developments have stimulated studies about the mechanisms by which UVA 1 phototherapy works.

The mechanism of action underlying the therapeutic effectiveness of UV phototherapy of atopic dermatitis patients has not been well understood. A number of recent observations, however, indicate that T-helper cells present in lesional skin of atopic dermatitis patients are important targets for UV phototherapy. Phototherapy of atopic dermatitis employing longwave UVA radiation (340-400 nm), which effectively penetrates the dermal layers of human skin and thus has the potential to directly affect intradermal T-cells, has been shown to be superior to short wavelength UVB radiation, which is almost exclusively absorbed by the epidermis. Accordingly, successful UVA phototherapy of atopic dermatitis was associated with significant downregulation of the in situ expression of T-helper cell derived cytokines as well as significant reduction in the number of intradermal $CD4^+$ T-cells. These observations led us to speculate that UVA phototherapy may act through depletion of skin-infiltrating T-helper cells.

It has therefore been of interest to learn that UVA radiation can induce apoptosis. In murine lymphoma cells, in vitro UVA irradiation induced apoptosis 4 hours after exposure by a process which did not require macromolecular synthesis, and also 24 to 48 hours after irradiation through a mechanism depending on de novo protein synthesis. We here demonstrate that UVA phototherapy induced apoptosis in T-helper cells present in eczematous skin of atopic dermatitis patients. We also show that in vitro UVA irradiation-induced apoptosis in atopen-specific human T-helper cells from atopic eczema was mediated by the FAS/FAS-ligand system and induced through the generation of singlet oxygen [2].

Result

In order to assess whether UVA phototherapy of atopic dermatitis patients was accompanied by the induction of apoptosis in skin-infiltrating T-helper cells, sequential biopsies were taken from eczematous skin of five patients before the 1st, and after the 1st, 2nd, 3rd, 4th and 10th exposure to UVA radiation. Skin specimens were analyzed for apoptotic cells using the TUNEL assay, followed by an anti-CD4 staining to detect T-helper cells. Before therapy, numerous $CD4^+$ cells were present intradermally in eczematous skin, This cell population did not contain a significant number of apoptotic cells, and this was in sharp contrast to skin specimens obtained from the same skin areas after UVA radiation therapy had been started. Already after the 1st UVA radiation exposure, $CD4^+$, apoptotic cells were detected. During subsequent UVA treatments, the number of double positive cells was further increased, whereas the total number of $CD4^+$ cells decreased. After ten exposures, the total number of intradermally located, $CD4^+$ T-cells had been significantly diminished, and remaining cells almost exclusively showed signs of apoptosis.

In order to assess whether UVA radiation is capable of directly inducing apoptosis in skin-infiltrating T-helper cells, atopen-specific T-cells, which had been derived from eczematous skin of atopic dermatitis patients, were employed. Apoptosis was detected in UVA-irradiated T-cells by the TUNEL method and by assessment of DNA laddering. In atopen-specific human T-helper cells, in vitro UVA irradiation induced apoptosis in a dose- and time-dependent manner. Significant apoptosis was already detectable 4 hours after exposure, reaching a maximum 24 hours after irradiation with 30 J/cm^2 UVA radiation. In order to determine the involvement of the FAS/FASL system in UVA radiation-induced human T-helper cell apoptosis, FAS and FASL expression was measured in irradiated cells using FACS analysis. Before UVA radiation exposure, FASL molecules were not present on the cell surface, but significant FASL surface expression was detected in UVA-irradiated cells already 4 hours after exposure. UVA radiation-induced expression of surface FASL expression was dose-dependent and maximal upon exposure of cells to 30 J/cm^2 UVA. In contrast to FASL, FAS surface expression remained essentially unaltered upon UVA irradiation. In order to assess the functional relevance of UVA radiation-induced FASL surface expression for UVA radiation-induced human T-helper cell apoptosis, cells were treated with the blocking anti-FAS-antibody ZB4. Addition of ZB4 antibodies, but not of equivalent concentrations of an isotype control antibody, significantly lowered UVA radiation-induced human T-helper cell apoptosis.

In the following experiments, reagents capable of quenching (sodium azide) or enhancing (deuterium oxide) singlet oxygen effects were assessed for their capacity to modulate UVA radiation-induced human T-helper cell apoptosis. Irradiation of cells in the presence of sodium azide significantly inhibited UVA radiation-induced FASL surface expression as well as apoptosis in human T-helper cells, whereas irradiation of cells in the presence of deuterium oxide resulted in a slight, but consistent increase in the percentage of FASL expressing as well as apoptotic cells.

We next assessed whether UVA radiation-induced apopotosis could be mimicked by stimulating unirradiated human T-helper cells with singlet oxygen. Singlet oxygen was generated by thermal decomposition of $NDPO_2$. Singlet oxygen increased FASL surface expression in unirradiated T-cells to an extent similar to that observed in UVA-irradiated cells. Similar to FASL surface expression, singlet oxygen generation also induced apoptosis in unirradiated cells. $NDPO_2$ induced FASL surface expression as well as apoptosis were significantly enhanced, if T-cells were stimulated in the presence of deuterium oxide. Treatment of cells with NDP did not induce FASL surface expression or apoptosis. Addition of anti-FAS-antibody ZB4 significantly inhibited apoptosis in unirradiated cells, which had been expsoed to $NDPO_2$.

Discussion

Phototherapy led to induction of apoptosis in intradermal T-helper cells, subsequent depletion of T-cells from lesional atopic skin and concomitant improvement of clinical symptoms. Taken together these results indicate that the therapeutic effectiveness of UVA phototherapy for atopic dermatitis results from induction of T-helper cell apoptosis. At least under in vitro conditions, human T-helper cell apoptosis may be induced by short wavelength UVB radiation as well. Depletion of T-cells from human skin via induction of apoptosis may thus represent a general mechanism relevant for phototherapy of inflammatory skin diseases.
UVA radiation-induced Fas-L expression was of functional relevance, because interference with Fas/Fas-L interaction through addition of a blocking anti-Fas antibody effectively prevented UVA radiation-induced human T-cell apoptosis. In this regard, UVA radiation-induced human T-helper cell apoptosis resembled T-cell apoptosis induced by anti-CD3 antibody, phorbol ester plus calcium ionophore, staphylococcal enterotoxin superantigen and cytotoxic drugs. The present study indicates that increased Fas-L surface expression is important for UVA radiation-induced T-cell apoptosis. Recently, it has been demonstrated that the generation of singlet oxygen is a primary mediator in UVA radiation-induced biological effects including increased expression of cell surface molecules. No direct methods exist to detect singlet oxygen in irradiated cells. Evidence for the involvement of singlet oxygen in biological

processes is therefore circumstantial and is based on indirect methods, including the use of singlet oxygen quenchers, enhancers of the life-time of singlet oxygen, and effects induced by singlet oxygen-generating systems. In the present work, a similar experimental approach was used to assess the role of singlet oxygen in UVA radiation-induced upregulation of Fas-L expression and induction of apoptosis in human T-helper cells. The capacity of sodium azide to suppress, of deuterium oxide to enhance, and of $NDPO_2$ to mimick UVA radiation-induced Fas-L expression and apoptosis indicated a prominent role for singlet oxygen in this system. Singlet oxygen-induced T-cell apoptosis involved the Fas/Fas-L system, because blocking anti-Fas antibodies effectively inhibited human T-helper cell apoptosis which was induced by UVA irradiation or stimulation with a singlet oxygen generating system. These studies demonstrate that singlet oxygen, by virtue of its capacity to activate the Fas/Fas-L system, serves an important role in control of human T-cell apoptosis. Ultraviolet A radiation-induced singlet oxygen generation is thought to occur within seconds and does not require protein synthesis. UVA radiation-induced singlet oxygen generation could thus account for the fast increase in Fas-L expression and induction of apoptosis, which was observed in the present study in irradiated T-cells.

2. CUTANEOUS T-CELL LYMPHOMA

Introduction

We have shown that UVA1 radiation phototherapy induces clinical improvement in patients with atopic dermatitis through induction of apoptosis in skin-infiltrating T-cells [2]. This indicated the possibility that UVA-1 might be beneficial for the treatment of other T-cell-mediated skin diseases as well. Accordingly, UVA-1 phototherapy was recently found to be highly effective for cutaneous T-cell lymphoma (CTCL) [3]. In this study, equivalent clinical responses could be achieved regardless of whether 130 J/cm^2, which is the dose routinely used to treat AD, or 60 J/cm^2 UVA-1 were given. We therefore speculated that malignant T-cells might be more sensitive to UVA-1 phototherapy as compared with normal T-cells. In this study, we therefore wanted to test the hypothesis that malignant T-cells are more sensitive to UVA-1 irradiation-induced apoptosis as compared with normal T-cells.

Methods and materials

Normal CD4 T-cells were prepared from healthy individuals by employing magnet cell sorting system. CD4 purity was over 95%. Cells from different malignat T-cell lines including Jurkat, MOLT4, ATL102, HPB-ALL, and HUT 102 (CTCL line) were employed as malignant T-cells. For UVA-1 irradiation UVA-1 Sellamed 2000 System Dr. Sellmaier (Sellas, Gevelsberg, Germany) irradiation device was used. The emission was filtered with a UVA-1 filter (Sellas, Gevelsberg, Germany) and an infrared absorbing filter UG1 (Schott, Mainz, Germany) and consisted exclusively of UV-wavelength greater than 340 nm and smaller than 450 nm. The UVA-1 irradiance was measured with a IL 1700 photometer (International Light, Newburyport, MA) and found to be 50 mW/cm^2 at 30 cm distance. Both freshly isolated normal and malignant T cells were irradiated in vitro with UVA1. After 24 hours, cells were harvested and subjected to FACS analysis by using Annexin V method [4].

Results

All cell lines underwent apoptosis after UVA-1 irradiation in a dose dependent manner (Fig 1). In comparison with the normal CD4 T-cells, cells from all the malignant T cell lines assessed showed a higher rate of UVA-1 induced apoptosis. At a dose of 20 J/cm^2, apoptosis rate in malignant cells was between 35% -65% as compared with normal T cells, where less than 18 % were apoptotic. This difference between malignant and normal T cells was statistically significant at a dose of 20 J/cm^2 UVA1 for all cell lines tested at a p level < 0.05.

We have shown that UVA-1 radiation induces T cell apoptosis through the FAS-FASL system via singlet oxygen generation [2]. In order to assess whether the differences in sensitivity to UVA-1 radiation-induced apoptosis are defined at the level of the FAS system, Jurkat cell lines which express different FAS level on the surface were analyzed. Three Jurkat cell lines were used. Wild type Jurkat cells had abundant FAS expression on the surface. The C5 cell line expressed less FAS level and upon treatment with anti-FAS-antibody did not undergo apoptosis. The R1 cell line had almost no FAS expression on the surface expression and as a result was resistant to FAS antibody or FASL treatment. The induction of apoptosis in these three cell lines after UVA-1 irradiation was assessed. Both C5 and R1 lines exhibited significantly lower apoptosis rate as compared with wild type Jurkat cells.

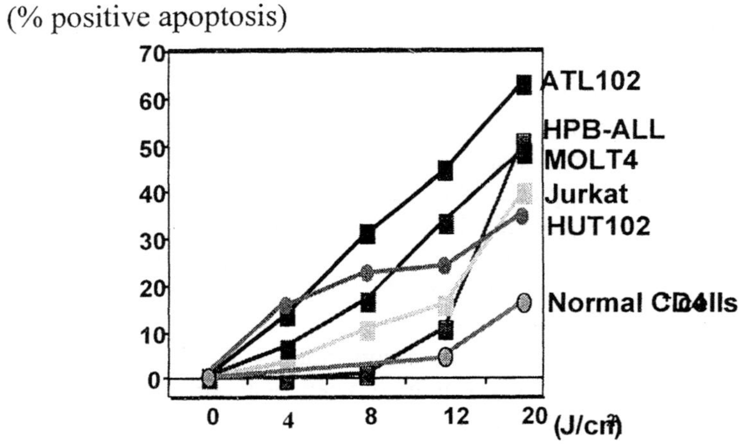

Fig 1. Malignant T-cells are Sensitive to Ultraviolet A-1 (UVA-1) Radiation-induced Apoptosis.

Discussion

Malignant T-cells showed a higher susceptibility towards UVA-1 radiation-induced apoptosis as compared with normal $CD4^+$ T-cells. These data indicate that malignant T-cells are characterized by an exquisite sensitivity to UVA-1 radiation-induced apoptosis. This difference might explain why UVA-1 can be used at lower doses for the therapy of CTCL, as compared with atopic dermatitis.

The susceptibility of T-cells towards UVA-1 radiation-induced apoptosis correlated with their FAS surface expression. This observation further supports the concept that the FAS/FAS-ligand system plays a critical role in UVA-1 radiation-induced T-cell apoptosis.

REFERENCE

1. Krutmann J: Therapeutic photomedicine phototherapy; In :Freedberg IM et al (eds): ***Fitzpatrick's Dermatology in general medicine***, 5th Ed, McGraw-Hill, New York, 1999, 2870-2879.
2. Akimichi Morita, Thomas Werfel, Helger Stege, Constanze Ahrens, Karin Karmann, Markus Grewe, Susanne Grether-Beck, Thomas, Ruzicka, Alexander Kapp, Lars-

Oliver Klotz, Helmut Sies, and Jean Krutmann. Evidence that Singlet Oxygen-induced Human T Helper Cell Apoptosis Is the Basic Mechanism of Ultraviolet-A Radiation Phototherapy , ***J. Exp. Med.*** 1997 186: 1763-1768.

3. Heidi Plettenberg, Helger Stege, Thomas Ruzicka, Yoko Hosokawa, Takuo Tsuji, Akimichi Morita, Jean Krutmann, Ultraviolet A-1 (340-400nm) phototherapy for cutaneous T cell lymphoma, ***J Am Acad Dermatol,*** 41: 47-50, 1999.
4. Akimichi Morita and Jean Krutmann: UVA radiation-induced apoptosis. In: Methods in Enzymology, special volume on "Singlet oxygen, UV-A & Ozone (L. Packer, H. Sies, eds), Academic Press, Orlando, Fl, 2000, 302-309, ***volume 319 of Methods in Enzymology***.

IN VITRO FUNCTION OF DOUBLE-DOSE PLATELETS TREATED WITH THE PATHOGEN INACTIVATION HELINX™ TECHNOLOGY

[1]K. Janetzko, [2]L. Corash, [2]L. Lin, [3]J. Flament, [1]H. Klüter

[1]Institute of Transfusion Medicine and Clinical Immunology, Red Cross Blood Service of Baden-Württemberg, Faculty of Clinical Medicine Mannheim, University of Heidelberg, Gemany, [2]Cerus Corporation, Concord, CA, USA, [3]Baxter R&D Europe, Nivelles, Belgium

Key words: double-dose platelets, pathogen inactivation, psoralen S-59, UVA

1. Abstract

Helinx™ technology uses the psoralen S-59 and ultraviolet A (UVA) light to inactivate a broad spectrum of bacteria, viruses, and leukocytes in platelet concentrates (PC). In a paired study we investigated the effect of the Helinx™ technology on platelet function over a storage period of 5 days. Double-dose apheresis PC (7.0×10^{11} platelets) resuspended in 35% plasma/65% platelet additive solution (PAS III) were collected ($n = 8$) and split into two equal size PC. One PC was not treated (Control), the other PC was treated with 150 μM S-59 and 3 Joules/cm^2 UVA on Day 1 (Test).

After 5 days of storage, there were no statistically significant differences ($p > 0.05$) between the Test PC and Control PC with respect to pH (7.0 vs 6.9), hypotonic shock response (HSR) (37% vs 36%), CD62 expression (29% vs 28%), morphology score (307 vs 301), cell lysis by electron microscopy (15% vs 14%) and platelet count (868 vs 900 x 10^9/L). Lactate dehydrogenase (LDH) was increased (175 vs 104 IU/L) for Test PC vs Control PC.

These results demonstrate that double-dose collection of apheresis platelets can be treated with the Helinx™ technology as two products. In vitro quality of pathogen-inactivated split-dose PC did not differ significantly from Control PC over the storage period of 5 days.

2. Introduction

The risk of transfusion-associated infection still represents a major concern in transfusion medicine. Although pre-donation evaluations of donors and sensitivity of serological testing have steadily improved, blood components continue to carry a substantial risk of transmitting pathogens. Diagnostic tests may be negative during the window period before seroconversion of an infected donor or insensitive for detection of pathogens or antibody responses. For unidentified pathogens, no routine tests are available. Platelet concentrates (PC) stored at 22°C for up to 5 days pose particular problems. The storage conditions allow any contaminating bacteria to multiply rapidly, causing clinically significant sepsis in the transfusion recipients. Estimation of the frequency of bacterial contamination ranges up to 0.4% in all PC transfused (1-3). Ensuring safe cellular blood products is particularly difficult because any effective treatment must inactivate a broad spectrum of cell-free and cell-associated viruses, bacteria, and leukocytes without inducing toxicity and carcinogenicity yet preserving the cell functionality and viability during the entire storage period.

There are several reports of photochemical treatment (PCT) having been used for pathogen inactivation. However, because of low psoralen reactivity (8 methoxypsoralen, 8-MOP) or a positive mutagenic response (4'-aminomethyl-4,5',8-trimethylpsoralen, AMT), these methods were not applicable for routine processes (4,5). PCT using Helinx technology with a new psoralen (S-59) in combination with a low dose of UVA light, however, has demonstrated encouragingly effective inactivation of high levels of certain enveloped and non-enveloped viruses, bacteria, and leukocytes. In the S-59 inactivation procedure, platelets are resuspended in autologous plasma and platelet additive solution (PAS III) and mixed with the psoralen S-59. This process targets nucleic acids by reversible intercalation into helical regions of DNA and RNA. After a short illumination with UVA light (approximately 3 minutes), S-59 reacts with the pyrimidine bases of the nucleic acids to form monoadducts or covalent crosslinks (6). These bonds block the process of transcription and translation and, therefore, effectively block replication of the genome leading to inactivation of pathogens (7). After illumination, residual S-59 and free photoproducts are removed by incubation with a compound adsorption device (CAD), and the PC can be stored under normal conditions for up to 5 days. Table 1 gives data confirming the efficacy of this procedure on viral and bacterial inactivation.

Table 1: *Efficacy of Viral and Bacterial Inactivation in Platelet Concentrates by PCT*

Virus	Model	Log inactivation[1]
HIV-1	Cell-free	>6.2
	Cell-associated	>6.1
	Clinical isolate[2]	>3.4
	Proviral	Below limit of detection
HIV-2	Clinical isolate[2]	>2.5
HBV	Human MS-2 strain[3]	>5.5
	Duck Hepatitis B virus	>6.2
HCV	Human Hutchinson strain[3]	>4.5
	Bovine Viral Diarrhea virus	>6.0
HCMV	Cell associated	>5.9

Gram-negative Bacteria	Log inactivation
Escherichia coli	>6.4
Serratia marcescens	>6.7
Klebsiella pneumonia	>5.6
Pseudomonas aeruginosa	4.5
Salmonella choleraesuis	>6.2
Yersinia enterocolitica	>5.9
Enterobacter cloacae	5.9

Gram-positive Bacteria	Log inactivation
Staphylococcus epidermidis	>6.6
Staphylococcus aureus	6.6
Streptococcus pyogenes	>6.8
Listeria monocytogenes	>6.3
Corynebacterium minutissimum	>6.3
Bacillus cereus	3.6–3.9

1. The log inactivation is calculated by subtracting the residual infectivity (expressed in log titer) after Helinx treatment from the initial infectivity (expressed in log titer). The ">" sign indicates inactivation to the limit of detection.
2. The level of log inactivation demonstrated for HIV clinical isolates was limited by the highest available viral inoculum.
3. The MS-2 strain HBV and the Hutchinson strain HCV are human hepatitis viruses. Their infectivities are measured using an assay in chimpanzees.

3. Materials and Methods

In a paired study, we investigated the effect of photochemical treatment (PCT) using Helinx technology with the psoralen S-59 in combination with UVA light on platelets over a storage period of 5 days. Using the AMICUS® cell separator (Baxter) we collected single donor platelet concentrates with a platelet yield of 7.0×10^{11}, resuspended in 35% autologous plasma and 65% platelet additive solution (PAS III). Each platelet concentrate was split into two, an untreated Control unit and a photochemically treated Test unit. PCT occurred on the day after collection. Test PC were treated with 150 μM S-59 and a 3 Joules/cm^2 UVA treatment (approximately 3 minutes). Following illumination, the levels of residual S-59 and free photoproducts were reduced by treatment of the platelet mixture with a compound adsorption device (CAD) for 4 hours. To evaluate platelet functionality and viability, we took platelet samples before, during, and after PCT, as well as at the end of the storage period on Day 5, and measured platelet count, pH, and lactate dehydrogenase (LDH). The extent of platelet activation was estimated by the expression of CD62 and hypotonic shock response (HSR). Morphological changes of platelets were evaluated by morphology score. The percent (%) platelet lysis was determined by electron microscopy (EM).

4. Results

After 5 days of storage, there were no statistically significant differences ($p > 0.05$) between the Test PC and Control PC with respect to pH (7.0 vs 6.9), HSR (37% vs 36%), CD62 expression (29% vs 28%), morphology score (307 vs 301), cell lysis by EM (15% vs 14%) and platelet count (868 vs 900 x 10^9/L). LDH was increased (175 vs 104 IU/L) for Test PC vs Control PC. Compared with the initial pre-photochemical treatment values, we observed a decrease in all parameters through the end of the observation period in all products; this decrease is a normal event (with the exception of CD62 expression) which results from the storage time. However, we did not observe any differences between treated and untreated PC that were due to the treatment procedure. During platelet storage the decrease in platelet morphology score correlated with an increase in the extent of platelet lysis (by EM). When we compared the Test product morphology values with pretreatment values, no significant differences were detected in the Test PC, while significant differences were detected in the Control PC (Table 2).

Table 2. *In Vitro Data of Storage Parameters before PCT and at the End of the Storage Period*

Parameters		Pre-photochemical treatment	Day 5 at the end of the storage period
Platelet count ($\times 10^9$/l)	PCT	964 ±109	868 ±127
	Control	973 ±74	900 ±116
CD62 expression (%)	PCT	38.4 ±8.6	29.3 ±12.3
	Control	39.1 ±9.2	27.9 ±10.6
pH (22°C)	PCT	7.17 ±0.09	6.95 ±0.11*
	Control	7.08 ±0.33	6.92 ±0.14*
Morphology score	PCT	309 ±61	307 ±27
	Control	305 ±60	301 ±26*
Platelet lysis (%)	PCT	12.0 ±8.0	14.8 ±4.9
	Control	13.3 ±5.3	13.8 ±5.2

**Day 5 results are significantly different from values obtained prior to photochemical treatment.*

In analyzing the hypotonic shock response (HSR), we found a significant difference between Control and Test platelets as a result of the compound adsorption device treatment (CAD) step. However, there were no statistically significant differences between the Control and Test platelets on Day 5 of storage; this indicated a reversible short-time effect on the S-59 treated platelets (Figure 1).

Figure 1: *Results of Hypotonic Shock Response over the 5 Day Storage Period (* indicates statistically significant differences)*

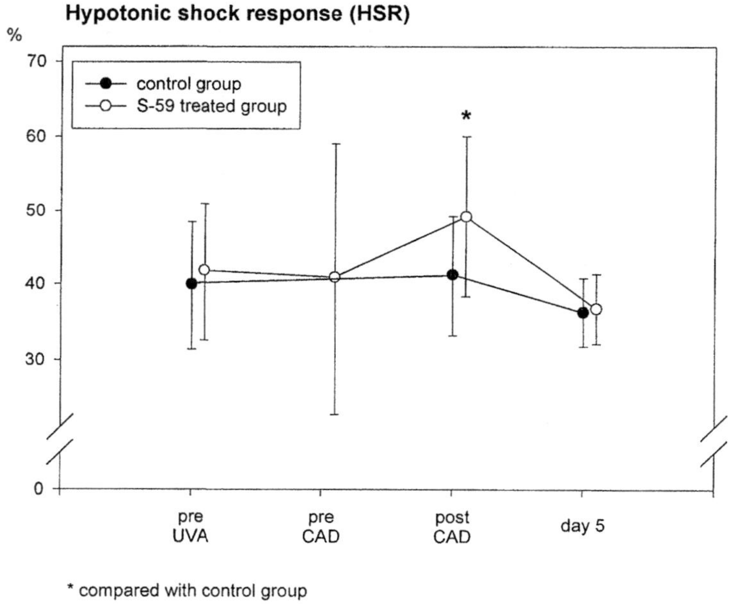

We found comparable results in both Control and Test platelets when measuring LDH concentration prior to CAD treatment. The CAD treatment caused a significant increase in extracellular LDH. Over the length of the storage period, however, we found no further changes as the extracellular LDH increased equally in Control and Test platelets; this again was due to the length of storage time (Figure 2).

Figure 2: *Results of Lactate Dehydrogenase (LDH) over the Storage Period of 5 Days (*, #, + indicate statistically significant differences)*

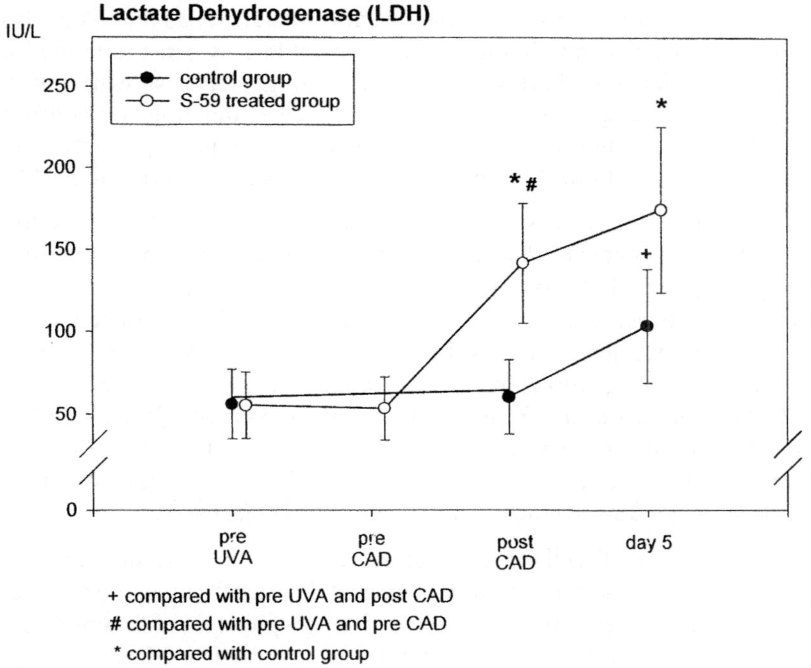

5. Discussion and Conclusion

These results demonstrate that double-dose collection of apheresis platelets can be treated with the Helinx™ technology as two products. In vitro quality of pathogen-inactivated split-dose PC did not differ significantly from Control PC over the storage period of 5 days. The results of our in vitro studies demonstrate an increase in extracellular LDH and a transient elevation in HSR as a result of the S-59 reduction (CAD) procedure. An increase in extracellular LDH is usually due to lysis of platelets. However, we did not detect any difference between the Control and Test platelets with respect to the extent of platelet lysis by EM analysis. In addition, platelet morphology was comparable between the treated and the control platelets. Further, previous findings have shown that this procedure has negligible effects on in vitro platelet function; we confirmed those findings through analysis of pH, blood gas, glucose, lactate, CD62 expression, aggregation, shape change, ATP release, and morphology (8).

Results of other studies demonstrate that PCT also leads to an inactivation of leukocytes by blocking proliferation and cytokine synthesis, namely IL-8 and IL-1β (9). Previous studies have demonstrated a strong correlation between these cytokine levels and the incidence of febrile non-hemolytic transfusion reactions (FNHTRs) (10). Cytokines generated in platelet units during storage have been implicated as possible mediators of FNHTRs when increasing from non-detectable to physiologically significant concentrations during the storage of PCs. Therefore PCT with S-59 may reduce the rate of adverse transfusion-associated reactions due to the inhibition of cytokine synthesis from white blood cells (WBCs).

Moreover, T cells contaminating cellular blood components have been implicated as the initiating agents for graft-versus-host disease (GVHD). At present the only effective prophylaxis for transfusion-associated (TA)-GVHD, which is fatal in 80% to 90% of cases (11), is irradiation of cellular blood products using a gamma source (12). In vitro and in vivo findings suggest that PCT with S-59 and UVA, like gamma irradiation treatment, inactivates high levels of T cells in platelet concentrates and thus has the potential to prevent TA-GVHD (13, 14).

Toxicology studies were completed and included single dose and multiple dose (up to 13 weeks duration) toxicity, safety pharmacology

(CNS, renal, and cardiovascular), reproductive toxicity, genotoxicity, p53 carcinogenicity, vein irritation, phototoxicity, and toxicokinetics (multiple dose pharmacokinetics). No specific target organ toxicity (clinical or histologic pathology), reproductive toxicity, or carcinogenicity was observed. S-59 and/or PCT formulations demonstrated CNS, electrocardiographic (ECG), and phototoxicity at non-clinical doses. Based on the extremely large safety margins (>30,000-fold expected clinical exposures), the CNS and ECG observations are not considered to be of toxicological relevance. Additionally, after a complete assessment, the results of mutagenicity and phototoxicity studies are not considered to be relevant for the proposed use of the photochemically-treated platelets. The results of an extensive series of studies have thus demonstrated no toxicologically relevant effects of platelets treated with Helinx™ technology. The results of the toxicokinetic analyses from 3-month rat and dog studies with S-59 PCT formulations +/-CAD treatment were compared with clinical data. These data indicate that the test articles used in the key toxicology studies provided large multiples of the clinical exposure to S-59, whether the comparison was based on dose, C_{max}, or AUC (Ciaravino V, McCullough T, and Dayan A. Pharmacokinetic and toxicology assessment of INTERCEPT™ (S-59 psoralen and UV-A treated) platelets. Human and Experimental Toxicology; Manuscript in preparation.).

Clinical studies carried out in healthy volunteers have demonstrated acceptable post-transfusion recovery and platelet life span for Helinx-treated platelets compared to standard platelets (16). Moreover, template bleeding time and hemostasis did not show a significant difference in patients receiving treated platelets compared with a control group. Additionally, the intervals to the next transfusion were not significantly different (17). Further studies are ongoing in the United States and in Europe (*euroSPRITE*) to investigate the therapeutic efficacy and safety of Helinx-treated platelets in support of thrombocytopenic chemotherapy patients. The results of the *euroSPRITE* trial show that hemostasis after transfusion of control and Helinx-treated platelets was comparable. Helinx platelets provided therapeutically adequate platelet count increment, corrected count increment, and transfusion intervals compared to control platelets. Bleeding, RBC transfusions, acute transfusion reactions, and refractoriness to platelet transfusions were similar between the two groups (18).

6. References

1. Goldman M., Blajchman M.A. Blood product-associated bacterial sepsis. Transfus Med Rev 1991; 5:73-83.

2. Morrow J.F., Braine H.G., Kickler T.S., et al. Septic reactions to platelet transfusions. A persistent problem. JAMA 1991; 266:555-8.

3. Blajchman MA, Ali AM. Bacteria in the blood supply: an over-looked issue in transfusion medicine. In: Nance ST, ends Blood safety: current challenges. Bethesda: American Association of Blood Banks, 1992:213-28.

4. Lin, L., G.P. Wiesehahn, P.A. Morel, and L. Corash. Use of 8-methoxypsoralen and long-wavelength ultraviolet radiation for decontamination of platelet concentrates. Blood 1989; 74: 517-525.

5. Wagner SJ, White R, Wolf L, et al. Determination of residual 4′-aminomethyl-4.5′8-trimethylpsoralen and mutagenicity testing following psoralen plus UVA treatment of platelet suspension. Photochem Photobiol 1993; 57:819-24.

6. Cimino G, Gamper H, Isaacs ST, Hearst JE. Psoralens as photoactive probes of nucleic acid structure and function: organic chemistry, photochemistry, and biochemistry. Ann Rev Biochem 1985; 54:1151-1193.

7. Lin L, Alfonso R, Behrman B, et al. Photochemical treatment of platelet concentrates with a novel psoralen and UVA to enhance the safety of platelet transfusion. Infusionsther Transfusionsmed 1998; 25:39-48.

8. Lin L, Cook DN, Wiesehahn GP, et al. Photochemical inactivation of viruses and bacteria in human platelet concentrates using a novel psoralen and long wavelength ultraviolet light. Transfusion 1997; 34:423-435.

9. Hei DJ, Grass J, Lin L, Corash L, Cimino G. Elimination of cytokine production in stored platelet concentrate aliquots by photochemical treatment with psoralen plus ultraviolet A light. Transfusion 1999; 39:239-248.

10. Heddle NM, Klama L, Singer J, Richards C, Fedak P, Walker I, Kelton JG: The role of the plasma from platelet concentrates in transfusion reactions. N Engl J Med 1995; 331: 625.

11. Spitzer TR: Transfusion induced graft-vs.-bone disease. "In *Graft-vs.-disease. Immunology, Pathophysiology and Treatment*", Burakoff SJ, Deeg HJ, Ferrara J, Atkinson K (eds): NY. Marcel Dekker. 1990, p 539.

12. Anderson KC: Clinical indications for blood components irradiation. "*In Irradiation of blood components*", Bethesda MD, Baldwin ML, Jeffries LC (eds): American Association of Blood Banks. 1992, p. 31.

13. Grass JA, Hei DJ, Metchette K, et al. Inactivation of leukocytes in platelet concentrates by photochemical treatment with psoralen plus UVA. Blood 1998; 91:2180-2188.

14. Grass JA, Wafa T, Reames A, et al. Prevention of transfusion-associated graft-versus-host disease by photochemical treatment. Blood 1999; 93:3140-3147.

15. Corash L, Behrman B, Rheinschmidt M, Wages D, Snyder E. Raife T, Kagen L, Baril L, Davis K, Metzel P, Walsh J, Smith D, Shafer S, Cimino G, Hei D, Lin L, and Buchholz DH. Post-transfusion viability and tolerability of photochemically treated platelet concentrates (PC). Blood 1997; 90: 267a.

16. Slichter, SJ, L. Corash, M. Grabowski, R. LeBlanc, J Sedgwick, D. Townsend-MaCall, K. Davis, B. Behrman, M Rheinschmidt, P. Metzel, L. Lin, and D.H. Buchholz. Viability and hemostatic function of photochemically treated (PCT) platelets (PLTS) in thrombocytopenic patients. Blood 1999; 94(S1): 376a.

17. Van Rhenen D, Gulliksson H, Pamphilon D, et al. S-59 (HELINX^{TM}) photochemically treated platelets (plt) are safe and effective for support of thrombocytopenia: Results of the *euroSPRITE* Phase 3 trial. Blood 2000; 96:819a (Abstract).

Photomedicine and Photobiologic Aspects In Skin

LIGHT IN THE TREATMENT OF SKIN DISEASE

Herbert Hönigsmann
Division of Special & Environmental Dermatology, University of Vienna Medical School, Vienna, Austria

Introduction

This review covers the current practice of phototherapy with ultraviolet (UV) radiation, of psoralen photochemotherapy (PUVA) and of photodynamic therapy (PDT) in the treatment of skin diseases and superficial skin tumors.

1. PHOTOTHERAPY

1.1. Principles and Mechanisms of Phototherapy

Phototherapy refers to the use of artificial UVB without the addition of exogenous photosensitizers. The radiation is absorbed by endogenous chromophores. Photochemical reactions of these absorbing biomolecules result in alterations of skin biology and thus lead to the treatment effect.

The best-characterized chromophore for UVB is nuclear DNA. Absorption of UV by nucleotides leads to DNA photoproduct formation, mainly pyrimidine dimers and the pyrimidine (6-4) pyrimidone photoproduct. UVB exposure is known to cause a reduction of DNA synthesis. Therefore, it can be used to suppress the accelerated DNA synthesis in psoriatic epidermal cells. UVB exposure up-regulates the tumor suppressor gene product p53, which is involved in the control of the cell cycle gene and may thus be relevant for the reversal of the shortened cell cycle in keratinocytes of psoriatic epidermis. The UVB-induced alteration of cell cycle control mechanisms is also important for apoptosis, which is seen in skin after UVB exposure ("sunburn cells"), and for the control of skin carcinogenesis.

Besides the effect on cell cycle, UV induces prostaglandin release and alters cytokine expression and secretion. Interleukins 6 and 10 seem to play important roles for systemic symptoms of UV phototoxicity (sunburn) and immune suppression. All these mechanisms may be equally important for therapeutic effectiveness and side effects.

Also, UV-induced cellular molecular effects occur which are independent of nuclear DNA damage. These involve membrane receptors and molecular signaling pathways that regulate transcriptional activity. Thus, gene expression may be altered also independently of DNA damage. The existence of such molecular regulatory mechanisms indicates additional or parallel molecular pathways but the actual cellular targets and effector mechanisms of phototherapy are not yet identified. One important aspect of the mechanisms of phototherapy is its effect on the cutaneous immune response. UV radiation suppresses contact allergy, delayed type hypersensitivity, and immune surveillance against UV-induced non-melanoma skin carcinoma in mice. Langerhans cells are very sensitive to UVB, which alters their antigen presenting function. Keratinocytes secrete soluble mediators such as interleukin 1 and 6, prostaglandin E2, tumor necrosis factor alpha, and several others under the influence of UV radiation. This may alter the cutaneous immune response by direct effects on T lymphocytes or via their influence on the function of epidermal Langerhans cells. In psoriasis both, epidermal keratinocytes and cutaneous lymphocytes may be targeted by UVB. Immune suppression, alteration of cytokine expression as well as cell cycle arrest may contribute to the suppression of disease activity in several other dermatoses.

1.2. Action spectrum and radiation sources

Parrish and Jaenicke reported an action spectrum for antipsoriatic activity demonstrating that 304 and 313 nm were optimally effective even with suberythemogenic doses. The addition of UVA does not enhance the therapeutic efficacy of UVB. The Philips TL01 lamp emitting narrow-band UVB (311-313 nm) was introduced to optimally meet the requirements for antipsoriatic activity.

1.3. Phototherapy regimens

The initial therapeutic UVB dose lies at 75 to 100 % of the minimal erythema dose (MED). Treatments are given two to five times weekly. The rate of dose increments depends on treatment frequency and the effect of the preceding therapeutic exposure. The objective of the dose increments is to maintain a minimally perceptible erythema as a clinical indicator of optimal dosimetry. Treatments are given until total remission is reached or no further improvement can be obtained with continued phototherapy.

1.4. Phototherapy for psoriasis

Primarily eruptive and seborrheic forms of psoriasis respond rapidly to broadband UVB. The results with chronic plaque-type psoriasis, however, are not so favorable. Although broadband UVB phototherapy can be very effective, it is second to PUVA both with clearing efficiency and remission

times. Narrow-band (312 nm) phototherapy is superior to conventional broadband UVB with respect to both clearing and remission times. In Europe, narrow-band UVB has mostly replaced conventional phototherapy. From own studies and other reports we consider narrow-band phototherapy nearly as effective as PUVA.
Upon clearing treatment is either discontinued or, as done in some institutions, the patients are subjected to maintenance therapy for one or two months. It is generally believed that maintenance therapy keeps patients clear of disease. However, the efficacy and requirements of maintenance are controversial because of the additional radiation load delivered to the patients' skin.

1.4.1. Combination Treatments

The use of adjunctive agents and combination therapies is directed at an improved efficacy and a reduction of the cumulative UVB burden in order to reduce the risk of long-term side effects. Narrow-band UVB has been successfully used in a variety of combination therapies. Systemic drugs such as retinoids increase efficacy, particularly in patients with chronic and hyperkeratotic plaque type psoriasis. Also, retinoids may possibly reduce the carcinogenic potential of UVB phototherapy.

1.4.2. Side Effects

Short-term side effects consist of erythema, dry skin with pruritus, occasional blistering, and an increased frequency of recurrent herpes simplex eruptions. Long-term side effects include photoaging and carcinogenesis. UVB is a known carcinogen; however, its carcinogenic potential seems to be less than that of PUVA. Stern and Laird in their 16-center study failed to show a relationship between UVB phototherapy and non-melanoma skin cancer but future studies might perhaps reveal the opposite.

1.5. Phototherapy for Mycosis Fungoides

Since it was known that patients with early stage MF benefit from exposure to natural sunlight, conventional UV lamps were long used for MF treatment before modern high-intensity lamps became available. Responses to phototherapy were better in patients with eczematous (patch) stage mycosis fungoides than in patients with the plaque stage. Thus, UVB phototherapy seems to be efficient in early eczematous -stage MF. Prolonged maintenance exposure is necessary to prevent early relapses. UVB phototherapy is a safe therapeutical option for stage I MF but is not indicated in patients with plaque stage or more advanced stages of MF.

1.6. Phototherapy for Vitiligo

Although PUVA may be the most commonly employed phototherapeutic modality in vitiligo therapy, consistent treatment with UVB irradiation for a sufficiently long period can also achieve acceptable results. There are only a few publications about UVB phototherapy for vitiligo. In a study comparing whole body narrowband UVB irradiations with PUVA using topical psoralen, narrowband UVB was as effective as PUVA but had fewer side effects.

1.7. Phototherapy for Atopic Dermatitis

It is known since decades that UV radiation is beneficial for patients with atopic dermatitis. Modern fluorescent lamps with defined emission spectra for phototherapy of atopic dermatitis have been continually used from the 1970s until today. In the recent past new phototherapeutic options, namely UVA-1 therapy and narrowband UVB therapy have been introduced. Dermatologists may now select from a diversified spectrum of distinct modalities the phototherapy of choice for their particular patient. In general, symptomatic phototherapy of acute, severe exacerbation of atopic dermatitis may be achieved with UVA-1 and PUVA, whereas conventional UVA/UVB and narrowband UVB therapy represent phototherapeutic modalities that are primarily indicated for treatment of chronic stages of this disease.

1.7.1. UVA-1 Phototherapy

The rationale for employing UVA-1 radiation (340-400nm) in the treatment of patients with atopic dermatitis was based on immunological studies in which it was demonstrated that exposure of human skin to UVA-1 radiation induced abrogation of epidermal Langerhans cell function to activate alloreactive T cells. Clinical studies comparing the efficacy of broadband UVB therapy versus UVA/UVB therapy in the management of patients with atopic dermatitis indicated that the therapeutic effectiveness of conventional UVB therapy may be significantly improved by increasing the UVA portion of the action spectrum.

The therapeutic effectiveness of UVA-1 irradiation in the management of patients with atopic dermatitis was first assessed in a pilot study in which patients exposed once daily to 130 J/cm2 UVA-1 (high-dose UVA-1 therapy) for 15 consecutive days. The therapeutic effectiveness of UVA-1 therapy was compared to that of a conventional UVA/UVB therapy by employing both modalities as a monotherapy. Assessment of clinical scores demonstrated that UVA-1 therapy was superior in inducing clinical improvement. In a subsequent multicenter trial, UVA-1 was compared with conventional UVA/UVB therapy and topical treatment with fluocortolone. After 10 treatments, patients in all three groups had improved, but clinical improvement was significantly greater in patients on either corticosteroid or

UVA-1 therapy, as compared to UVA/UVB therapy. UVA-1 therapy may thus be used as an alternative to glucocorticosteroids to treat severe atopic dermatitis.

Except for eczema herpeticum, no acute side effects have been observed in any of the atopic dermatitis patients treated with UVA-1. No other side effects have been observed, although its potential carcinogenic risk is a theoretical concern. The actual contribution of UVA-1 radiation to the development of malignant melanoma in humans is unknown but cannot be excluded.

1.7.2. Mechanisms

Several studies show that UVA-1 is capable of downregulating expression of interferon-alpha in lesional skin of patients with atopic dermatitis. Inhibition of interferon- alpha expression in atopic eczema may not only be achieved by UVA-1 therapy, but could also be seen after topical application of glucocorticosteroids and most likely represents a general mechanism by which various treatment forms induce symptomatic relief. In vivo studies have revealed that UVA-1 phototherapy induced apoptosis in skin-infiltrating T cells and thereby caused a gradual reduction of the inflammatory infiltrate and a concomitant improvement of the disease. These observations have stimulated interest in the use of UVA-1 phototherapy for other T-cell-mediated skin diseases as well.

1.8. Phototherapy for Polymorphous Light Eruption (PLE)

The PLE rash occurs less frequently or entirely disappears during the course of the summer months. This gradually increasing tolerance towards sunlight is termed 'hardening effect'. The presumed mechanisms of action underlying this effect are an enhanced filtering effect of the skin as a result of increased melanin formation and thickening of the stratum corneum and an altered immunological reactivity of the skin. Photoprevention of PLE is likely to be based on the same mechanisms. Desensitization with broadband UVB is especially suitable for patients whose PLE is precipitated by UVA. Several studies have shown that 12 to 15 treatments at a frequency of 3 to 5 exposures per week increase the patients' tolerance to sunlight and are effective in preventing or attenuating the occurrence of PLE. Treatment induced pigmentation is not a reliable indication of therapeutic effect since patients with skin types I and II who only tan poorly may experience good PLE prevention.

Despite widespread use, treatment regimens vary greatly between centers. In some units, narrowband UVB has gradually replaced broadband UVB and PUVA and is considered the treatment of first choice for PLE desensitization.

In a recent controlled study UVB and PUVA were found to be equally effective in preventing PLE.

2. PHOTOCHEMOTHERAPY WITH PSORALENS

Psoralen photochemotherapy (PUVA) is the combined use of psoralens (P) and long wave UV radiation (UVA). The combination of drug and radiation results in a therapeutic effect, which is not achieved by the single component alone. Remission of skin disease is induced by repeated controlled phototoxic reactions.

2.1. Principles and Mechanisms

PUVA is performed by the administration of a fixed dose of a psoralen preparation at a constant interval before UVA exposure. These parameters are kept unchanged in order to make psoralen plasma levels as reproducible as possible. Initial dose finding (phototoxicity testing) and dose adjustments are done by varying the UVA fluence.

Absorption of photons is confined to the skin according to the penetration characteristics of UVA. PUVA-induced phototoxic reactions manifest as delayed sunburn-like erythema and inflammation that upon overdosage can progress to blister formation and even superficial skin necrosis. Therefore, UVA dosimetry is most critical for safe and efficient PUVA therapy.

Psoralens are linear furocoumarins, which were originally derived from plants. Currently, 8-methoxypsoralen (8-MOP, methoxsalen) is most commonly prescribed, but 5-methoxypsoralen (5-MOP, bergapten) is also used where available. The synthetic furocoumarin, 4,5',8-trimethylpsoralen (TMP) is utilized for bathwater-delivered PUVA mainly in Scandinavia. Oral psoralens are metabolized in the liver and excreted with the urine within 12 to 24 hours. Psoralens are only reactive when activated by UV radiation and the activity is confined to those layers of skin, which are reached by UVA. The radiation penetrates epidermis and papillary dermis and reaches also the superficial vascular plexus.

Psoralens react with DNA in three steps. First the psoralen intercalates into the DNA double strand in the absence of UV radiation. Upon irradiation either reactive double bond may form a cyclobutane monoadducts (MA) with a pyrimidine base. One of these, the 4',5' MA can form a psoralen-DNA cross-link after absorbing a second photon. DNA-psoralen crosslinks inhibit DNA replication and cause cell cycle arrest. Psoralen photosensitization leads to an alteration of cytokine and cytokine receptor expression as well as cytokine secretion. PUVA can reverse the pathologically altered pattern of keratinocyte differentiation markers and reduce the number of proliferating epidermal cells. Infiltrating lymphocytes are strongly suppressed by PUVA, with variable effects on different T-cell subsets. It has been also demonstrated that PUVA is far more potent in inducing apoptosis in lymphocytes than in keratinocytes at a similar level of the antiproliferative activity in both cell types. This specific response pattern may explain the high efficacy in mycosis fungoides (cutaneous T cell lymphoma) as well as in inflammatory skin diseases.

2.2. Action Spectrum

The action spectrum for 8-MOP-induced delayed erythema has its maximum activity in the 330 nm range and it seems that the therapeutic activity parallels the erythema action spectrum for 8-MOP. The psoralen action spectrum is well covered by the conventional therapeutic UVA fluorescent tubes and broad-spectrum metal halide lamps.

2.3. Photochemotherapy for psoriasis

Practically all forms of psoriasis respond to PUVA, although the management of erythrodermic or generalized pustular psoriasis is more difficult. Treatments are given 2-4 times weekly. Maintenance therapy is performed quite commonly and consists of four weeks of twice-weekly treatments at the last UVA dose used for clearing, followed by four once weekly exposures. As with UVB the impact of maintenance therapy that leads to a higher cumulative number of treatments and cumulative UVA doses is unknown.

Topical application of TMP or 8-MOP by bath water has been popular in Scandinavia for many years, but only recently has attracted worldwide interest. Bathwater delivery helps to avoid gastrointestinal and ocular side effects because there is no systemic photosensitization. Skin psoralen levels are more reproducible and photosensitivity is less persistent than after oral administration.

2.3.1. Combination Treatments

As with phototherapy, PUVA can be combined with other treatments to improve efficacy and reduce possible side effects. Topical therapy does not find all patients' acceptance since it is considered a step backward to conventional treatment. The combination of PUVA with systemic retinoids is one of the most potent therapeutic regimens for psoriasis. RePUVA can reduce the number of exposures by one third and the total cumulative UVA dose by more than one half and may clear also "poor PUVA responders". The mechanism of the synergistic action of retinoids and PUVA is unknown but it may be due to the accelerated desquamation of the psoriatic plaques that optimizes the optical properties of the skin and to the reduction of the inflammatory infiltrate. As an additional beneficial effect retinoids may theoretically protect against long-term carcinogenic effects of PUVA by reducing the number of exposures and by its potential cancer protective effect.

2.3.2. Side effects and long-term hazards

Redness, swelling, and occasionally blister formation can occur as seen with excessive sunburns. While UVB-induced erythema peaks before 24 hours, the peak PUVA reaction is not reached before 72 hours. Generalized pruritus or tingling sensations may herald phototoxic side effects. Besides the phototoxic episodes, psoralens can induce systemic side effects in the absence of light. Most notorious is nausea and vomiting induced by oral 8-MOP. Accidental ocular UVB overexposure may induce photokeratitis due to its absorption in conjunctiva and cornea. UVA penetrates into the ocular lens and is cataractogenic by itself with high fluences. Despite experimental data, which indicated a risk of premature cataract formation, clinical evaluation shows no increase in lens opacities even in patients who neglected careful eye protection.

The long-term side effects generally resemble those known as photoaging. High cumulative doses result in pigmentary changes, xerosis, loss of elasticity, wrinkle formation and actinic keratoses. Additionally, PUVA may induce profuse formation of dark lentigines, termed PUVA lentiginosis. The major concern with prolonged and repeated phototherapeutic regimens is the induction or promotion of skin cancers. Almost all data were obtained from psoriatics since they are the largest group of patients receiving PUVA. The risk is related to DNA damage but PUVA induced down-regulation of immune responses could play an additional role. In PUVA patients the risk of squamous cell carcinoma, but not of basal cell carcinoma, is significantly increased in comparison with matched controls and the magnitude of the increase appears to be dose-dependent. However, there is uncertainty about the contribution of PUVA to these observations: many of the reported patients had previous exposure to excessive sunlight and to treatments of known carcinogenicity (e.g. arsenic, UVB, antimetabolites).

No increased risk of malignant melanoma was found in all large-scale studies reported so far. However, more recently, Stern et al reported 11 melanomas in 9 patients from the cohort of 1380 patients enrolled in the PUVA 16-Center Study. Since there are no data available on dosimetry it is unclear whether melanomas are associated with a certain cumulative UVA dose or with episodes of phototoxic burns. Moreover, no information is given on other co-carcinogenic factors, on the clinical type of melanoma and, most importantly,

on the patients' history (dysplastic nevi, family history of melanoma). Besides this lack of important information this report emphasizes that the guidelines for PUVA treatment must be rigidly observed. Patients with long-term PUVA must receive lifetime monitoring.
Based on recent studies bath-PUVA with TMP or 8-MOP seems to bear no relevant risk of carcinogenesis. These data on the long-term safety of bath-PUVA are encouraging but no premature conclusions should be drawn.

Animal studies show that UVB is carcinogenic but model studies indicate that the observed risk of squamous cell carcinoma is much higher with PUVA than the calculated risk with UVB. The carcinogenic risk of UVB in comparison with PUVA is not known, and it will be crucial to monitor its long-term effects in psoriatics.

2.4. Photochemotherapy for Mycosis Fungoides

Gilchrest et al. in 1976 were the first to report the successful use of PUVA in treating MF. Nine patients with unsatisfactory responses to other treatments and histologically confirmed plaque- or tumor-stage or the erythrodermic form of MF were subjected to photochemotherapy. Of these nine patients, four achieved complete remissions and five experienced significant improvement.
Several studies on large patient cohorts have provided information on the response rates of initial PUVA treatment in relation to the disease stage Herrmann et al. have summarized the outcome of five of these studies, comprising a total of 244 patients. They have calculated the following average complete initial response rates: stage IA (90 %), stage IB (76 %), stage IIA (78 %), stage IIB (59 %), and stage III (61 %).
Of major relevance to predicting long-term outcome are data on relapse rates and mean disease-free intervals. Published data thus far indicate that PUVA is an excellent treatment option for early stage (IA-IIA) MF, where high rates of complete clearing can be achieved, and a substantial percentage of patients remain free of disease for many years. In advanced stages (IIB - IVB), PUVA is not sufficient as monotherapy, but as adjunctive treatment it can reduce the tumor burden in the skin and can increase the quality of life for patients.

2.4.1. Short-Term and Long-Term Hazards

The range of acute side effects of PUVA in patients with MF is, in principle, the same as in patients with other PUVA-responsive diseases. Increased phototoxicity can be occasionally observed in plaques and tumors of MF, sometimes leading to erosions and superficial ulcerations.
The most serious concern relates to the development of cutaneous malignancies. There is clear evidence that the combined or sequential use of

different therapies for MF may increase synergistically the risk of squamous cell carcinomas, whereas it is presently not possible to accurately estimate the risk for MF patients having been treated with PUVA monotherapy. The impact of this risk has to be balanced against the therapeutic benefit, but this may be meaningless in a potentially life-threatening disease such as MF.

2.5. Photochemotherapy for Vitiligo

PUVA represents the gold standard for vitiligo treatment. However, the patients must be aware that the treatment may require months or even years. 8-MOP, 5-MOP and TMP have been successfully used in several studies. The MPD test should preferentially be carried out in a non-sun exposed vitiliginous area. Treatments are given three times a week. The UVA dose is increased until a light pink erythema is induced in the vitiliginous areas. Further increases are not useful. There are no parameters that allow a reliable prognosis for an individual patient. Darker-skinned patients are more likely to achieve an acceptable degree of repigmentation. Treatment is more efficient in generalized vitiligo than in the focal or segmental variant, which often do not respond satisfactorily. Lesions of the face, neck and trunk respond more quickly than those in acral sites.

The course of repigmentation in phototherapy and photochemotherapy is very similar. The actual improvement of vitiligo can be first seen as early as after 15-20 irradiations, when perifollicular and marginal repigmentation appears. This may be followed by a spread of the repigmentation, which after several months to years can result in complete confluence of the follicular areas. Not many reports are available on the long-term efficacy of phototherapy. After PUVA treatment the majority of the patients retain PUVA-induced repigmentation for many years.

2.6. Photochemotherapy for Atopic Dermatitis

Systemic PUVA has been found to be effective for the treatment of atopic dermatitis. Although there is no doubt that PUVA therapy can be successfully used not only for moderate, but also severe and even erythrodermic forms of atopic dermatitis, it has some disadvantages. As compared to PUVA therapy of psoriasis, the actual number of treatments to clear atopic dermatitis was found to be relatively high. Even more important, cessation of PUVA therapy was associated with the occurrence of rebound phenomena in a high percentage of patients if photochemotherapy was not combined with systemic corticosteroids or if maintenance therapy was not continued for longer time intervals.

2.7. Photochemotherapy for Polymorphous Light Eruption

Since photochemotherapy of PLE involves the administration of a photosensitizer only small doses of UVA are required. Such doses are in general much lower than those required to cause PLE but still have the potential to induce pigmentation and to modulate immune responses in the skin. The efficacy of photochemotherapy in the prevention of PLE was confirmed in many clinical studies with comparably good therapeutic results. PUVA provides for almost complete or complete prevention of symptoms in 90 to 100% of patients.

PUVA hardening is indicated in selected patients with severe PLE that cannot be prevented by the use of sunscreens or UVB phototherapy. Treatment is given 3 times weekly over a period of 4 weeks. For the therapeutic effect suberythemogenic UVA doses are generally sufficient.

2.8. Phototherapy & Photochemotherapy of Other Diseases

In addition to its confirmed efficacy for classical indications as discussed above, phototherapy has also shown promise in the therapy of a variety of other dermatoses. However, since these conditions have been treated mostly on a tentative or trial basis, success rates have been poorly defined. Nevertheless, there has been a steady increase in the number of dermatoses for which phototherapy can now be regarded as a standard therapeutic option, and such disorders include in particular lichen planus, urticaria pigmentosa, granuloma anulare, pityriasis lichenoides, lymphomatoid papulosis and localized scleroderma. In addition, a large number of case reports also suggest the efficacy of the treatment in a variety of other dermatoses; however, in these last instances, the reports are based mostly on single cases and no definite positive conclusions can be drawn. In these instances the reader is strongly recommended to consult the original literature.

3. PHOTODYNAMIC THERAPY (PDT)

Photodynamic therapy aims to destroy selectively the desired target tissue and to avoid or minimize damage to vital structures. The photodynamic reaction consists of the excitation of photosensitizers (usually porphyrins) by visible light in the presence of oxygen resulting in the generation of reactive oxygen species (ROS). These ROS mediate cellular and vascular effects depending on the tissue localization of the photosensitizer. This results in a direct or indirect cytotoxic effect on the target cell.

3.1. Principles of Photodynamic Therapy

3.1.1. Photosensitizers

For dermatological use only "endogenous" sensitizers such as 5-aminolevulinic acid (ALA) are currently available. 5-ALA is a metabolite of the heme biosynthesis pathway, which induces the synthesis of the actual photosensitizer (protoporphyrin IX, PpIX) in the target tissue. The concentration of the PpIX depends on the metabolic status of the diseased tissue.

3.1.2. Light Sources and Dosimetry

Light penetration into skin increases with longer wavelengths. To match the absorption maxima of PpIX, wavelengths around 630nm are necessary. As opposed to other disciplines dermatological applications need no fiberoptics for endoscopic treatment and thus simple incoherent light sources can be used. With regard to therapeutic results there is no difference between coherent and incoherent light. A standardized light dosimetry for PDT does not exist. For PDT of skin tumors and actinic keratoses light doses of 100-150 J/cm^2 are necessary. With current incoherent light sources the treatment duration is about 20 to 30 minutes.

3.1.3. Mechanism of Action

PDT-induced effects are mediated by photooxidative reactions. During irradiation, the photosensitizer is absorbing light and is converted to an excited (triplet) state. The energy can be then transferred to molecular oxygen resulting in the generation of reactive oxygen species, mainly singlet oxygen. Visible damage consists of cell membrane defects as a consequence of lipid peroxidation with consecutive cell lysis. Depending on the intracellular localization of the photosensitizers also damage to subcellular structures such as mitochondria, lysosomes or endoplasmic reticulum occurs whereas DNA is not a primary target.

3.2. Photodynamic Therapy in Dermatology

Topical application of photosensitizers avoids generalized cutaneous photosensitivity. However, only small hydrophilic molecules such as ALA penetrate well into the skin. ALA application induces the production of intracellular porphyrins selectively in epidermal cells and the pilosebaceous unit. These structures synthesize porphyrins to a much higher amount than fibroblasts, myocytes or endothelial cells. Epithelial tumors synthesize much higher amounts of protoporphyrin IX than the surrounding tissue and can therefore be destroyed without damage to normal skin. Topical ALA-induced photosensitivity is restricted to the target area. Systemic porphyrin induction is not observed following topical application. The only significant acute side effect of ALA-PDT is a stinging pain during and shortly after irradiation.

3.2.1. Oncologic Indications

Kennedy et al. reported the first clinical results with topical ALA-PDT in 1990. After application of 20% ALA in a cream formulation and an incubation time of 3-6h, lesions were irradiated with a 500 W slide projector. Out of 80 BCC 90% showed complete remission within 2-3 months after therapy. Wolf et al treated 70 skin tumors with topical ALA-PDT. Five of six superficial squamous cell carcinomas, nine actinic keratoses, and 36 of 37 superficial BCC showed complete remission. Topical ALA-PDT is effective with excellent cosmetic results in patients with nevoid BCC syndrome, arsenic-induced skin cancer or with skin cancer due to immunosuppression after kidney transplantation. Nodular BCC also does not respond sufficiently (response rate 64% in 25 lesions) probably because of insufficient light penetration. Tumor thickness appears to be crucial and should not exceed 2-3 mm when a single treatment session is planned. Treatment efficacy can be enhanced by repeated treatment sessions.

3.3. Perspectives

The efficacy of PDT in the treatment of superficial neoplastic skin lesion, in particular, actinic keratoses, Bowen's disease, and basal cell carcinomas has been sufficiently documented. PDT may also find a place in the treatment of selected patients with psoriasis. Crucial parameters are the depth of the penetration of light as well as of the sensitizer into the skin. Moreover, for the treatment of skin cancers with metastatic potential patients have to be selected carefully and histological diagnosis and determination of tumor thickness are a prerequisite. There is still need for controlled studies and for standardized treatment protocols.

Bibliography

Honig B., Morison W. L., Karp D. Photochemotherapy beyond psoriasis. J Am Acad Dermatol 1994; 31:775-790

Hönigsmann, H. "Polymorphous light eruption." In *Clinical Photomedicine*, eds. H. W. Lim, N. A. Soter, Marcel Dekker, New York, 1993; pp 167-179

Hönigsmann, H., Szeimies, R.-F., Knobler, R., Fitzpatrick, T. B., Pathak, M.A., Wolff. K. "Photochemotherapy and Photodynamic Therapy." In *Dermatology in General Medicine 5th Edition*, eds. T. B. Fitzpatrick et al, McGraw-Hill Inc. New York 1999; pp 2880-2900

Hönigsmann H. Phototherapy for psoriasis. Clin Exper Dermatol 2001; 26:1-9

Hönigsmann H., Tanew A., Bergstresser P. R. "Photo(Chemo)Therapy for Cutaneous T-Cell Lymphoma." In *Dermatological Phototherapy and Photodiagnostic Methods*, eds. J. Krutmann H. Hönigsmann C. A. Elmets P. R. Bergstresser, Springer, Berlin-Heidelberg-New York, 2001; pp 122-134

Krutmann, J., Morita A. "Photo(chemo)therapy for Atopic Dermatitis" eds. J. Krutmann, H. Hönigsmann, C. A. Elmets, P. R. Bergstresser, Springer, Berlin-Heidelberg-New York, 2001; pp 93-108

Morison W. L. Phototherapy and Photochemotherapy of skin disease" 2nd edn. Raven Press, New York, 1991

Morison W. L., Baughman R. D., Day R. M., Forbes P. D., Hönigsmann H., Krueger G. G., Lebwohl M., Lew R., Naldi L., Parrish J. A., Piepkorn M., Stern R. S., Weinstein G. D., Whitmore S. E. Consensus workshop on the toxic effects of long-term PUVA therapy. Arch Dermatol 1998; 134:595-598,

Ortel B., Hönigsmann H. "Phototherapy and Photochemotherapy" In *Photodermatology*, ed. Hawk J. L. M., Arnold, London 1999; pp 223-240

Ortel B., Alge C., Pandy A. "Phototherapeutic Options for Vitiligo" In *Dermatological Phototherapy and Photodiagnostic Methods*, eds. J. Krutmann, H. Hönigsmann, C.A. Elmets, P. R. Bergstresser, Springer, Berlin-Heidelberg-New York, 2001; pp 136-161

Stern R. S., Liebman E. J., Vakeva L. Oral psoralen and ultraviolet-A light (PUVA) treatment of psoriasis and persistent risk of nonmelanoma skin cancer. PUVA Follow-up Study. J Natl Cancer Inst 1998; 90:1278-1284

Szeimies R.-M., Karrer S., Abels C., Landthaler M. & Elmets C. A. "Photodynamic Therapy in Dermatology." In *Dermatological Phototherapy and Photodiagnostic Methods*, eds. J. Krutmann, H. Hönigsmann, C. A. Elmets, P. R. Bergstresser, Springer, Berlin-Heidelberg-New York, 2001; pp 209-247

DEVELOPMENT OF PERSONAL UVB SENSOR: DETECTION OF PREVITAMIN D PHOTOSYNTHESIS

Igor Gvozdovskyy and Irina Terenetskaya
Institute of Physics, National Academy of Sciences of Ukraine, 03039, Kiev-39, Ukraine. E-mail: teren@iop.kiev.ua

INTRODUCTION

Although health-promoting properties of sunlight have been recognized from the beginning of civilization, predicted increase in solar UVB radiation (280-315nm) produced by ozone depletion has raised concerns on the resulting effects on human health and ecosystem. While there are both beneficial and harmful UVB effects, it is an increase in the harmful effects, which attracts most attention.

Excessive UV exposures are generally associated with acute and chronic health effects, such as erythema, skin cancer, immune system suppression, cataract. To raise public awareness and to provide information about potential risks of solar UV exposure, the use of a simple universally applicable solar UV index related to minimum erythema dose has been recommended by WHO in 1994 (1 MED = 200 J/m^2).

A number of personal UV dosimeters have been designed to measure ambient solar UV levels and give an indication of one's accumulated UV dose. At the moment variety of small electronic instruments are available on the market. For the most part the spectral sensitivity corresponds to the CIE erythema action spectrum [1], and anyone can easily determine exposure for healthy sunbathing according to the individual skin type.

Worthy of mention is passive chemical UV dosimeters that change optical characteristics when sunlight is applied to the materials. They have been developed for public use and calibrated for the erythemally effective exposure. The SUNTEST is a chemical UV dosimeter with silver-mercury-oxalate suspension [2]. The accumulated UV dose can be evaluated *in situ* from the darkening of the dosimeter, from white to dark brown, comparing it with the calibration scale. Another chemical dosimeter - polysulphone film, has been used extensively for measurements of human exposure in epidemiological studies [3-5]. However, it is unable to provide UV dosimetry *in situ* because the laboratory analysis of an exposed material is needed with the use of UV absorption spectroscopy.

Furthermore, several biological dosimeters have been developed to indicate the effect of UV irradiance on a specific biological system over a designated period of time. They use biological material as a UV-sensitive target for assessing UV biological effectiveness, primarily its DNA-damaging capacity. But the result is (often) not available immediately after an exposure, as further (sometimes time-consuming) laboratory analysis of the dosimeter material is required to determine the degree of UV-initiated change. The examples are a Biofilm consisted of immobilized bacterial spores *Bacillus subtillis* [6,7], bacteriophage T7 in buffer solution [8] and solution of DNA molecule [9].

While excessive UV exposures are commonly associated with adverse health effects, small amounts of UV are beneficial for people and essential in the natural production of vitamin D_3 in skin. Natural exposure to sunlight is responsible for maintaining adequate vitamin D[1] nutrition for most of the population in the world, and the importance of sunshine as a therapeutic measure for preventing the bone deforming diseases is now appreciated. Moreover, in addition to the well-established effects of vitamin D on maintenance of mineral homeostasis, recent evidence has suggested a wider biologic role of vitamin D and its active metabolites in tissues not primarily related to mineral metabolism [10,11].

In particular, as reported by latest epidemiologic study [12], several types of cancer (breast, prostate, kidney, colon, ovary, and pancreas) have a strong association with low levels of vitamin D and lack of sunlight. It has been noted that northern countries have higher levels of cardiovascular disease and more heart attacks occur in winter months [13,14]. Vitamin D and/or sunlight (UV-B) have been shown to lower blood pressure, restore insulin sensitivity, lower cholesterol, and normalize food intake and blood sugar [15]. Low vitamin D is associated with several autoimmune diseases including Multiple Sclerosis, rheumatoid arthritis, thyroiditis and Crohn's disease [15].

---

[1] Two principal chemical species of vitamins - D_2, or ergocalciferol ($C_{28}H_{44}O$) and D_3, or cholecalciferol ($C_{27}H_{44}O$) are available. The terminology vitamin D is employed here in general sense. Both vitamins D_2 and D_3 are photochemically produced from their precursors ergosterol and 7-dehydrocholesterol (7-DHC). It is significant that basic monomolecular isomerizations of the two steroid species occur in perfect analogy.

Hence it follows that UV dosimetry from the standpoint of its vitamin D synthetic capacity deserves particular attention, especially in the context of dramatic influence of seasonal and latitudinal changes in solar UVB radiation on the ability of sunlight to provide the human requirements for vitamin D [16].

UV DOSIMETRY USING AN *IN VITRO* MODEL OF VITAMIN D_3 SYNTHESIS

As is known vitamin D_3 is synthesized from 7-dehydrocholesterol through two consecutive monomolecular isomerizations catalyzed by UVB radiation. At the first step previtamin D is formed by photoinduced hexadiene ring opening. Further it is converted into vitamin D by thermoinduced intramolecular hydrogen shift. Consequently, concentration of accumulated previtamin D in a definite exposure time can serve as a measure of biologically effective 'antirachitic' UV dose. The problem is that side photoconversions of previtamin D give rise to multicomponent photoisomer mixture that drastically complicate the concentration analysis.

In the pioneering applications of an *in vitro* model of vitamin D_3 synthesis (ethanol solution of 7-dehydrocholesterol) the analysis was conducted using high performance liquid chromatography (HPLC) [17]. Obviously time consuming HPLC analysis is unable to provide the UV measurements *in situ*.

A major step toward an *in situ* dosimetry was made with development of original spectrophotometric analysis [18,19]. To measure the 'antirachitic' UV dose, the solution of 7-DHC in ethanol ($C = 20$ μg/ml) is exposed in rectangular quartz cuvette of 0.5÷1cm thickness, and UV absorption spectra of the solution are recorded within 230-330nm before and after an exposure to sunlight (or artificial UV light). Concentration of accumulated previtamin D is derived from the spectra by computer processing using specially designed program. This enables performance of the 'antirachitic' UV dosimetry *in situ* if portable spectrophotometer is available on site.

Besides, it is exceptionally important that adequate mathematical model has been developed which links biological and physical units and enables calculation of the photoreaction kinetics for any monochromatic or polychromatic UV source with the intensity distribution $I_0(\lambda)$ [18]. It has been found that action spectrum of previtamin D photosynthesis (*in vitro*) calculated with the use of the model correlates well with the experimentally measured one.

Using a number of narrow band filters ($\Delta\lambda = 2$ nm) and calibration facility at the Institute of Space Aeronomy (Brussels) the physical and erythemal doses required for accumulation of 5% previtamin D have been determined for different wavelengths [21] (Table 1). It is clear from the Table that the relation between physical, erythemal and 'antirachitic' doses is not constant with the wavelength, and because of this, may vary over a wide range with the constantly changing spectral composition of sunlight.

Table 1. Physical and erythemal doses required for accumulation of 5% of previtamin D under quasi-monochromatic UV irradiation at different wavelengths.

Wavelength (**nm**)	260	270	280	290	300	310
Previtamin D (%)	5	5	5	5	5	5
Erythemal dose (**MED**)	0.28	0.21	0.2	0.48	0.70	3.87
Physical dose (**J/m^2**)	55	42.7	40	95	205	10400

From the above discussion it appears that while the method is workable, it is not easily accessible in personal UV dosimetry in view of the use of liquid medium, UV spectrophotometer and computer processing of the absorption spectra.

Hence our task was to find a method for easier monitoring the previtamin D photosynthesis or even to find a possibility to make the process visible. With this aim we concentrated our attention on the nematic liquid crystals with intended use of provitamin D molecule as a chiral dopant [22].

It is believed that peculiar features of the structure endow the 7-dehydrocholesterol molecule with this ability. The molecule can be considered 'lath shaped' with dimensions approximately 4Å × 7Å × 20Å [23]. Actually, the molecule is not flat, but presents a corrugated appearance with a small twist about the long axis. In addition the methyl groups at carbons 10 and 13 protrude out of the molecular plane. By this is meant that the parallel layers in nematic liquid crystal will be twisted with respect to one another by the action of 7-DHC dopant.

LIQUID CRYSTALLINE SENSOR

Basic features of nematic crystal with chiral dopant

As is known, chiral dopant is optically active substance that induces cholesteric phase when added in small amount into nematic liquid crystal (LC). The director of the LC molecules adopts a helically twisted orientation under the action of chiral dopant (Fig.1).

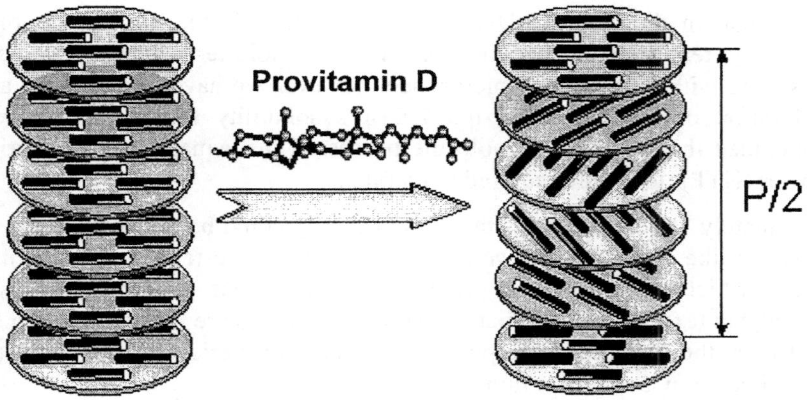

*Figure 1. Schematic representation of transformation of nematic LC into cholesteric phase induced by chiral dopant of provitamin D and an idealized representation of the order of a cholesteric liquid-crystalline phase. Each layer exhibits nematic-like order and the layers are twisted with respect to one another to form a macrohelix. The solute molecules are aligned such that their long axes lie parallel to the axis described by the solvent molecules. The distance it takes the director to complete a turn of 360^0 is called the cholesteric pitch **P** of the liquid crystal.*

If unpolarised white light is incident parallel to the helical axis, a narrow wavelength band will be reflected while all the others are transmitted. The critical wavelength λ_0 is determined by the helix pitch $\boldsymbol{P}$ and the average index of refraction $\boldsymbol{n}$ so that $\lambda_0 = \boldsymbol{P} \cdot \boldsymbol{n}$. This can introduce various colors of cholesteric LC if the wavelength of selectively reflected light λ_0 is in the visible region of the spectrum.

Since the molecular conformation of steroid moiety of provitamin D molecule is altered by the photochemical conversion into previtamin D, it is expected that cholesteric pitch and the wavelength for selective reflection will be affected and, as a result, the LC will vary in color. This could provide the easiest detection of previtamin D synthesis and evaluation of the accumulated UV dose *in situ* by comparison of the LC cell color with the calibration scale.

To fulfill the desired requirements for host matrix, the LC should be transparent in the UVB range, thermally stable to the temperature variations within 10-40^{0}C, be a good solvent for 7-DHC and be stable in relation to UV and visible light. Two nematic liquid crystals LC-805 (NIOPIK, Russia) and ZLI-1695 (Merck) have been selected as host matrices, and 7-DHC (Sigma) was dissolved in the LC material in concentration 5÷10 wt.%.

Although on dissolving 7-DHC in the nematic LC the appearance of a papillary texture characteristic for induced cholesteric phase has been observed with a polarizing microscope, however we have not detected any coloration of the LC up to the 7-DHC's solubility limit. Thus, it was concluded that the 7-DHC molecule possessed inadequate helical twisting power (HTP)[2] to ensure required helix pitch.

Fortunately, it occurred to us that this difficulty could be obviated by use of a wedge-like cell. As is known a chiral nematic (cholesteric) LC, filled into a cell which was treated to provide a homogeneous alignment, forms a Grandjean texture [24] in which quasi-nematic layers are twisted (see Fig.1), and over the distance corresponding to the cholesteric pitch the director would perform a complete turn.

If the walls of the cell are not parallel to each other, but form a wedge, the width of the cell is proportional to the distance from the origin of the wedge angle. At certain distances the spacing between the cell surfaces equals multiple values of half of the natural pitch of the cholesteric helix (***P***/2) (see Fig.2). If the LC molecules are oriented parallel at both surfaces of the cell, the helix is distorted between these positions. With increasing cell spacing the cholesteric helix is at first elongated until a new half turn of the director is added and a compressed helix with an additional half turn is formed. This change in total twist leads to the formation of disclination lines (Cano) [24]. The disclinations occurring at equal distances from one another can be easily observed with a polarizing microscope.

---

[2]At the dopant concentrations $\boldsymbol{C} < 10$ % (by the mass fraction) the pitch is inversely proportional to ***C***, and the constant of proportionality is called the helical twisting power (***HTP***) of the chiral dopant. The HTP is defined as: $\boldsymbol{HTP} = (\boldsymbol{P} \cdot \boldsymbol{C})^{-1}$

EXPERIMENTAL

The wedge-like cell was prepared using Mylar spacer between two quartz plates (15 mm × 20 mm) as substrates. The cells thickness was measured by interferometer (***L*** = 63μm). To provide planar boundary conditions the quartz substrates were spin coated with polydimethylsiloxane annealed on the hot plate at 200-250^{0}C over 4 hours and mechanically rubbed to ensure planar uniform alignment of the LCs. To avoid oxidation of 7-DHC the LC cells were carefully stuck along the perimeter. With these cells it was possible to follow the photoreaction course of provitamin D photoisomerization by recording the UV absorption spectra and, in parallel, by visual observations of the cells using polarizing microscope.

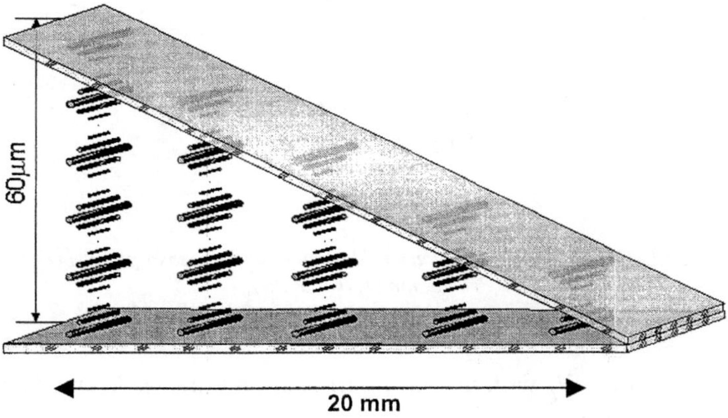

Figure 2. Schematic representation of the wedge-like cell and formation of disclination lines (Cano) in chiral nematic (cholesteric) liquid crystal.

Transformation of the nematic LCs into cholesteric phase was evidenced by observation the Cano lines when the cell was sandwiched between crossed polarizers. Well known dependence of the number of Cano disclinations (N_C) on the dopant concentration (C) has been detected. N_C =10 has been found at C =10 wt.% that corresponded to maximum solubility of 7-DHC. The cholesteric pitch P was calculated by equation $P = 2L/N_C$ [24], and the changes of P were detected within the range of 9 ÷ 24 μm depending on the concentration of 7-DHC. What this means is that the selective reflection wavelength lies outside the visible spectrum (as it has been suggested earlier).

UV lamp irradiation

The sample with initial number of the Cano lines N_C=5 was irradiated with fluorescent lamp (EL-30) of integral UV intensity 0.5 mW/cm^2 within spectral range 260-350nm (Fig.3).

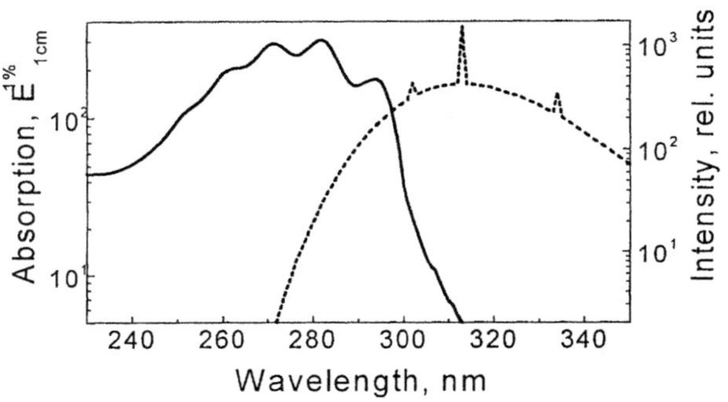

Figure 3. The EL-30 lamp radiation spectrum (dash line) and the absorption spectrum of Provitamin D_3 (solid line).

The decrease in the N_c with increasing irradiation time has been detected by placing the sample between crossed polarizers (Fig.4). This indicates that 7-DHC undergoes photoinduced transformation into the photoisomer possessing considerably lesser helical twisting power (HTP). Obviously, HTP of 7-DHC molecule with its rigid steroid skeleton may be significantly decreased by hexadiene ring-opening into conformationally flexible molecule of previtamin D. This gives rise to the increase in the cholesteric pitch and, as a result, is accompanied by the N_c decrease.

Consequently, by this means our findings hold promise for dosimetry of biologically active UVB solar/artificial radiation (280 – 315nm) by visual observation the Cano lines decrease with UV dose. This distinctly simplifies previously suggested method of UV dosimetry based on the spectral monitoring previtamin D photosynthesis in ethanol solution [19,20].

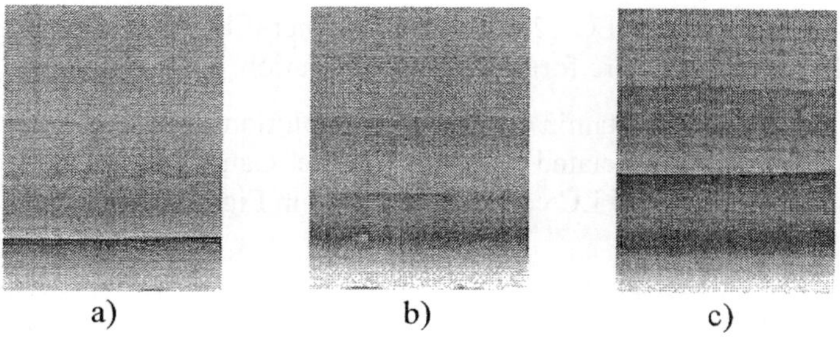

Figure 4. Photographs of the LC wedge cell sandwiched between crossed polarizers that are illustrative of the Cano lines decline under irradiation with the UV lamp: a) $N_C = 5$ at $t = 0$; b) $N_C = 4$ at $t = 25min$; c) $N_C = 3$ at $t = 60min$.

Solar irradiation

It is important for the UV biodosimetry to have great confidence that the decrease in the Cano stripes number in the LC cell is caused by previtamin D formation. To be certain that the N_C decrease closely corresponds to previtamin D photosynthesis *in vitro*, the LC cell (ZLI-1695 with 7-DHC dopant) and the cuvette with ethanol solution of 7-DHC (C = 20μg/ml) were simultaneously exposed to sunlight around noon (clear day).

Before the exposure the UV absorption spectra of both samples had been recorded by the spectrophotometer KSVU-23 (LOMO, Russia), and the initial number of the Cano lines in the LC cell had been fixed ($N_C = 10$ at $t = 0$). In the course of solar irradiation the number of the Cano lines in the LC cell was checked time by time by placing the cell between crossed polarizers. At the moment when one Cano line was disappeared the absorption spectra of both samples were recorded again. The same procedure was repeated to the total exposure time of 200 min. During the night the samples were stored at the refrigerator, and on the next day were exposed 160 min in total until Cano lines number reduced to $N_C = 5$.

To derive the concentration of accumulated previtamin D the absorption spectra of the solution were processed with computer according to the original program in which known individual spectra of the photoisomers were used as the reference ones [19]. However, spectral analysis of the LC samples was impeded because of small shift and distortion of the 7-DHC spectrum in the LC matrix. Nevertheless, HPLC analyses of 7-DHC photoisomerization in LC [25] and in polymer film [26] showed that the same photoproducts were formed as in the solution.

Accumulation of previtamin D in ethanol solution during the exposure to sunlight and the associated changes in the Cano lines number in the simultaneously exposed LC cell are presented in Fig. 5.

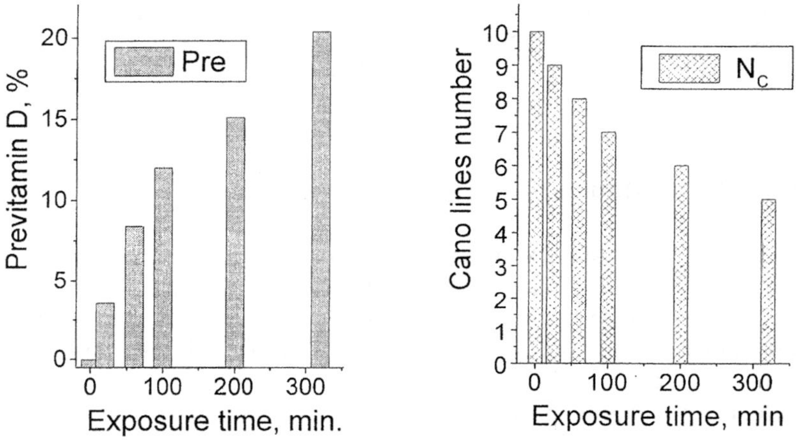

Figure 5. Accumulation of previtamin D in ethanol solution (left) and decrease in the Cano lines number in the LC cell during simultaneous exposure to sunlight
(August 21, 2000, local time 12.00-15.20 and August 22, 2000, local time 12.40-15.00)

From the result obtained it may be deduced that dependence of the Cano lines number on the accumulated UV dose proceeds in close agreement with photosynthesis of previtamin D in solution. This conclusion is more convincingly demonstrated in Fig.6, where one can see wholly satisfactory linear relationship between the N_C and previtamin D_3 concentration. From the data presented in Fig.6 one can determine that $\Delta N_C = -1$ corresponds to ~4% concentration of previtamin D, that is ~0,8 μg/ml of previtamin D accumulation *in vitro*.

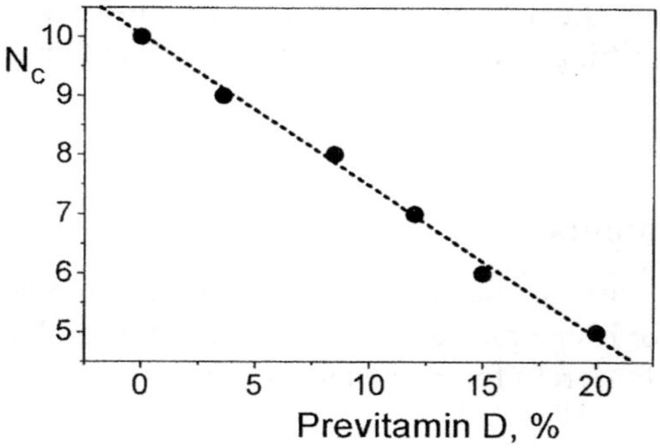

Figure 6. Correlation between the Cano lines number N_C in the LC cell and Previtamin D_3 accumulation in ethanol solution.

CONCLUSIONS

A new biosensor of vitamin D synthetic capacity of solar/artificial UVB radiation has been developed that is based on nematic liquid crystal with provitamin D_3 dopant. For the first time it has been experimentally demonstrated that nematic liquid crystals (LC-805, ZLI-1695) are converted into cholesteric phase by doping with chiral steroid biomolecules of provitamin D_3.

During UV exposure of the admixed LC in a wedge-like cell the decrease in the Cano lines number has been observed as a result of UV initiated photoisomerization which changed helical twisting power of the dopant molecules of 7-DHC. Excellent correlation between this clearly defined N_C dependence and previtamin D accumulation *in vitro* has been established. This suggests that the decrease in the N_C on the exposure time is well representative of the accumulated dose of biologically active “antirachitic” UV radiation, and after calibration the sensor can be used as personal UV-B biodosimeter.

In this manner the doped LC film can ensure personal monitoring the vitamin D synthetic capacity of sunlight by visual observation with crossed polarizers. The sensor great advantage is that it can be used to evaluate UV exposure in realistic conditions of changing orientation and surface position, e.g. for investigating exposure of different body parts during various activities. Further attempts will be directed to greater simplification of the procedure by shifting the selective reflection band into visible spectral region.

Acknowledgments

The authors wish to thank Prof. G.Chilaya for providing the LC-805, Prof. R.Weiss for his helpful advice on the ZLI-1695 and to Drs. R. Wildner and W.Becker (E.Merck) for the generous gift.

REFERENCES

1. Commission Internationale d'Eclairage. A reference action spectrum for ultraviolet induced erythema in human skin. C.I.E. J. 1987; 6:17-22.
2. Horkay I, Wikonkal N, Patko J, Bazsa G, Beck M, Ferenczi A, Nagy Z, Racz M, Szalay T. SUNTEST: a chemical UVB radiation dosimeter. J. Photochem. Photobiol. B:Biol. 1995; 31:79-82.
3. Davis A, Deane GHW, Diffey BI. Possible dosimeter for ultraviolet radiation. Nature 1976; 261:169-179.
4. Airey DK, Wong JCF, Fleming RA, Meldrum IR. An estimate of the total UVB exposures for outdoor workers during a south-east Queensland summer. Health Phys. 1997; 72(4):544-549.
5. Krins A, Dorschel B, Henniger J, Knuschke P, Bais A. Readings of polusulphone film after fractionated and continuous exposures to UV radiation and consequences for the calculation of the reading resulting from polychromatic UV radiation. Radiat. Prot. Dosim. 1999; 83:303-308.
6. Quintern LE, Puskeppeleit M, Rainer P, Weber S, ElNaggar S, Eschweiler U, Horneck G. Continuous dosimetry of the biologically harmful UV radiation in Antarctica with the biofilm technique. J. Photochem. Photobiol. B:Biol. 1994; 22:59-66.
7. Horneck G. Quantification of biological effectiveness of environmental UV radiation. J. Photochem. Photobiol. B: Biol. 1995; 31:43-49.
8. Ronto G, Gaspar P, Berces A. Phages T7 in biological UV dose measurement. J. Photochem. Photobiol. B: Biol. 1992; 12:285-294.
9. Regan JD, Yoshida H. DNA UVB dosimeters. J. Photochem. Photobiol. B: Biol.1995; 31:57-61.
10. Mayar Anita, Norman Anthony. "Vitamin D", in: *Encyclopedia of Human Biology*, Academic Press, Inc. 1991; Vol. 7:859-871.
11. Reichel H, Koeffler HP, Norman AW. The role of the vitamin D endocryne system in health and disease. N. Engl. J. Med. 1989; 320:980-991.
12. Grant WB. An estimation of excess cancer mortality in the United States due to inadequate doses of solar ultraviolet-B radiation and/or vitamin D.
13. Segall JJ. Latitude and ischemic heart disease (letter). Lancet 1989; 1:1146.

14. Williams FL, Lloyd OL. Latitude and heart disease (letter). Lancet 1989; 1:1072-1073.
15. Sullivan K. Preliminary Report on the importance of sunlight and vitamin D. http://www.sunlightandvitamind.com
16. Webb AR, Kline LW, Holick MF. Influence of season and latitude on the cutaneous synthesis of vitamin D3: exposure to winter sunlight in Boston and Edmonton will not promote vitamin D3 in human skin. J. Clin. Endocrinol. Metabol. 1988; 67:373-378.
17. McLaughlin JA, Anderson RR, Holick MF. Spectral character of sunlight modulates photosynthesis of previtamin D3 and its photoisomers in human skin. Science 1982; 216:1001-1003.
18. Terenetskaya IP. Provitamin D photoisomerization as possible UVB monitor: kinetic study using tunable dye laser. *SPIE* 1994; 2134B:135-140.
19. Terenetskaya IP. Spectral monitoring of biologically active solar UVB radiation using an *in vitro* model of vitamin D synthesis. Talanta 2000; 53:195-203.
20. Galkin ON, Terenetskaya IP. 'Vitamin D' biodosimeter: basic characteristics and potential applications. J. Photochem. Photobiol. B:Biology 1999; 53:12-19.
21. Bolsée D, Webb AR, Gillotay D, Dörschel B, Knuschke P, Krins A, Terenetskaya I. Laboratory facilities and recommendations for the characterization of biological ultraviolet dosimeters. Applied Optics 2000; 39: 2813-2822.
22. Leder LB. Rotatory sense and pitch of cholesteric liquid crystals. J. Chem. Phys. 1971; 55:2649-2657.
23. Carlisle CH, Crowfoot D. Proc. Roy. Soc. (London) 1945; A184:64-69.
24. De Gennes, Paul, *The Physics of Liquid Crystals*, Oxford: Clarindon Press, 1974.
25. Cassis EG, Weiss RG. Liquid-crystalline solvents as mechanistic probes--V. An investigation of the effect of cholesteric order on the formation rates of vitamin D_3 from pre-vitamin D_3 and of pre-vitamin D_3 from vitamin D_3. Photochem.&Photobiol. 1982; 35:439-444.
26. Majima T, Nakamura Y, Sato T. Waseda University Bulletin 1991; 13:13-15.

LIGHT EMITTING DIODES FOR PHOTOTHERAPY FOR THE CONTROL OF JAUNDICE

Hendrik J. Vreman, Ronald J. Wong, and David K. Stevenson
Stanford University School of Medicine, Stanford CA, 94305, USA

INTRODUCTION

The mission of our laboratory is the study of neonatal jaundice. An important focus of our research is to bring technology related to the diagnosis, prevention, and treatment of jaundice from the laboratory to the bedside. The following report summarizes our studies related to the efficacy and safety of light in the treatment of jaundice.

Phototherapy (PT) is the most frequently used treatment for jaundice in neonates.[1] It is the only major therapy, except for transplantation, presently available for the control of excessive hyperbilirubinemia and jaundice in patients with Type 1 Crigler-Najjar Syndrome (CNS)[2], who require life long nightly PT. The efficacy of PT is dependent upon irradiance (light intensity), quality of the light (optimal in the blue-green, 400 to 525 nm)[3], exposed body surface area, and duration of exposure. The most commonly used light sources are: fluorescent tubes (daylight, broad and narrow spectrum blue); spotlights with halogen bulbs; and fiberoptic pads for contact use. All, but the fiberoptic devices, use broad wavelength light sources, which can emit UV and/or IR radiation in addition to visible light outside the bilirubin absorption spectrum (460 ± 60 nm).[4] These wavelengths are primarily responsible for the reported side effects of PT, which are mediated through interactions with endogenous or administered photosensitizers (see Table 1). The removal of bilirubin via PT is dependent upon: (1) the rate of bilirubin transport from the circulation to the skin; (2) the rate of bilirubin photoalteration (rate limiting step); (3) the transport of photoproducts from skin to the circulation; and (4) the excretion of these products by the liver and kidneys. The major mechanisms of bilirubin photoalteration have been identified and are shown in Figure 1.[5]

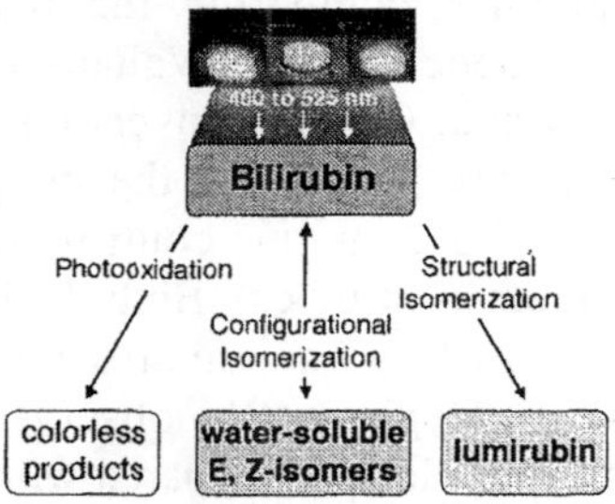

Figure 1. The major mechanisms of bilirubin photoalteration.

Currently we are developing a new PT light source based upon the recently developed high intensity gallium nitride light emitting diodes (LEDs).[6,7] These semiconductor devices have several potential advantages over conventional light sources for PT. The blue LEDs emit high intensity, narrow spectrum light, which overlaps the absorption spectrum of bilirubin, the jaundice pigment.[8] They have no glass parts, are small in size (< 5 to 9 mm), and lightweight (< 300 mg). The lens type and, especially, surface mount LEDs can be mounted on rigid or flexible low voltage circuit boards to form devices of unlimited sizes and shapes, such as canopies, pads, blankets, and even clothing. In addition, they have very long life (> 3 yrs). Finally, their emission spectrum can be selected for optimizing *in vivo* bilirubin photodegradation while minimizing activation of endogenous or administered photosensitizers (e.g. riboflavin[9], drugs[10]).

Table 1. *Documented Adverse Effects of Phototherapy*[11-13]

Skin: bronzing, tanning, transient rash, poor healing, scarring, cancer, insensible water loss, photosensitization, hyperthermia

Gastrointestinal: gallstone formation, increased motility, loss of nutrients and vitamins, hypocalcemia, electrolyte imbalance

Eye: retinal and ocular injury

Generalized: circadian rhythm disturbances, growth retardation, changes in gonadotropin levels, disturbances of blood flow, increased incidence of intracranial and pulmonary hemorrhages, apnea, etc

The objectives of our ongoing studies are: (1) to determine and compare the efficacy of LED and conventional light sources to degrade bilirubin *in vitro* and *in vivo*, and (2) to examine LED light for limiting some of the (potential) side effects of PT, which have been observed with conventional light devices (Table 1).

Early PT Studies

Our earliest work evaluated the use of light for the treatment of neonatal jaundice comparing the feasibility and efficacy of a newly developed fiberoptic blanket (Wallaby Phototherapy System, Healthdyne Technologies, Marietta, GA) to conventional PT on 42 full-term infants with nonhemolytic jaundice.[14] Infants in the study group were treated with the fiberoptic blanket and those in the control group were placed under a standard PT unit (fluorescent tubes). Both devices delivered an average irradiance of 7.0 ± 0.5 $\mu W/cm^2/nm$ as measured with the Joey Dosimeter (Healthdyne Technologies). We observed that incremental changes in serial plasma bilirubin levels compared with the initial bilirubin concentration did not differ between the study and control groups. However, for infants whose initial plasma bilirubin concentration exceeded 200 μmol/L (or 11.7 mg/dL, i.e. 17.1 μM = 1 mg/dL), a statistically significant difference was found between the average bilirubin level (μmol/L) at initiation of PT and the average bilirubin level at termination of treatment in both study (266 ± 22 vs 206 ± 42, respectively) and control groups (265 ± 28 vs 223 ± 26, respectively). We concluded that PT delivered by the fiberoptic blanket is safe and has an efficacy comparable to that of conventional PT. It also provides a convenient alternative PT treatment, which obviates the need for eye patches and facilitates maternal bonding to the infant during therapy.

Several other studies have been performed by others evaluating this device and another optical fiber device, the BiliBlanket (Ohmeda Inc., Columbia, MD), which had become available for clinical use.[15] The major drawback of these light sources is their limited ability to deliver high intensity light to a large body surface area. However, these devices are particularly attractive and safe for use in home PT of newborns.[15,16]

Photooxidative and Phototoxicity Studies

Natural and synthetic derivatives of heme (metalloporphyrins, Mps) are being evaluated as potential drugs for the prevention of neonatal hyperbilirubinemia.[17] These compounds are competitive inhibitors of heme oxygenase, the rate-limiting step in the production of bilirubin from heme.[18] Because carbon monoxide (CO) is also produced in equimolar amounts from this pathway, the quantitation of CO can be used as an index of the rate of heme degradation and bilirubin production *in vitro* as well as *in vivo*.[19] However, some Mps, such as tin proto- and mesoporphyrin (SnPP and SnMP), have been shown to be potent photosensitizers presumably through their ability to generate singlet oxygen in the presence of light. Singlet oxygen, in turn, oxidizes numerous molecules such as membrane proteins, lipids as well as bilirubin, to various products, which cause alterations and damage to the physiological and biochemical integrity of the organism. This mechanism is the basis for the use of porphyrin analogs in photodynamic destruction of certain tumors.[20]

In our studies (see below), we have established that CO is also a predictable and measurable product of both *in vitro*[21] and *in vivo*[22] photosensitizer-mediated photooxidative reactions. *In vitro* reactions are usually performed at ambient temperature in 2-mL clear glass vials with septum-sealed caps. Prior to the application of light to the bottom of vials (containing < 100 μL reaction mixture) the headspace is purged with CO-free air. After light exposure, the vial headspace gas is injected into a gas chromatograph connected to a sensitive Reduction Gas Detector (Trace Analytical Inc., Menlo Park, CA) via a double needle assembly for the quantitation of CO.[23] The instrument has a sensitivity of 1 pmol CO/vial.[24] CO generation is expressed as pmoles CO generated per h.

For *in vivo* CO generation studies, neonatal rats or mice or adult hairless mice are placed in specially constructed clear plexiglass chambers (7 to 30 mL volume) surrounded by light sources and supplied with CO-free air (20 to 50 mL/min). Total body CO excretion (VeCO) is calculated from the measurement of the flow rate and CO concentration of the chamber outlet gas and the animal's body weight. CO concentrations are quantitated by gas chromatography as described above. VeCO is expressed as μL CO excreted per h per kg body weight.[19,25]

In Vitro Studies

Because we routinely measure the production of CO as an index of bilirubin production in whole animals and isolated tissue preparations[26,27], which may be exposed to various Mps including SnPP, we investigated the possible production of CO from the oxidation of organic molecules in the presence of Mps and light.[28] We found that the *in vitro* generation of CO occurred in the presence of SnPP (5 μM) and cool white light (19 $\mu W/cm^2/nm$ or 10 W/m^2) from SnPP alone, human serum albumin (HSA), glucose, histidine, ethanolamine, medium-chain triglycerides, NADPH, and human plasma. Further detailed studies with HSA and NADPH established that the photogeneration of CO is not only nearly linear with time and irradiance; but also, curvilinear with respect to the SnPP concentration and dependent upon the presence of oxygen in the reactor headspace. In addition, cool white light generated less CO from HSA and NADPH than equidistantly placed blue or green fluorescent PT light tubes. Comparison of SnPP with other Mps indicated that SnMP is the most photoreactive, while zinc protoporphyrin (ZnPP) is the least.

In a subsequent study, our collaborators and we investigated the phototoxicity of six protoporphyrin dimethylesters [i.e. cobalt (Co), copper (Cu), manganese (Mn), nickel (Ni), Sn, and Zn]. Hemolysis of human erythrocytes and inactivation of two enzymes (acetylcholinesterase and β–galactosidase) were used to assess the phototoxicity of these metal chelates. SnPP, the only porphyrin found to hemolyze erythrocytes at a concentration of 40 μM (radiation dose, 230 kJ/m^2), was much less efficient than either free protoporphyrin IX or hematoporphyrin. SnPP completely inactivated β-galactosidase at concentrations above 15 μM (75 kJ/m^2) and drastically interfered with acetylcholinesterase activity at a concentration of 150 μM (75 kJ/m^2). CoPP, CuPP, MnPP, NiPP, and ZnPP were ineffective photohemolytic agents at 40 μM (230 kJ/m^2), but inactivated acetylcholinesterase and β-galactosidase activities to varying degrees. The results suggested that (i) metal ions reduce the phototoxicity of protoporphyrin IX, (ii) different metal ions reduce the phototoxic activity of protoporphyrin IX to different degrees, and (iii) the biological activities of the various metal complexes vary in different assay systems.

In Vivo Studies

In order to determine if the *in vitro* photogeneration of CO from photoreactive Mps correlated with *in vivo* phototoxicity, we studied the phototoxic effects of SnPP, SnMP, and the zinc analogs, ZnPP and ZnMP, in the neonatal rat model.[29] Wistar rat pups (24 to 36 h old) were injected with up to 60 μmol Mp i.p./kg body weight. The animals were placed over cool white fluorescent tube light (20 $\mu W/cm^2/nm$) for 12 h. This exposure resulted in a SnPP dose-dependent mortality with an LD_{50} of 11.7 μmol/kg. Treatment with 30 μmol SnMP/kg caused a 40% mortality rate. No fatalities were observed among light-exposed ZnPP- or ZnMP-treated pups. No deaths were observed among control pups treated with the highest Mp doses but kept in the dark. Similarly, no mortality was observed in untreated light-exposed control animals. We concluded that SnPP and SnMP are potentially fatal phototoxic substances in the neonatal rat and that ZnPP and ZnMP may be safer drugs for patients receiving light exposure. Furthermore, these results qualitatively paralleled those observed in previous *in vitro* assays.

Photosensitizer Studies

Early work by others have demonstrated that riboflavin (RF, Vitamin B_{12}), a natural photosensitizer, is degraded during PT to the point of deficiency.[30] It also accelerates the rate of bilirubin photodecomposition *in vitro* and *in vivo* in Gunn rats.[31] These results suggested that the systemic administration of an appropriate agent (photosensitizer) could potentially improve the efficacy of PT. A number of other studies have also investigated this deceptively attractive concept[9], but ignored, however, that singlet oxygen is not a magic bullet, which degrades only bilirubin. It was then found that co-administration of RF, through interactions with light, can lead to serious phototoxic reactions and eventually to skin necrosis in Gunn rats.[30,31]

Similarly, we also found that bilirubin photodegradation was enhanced by the presence of known photosensitizing Mps, such as SnPP and SnMP during exposure to light from clinically used fluorescent tubes.[28] As expected, different light qualities (colors) yielded different results. Solutions of bilirubin (92 mg/L) in HSA were exposed *in vitro* in the presence of tin and zinc Mps (10 μM) at 37°C during 30 min exposure to light. We found that bilirubin was photodegraded to approximately 60% (55/92 mg/L) of the dark control under cool white light (17 $\mu W/cm^2/nm$). The presence of ZnPP or ZnMP barely reduced the bilirubin concentration [57% (52/92) and 54% (50/92) of dark control], but the presence of photosensitizing SnPP or SnMP caused a significant enhancement of degradation down to 35% (33/92) and 25% (22/92) of dark control, respectively. The results demonstrate that the tin Mps are the potent photosensitizers capable of potentiating the native photodegradation of bilirubin in biologic matrices.

LED Studies

Because of the potential advantages of LEDs for use in PT, we compared the efficacy for photodegrading bilirubin between single LEDs of different color (Nichia Chemical Industries Ltd., Tokushima, Japan). Based on these results, we then constructed a prototype PT device with three focused arrays, each with 100 15° blue LEDs (see Fig. 1) and compared its efficacy with that of conventional PT devices by measuring the *in vitro* photodegradation of bilirubin containing HSA.[28] When blue, blue-green, green, and white LEDs were compared, blue light was found to be the most effective in degrading bilirubin 28% of the dark control, followed by blue-green (18%), and then white light (14%). Green light was least effective (11%). The prototype device was found to generate a greater irradiance (> 200 $\mu W/cm^2/nm$) than any of the conventional devices tested. It also supported the greatest rate of bilirubin photodegradation. We concluded that light from LEDs might be more effective for the treatment of hyperbilirubinemia than light from presently used PT devices. Furthermore, the unique characteristics of this light source may make it ideally suitable for use in safe and lightweight home PT devices.

On the basis of these *in vitro* results, we developed a prototype clinical device containing six focusable panels with 100 blue LEDs each. The efficacy of this device was compared to conventional PT devices at similar light intensities. A prospective open randomized study was performed at two university-affiliated community hospitals in Jerusalem.[32] Sixty-nine jaundiced, but otherwise healthy, term infants met the entry criteria for PT set by the American Academy of Pediatrics' Practice Parameter.[33] The main outcome measures were the duration of PT and the rate of decrease in serum total bilirubin (STB) concentration (expressed as μmol/L/h). The mean STB concentrations at initiation and termination of treatment did not differ between newborns receiving LED or conventional PT. The duration of PT and the rate of decrease in STB concentration were not statistically different between the two groups. The average rate of decrease in STB was -3.16 μmol/L/h (95% confidence limits -4.81, -1.51) in newborns receiving LED PT compared with -2.19 μmol/L/h (-3.99, -0.40) in those treated with conventional PT ($p < 0.14$). No side effects were noted in any of the newborns. The blue LED device was found to be as effective as conventional PT. It also was readily accepted by nursing staff because it did not produce the sensation of vertigo, which is often associated with exposure to the special blue fluorescent tubes. We concluded that LED PT devices could provide much higher irradiances, and thus greater efficacy.

Previous studies by others[4,34] have reported conflicting results concerning the efficacy of blue versus green light PT. This disparity may be attributed to the use of light sources, which provided relatively broad-spectrum non-standardized light. Similar to the above study[32] where we used a blue LED prototype device, we produced a second prototype LED device with six focusable panels with 100 15° blue-green LEDs. Because LEDs have narrow luminous spectra, we can compare, for the first time, the efficacy of blue- (459 ± 22 nm) vs blue-green (505 ± 38 nm) PT.[35] One hundred and fourteen jaundiced but otherwise healthy term infants were exposed to either LED or conventional PT (Microlites, Air Shields, Hatboro, PA, contains 3 halogen bulbs) with irradiances ranging from 5 to 8 $\mu W/cm^2/nm$ at a distance of 50 cm. Infants receiving LED PT were further randomized to either blue or blue-green light exposure. The average rate of decrease in STB was –2.82 μmol/L/h (95% confidence limits -4.01, -1.62) in newborns receiving blue LED PT compared with -1.55 μmol/L/h (-2.86, -0.25) and –2.42 (-3.22, -1.63) in those treated with blue-green and conventional PT, respectively. We concluded that blue LED and conventional PT were more effective than blue-green LED PT, but the difference did not reach statistical significance. The lower efficacy of blue-green PT was consistent with our previous findings.[8]

Using genetically jaundiced, neonatal Gunn rats, we evaluated LED light efficacy through *in vivo* measurements of plasma bilirubin (PBR).[22] Light safety was assessed through measurements of VeCO, as described above. Heme catabolism is the major component of VeCO, but we hypothesized that because photooxidative injury could also generate CO, this would be reflected in elevated VeCO levels. Pups, 36 to 60 h old, were studied chambers closely surrounded by each light source and supplied with air for VeCO measurements. Control pups were kept in the dark, while those in the study group were exposed for 4 h to the following light delivery devices: 40 blue, blue-green, green, or white LEDs; BiliBlanket; or two cool-white and one special blue (BB) fluorescent tubes. Post-PT, animals were sacrificed and blood was collected for PBR quantitation. Efficacy was expressed as the % decrease in PBR levels compared to controls. Safety was expressed as the % change in VeCO before PT and 30 min after commencement of PT. We found that the decreases in PBR levels in comparison to controls after exposure to LED light in increasing order were 14 ± 7 for green, 22 ± 9 for white, 31 ± 7 blue-green, and 52 ± 6 for blue. The decrease in PBR was 48 +5 and 45 ± 4% after exposure to BiliBlanket and fluorescent tubes, respectively. The change from baseline VeCO levels was greatest during PT with fluorescent tubes (105 ± 2%), followed by BiliBlanket (41 ± 13%), then blue LED (11 ± 15%), and least for white (-4 ± 22), green (-4 ± 6), and blue-green LEDs (-4 ± 5%). These *in vivo* results support our previous *in vitro* findings[8] that for LEDs, blue light was most effective for PBR removal. Even though the efficacy of the BiliBlanket and fluorescent tubes (360° light exposure, close contact) approached that of the blue LED, it is inappropriate to relate this finding to the clinical setting because these conventional devices are never used as contact devices. LEDs, in contrast, can be used for close body contact to deliver high intensity light to an unlimited surface area and shape. The higher VeCO values indicated greater photosensitization with the conventional PT devices. Consequently, we concluded that blue LED light is both safer and more effective than conventional PT devices.

To identify the origin of the light-mediated increases in VeCO observed in light-exposed neonatal rats[22], we monitored, in modified chambers, total body, head, and trunk VeCO of 1 to 7 day-old Wistar rat pups kept in the dark and during subsequent 30 min of 360° exposures to the following light sources [device, wavelength ranges (nm)]: fluorescent (BiliLite: 350 to 600), fiberoptic (BiliBlanket: 400 to 550), and LEDs (blue: 459 ± 22; green: 525 ± 40).[36] VeCO (μL CO/h/kg, % of dark, n) increased the most after exposure to BiliLite (69 ± 10, 83%, 5), followed by blue LED (55 ± 8, 39%, 5), then by BiliBlanket (56 ± 10, 35%, 5), and least for green LED (46 ± 12, 21%, 3) as compared to dark controls (40 ± 9, 0%, 18). Subsequent separate CO measurements (n = 2) of the head and trunk demonstrated that only the head

(22 vs 5 μL CO/h/kg), representing pulmonary CO excretion, excreted significant rates of CO in the dark. Upon exposure to light, CO excretion of the head (~15% of body surface area) and the trunk increased quickly (within 5 to 30 min) and significantly, ranging from 27 to 37 and 8 to 26 μL CO/h/kg, respectively, for all light sources, and returned to dark CO levels within 30 min after cessation of light exposure. We concluded that the light-induced elevation of VeCO is caused by transdermally excreted CO. This CO is most likely produced through endogenous photosensitizer-mediated photooxidation of dermal biomolecules, although dissociation of carboxyhemoglobin in circulating re blood cells cannot be entirely excluded.[37]

We have described above that exposure of rat pup skin to light leads to increased excretion of CO from the dermis. We hypothesized that this phenomenon resulted from the photooxidation of organic compounds in the presence of photosensitizers, such as RF, in the skin. In order to test our hypothesis, we compared how blue LED or fluorescent light affected VeCO levels in RF-treated Gunn rat pups.[38] Homozygous Gunn rat pups (n = 6) were identified by transcutaneous bilirubinometry (TcB, mg/dL) using the BiliCheck System (SpectRx Inc, Marietta, GA). Once baseline VeCO levels were established, pups were subjected to 360° exposure for 30 min to 91 $\mu W/nm/cm^2$ of either blue LED (n = 40) or fluorescent (2 cool white/2 BB tubes) light. Once VeCO returned to baseline levels, 10-μmol RF i.p./kg BW was injected and the VeCO was monitored for another 30 min followed by 30 min exposure to each light source. We found that exposure to fluorescent light resulted in an 8-fold increase from baseline VeCO levels when compared to exposure to LED light (95 ± 6% vs 12 ± 7%). Furthermore, in the presence of RF, VeCO increased to 181% of basal rates during exposure to fluorescent light in contrast to an increase of only 89% during blue LED light exposure. Finally, fluorescent and blue LED light exposure decreased transcutaneous bilirubin levels to 73% and 61%, respectively. Because of the greater degree of photosensitization during exposure to fluorescent light as indexed by the level of light-mediated CO production, we concluded that blue LED light appears to be safer for use in PT.

A Portable PT Bed

Infants born with the CNS lack the hepatic enzyme uridine diphosphate glucuronyl transferase, which conjugates bilirubin to mono- and diglucuronides. This enzyme deficiency leads to the accumulation of bilirubin in circulation and tissues to potentially toxic levels.[2] To control bilirubin levels, nightly PT treatments with blue light are needed to convert bilirubin to excretable products. Because of the need for chronic PT, patients are limited in travel because current PT devices are usually large and not transportable. We have constructed a PT bed/light source combination, which is compact, lightweight (~ 30 kg), and transportable via air and surface modes. Assembly or disassembly requires < 20 min. The bed consists of a modified aluminum camping cot (80 x 170 cm, EZ Sales Mfg, Inc., Gardena, CA) fitted with a set of additional head, foot, and side fabric stretching bars to allow for tautness adjustment of the bed fabric using turnbuckles. The fabric can consist of 3-mm mesh white nylon netting, but sheer polyester fabric has been found to provide the most comfortable, and yet relatively transparent, sleeping surface. The light source consists of two polyethylene cases (91 x 61 x 86 cm) each fitted, at present, with a bank (81 x 61 cm) of eight TL20W/52 (Philips, The Netherlands) fluorescent tubes, which emit blue light with an emission spectrum of 400 to 500 nm, with a peak of 460 nm. These tubes are used throughout Europe for PT of neonates and CNS patients. They are similar in spectral characteristics and efficacy to the special blue (BB) fluorescent tubes used in the US. The light banks with electronic ballasts, powered by 110V AC, are placed under the bed, 10 cm from the body, and deliver 40 to 60 $\mu W/cm^2/nm$ across the entire bed. This irradiance is at least twice that delivered by stationary overhead lamps. The first prototype bed is being used by a CNS patient at home and during travel. His jaundice responds well to PT with this bed and his STB levels are generally well controlled, as demonstrated by a decrease in TcB measurements after nightly PT. Our ultimate goal is to replace the broad light spectrum fluorescent tubes with narrow spectrum (460 ± 22 nm) 15° lens type blue LEDs or lensless surface mount LEDs.

Summary and Future Directions

To date, our studies have demonstrated that blue LEDs appear to be most effective in degrading bilirubin, in part because of the high light intensity, small size, and low heat output of these light sources, which allows them to be used in close proximity to patients needing PT.

We are now developing blue LED containing panels for the treatment of neonatal jaundice for use in newborn basinets to replace the presently used bulky overhead and therapy limiting fiberoptic devices. Although the price of a single blue LED is approximately $0.75 and many are needed for high intensity devices, the increasing use of LEDs in advertising panels, traffic signals, etc as well as competition between the different manufacturers is expected to decrease prices. Thus, the use of LEDs in PT devices will soon become an economic possibility.

Although blue LED light is associated with lower rates of dermal CO production and thus appears to be a safer light source than fluorescent tube light, far more work needs to be conducted to determine what photosensitizer(s) mediate this process, what dermal components are altered by this oxidation process, if this alteration is injurious, and how it can be prevented. The various light side effects (see Table 1), minimal or absent during temporary neonatal PT, are exacerbated during the intense and chronic CNS PT. Therefore, more effort needs to be directed to investigate and minimize these effects in this population. It is most likely that answers to these questions are more relevant to chronic PT of CNS patients than to jaundice of other etiologies.

Despite the development of several promising, more permanent treatments for CNS, which are on the horizon, PT will continue to be needed in the future. The use of LED light, which has a selectable wavelength range, may be the key. In addition, the identification of protective compounds may be another approach.

Furthermore, lightweight PT beds need to be developed to improve the quality of life in this population, which is healthy, except for the lack of only one liver enzyme. We envision the ultimate PT device to be a LED-containing blanket or sleeping bag device, which can be taken and used anywhere.

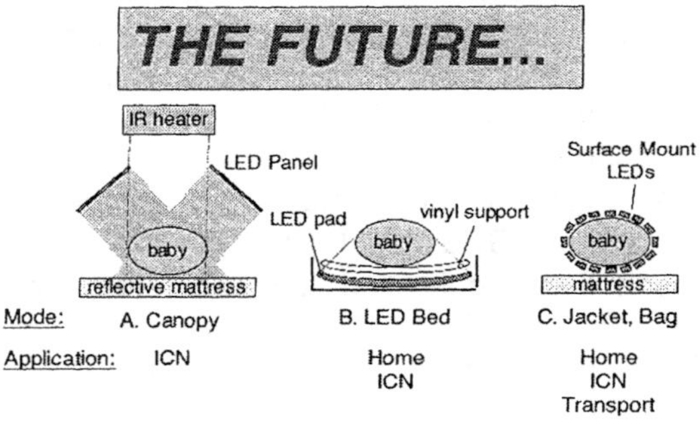

Figure 2. LED devices of the future.

ACKOWLEDGMENTS

This work was supported by the National Institutes of Health, grants HD14426 and unrestricted gifts from the H. M. Lui Research Fund, the Hess Research Fund, and the Mary L. Johnson Research Fund.

REFERENCES

1. Brown AK, Kim MH, Wu PY, Bryla DA. Efficacy of phototherapy in prevention and management of neonatal hyperbilirubinemia. Pediatrics 1985;75:393-400.
2. Crigler JFJ, Najjar VA. Congenital familial non-hemolytic jaundice with kernicterus. Pediatrics 1952;10:169-180.
3. Ennever JF, McDonagh AF, Speck WT. Phototherapy for neonatal jaundice: Optimal wavelengths of light. J Pediatr 1983;103:295-9.
4. Ennever JF. Blue light, green light, white light, more light: Treatment of neonatal jaundice. Clin Perinatol 1990;17:467-81.
5. McDonagh AF, Lightner DA. Phototherapy and the photobiology of bilirubin. Semin Liver Dis 1988;8:272-83.
6. Nakamura S, Fasol G. InGaN single-quantum-well LEDs. The Blue Laser Diode. Berlin: Springer-Verlag, 1997:201-221.
7. Fasol G. Longer life for the blue laser. Science 1997;278:1902-1903.
8. Vreman HJ, Wong RJ, Stevenson DK, et al. Light-emitting diodes: A novel light source for phototherapy. Pediatr Res 1998;44:804-9.
9. Sanvordeker DR, Kostenbauder HB. Mechanism for riboflavin enhancement of bilirubin photodecomposition in vitro. J Pharm Sci 1974;63:404-8.
10. Redmond RW, Gamlin JN. A compilation of singlet oxygen yields from biologically relevant molecules. Photochem Photobiol 1999; 70:391-475.
11. Ente G, Klein SW. Hazards of phototherapy. N Engl J Med 1970;283: 544-5.
12. Rao HKM, Elhassani SB. *Primum non nocere*: Iatrogenic complications of procedures performed on the newborn Part 3. Phototherapy, tube feedings, and others. P-N 1981;March/April:23-25.
13. Aouthmany MM. Phototherapy increases hemoglobin degradation and bilirubin production in preterm infants. J Perinatol 1999;19:271-4.
14. Gale R, Dranitzki Z, Dollberg S, Stevenson DK. A randomized, controlled application of the Wallaby phototherapy system compared with standard phototherapy. J Perinatol 1990;10:239-42.
15. George P, Lynch M. Ohmeda BiliBlanket vs Wallaby Phototherapy System for the reduction of bilirubin levels in the home-care setting. Clin Pediatr (Phila) 1994;33:178-80.
16. Stevenson D. Home phototherapy: Risks versus benefits. Clin Pediatr 1986;25:300-2.
17. Stevenson DK, Rodgers PA, Vreman HJ. The use of metalloporphyrins for the chemoprevention of neonatal jaundice. Am J Dis Child 1989; 143:353-6.
18. Tenhunen R, Marver HS, Schmid R. The enzymatic conversion of heme to bilirubin by microsomal heme oxygenase. Proc Natl Acad Sci USA 1968;61:748-55.

19. Vreman HJ, Ekstrand BC, Stevenson DK. Selection of metalloporphyrin heme oxygenase inhibitors based on potency and photoreactivity. Pediatr Res 1993;33:195-200.
20. Land EJ. Porphyrin phototherapy of human cancer. Eur J Radiat Biol 1984;46:219-23.
21. Vreman HJ, Gillman MJ, Downum KR, Stevenson DK. *In vitro* generation of carbon monoxide from organic molecules and synthetic metalloporphyrins mediated by light. Dev Pharmacol Ther 1990; 15:112-24.
22. Chan ML, Vreman HJ, Wong RJ, Young BW, Stevenson DK. *In vivo* study of light-emitting diodes (LEDs) as a light source for photo-therapy in neonatal jaundiced rats. Pediatr Res 1999;45:1104-1104.
23. Vreman HJ, Stevenson DK. Detection of heme oxygenase activity by measurement of CO. In: Maines MD, Costa LG, Reed DJ, Sassa S, Sipes IG, eds. Current Protocols in Toxicology. New York: John Wiley & Sons, Inc., 1999:9.2.1-9.2.10.
24. Vreman HJ, Kwong LK, Stevenson DK. Carbon monoxide in blood: An improved microliter blood sample collection system with rapid analysis by gas chromatography. Clin Chem 1984;30:1382-6.
25. Vreman HJ, Zentner AR, Wong RJ, Stevenson DK. Carbon monoxide production and upregulation of heme oxygenase activity in mice after heme administration. Pediatr Res 1999;45:1358-1358.
26. Dennery PA, McDonagh AF, Spitz DR, Rodgers PA, Stevenson DK. Hyperbilirubinemia results in reduced oxidative injury in neonatal Gunn rats exposed to hyperoxia. Free Radic Biol Med 1995;19:395-404.
27. Vreman HJ, Mahoney JJ, Stevenson DK. Carbon monoxide and carboxyhemoglobin. Adv Pediatr 1995;42:303-25.
28. Vreman HJ, Stevenson DK. Metalloporphyrin-enhanced photodegradation of bilirubin *in vitro*. Am J Dis Child 1990;144:590-4.
29. Hintz SR, Vreman HJ, Stevenson DK. Mortality of metalloporphyrin-treated neonatal rats after light exposure. Dev Pharmacol Ther 1990; 14:187-92.
30. Sisson TR. Photodegradation of riboflavin in neonates. Fed Proc 1987; 46:1883-5.
31. Ballowitz L, Bunjamin A, Hanefeld F, Lietz L, Stüttgen G, Wirjadi D. Effects of riboflavin on Gunn rats under phototherapy. Pediatr Res 1979;13:1307-15.
32. Seidman DS, Moise J, Ergaz Z, et al. A new blue light-emitting phototherapy device: A prospective randomized controlled study. J Pediatr 2000;136:771-774.
33. American Academy of Pediatrics. Practice parameter: Management of hyperbilirubinemia in the healthy term newborn. Provisional Commit-tee for Quality Improvement and Subcommittee on Hyperbilirubin-emia [published erratum appears in Pediatrics 1995 Mar;95(3):458-61]. Pediatrics 1994;94:558-65.

34. Donzelli GP, Pratesi S, Rapisardi G, Agati G, Fusi F, Pratesi R. 1-day phototherapy of neonatal jaundice with blue-green lamp [letter]. Lancet 1995;346:184-5.
35. Moise J, Seidman DS, Ergaz Z, et al. Efficacy of phototherapy with blue and blue-green light emitting diodes (LEDs) and conventional devices: A prospective randomized controlled study. Pediatr Res 1999; 45:212A.
36. Vreman HJ, Wong RJ, Chan ML, Young BW, Stevenson DK. Dermal carbon monoxide (CO) production and excretion in neonatal rats during light exposure. Pediatr Res 2000;47:463-463.
37. Coburn RF, Danielson GK, Blakemore WS, Forster RE. Carbon monoxide in blood: Analytical method and sources of error. J Appl Physiol 1964;19:510-15.
38. Vreman HJ, Wong RJ, Chan ML, Young BW, Stevenson DK. Effect of fluorescent- and light emitting diode (LED) blue light exposure on riboflavin-treated Gunn rat pups. Pediatr Res 2001;49:325A.

DIFFERENTIAL EFFECTS OF ULTRAVIOLET A AND ULTRAVIOLET B RADIATION ON RISK OF MELANOMA: EPIDEMIOLOGICAL EVIDENCE

Cedric F. Garland, Frank C. Garland, and Edward D. Gorham
University of California, San Diego, La Jolla, California 92093

INTRODUCTION

Exposure to solar radiation is thought to influence the risk of melanoma, however the action spectrum that is most important in induction of melanoma is not well understood. Mortality rates from melanoma are highest in the southeastern U.S., where solar radiation levels are higher than in the northeastern U.S. Paradoxically, mortality rates are relatively low in the area of the U.S. with the highest level of solar radiation, the southwest. Regional differences in skin pigmentation among Caucasians may play a role in this geographical pattern. Mortality rates are markedly lower in Asians and Blacks than in Caucasians. This analysis is primarily concerned with variations in melanoma rates among Caucasian populations, although data on other ethnic or racial groups were included to the extent available.

BACKGROUND

More than 50 yèars have elapsed since it was hypothesized that malignant cutaneous melanoma was somehow associated with exposure to sunlight. Studies of this association have been extremely mixed. It is known that people engaged in outdoor occupations generally have lower incidence rates of melanoma than those engaged in indoor work (1, 2).

Historically most populations have spent much of the day outdoors in the sun, working to raise agricultural crops since the agricultural revolution, which is estimated to have started in approximately 10,000 B.C. Presumably outdoor laborers on farms or working in construction were exposed to an environment containing massive amounts of ultraviolet radiation. Much of the cutaneous exposure was unavoidable, as ultraviolet energy is reflected from many surfaces and attempts at shading are unlikely to be completely successful. Sunburn has been a common clinical problem in Caucasian populations for hundreds and probably thousands of years. Children may have had great exposures to ultraviolet throughout history, as they lived and often worked outdoors. Even today it is hard to make children wear hats or the amount and type of clothing that would be needed to prevent solar overexposures.

Melanoma incidence and mortality rates have increased at an appreciable rate since 1955 in virtually all Caucasian populations, although the increase was delayed and the rate of increase has been far more modest in some countries, such as Italy. By contrast melanoma was so rare before 1955 that it was not tabulated separately as a cause of death in vital statistics reports from the US National Center for Health Statistics, but combined with deaths from squamous and basal cell carcinomas (3). The rate of increase in melanoma has been reported to be at least 2% per year in recent years (4). There is a strong association between sex and site, with the greatest rise in melanoma on the trunk in males, and on the lower limbs and, increasingly in recent years, on the trunk in females. The highest rates of melanoma incidence and mortality in the world by far occur in New Zealand and Australia (3). Melanoma is now the fourth most commonly occurring cancer in Australia and New Zealand, where it is expected to afflict 1 in 25 persons (4). In Europe the highest rates of melanoma occur in Norway, Denmark, Slovenia, the former Czech republic, and Sweden. The increased rates in melanoma occurred first in Northern European countries of Europe, and later in the more southern areas. There are a number of correlated factors associated with melanoma risk, including Fitzpatrick-Pathak skin phototype (range I-VI), skin type, hair color, and predominant melanin type (eumelanin or pheomelanin). Risk is also associated with the amount and types of clothing worn. Some occasionally reported correlates of melanoma incidence, such as number of childhood sunburns, are so closely associated with skin phototype that they probably convey no independent information. People with fair skin tend to be more likely than those with more pigmented skin to experience solar erythema.

While the association of melanoma risk with sunlight is uncertain, there are facts that suggest that exposure to ultraviolet radiation or light is a factor. Most attention to date has centered on ultraviolet B (UVB), radiation since UVB is the cause of solar erythema and is readily absorbed by DNA, resulting in certain "signature" mutational photoproducts, such as thymidine dimers. It generally has been thought that the action spectrum for human melanoma is either UVB or unknown, although a role for ultraviolet A (UVA) has been suspected (5-8). Approximately 100 UVA photons reach the melanocyte for each ultraviolet B photon (9). Based on quantum calculation of wavelength dependence of photons and a UVA/UVB photon ratio of 100 (9), the total calculated amount of ultraviolet energy in the UVA reaching the melanocyte is 75 times greater than the amount of UVB energy. While DNA is a UVB chromophore that allows direct mutation, UVA mutates DNA indirectly through free-radical mediated oxidative damage to guanine bases (9).

One of the most compelling questions in melanoma is the question of the action spectrum. The action spectrum for a fish model of a melanocytic tumor resembling or possibly the same as melanoma includes three peaks in the UVA, with a prominent peak at 365 nm (6-8). Convolution of the solar spectrum with the action spectrum for the fish model indicates that 90% of melanoma is due to UVA, if the fish model is applicable to humans (6-8). The fish model is *Xiphophorus,* a swordtail-platyfish hybrid. The only other known animal model of melanoma is the miniature opossum, *Monodelphis domestica.* Some adults of these animals develop melanoma upon prolonged exposure to a mixed UVB-UVA radiation source (10). A study of infant opossums revealed a single melanoma precursor lesion after exposure to UVA (10), in one of 70 animals exposed on 9 occasions to 250 MJ/m^2 UVA.

Since the action spectrum in humans is still uncertain, this study examined two natural experiments for information that might be supportive of either the UVA or the UVB hypothesis of melanoma promotion, or of a combination. One is the residence of different populations at different latitudes and the other is the discovery, in 1922, of chemical sunscreens, and their diffusion throughout the population of industrialized countries in the 1950's and subsequently. The epidemiological evidence suggests that the ratio of UVA to UVB might be an important correlate of melanoma risk. This ratio varies by latitude and is altered when chemical sunscreens are applied to the skin.

METHODS

The ultraviolet photon flux for each nanometer in the UVA (310-400 nm) and terrestrial UVB (300-310 nm) were modelled as a function of solar noon zenith angle at different latitude at the equinoxes using clear sky data collected by a high-resolution scanning spectrophotometer in San Diego CA (San Diego CA: Biospherics, Inc.). The solar noon zenith angle is the number of degrees that the sun, at its highest point during the day, deviates from the zenith. The zenith is a point in the sky on a line perpendicular to the earth's surface. A solar zenith angle of 0 degrees, for example, denotes a position of the sun directly overhead. The solar zenith angle was calculated for April 21 and September 21, the dates of the equinoxes. These are the dates when the equator is parallel to the plane of rotation of the earth around the sun. On these dates, the highest point the sun reaches in the sky is equal to 33 degrees below the zenith (in the southern sky). At the equinoxes, the noon solar zenith angle is equal to the latitude. For example, in San Diego CA (33 degrees North), the maximum elevation of the sun above the horizon at the equinoxes is 57 degrees, corresponding to a solar zenith angle of 33 degrees. In Boston (42 degrees North) the maximum equinoctal elevation of the sun above the horizon is only 48 degrees. The method used for this study provides an approximate estimate of the photon flux in nanometer or smaller increments. The model did not take into account the effect of variations in thickness of the Hartley and Huggins bands whose ozone absorbs UVB. Instead, for simplity in calculations, a uniform ozone thickness of 300 Dobson Units (DU) was used. This is considered reasonable for the equinoxes, when variations in ozone thickness are minimal. The model also did not take into account the possible effects of cloud cover.

Age-adjusted mortality rates of melanoma were obtained from the World Health Organization using the website of the International Agency for Research on Cancer (www.iarc-dep.fr). Owing to delays in reporting, the earliest year for which reasonably complete data were available was 1990, and that year was used for the analyses. The rates were reported by WHO by 50 countries. The rates were age-adjusted by WHO to the world standard population and were provided specific for sex.

The data were entered using Microsoft Excel (Redmond WA: Microsoft, Inc.). Trend lines and coefficients of determination were calculated using the Excel Chart subroutine. The best fit to the data points was obtained using an exponential trend line. Age-adjusted incidence rates of melanoma were obtained from the Connecticut Tumor Registry-SEER and were plotted. Dates of introduction of sunscreens were obtained from a standard resource (5,11) and archived newspaper and magazine reports and advertisements. The action spectrum of para-aminobenzoic acid (PABA) was obtained from Sadtler Spectra.

RESULTS

Melanoma mortality rates are shown by country in Figure 1. The highest rate in men was 5.9 per 100,000 in New Zealand (42 degrees South) and 5.0 per 100,000 in Australia (38 degrees South). Forty-three of the 50 countries had predominantly Caucasian populations. Nine had significant Hispanic populations. Five had predominantly Asian populations, and two had predominantly black populations. The lowest rates in a Caucasian population were in both sexes in Malta, a country with a predominantly Caucasian population with many fair-skinned and "olive" skinned residents living on an island in the Mediterranean, south of Sicily. Similar or slightly lower rates were observed in four populations of oriental origin and one with a predominantly black population, with the lowest rate in the Republic of Korea (0.1 per 100,000 population per year in both sexes).

Melanoma mortality rates increased with increasing latitude in men (Figure 2) and women (Figure 3). Latitude accounted for 28 percent of the variation in melanoma mortality rates in men and 47 percent of the variation in melanoma mortality rates in women.

Paradoxically, melanoma mortality rates appeared to decrease with with increasing UVB (Figure 4) and UVA flux (Figure 5). However, the rates increased with increasing UVA/UVB ratio in men (Figure 6) and women (Figure 7).

The natural experiment of sunscreen adoption in the U.S. is illustrated by a plot showing the acceleration in melanoma incidence in the Connecticut Cancer Registry following the introduction in 1935 of suntan lotions containing PABA or derivatives in low SPF formulations and more rapid acceleration following the introduction of high SPF sunscreens in 1961 (Figure 8). Common chemical sunscreens such as PABA, which has peak absorption at 280 nm (Figure 9) absorb UVB well but UVA poorly or not at all, dramatically raising the UVA/UVB ratio at the surface of the skin.

DISCUSSION

This study and its UVA/UVB model have several important limitations and strengths. Some of the most evident of these are described in the following sections.

Strengths

The UVA/UVB ratio model accounts for much of the paradoxical global variation in mortality rates from melanoma, specifically the high rates in latitudes relatively distant from the equator.

The ratio model accounts for the unexpected and paradoxical increase in risk of melanoma in chemical sunscreen users reported in 6 epidemiological studies, since sunscreens markedly increase the UVA to UVB ratio.

The ratio model accounts for the significantly higher prevalence of melanocytic nevi in children whose parents regularly applied sunscreen to them (12) and the poor effectiveness of sunscreen in preventing melanocytic nevi in a clinical trial (13).

The model accounts for the higher melanoma mortality rates in New Zealand than in Australia (11), which is paradoxical since New Zealand is more distant from the equator than Australia.

The mortality rates were adjusted for age differences among the populations being compared and the findings were consistent in both sexes

Limitations

The UVA/UVB ratio model does not attempt take into account variations in the pigmentation of the skin in different areas of the world, since these differences have not been quantified. However, 43 of the 50 countries reporting to WHO have predominantly Caucasian populations, and some of the countries with the lowest mortality rates have predominantly Caucasian populations (Malta, Greece, Italy, Spain).
An upward trend exists, albeit over a narrower range of mortality rates, when the 7 countries with non-Caucasian populations are excluded.

The ratio model when applied to geographic variation does not account for variations in the thickness of the ozone layer, since it assumes uniform ozone density of 300 DU, or of variations in cloud cover (not necessarily serious since the ozone distribution is rather uniform at the equinoxes and the predominant attenuation by clouds applies equally to UVA and UVB photons, leaving the ratio similar.

The model is based on mortality rates, which are the only rates that are available from countries covering such a broad range of latitudes.

Incidence rates might give a better approximation of risk estimates, although mortality rates are a more certain endpoint.

The equnoctal UVA flux at noon is about half as high at the arctic circle as at the equator. Therefore there is less UVA present in an absolute sense in the higher latitudes than the lower. However, the difference between UVA fluxes at the equinoxes is not massive.

MECHANISMS

UVB-activated signals in the normal, solar unaccommodated epidermis (Figure 10) are thought to stimulate a photoprotective epidermal response (PER) that includes growth in the epidermis, leading over a period of days to weeks to thickening of the stratum corneum (Figure 11), an adaptation that tends to help scatter ultraviolet radiation, somewhat reducing the penetration of ultraviolet radiation (Figure 12).

UVB-activated signals are thought to trigger synthesis and release of melanosomes by melanocytes (Figure 13) , further reducing the penetration of UVR to the basal epidermis and melanocytes (Figure 14). It has been hypothesized that one of these signals may be triggered by certain UVB-induced mutations of the DNA (Figure 11).

When only UVA is present in normally encountered solar dosages, the epidermis does not thicken appreciably and pigmentation is relatively weak. A relatively high dosage of UVA (8-25 J/cm2 for one hour or more) is required, for example, to produce visible persistent pigment darkening in most people, due to melanin photo-oxidation 2-24 hours after irradiation (14)

When ultraviolet radiation excludes most UVB and consists of virtually all UVA, as with chemical sunscreens or solar exposures at high latitudes, the natural photoprotective epidermal response is diminished or absent. As a result the basal epithelium, including the melanocytes, is exposed to a relatively massive flux of UVA photons.

UVA photons activate free radicals that damage the DNA through oxidative damage to guanine bases (14). The ultimate result is mutation of the DNA and promotion of melanoma through a series of mutations of proto-oncogenes or tumor supressor genes.

CONCLUSIONS

The current model of a possible positive, if mixed, relationship between ultraviolet radiation and risk of melanoma cannot readily account for the inverse association observed in this study between latitude and melanoma mortality rates observed in this study.

The UVA/UVB ratio can explain more of the epidemiology in quantitative terms than any other existing model. Further research is needed to improve the model, including the building in of an adjustment for the proportion of the population by degree of fairness of the skin in each country. As incidence rates become available for more countries, similar analyses should be conducted using these rates to confirm the results of these analyses..

In the meantime, caution should be associated with any exposure of the skin to radiation where the ratio of UVA to UVB is higher than normally encountered in ordinary sunlight. Such exposures commonly are produced by application of chemical sunscreens, which should normally be avoided except for sunscreen containing true boroad-spectrum blockers in effective concentrations, such as titanium dioxide or zinc oxide. (A few individuals may become sensitized to these agents.)

Caution should also be exercised by fair-skinned persons living at high latitudes with respect to moderation of duration of solar exposure, since UVA fluxes are potentially high even at high latitudes. In persons who skin can form pigment readily, allowing normal seasonal accommodation to sunlight by regular solar exposure for brief periods of time (15-30 minutes per day) is probably the wisest course of action. Individuals regularly exposed to moderate amounts of sunlight in the course of their occupation have the lowest incidence rates of melanoma (1).

Acknowledgements. The authors would like to acknowledge the contributions of several sciemtists who advised on several aspects of this report. These include Germar Bernhard, Ph.D., of Biospherics, Inc., San Diego, California, for scanning spectrophotometer UV data and latitude modeling; Terence O'Grady, M.D., Associate Clinical Professor, University of California San Diego, Department of Dermatology (Dermatopathology) for review of diagrams of the epidermis; and William B. Grant, Ph.D., for advice and consultation concerning ozone. We would also like to thank the NASA Goddard Space Flight Center Total Ozone Monitoring Program for data on global ozone thickness and erthemal UV flux.

REFERENCES

1. Garland F, White M, Garland C, Shaw E, Gorham E. Occupational sunlight exposure and melanoma in the U.S. Navy. Arch Environ Health 1990;45: 261-7.
2. Franceschi S, La Vecchia C, Lucchini F, Cristofolini M. The epidemiology of cutaneous malignant melanoma: aetiology and European data. Eur J Cancer Prev 1991;1(1):9-22.
3. U.S. National Center for Health Statistics. Vital statistics of the United States, 1986: Mortality, Part A. Washington DC: U.S. Public Health Service; 1989.
4. Marks R. Epidemiology of melanoma. Clin Exp Dermatol 2000;25:459-63.
5. Garland CF, Garland FC, Gorham ED. Rising trends in melanoma: an hypothesis concerning sunscreen effectiveness. Annals of Epidemiology 1993;3: 103-10.
6. Setlow R, Woodhead A, Grist E. Animal model for ultraviolet-radiation-induced melanoma:Platyfish-swordtail hybrid. Proc Nat Acad Sci 1989;86:8922-6.
7. Setlow R, Grist E, Thompson K, Woodhead A. Wavelengths effective in induction of malignant melanoma. Proc Nat Acad Sci 1993;90:6666-70.
8. Setlow R, Woodhead A. Temporal changes in the incidence of malignant melanoma: explanation from action spectra. Mutation Res 1994;307:365-74.
9. Nim H, Honigsman H, Gilchrest B, et al. American Academy of Dermatology Consensus Conference on UVA Protection of Sunscreens. J Am Acad Dermatol 2000;44:505-8.
10. Robinson K, Kripke M, Ley R. Ultraviolet A and melanoma in Monodephis domestica sucklings. JNCI 2000.
11. Packaged Facts. Suncare products market. New York: Information SVP; 1990.
12. Autier P, et al. Use of sunscreens and prevalence of melanocytic nevi in children. JNCI 1998;90:1873-80.
13. Gallagher et al. Clinical trial of broad-spectrum sunscreen in prevention of nevi in children in Vancouver, Canada. JAMA 2000;283:2955-60.
14. Lim H, Naylor M, Honigsman H, et al. AAD consensus conference on UVA protection of sunscreens: summary and recommendations. J Am Acad Dermatol 2001;44:505-8.

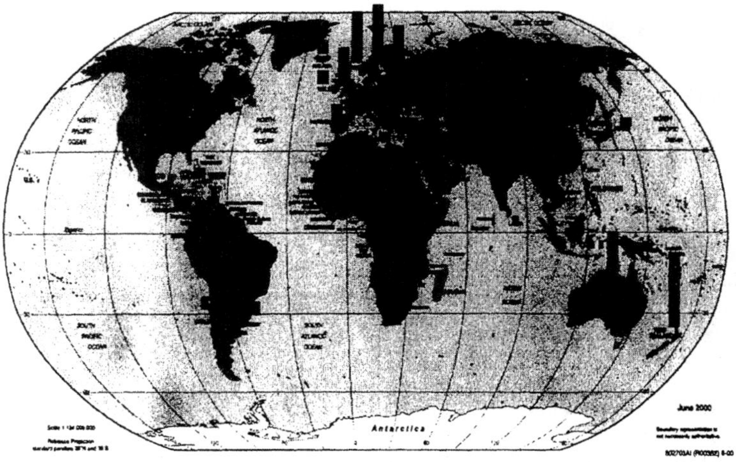

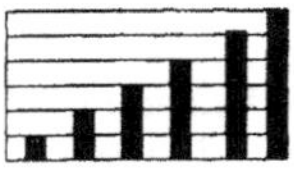

Figure 1. Age-adjusted melanoma mortality rates, men

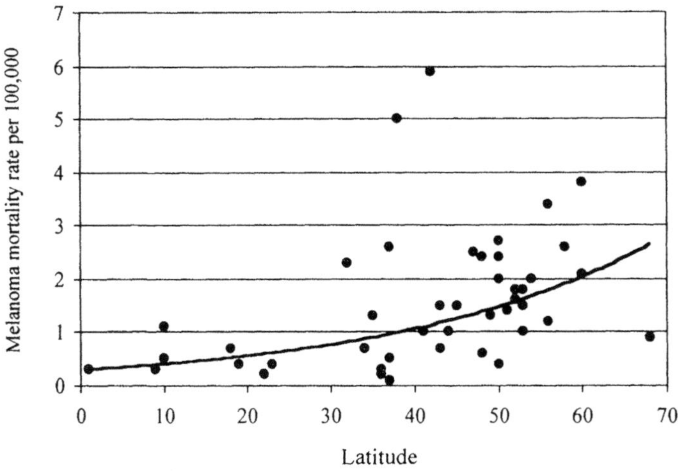

Figure 2. Age-adjusted melanoma mortality rates by latitude, men

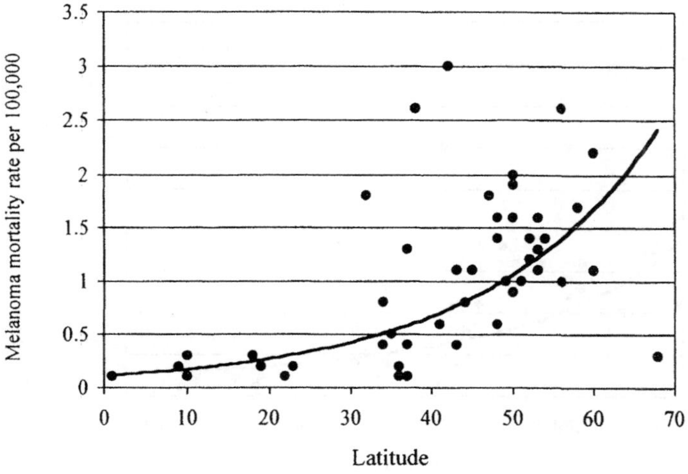

Figure 3. Age-adjusted melanoma mortality rates by latitude, women

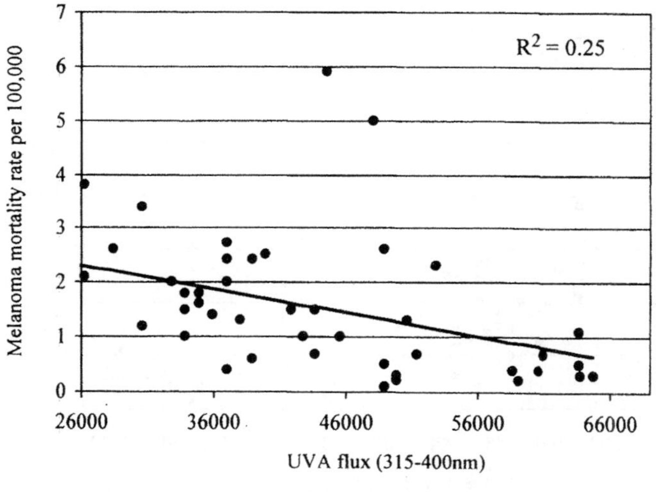

Figure 4. Age-adjusted melanoma mortality rates by UVA flux, men

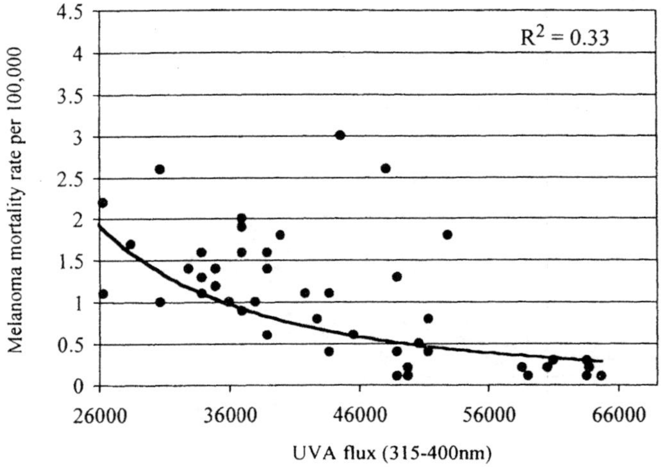

Figure 5. Age-adjusted melanoma mortality rates by UVA flux, women

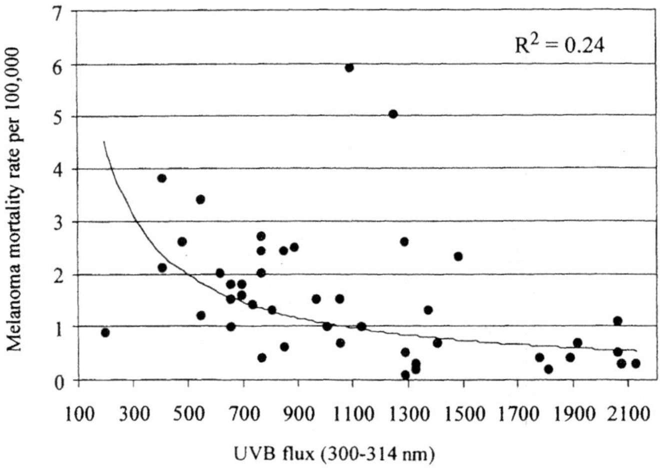

Figure 6. Age-adjusted melanoma mortality rates by UVB flux, men

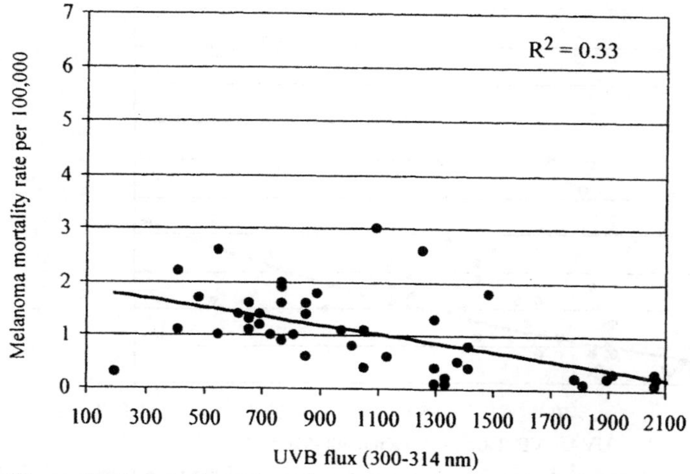

Figure 7. Age-adjusted melanoma mortality rates by UVB flux, women

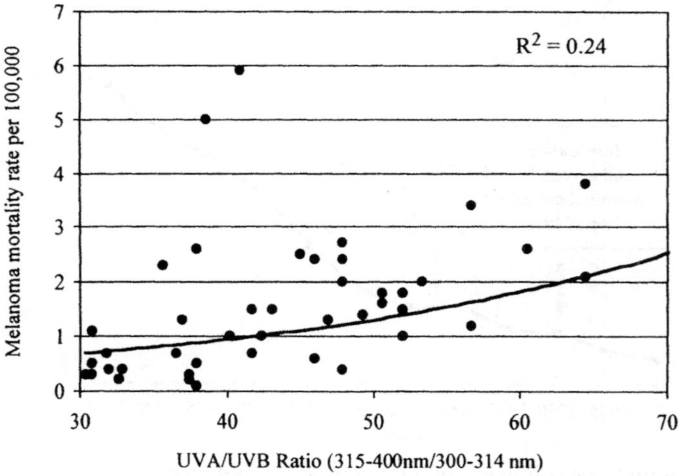

Figure 8. Age-adjusted melanoma mortality rates by UVA/UVB ratio, men

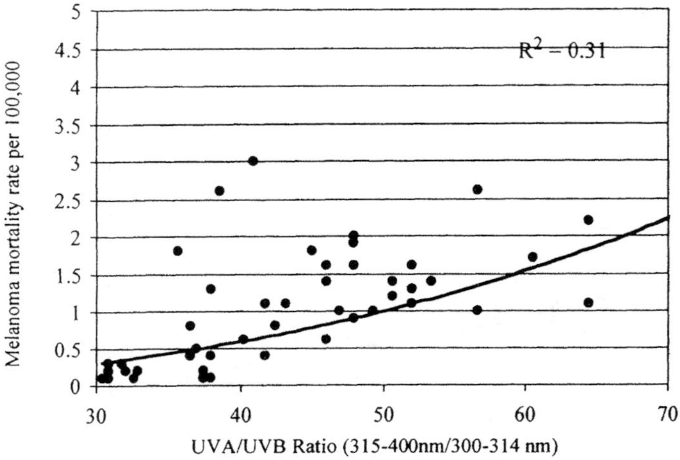

Figure 9. Age-adjusted melanoma mortality rates by UVA/UVB ratio, women

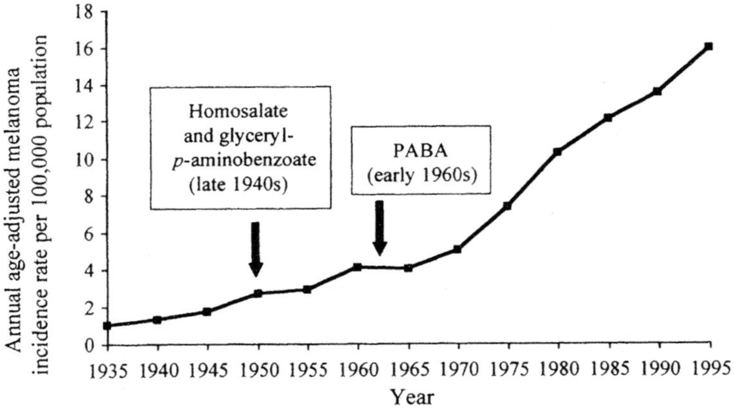

Figure 10. Annual age-adjusted incidence rates of melanoma, Connecticut, and dates of introduction of suntan lotions and sunscreens

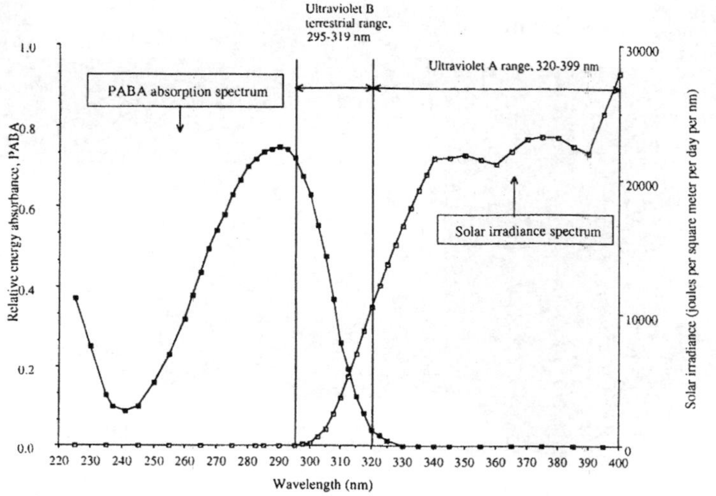

Figure 11. Solar ultraviolet irradiance and relative energy absorbance by para-aminobenzoic acid

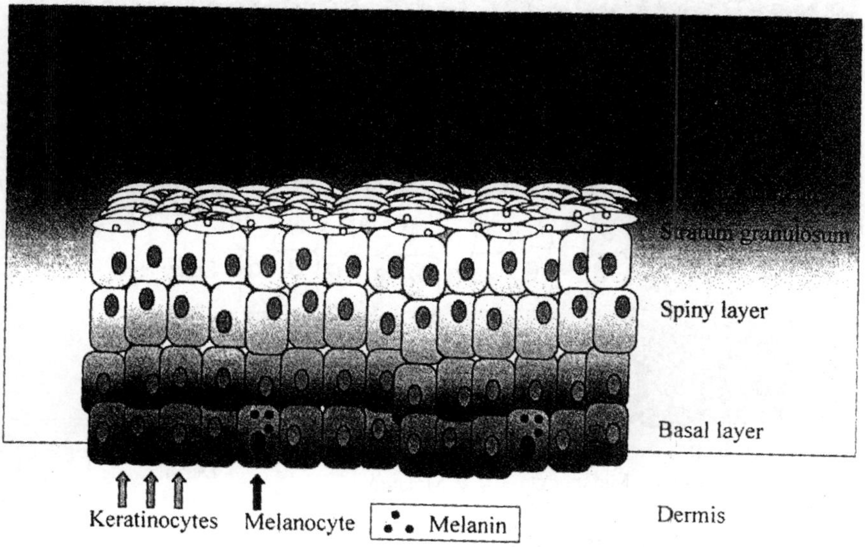

Figure 12. Normal epidermis with no solar accommodation

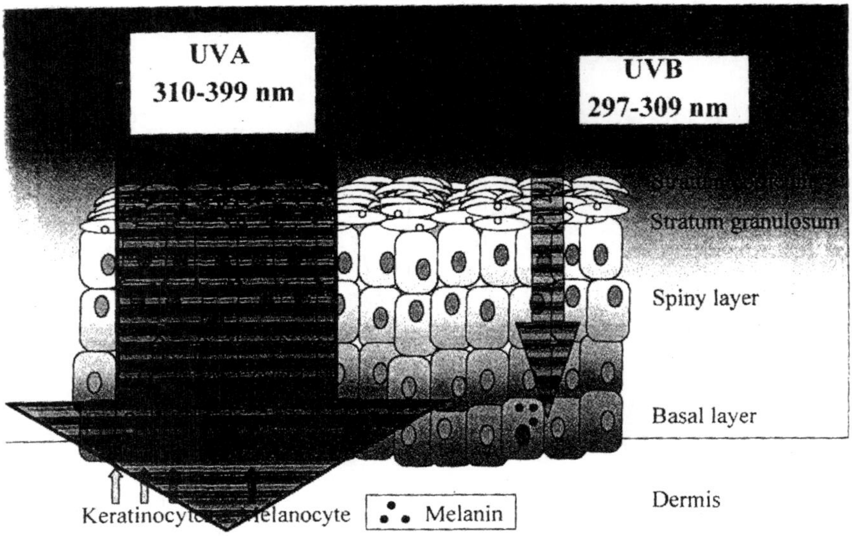

Figure 13. Normal epidermis with no solar accommodation, ultraviolet penetration shown

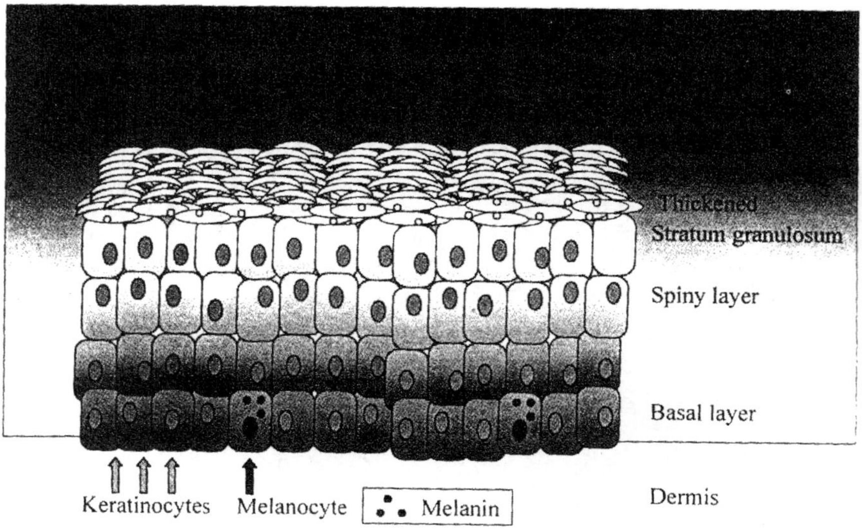

Figure 14. Thickening of epidermis in response to UVB

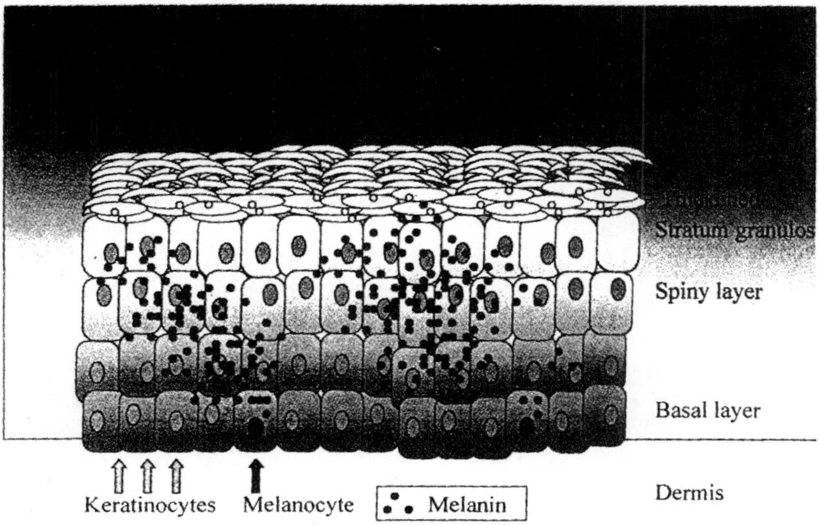

Figure 15. Pigmentation of epidermis mainly in response to UVB

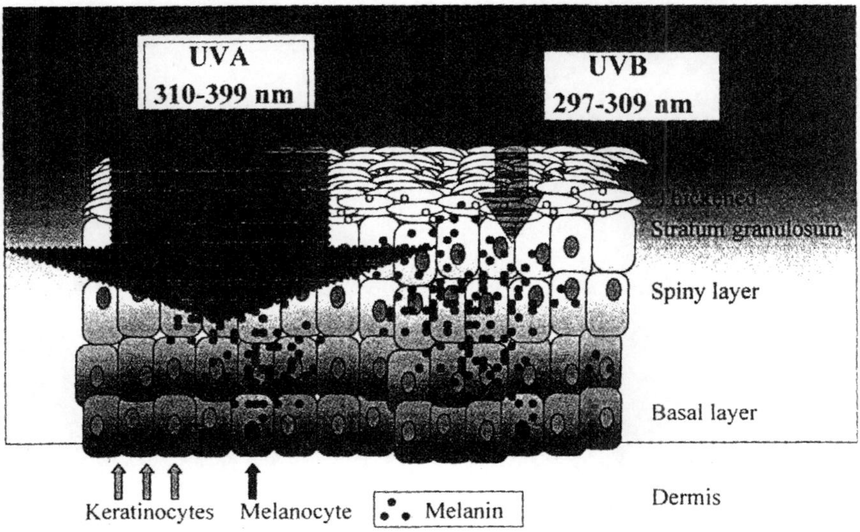

Figure 16. Accommodated and pigmented epidermis, reduced ultraviolet penetration

COMBINATION THERAPY OF MELANOMA CELLS WITH PHOTORADIATION AND SONOLYSIS

Norio Miyoshi, Toshiyuki Ogasawara*, Toru Ogawa*, Kazuo Sano*, Shin-ichi Higashi, Masakatsu Watanabe**, Masaru Fukuda and Haruo Hisazumi*****

*Department of Pathology, *Dentistry and Oral Surgery, Fukui Medical University, Matsuoka, Yoshida-gun, Fukui 910-1193; **National Giant Spectrograph Lab., Okazaki; and Department of Urology, Kanazawa University, Kanazawa Japan*

INTRODUCTION

Treatment of a melanoma has been a very important problem to be solved in dermatology, especially considering that melanoma cells were resistant to radiation therapy and were easily metastasized. It was reported that the target of these treatments are membrane in both cases of PDT (chemical attack) and of SDT (physical attack).

In this study, we applied to treat the cultivated melanoma (HMF) cells by PDT and SDT with a sensitizer, Chlorine e6 derivative, ATX-70. The aim of this study is (1) To find where the ATX-70 molecule is active, out-side or in-side of the cells, following treatment with the following combined therapies; hyperthermia, radiation, chemotherapy, photodynamic therapy (PDT) and sonodynamic therapy (SDT). (2) Which sequence of PDT or SDT is best as a first treatment for the combination therapy? We examined the cytotoxicity of HMF cells for the combination and single treatments of PDT and SDT using flow cytometry after the double staining of the nuclear (Propidium iodide) and membrane (Annexin-V/FITC).

MATERIALS AND METHODS

(1) *Cells and photosensitizer*: A human melanoma cultivated (HMF) cells were cultured in RPMI1640 with 10% calf-serum for 3 days. The sensitizer used in this study was a chlorine-e6 derivative, ATX-70.

(2) *Changing wavelength of photoirradiation for single PDT*: We used the Giant Spectrograph in National Institute for Basic Biology as a light source (400, 450, 550, 570 and 590 nm; 2.88 J/cm2) against 8.25 μM ATX-70 in HMF cell suspensions.

(3) *Combination of PDT and SDT*: The light source of the PDT was a semiconductor laser light (405 nm, 2.88 J/cm2). The sonicator was product by Bronsonic (47 kHz, 50 W, 10 sec).

(4) *Detection method of the cytotoxicity of HMF cells treated the combination-treatments*: Flow cytometry (Epics-type, Coulter Co.) analysis was done after the double staining using an apoptosis kits (Annexin-V labeled by FITC and Propidium iodide; MBL).

RESULTS AND DISCUSSION

(1) *Measurement of action spectra of ATX-70 out-side of HMF cells*: The fluorescence excitation spectra and the emission spectra of ATX-70 were measured the fluorescence intensity (Left Y-axis) of the out-side and the in-side of HMF cells) as shown in Fig. 1. One dot broken lines were the excitation and emission spectra of 8.25 μM ATX-70 (out-side of the cells) in the medium and the solid lines were those of 0.065 μM ATX-70 (in-side of the cells after 24 hr incubation: The ATX-70 molecules were incorporated 0.8 % into the lysosome of the cells where from the fluorescence microscopical photograph).

Both the ATX-70 molecules out-side and inside of the cells after 24 hr incubation were irradiated by light of various wavelengths at the same dose in order to measure of the action spectrum for the single PDT. The cytotoxicity (%; Right Y-axis) pointed on the excitation spectrum of the extracellular ATX-70 molecules from the two-dimensional graph (insertion; Y-axis: Front scattering light intensity; X-axis: Side scattering light intensity: the cell particle size decreased with increasing the damages of the cells) of the light scatters intensities of cells.

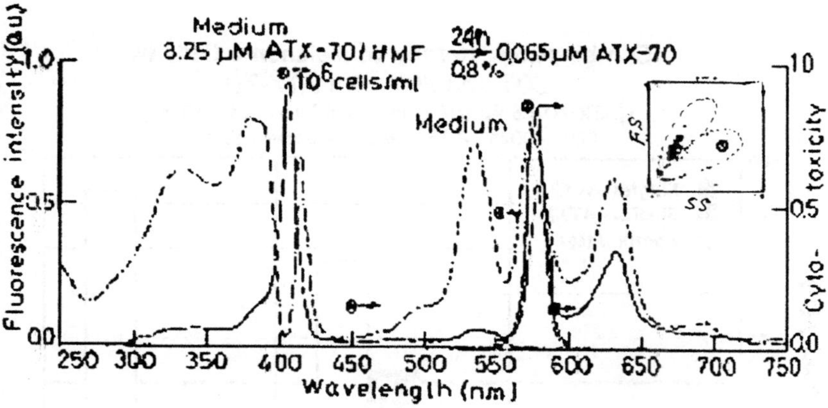

Fig.1. Action spectrum of PDT cytotoxicity on the excitation spectrum of 8.25 uM ATX-70 out of HMF cells.

It was considered that the aggregate form of ATX-70 which was bound the membrane surface were more active in comparison to the monomer from incorporated into the lysosome of the HMF cell in this single PDT conditions.

(2) *Cytotoxicity of the combined treatments of PDT and SDT of HMF cells*: Figure 2 shows the cytotoxicity (Left Y-axis) of HMF cells against the control (black Pole), single (slash Pole) and combination therapy (white Pole) changing the sequences of the SDT and PDT. At the control section, there were no effects of ATX-70 only. The enhancement (3.25 times) of the cytotoxicity by sonolysis with ATX-70 was found in the case of single SDT. On the other hand, sensitization of ATX-70 in PDT was not so large (1.71 times) and cytotoxicity was similar to that in the case of SDT. In the case of the combination, the cytotoxicity was synergistically enhanced by ATX-70 when the cells were first treated by PDT before the SDT. On

the other hand, only an adding effect was observed when the sequence was reversed as sown in Fig. 2.

It was considered that these sequences of the combination therapy of melanoma might be dependent on the chemical attack or physical effects against the cell membrane as the first treatment.

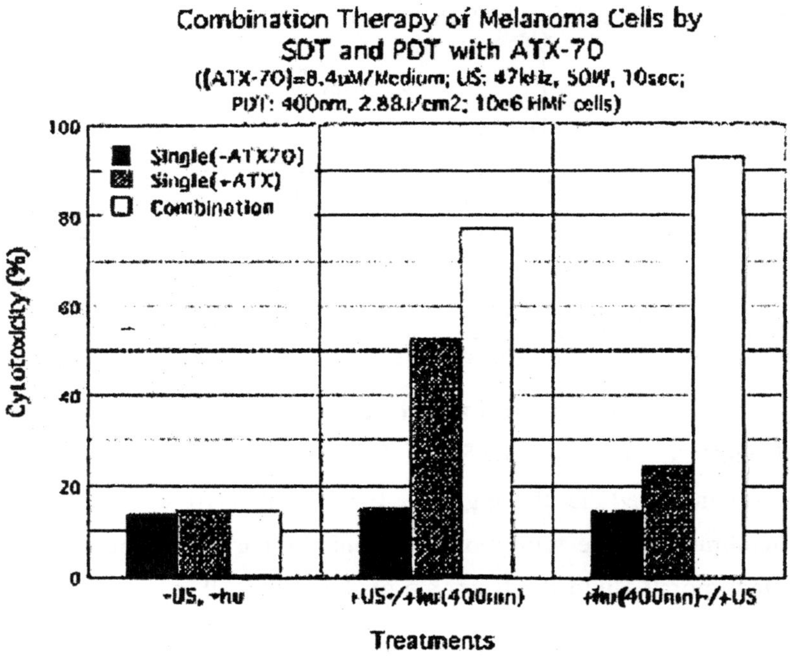

Fig.2. Combination therapy of melanoma cells by PDT and SDT with ATX-70.

In conclusion, it was found that the sequences of PDT and SDT for these combination therapy of human melanoma cells were very important to treat the cells; that was PDT (chemical membrane attack) first before SDT (physical membrane attack). Furthermore it was found that the synergistic cytotoxicity was almost due to the necrotic cell death by a flow cytometry analysis.

DNA DAMAGE CAUSED BY PHOTOEXCITED KETOPROFEN MEASURED BY THE COMET ASSAY

A.L. Vinette[1], J.R.N. McLean[2], J. McNamee[2], P. Belier[2] and J.C. Scaiano[1]
1) University of Ottawa, Ottawa, ON, Canada
2) Radiation Protection Branch, Health Canada, Ottawa, ON, Canada

1. INTRODUCTION

Since most of the studies looking at DNA damage from Non-Steroidal Anti-Inflammatory Drugs (NSAIDs) focus mainly on DNA in solution, without the DNA in its natural environment, it is difficult to extrapolate the noxious effects of NSAIDs to cellular DNA. The Comet Assay is a novel technique which allows for the detection of DNA damage at the single cell level as well as allowing for the cellular functions to go uninterrupted. Ketoprofen was the drug of choice because its photochemistry is fairly well understood, it is one of the most frequently prescribed drugs in clinical practice and it is one of the most frequently reported in phototoxic reactions.[1] The mononuclear fraction of human white blood cells isolated from human whole blood was used as the source of DNA in order to judge the effects on the human immune system. Solar simulated radiation emitted from a Xenon arc lamp consisting of a UVC filter was used as a UVA/UVB source. The photochemistry of Ketoprofen will be explained, the details of the Comet Assay and the reasoning leading up to the different experiments performed to quantify the amount of DNA damage caused from the interaction of Ketoprofen with UVA/UVB and DNA.

2. PHOTOCHEMISTRY OF KETOPROFEN

Ketoprofen is an arylpropionic acid within the NSAID class of compounds, which absorbs in the UVA/UVB region of the electromagnetic spectrum. Using laser-flash photolysis, many groups have studied its photochemistry in order to elucidate its mechanism for photodecomposition by identifying transient intermediates. The generally accepted photochemical mechanism (*Figure 1*), is as follows: upon UVA/UVB irradiation, Ketoprofen **1** absorbs a photon generating its singlet excited state **2**, decarboxylation leads to a ground state carbanion **3** where subsequent protonation generates 3-ethylbenzophenone **4**. Upon absorption of a second photon, **4** reaches a triplet excited state **5** from which deactivation can occur by various pathways.

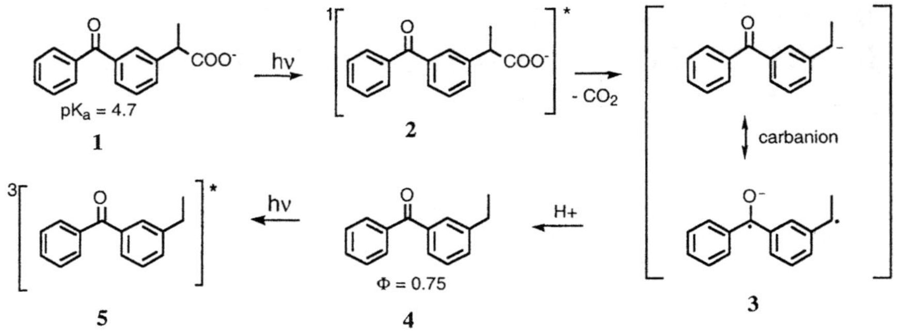

Figure 1. Reaction scheme for Ketoprofen photochemistry.

These deactivation pathways include: an energy transfer to a substrate or oxygen (generating Reactive Oxygen Species (ROSs)), formation of radicals and photoproducts, and hydrogen abstraction from a molecule. If the generation of the triplet excited state takes place within a cell nucleus, DNA, lipids and proteins are all target molecules ready to participate in the deactivation process and the results can be detrimental to cell components as well as cell functions.

3. THE COMET ASSAY

Developed by Singh et al.,[2] this somewhat revolutionary technique, also known as the "single cell gel electrophoresis" (SCGE) technique, is unique in that the observations are made at the single cell level allowing for the technique to be used with virtually any eukaryotic cell. The DNA damage detected by the Comet Assay can arise through various mechanisms, including DNA double- and single-strand breaks, DNA interstrand cross-linking, alkali labile sites, and incompletely repaired excision sites present at the time of lysis. A positive response in the assay means that the above events have been detected.[3]

We have adapted the technique in order to detect Ketoprofen sensitized DNA damage. Briefly, Ketoprofen is injected into the isolated mononuclear white blood cell culture. The cells are then incubated and subsequently mixed with agarose and casted onto a thin film (from here on referred to as 'gels'). The gels are first washed with a saline and detergent lysis buffer at pH 10 to release the DNA from the cellular and nuclear membranes. The exposed DNA within the gels is then washed with an alkaline unwinding buffer at pH 13 to allow DNA unwinding and alkali labile sites to convert to single-strand breaks. The gels, now containing the unwound DNA, are electrophoresed where DNA migration occurs leading to comet formation. After the electrophoresis step, the gels are washed in a neutralizing buffer, then allowed to dry overnight in ethanol. To visualize the damaged DNA, the gels are individually stained with a fluorescent nucleic acid dye and the comets are observed through a fluorescent microscope.

4. RESULTS AND DISCUSSION

Viability experiments on isolated human white blood cells were conducted to ensure that the Ketoprofen concentrations used in the experiments were not harming the cells. The white blood cells were isolated and cultured for a week with varying concentrations of Ketoprofen (10^{-6}, 10^{-5} and 10^{-4} M) and were counted every day for viability. The white blood cells showed to be viable throughout the whole week at all the concentrations of Ketoprofen applied (results not shown).

Since DNA absorption of UV radiation results in DNA crosslinks,[4] we chose a UV dose low enough to allow for an increase in DNA damage caused by the photoexcited drug. Human white blood cells were irradiated with solar simulated radiation at various doses (0, 0.25, 0.5, 0.75 and 1 MED) then incubated for 0, 2 and 4 hours and assayed with the Comet Assay. Results showed that UV damage rose quickly after ~0.2 MED; therefore, we chose 0.2 MED as the maximum UV dose for our experiments (results not shown).

With the Comet Assay, four parameters are generated by which DNA damage can be quantified, these are: head-to-tail ratio, tail moment, comet length and tail length. The tail moment is the preferred parameter in quantifying DNA damage since it is a measure of the difference in the center of gravity between the head and the tail of the comet. This parameter is more reliable in that it gives a clearer description of the location of the DNA within the comet; the results will be shown using the tail moment parameter.

Interesting results arose from a Comet Assay experiment where human white blood cells were incubated with varying Ketoprofen concentrations (10^{-6}, 10^{-5} and 10^{-4} M) and exposed to various doses of UV (0, 0.1 and 0.2 MED) (results not shown). In comparing to the control, at 0.1 MED, only when Ketoprofen was present at 10^{-5} M did we observe an increase in DNA damage; Ketoprofen at 10^{-6} M and 10^{-4} M showed a little more or the same amount of DNA damage respectively. At 0.2 MED, Ketoprofen at 10^{-6} M showed the same amount of DNA damage as the control while Ketoprofen at 10^{-5} M and 10^{-4} M showed a decrease in DNA damage compared to the control.

In a repeat experiment (*Figure 2*), somewhat similar results were observed. At 0.1 and 0.2 MED, all Ketoprofen concentrations showed a decrease in DNA damage compared to the control. These results lead us to believe that Ketoprofen has a protective effect on the DNA. When Ketoprofen is present, it filters the UV light, interfering with the DNA absorption of the radiation, hence preventing it from being damaged. We predict that this effect may be concentration dependent. Concurrent with the two-photon photochemistry of Ketoprofen, there would appear to be a lag time between the initial irradiation of the drug and the subsequent generation of noxious species.

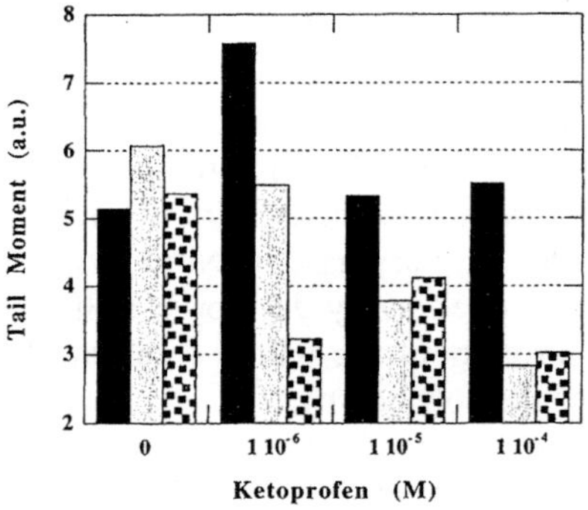

Figure 2. Tail moment parameter for quantification of DNA damage caused by photoexcited Ketoprofen at various concentrations (0, 10^{-6}, 10^{-5} and 10^{-4} M); UV doses range from 0 MED (solid black), 0.1 MED (solid grey) and 0.2 MED (dotted black).

At higher concentrations, the ratio of Ketoprofen to 3-ethylbenzophenone is high; therefore, even though 3-ethylbenzophenone is being formed, Ketoprofen molecules are absorbing most of the UV radiation, preventing 3-ethylbenzophenone molecules from absorbing a second photon and generating DNA damaging species. In this case, the Ketoprofen has a double protective effect, where it first filters the DNA from radiation and second, generates noxious species at a reduced rate by filtering 3-ethylbenzophenone from radiation.

At lower concentrations, the ratio of Ketoprofen to 3-ethylbenzophenone is smaller; therefore, the conversion of 3-ethylbenzophenone to noxious species is more efficient since less Ketoprofen molecules are present to absorb 3-ethylbenzophenone's second potential photon. In this case, lower concentrations of Ketoprofen may have a greater potential for damaging DNA.

In an attempt to explain these results with Ketoprofen's two-photon process, we decided to repeat the experiment with a more toxic drug, Tiaprofenic Acid,[5] whose photochemistry allows for the generation of noxious species upon the absorption of a single photon (*Figure 3*). This experiment consisted of incubating human white blood cells with Ketoprofen and Tiaprofenic Acid at physiological concentration (10^{-5} M), subsequently irradiating at various UV doses (0, 0.05, 0.1, 0.15 and 0.2 MED) and performing the Comet Assay. At all but one UV dose (0.05 MED), Ketoprofen and Tiaprofenic Acid were similar in that they both show an increase (0.15 MED) or a decrease (0.1 and 0.2 MED) in DNA damage. With the exception of 0.05 MED, Tiaprofenic Acid repeatedly shows more DNA damage than Ketoprofen reinforcing the fact Ketoprofen's two-photon process is preventing the observation of a clear trend in its sensitization of DNA damage.

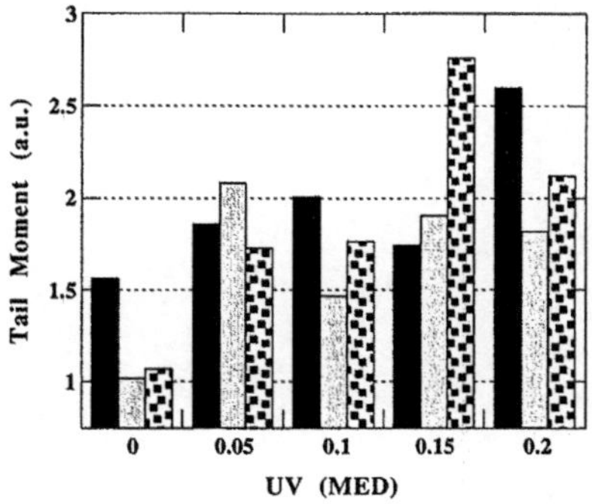

Figure 3. Tail moment parameter for quantification of DNA damage caused by photoexcited Ketoprofen (solid grey) and Tiaprofenic Acid (dotted black) at the physiological concentration (10^{-5} M) compared with the control (solid black); UV doses range from 0, 0.05, 0.1, 0.15 and 0.2 MED.

5. CONCLUSION

In our attempt to show that photoexcited Ketoprofen causes DNA damage in human white blood cells, while we did not observe a clear trend, we may have uncovered an additional element to NSAIDs' phototoxic mechanism. The DNA damage seems to be sensitive to the nature of the photoproducts generated at specific drug concentrations and UV doses. It will be worthwhile to investigate and identify the different photoproducts responsible for the protective/damaging effect of photoexcited Ketoprofen.

6. REFERENCES

1. Condorelli G, Costanzo LL, DeGuidi G, et al. Photosensitization induced by non steroidal anti inflammatory drugs: an overview of molecular mechanisms in biological systems. Newsletter 1996;58:66-77.

2. Singh NP, McCoy MT, Tice RR, Schneider LE. A simple technique for quantitation of low levels of DNA damage in individual cells. Exp Cell Res 1988;175:184-191.

3. Rojas E, Lopez MC, Valverde M. Single cell gel electrophoresis assay: methodology and applications. J Chromatog B 1999;722:225-254.

4. Chouini-Lalanne N, Defais M, Paillous N. Nonsteroidal antiinflammatory drug-photosensitized formation of pyrimidine dimer in DNA. Biochem Pharmacol 1998;55:441-446.

5. Agapakis-Caussé C, Boscá F, Castell JV, et al. Tiaprofenic acid-photosensitized damage to nucleic acids: a mechanistic study using complementary *in vitro* approaches. Photochem Photobiol 2000;71:499-505.

PHOTOSENSITIZATION OF HEPATITIS A VIRUS AND OTHER NON-ENVELOPED RNA VIRUSES BY *MESO*-SUBSTITUTED PORPHYRINS IN WATER AND HUMAN PLASMA

Michael J. Casteel[1], Avram Gold[1], Mark D. Sobsey[1], and Mark E. Brecher[2]
Department of Environmental Sciences & Engineering, School of Public Health, University of North Carolina at Chapel Hill[1]
Department of Pathology and Laboratory Medicine[2]*, University of North Carolina Hospitals Chapel Hill, NC 27599-7400, USA*

Some viruses that cause disease in humans are transmissible by consumption of contaminated water or through the transfusion of blood or its products. Human viruses such as hepatitis A virus (HAV) and parvovirus B19 (PVB19) have caused disease or have been detected in drinking water[9] or in blood products such as pooled plasma[15] and clotting factor[11,18,21]. HAV and PVB19 are small (22-27 nm), non-enveloped human viruses that contain ribonucleic acid or deoxyribonucleic genomes, respectively. A residual risk of infection from these agents is always present, because the effectiveness of barriers (*e.g.*, blood donor exclusion[17], microbiological testing, inactivation or treatment techniques[14,16]) against HAV and other viral pathogens for the prevention of contaminated blood products and water from reaching patients and consumers is finite. Non-enveloped viruses such as HAV are resistant to the removal and inactivation procedures used in both water safety[19,25] and transfusion medicine[16]. Procedures in transfusion medicine with demonstrated efficacy against lipid-enveloped viruses such as the human immunodeficiency virus, and hepatitis B and C viruses[3] are ineffective against HAV, PVB19, and perhaps other viruses with similar biological features. Hence, alternative methods for virus inactivation in water and in blood components are needed. One such method or approach, which utilizes photochemical or photodynamic processes, can be used for the inactivation of viruses in blood products and water[2,13]. These processes, which require the use of an optically excited dye or other molecule and a source of light have demonstrated ability for the inactivation of viruses and other microbes in blood[22] and water[13]. However, no information is available for the photosensitization of HAV and other non-enveloped viruses using the chemically well-defined, second-generation porphyrins. Porphyrins are any of a group of compounds containing the porphine structure of four pyrrole rings connected by methine bridges in a cyclic configuration, to which a variety of side chains are attached[5].

We conducted experiments to determine the virucidal effect of porphyrin photosensitizers in various solutions (solvent = water; phosphate or saline buffers; human plasma) on hepatitis A virus (HAV), poliovirus type 1 (strain LSc; PV1) and coliphage MS2 (bacterial virus). PV1 and MS2 are both non-enveloped viruses having single-stranded RNA genomes. Two porphyrins (Porphyrin Products, Logan, UT) were tested at 10^{-3} to 10^{-5}M concentrations: tetrakis (N-methyl-4-pyridiniumyl) porphine tetratosylate), abbreviated TMPyP4 and *meso*-tetra-(4-sulfonatophenyl)-porphine dihydrochloride ($C_{44}H_{32}N_4O_{12}S_4Cl_2$; MW 1007.69), abbreviated TPPS4. TPPS4 is a sulfonated, negatively charged compound while TMPyP4 has an overall positive charge. Photosensitization was tested on HAV, PV1, and MS2 in porphyrin-containing buffers (0.2M phosphate buffered saline (PBS), pH 7.2, or 0.14M NaCl, pH 7.0), in water, or in human plasma (Rh positive; Type O; received from University of North Carolina Hospitals). TMPyP4 or TPPS4 compounds were adjusted to the desired final concentration in the test solvents.

The cytopathic variant of HAV, strain HM175, was grown and assayed by the plaque technique in newly confluent layers of FRhK-4 (fetal rhesus kidney-derived) cells as previously described[8]. Poliovirus-1 and coliphage MS2 were propagated and assayed as previously described[1,24]. Toxicity assays consisted of adding porphyrins to dishes containing Chinese Hamster Ovary (CHO) cells. CHO cells were seeded into a six-well tissue culture plates and maintained in Eagle's Minimum Essential Medium (1X) containing heat-inactivated fetal calf serum, amino acids, and antibiotics. Concentrated porphyrin solutions were added to the existing maintenance media to the desired final concentration, incubated for 5 days at 37°C and 5% CO_2, and the cells were stained using crystal violet dye. Colony-forming units (cfu) of cells were enumerated and compared to the numbers of cells not exposed to porphyrins. Photosensitization experiments were conducted at room temperature under a laminar-flow hood. A UV lamp (Spectroline ENF260C; Spectronics Corp., NY) was the light source in these experiments and was set for long-wavelength (365 nm) UV irradiation. Porphyrin-containing solutions were magnetically mixed in petri dishes containing stir bars. The UV lamp unit was placed 3.5 cm above the dishes, and the test viruses were added to the porphyrin solutions and petri dishes to a final concentration of 10^4 to10^6 plaque-forming units (pfu) per milliliter. The UV light was turned on to begin an experiment. At 1, 10, 30, 60, or 90 minutes, aliquots were withdrawn from the petri dishes and then stored on ice in the dark. At the beginning and end of each experiments, aliquots of viruses in the porphyrin solution not exposed to light were withdrawn and kept on ice. Timed samples were serially diluted and assayed. Results are expressed as the percent reduction in $\log_{10}$ plaque-forming units (pfu) of infectious virus at time (t) when compared to numbers of viruses in porphyrin solutions not exposed to treatment with UV light.

When coliphage MS2 was added to 10^{-4}M TMPyP4 dissolved in NaCl (TMPyP4/NaCl) and irradiated with long-wavelength UV (LWUV) light, $\log_{10}$ (%) reductions were 0.3 (50%) in 1', 0.9 (87%) in 10', and >3.8 (>99.98%) in 60', while $\log_{10}$ reductions of MS2 in 10^{-4} M TPPS4 / PBS were 0.2 (37%), 2.5 (99.7%), and >3.5 (>99.97%) in 1, 10, and 60 minutes, respectively. In contrast to the declining inactivation observed for MS2, poliovirus was inactivated by >99.99% in just 1 minute in both 10^{-5} M TMPyP4 and TPPS4 in water. Hepatitis A virus suspended in 10^{-5} M TMPyP4/water was inactivated by 98% in 1 minute, and by >99.97% in 10 minutes. HAV suspended in 10^{-5} M TPPS4 / water was inactivated 0%, 92%, 99.2%, and 99.7%, after 1, 10, 30, and 90 minutes, respectively. Test microbes mixed with either molar concentrations of TPPS4 / PBS, or with TMPyP4 / NaCl or water, and which were not exposed to UV light, showed no loss of infectivity; essentially 100% of the initial infectious test microbes remained at the end of each experimental period. HAV and MS2 added to 10^{-4} or 10^{-5} M TPPS4 in plasma also were inactivated after exposure to LWUV light. When TPPS4 was added to a final concentration of 10^{-5} M in plasma, reductions of HAV were 37%, 68%, 80%, and 1 $\log_{10}$ (90%) in 1, 10, 30, and 90 minutes, respectively. At a higher concentration of 10^{-4} M TPPS4 in plasma, reductions were similar to those observed at the 10^{-5} M concentration in plasma, but was more extensive overall (>1.8 $\log_{10}$ (>98%) at 90 minutes). While a concentration of 10^{-5} M TPPS4/plasma irradiated with LWUV light had no effect on coliphage MS2 after 10 minutes of contact, MS2 was completely inactivated (by >99.8%) in 30 minutes. In contrast, MS2 in 10^{-5} M TMPyP4/plasma was completely inactivated (by >99.5%) in 1 minute. Levels of inactivation observed for HAV, PV1, and MS2 in porphyrin-containing solutions may have actually been higher than observed, as the detection limits were reached (*i.e.*, infectious virus was not detectable using the chosen assay).

To test the toxicity of porphyrins on mammalian cells, TMPyP4 and TPPS4 dissolved in PBS were added to cell culture media to final concentrations of 10^{-3}, 10^{-4}, or 10^{-5}M. Controls consisted of culture media only. After incubation for six days followed by crystal violet staining, the average number of cfu/well for the controls was 271 (range, 256-308). In contrast, a concentration of 10^{-3} M of either TPPS4 or TMPyP4 were completely toxic to CHO cells (*i.e.*, <1 cfu / well). However, counts of 320 and 310 cfu/well were recorded for concentrations of 10^{-4} or 10^{-5} M of TPPS4, respectively. TMPyP4 at a molar concentration of 10^{-5} yielded counts of 322 cfu/well, but at a concentration of 10^{-4}M, no cells could be detected. In another experiment, the average number of cfu/ well for controls was 193 cfu/well (range, 190-196).

Alternative methods for virus and other pathogen inactivation in blood and its components are needed[23], and photochemical processes seem to be the most promising[2]. Alternative methods to inactivate and/or remove viruses and other microbes also are needed for the treatment of water. Yet, there is relatively little data on photoinactivation methods for viruses such as HAV and other small, non-enveloped RNA viruses in blood products or in water, and to our knowledge there have been no reports on the use of porphyrins for this purpose. Porphyrins are photosensitizers used in human photodynamic tumor therapy (PDT), and porphyrin chemistry has evolved quickly in recent years, due to its use for the treatment of cancers and advancements made in the field of inorganic chemical synthesis. While other photosensitizers, such as methylene blue, psoralens, and others have been investigated for their ability to inactivate viruses and bacteria, the use of porphyrins has not been fully explored for the photosensitization of viruses and other microbes. Porphyrins have a multitude of unique features that could be exploited for inactivation of human pathogens in the clinical or environmental setting. For example, it is possible to change the overall chemical features of porphyrins by varying the functional groups that can be attached to the base structure of the porphyrin molecule. Porphyrins can be added directly into tubing or bags, or immobilized onto microbeads, resins, membranes, or other solid media. Porphyrins can be conjugated with monoclonal antibodies, or to oligonucleotide probes for specific adsorption to target nucleic acids or certain tissues or cell types[4,10,12]. Finally, porphyrin compounds have been shown to inhibit accumulation of the protease-resistant proteins (prions) that cause transmissible encephalopathy in animals and humans[6,20], a potential area of concern in blood safety[7].

Our data show that irradiation of porphyrin-containing solutions with long-wavelength UV light can inactivate infectious, non-enveloped viruses. The anionic porphyrin TPPS4 was able to inactivate infectious HAV in human plasma by almost 99% in a relatively short time period. Concentrations of porphyrins that inactivated viruses were not toxic to Chinese Hamster Ovary cells. Additional experiments are needed to optimize the ability of these and other porphyrins to inactivate other pathogenic viruses and bacteria that can contaminate water or human blood. Optimization of test conditions and parameters will enable the design and use of a more effective, rapid, and extensive means for the inactivation of hepatitis A virus, other small, non-enveloped RNA-containing viruses, and bacteria in water, blood, and other solutions. Additional experiments are needed to determine effects of dark toxicity and choice of solvent on the ability of porphyrins to inactivate microbes.

1. Adams, MH. Bacteriophages. Wiley-Interscience, New York, 1959.

2. Ben-Hur, E, Horowitz B. 1995. Advances in photochemical approaches for blood sterilization. Photochem Photobiol 1995; 62:383-388.

3. Ben-Hur, E, Horowitz B. Virus inactivation in blood. AIDS 1996; 10:1183-1190.

4. Benimetskaya, L, Takle GB, Vilenchik, M, Lebedeva I, Miller, P, Stein, CA. Cationic porphyrins: novel delivery vehicles for antisense oligodeoxynucleotides. Nucleic Acids Research 1998; 26(23):5310-7.

5. Bonnett, R. Photosensitizers of the Porphyrin and Phtalocyanine Series for Photodynamic Therapy. Chem Soc Reviews 1995; 19-33.

6. Caughey, WS, Raymond LD, Horiuchi, M, Caughey, B. Inhibition of protease-resistant prion protein formation by porphyrins and phtalocyanines. Proc Natl Acad Sci USA 1998; 95:12117-12122.

7. Chamberland, ME, Alter, HJ, Busch, MP, Nemo, G, Ricketts, M. Emerging Infectious Disease Issues in Blood Safety.Emerg Infect Dis 2001; 7(3).

8. Cromeans, T, Sobsey MD, Fields, HA. Development of a Plaque Assay for a Cytopathic, Rapidly Replicating Isolate of Hepatitis A Virus. J Med Virol 1987; 22:45-56.

9. DeSerres, G, Cromeans TL, Levesque B, Brassard, N, Barthe, C, Dionne M, Prud'homme H, Paradis D, Shapiro CN, Nainan OV, Margolis HS. Molecular confirmation of hepatitis A virus from well water: epidemiology and public health implications. J Infect Dis 1999; 179(1):37-43.

10. DiMauro, E., R. Saladino, P. Tagliaesta, V. De Sanctis, and R. Negri. 1998. Manganese Water-soluble porphyrin Senses DNA Conformation. J Molecular Biol 282:43-57.

11. Erdman DD, Anderson BC, Torok TJ, Finkel TH, Anderson LJ. Possible transmission of parvovirus B19 from intravenous immune globulin. J Med Virol 1997; 53(3):233-6.

12. Flynn SM, George ST, White L, Devonish W, Takle GB. Water-Soluble, Meso-Substituted Cationic Porphyrins - a Family of Compounds for Cellular Delivery of Oligonucleotides. BioTechniques 1999; 26:736-746.

13. Gerba CP, Wallis C, Melnick JL. Application of photodynamic oxidation to disinfection of tapwater, seawater, and sewage contaminated with poliovirus. Photochem Photobiol 1977; 26:499-504.

14. Hilfenhaus J, Nowak, T. Inactivation of Hepatitis A Virus by Pasteurization and Elimination of Picornaviruses during Manufacture of Factor VIII Concentrate. Vox Sang 1994; 67(suppl 1):62-66.

15. Koenigbauer UF, Eastlund T, Day JW. Clinical illness due to parvovirus B19 infection after infusion of solvent/detergent-treated pooled plasma.Transfusion 2000; 40(10):1203-6.

16. Lemon SM, Murphy PC, Smith A, Zou J, Hammon J, Robinson S, Horowitz B. Removal / neutralization of hepatitis A virus during manufacture of high purity, solvent/detergent factor VIII concentrate. J Med Virol 1994; 43:44.

17. Moor ACE, Dubbelman TMAR, VanSteveninck J, Brand, A. Transfusion-transmitted diseases: risks, prevention and perspectives. Eur J Haemotl 1999; 62:1-18.

18. Normann A, Graff J, Gerritzen A, Brackmann HH, Flehmig B. Detection of hepatitis A virus RNA in commercially available factor VIII preparation. Lancet 1992; 340(14):1232.

19. Peterson DA, Hurley TR, Hoff JC, Wolfe, LG. Effect of chlorine treatment on infectivity of hepatitis A virus. Appl Envr Microbiol 1983; 45(1):223-7.

20. Priola SA, Raines A, Caughey WS. Porphyrin and Phtalocyanine Antiscrapie Compounds. Science 2000; 287:1503-1506.

21. Robertson BH, Alter MJ, Bell BP, Evatt B, McCaustland KA, Shapiro CN, Sinha SD, Souci JM. Hepatitis A Virus Sequences Detected in Clotting Factor Concentrates Associated With Disease Transmission. Biologicals 1998; 26:95-99.

22. Santus R, Gellier P, Schrevel J, Maziere JC, Stoltz JF. Photodecontamination of blood components: advantages and drawbacks. Clin Hemor Microcirc 1998; 18:299-308.

23. Sloand EM. Viral risks associated with blood transfusion. Photochem Photobiol 1997; 65(3):428-431.

24. Sobsey MD, Carrick RJ, Jensen HR. Improved Methods for Detecting Enteric Viruses in Oysters. Appl Envr Microbiol 1978; 36:121.

25. Sobsey MD, Fuji T, Hall RM. Comparing Inactivation of Cell-Associated and Dispersed Hepatitis A Virus in Water by Free Chlorine and Monochloramine. J Am Water Works Assoc 1991; November issue

Real-Time Quantitative Reverse Transcriptase-Polymerase Chain Reaction (RT-PCR) Analysis of the Vitamin D Pathway in UV Irradiated Keratinocytes

John N. Flanagan[a], Daniel W. Rust, Vin Tangpricha, Tai C. Chen, and Michael F. Holick[a]
Vitamin D, Skin and Bone Research Laboratory and Endocrine Section, Department of Medicine and [a]Department of Biophysics and Physiology, Boston University Medical Center, Boston, Massachusetts 02118.

INTRODUCTION

Humans have developed many mechanisms for detecting and responding to solar ultraviolet (UV) and visible radiation. The primary UV responses are localized to the skin because of its direct interface with sunlight. There is a complex underlying network of response pathways initiated through hormonal and other biologic systems in the skin after UV irradiation which augment gene expression. It is well established that, during exposure to sunlight, the UVB portion (290-315nm) of the solar spectrum is responsible for the photolysis of epidermal and dermal stores of 7-dehydrocholesterol [provitamin D_3] to previtamin D_3 (1). Once formed, it is then isomerized by a thermally induced mechanism to vitamin D_3 (1). Vitamin D_3 a biologically inert form, enters the circulation and requires two successive hydroxylations before becoming biologically active. It is first hydroxylated in the liver by the vitamin D-25-hydroxylase [25-OHase] to form 25-hydroxyvitamin D_3 [25(OH)D_3] and then in the kidney to its biologically active form 1α,25-dihydroxyvitamin D_3 [1α,25$(OH)_2D_3$] by the 25-hydroxyvitamin D_3 1α-hydroxylase [1α-OHase] or to 24,25-dihydroxyvitamin D_3 [24,25$(OH)_2D_3$], a catabolic metabolite, by 25-hydroxyvitamin D_3 24-hydroxylase [24OHase] (2). Once formed, 1α,25$(OH)_2D_3$ acts on its target tissues by interacting with a specific nuclear receptor, the vitamin D receptor (VDR), to bring about its biological effects on gene transcription (3).

The major phyiologic role of 1α,25$(OH)_2D_3$ is to enhance the efficiency of intestinal absorption of dietary calcium and phosphorus and the mobilization of calcium and phosphorus stores from bone (2). Another physiologic role of 1α,25$(OH)_2D_3$ is to modulate cell proliferation and differentiation in cell types that express the VDR. It has been reported that the epidermis expresses VDR and the three major hydroxylases (4) involved in the metabolism of vitamin D (5). We were interested in the interaction between ultraviolet irradiation and 1α,25$(OH)_2D_3$ on the transcriptional activity of these genes in keratinocytes, a cell type which is sensitive to ultraviolet irradiation and responsive to $1\alpha,25(OH)_2D_3$. Thus, we examined the effects of exposing keratinocytes to UV radiation in the presence or absence of $1\alpha,25(OH)_2D_3$ on the expression of these three hydroxylase genes and the VDR gene by using real-time quantitative reverse transcriptase-polymerase chain reaction (RT-PCR).

MATERIALS AND METHODS:

Cell Culture

Normal human keratinocytes (NHK) were obtained from neonatal foreskin and maintained in serum-free defined medium as previously described (6). Cells at 60-70% confluency were treated with 10^{-8} M $1,25(OH)_2D_3$ or vehicle alone for twelve hours. At the end of treatment, the medium was replaced with phosphate-buffered saline (PBS) and cells were then exposed to either 10 mJ/cm^2 or 20 mJ/cm^2 using a solar simulator (Spectral Energy, Westwood, NJ) with a one kW xenon arc lamp (XMN 1000–21; Optical Radiation, Azuza, CA). The solar simulator used a 305-nm filter (Schott Glass Technologies, Duryea, PA) with an average irradiance of 100 μW/cm^2. The amount of UV irradiation was monitored with a research radiometer fitted with a UVB probe, at 285 $\pm$ 5 nm (model IL1700 A; International Light, Newburyport, MA). After irradiation, cells were incubated for an additional eighteen hours in the presence of either $1,25(OH)_2D_3$ or vehicle.

Real-Time Quantitative RT-PCR

Total RNA was harvested from NHK with the SV Total RNA Isolation System (Promega, Madison, WI, USA). 1α-OHase, 24-OHase, 25-OHase, and VDR mRNA levels were determined by quantitative RT-PCR assay based on a sequence specific dual label fluorogenic probe and the TaqMan technology using an ABI Prism™ 7700 Sequence Detector (Perkin Elmer Corp. Foster City, CA, USA). Two hundred nanograms of total RNA was analyzed for each reaction. Sequence specific primers and probes were designed with Primer Express software (Perkin Elmer Corp. Foster City, CA, USA) according to manufacture recommendation (Table 1).

The thermal cycling one-step RT-PCR included three stages. First, the reverse transcription was performed at 48° C for 30 minutes. Second, the Taq polymerase activation was performed at 50° C for 2 minutes followed by 95° C for 10 minutes. Third, the amplification thermocycle went through 40 cycles of the following sequence: melting at 95° C for 15 seconds and anneal/extend at 60° C for 1 minute. Quantitation of PCR products was determined using the ABI Prism™ Sequence Detection System software (Perkin Elmer Corp. Foster City, CA, USA). GAPDH was used as an endogenous reference gene for normalization.

Table I: Sequence Specific Primers and Probes

Gene	5′-Forward Primer-3′	5′-Reverse Primer-3′	5′-Probe-3′
1α-OHase	CCCAAGAGAG CGTGTTGGA	CTTCCTGTTCC GGGAAGAGTT	CGTGTCCACTC CCGCCAATAGC A
24-OHase	GGCCTGGATG TCGTATTTGC	TGAACGTTGGC TTCAGGAGAA	ACAATCCAACA AAGAGCCAAAT GCAGTTGAA
25-OHase	CTGCCTTCTC TGAGCCTGAA A	AAAGGGCACA GAGCCAAATG	CAGCCTGCTAC CCCCAGGATCC A
VDR	GACCCCACCT ACTCCGACTT C	GGCTCCCTCCA CCATCATT	CCAGTTCCGGC CTCCAGTTCGT GT

RESULTS

As demonstrated in figure 1, 1α-OHase, 25-OHase, and VDR all had moderate decreases in expression in NHK irradiated with 10 mJ/cm^2 as compared to nonirradiated cells. However, 24-OHase showed an 8-fold increase in expression in NHK when irradiated with 10 mJ/cm^2 as compared to nonirradiated cells. At 20mJ/cm^2, irradiated NHK had an increased expression in all four genes as compared to control, especially the 24-OHase gene which was up-regulated 50 fold. In the presence of 1α,25$(OH)_2D_3$, 1α-OHase, 25-OHase, and VDR gene expression showed decreased expression at both 10 and 20 mJ/cm^2 doses as compared to nonirradiated cells. On the other hand, 24-OHase had a dramatic 32- and 80-fold induction in gene expression in NHK relative to nonirradiated control cells when they were irradiated at 10 and 20 mJ/cm^2, respectively.

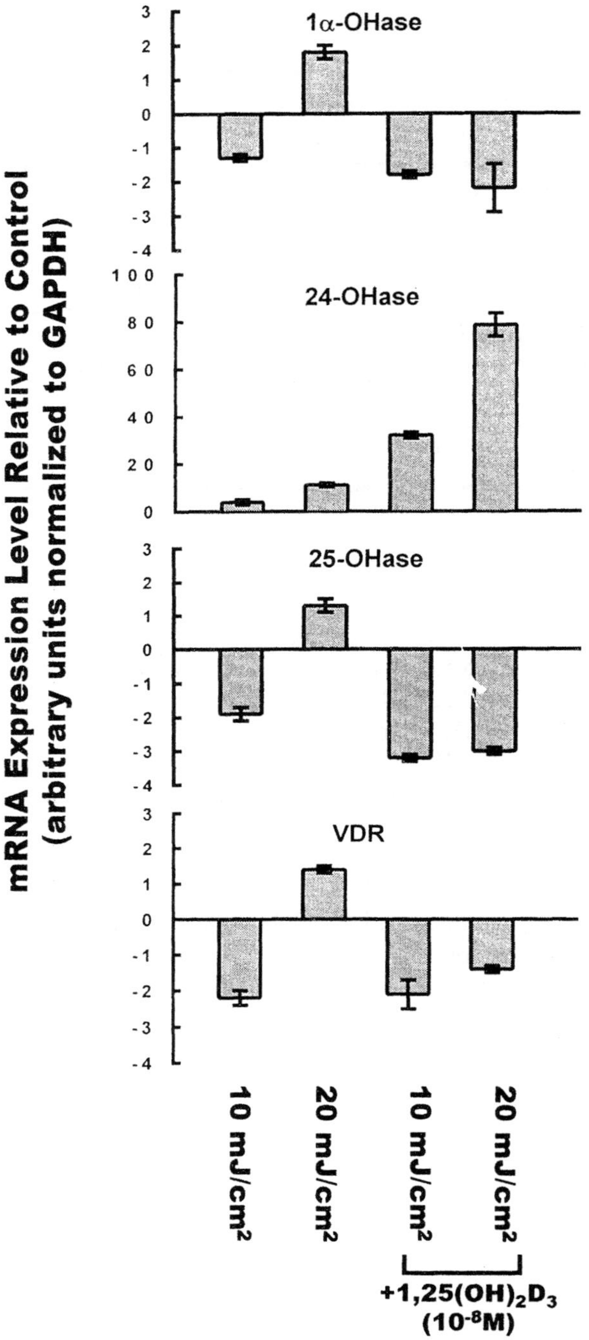

Figure 1. THE EXPRESSION OF GENES INVOLVED IN THE VITAMIN D METABOLIC PATHWAY IN KERATINOCYTES AFTER UV IRRADIATION IN THE PRESENCE AND ABSENCE

OF 1α,25(OH)$_2$D$_3$. Keratinocytes were treated with 1α,25(OH)$_2$D$_3$ or vehicle alone for 18 hours followed by UV irradiation at 0, 10 or 20mJ/cm^2 dosage. RNA was isolated 18 hours after UV irradiation.. Quantitative RT-PCR analysis for the 1α-OHase, 24-OHase, 25-OHase, and VDR are expressed as fold difference to nonirradiated cells at 0 hours and normalized to GAPDH.

DISCUSSION

Results from this study demonstrated the expression of genes involved in the vitamin D metabolic pathway in response to UV irradiation and 1α,25(OH)$_2$D$_3$ treatment in keratinocytes. Using a sensitive technique, such as real-time quantitative RT-PCR, we observed a moderate, but significant change in the expression of 1α-OHase, 25-OHase, and VDR. The much greater response in 24-OHase gene expression in UV irradiated-keratinocytes regardless of 1α,25(OH)$_2$D$_3$ treatment reveals an interesting observation. It is well established that UVB irradiation is responsible for the photolysis of provitamin D$_3$ stores to previtamin D$_3$ (1). Several groups have demonstrated that keratinocytes have the ability to convert vitamin D$_3$ to 1α,25(OH)$_2$D$_3$ through a two-step hydroxylation by 25-OHase then by 1α-OHase (4,8). Recently, Schuessler *et al* reported that 24-OHase was a good indicator of 1α,25(OH)$_2$D$_3$ production through 1α-OHase which induced the 24-OHase gene expression when treated with vitamin D$_3$. Here, we present preliminary evidence indicating that UV irradiation resulted in an induction of 24-OHase that in turn indicated production of 1α,25(OH)$_2$D$_3$ via photolysis of provitamin D$_3$. We also showed that the level of 24-OHase in the presence of 1α,25(OH)$_2$D$_3$ was further induced by increasing the amount of UV irradiation, suggesting that the increased levels of 24-OHase in response to UV irradiation can be correlated with turning on the vitamin D metabolic pathway from provitamin D$_3$.

It has been shown that keratinocytes, which express the VDR, are responsive to active 1α,25(OH)$_2$D$_3$ compounds to decrease proliferation and enhance terminal differentiation (7). In addition, active 1α,25(OH)$_2$D$_3$ compounds and UV phtotoherapy have been used as effective therapies for treating hyperproliferative disorders such as psoriasis (1). The preliminary data presented here provide evidence that UV phototherapy can increase the local production of 1α,25(OH)$_2$D$_3$ via greater conversion of provitamin D$_3$ to previtamin D$_3$ and ultimately leading to the synthesis of 1α,25(OH)$_2$D$_3$. There is increasing evidence showing that keratinocytes contains all the metabolic machinery necessary to convert provitamin D$_3$ to 1α,25(OH)$_2$D$_3$ by expressing 1α-OHase, 24-OHase, 25-OHase, and VDR. Here we present one more piece of evidence indicating that UV is capable of increasing 1α,25(OH)$_2$D levels by using 24-OHase gene expression as an indicator of its synthesis.

Acknowledgements
This work was supported in part by NIH Grant# MM01RR-00533 and the Indoor Tanning Association.

REFERENCES

1. Holick, M.F. Mc Collum Award Lecture, 1994: Vitamin D: new horizons for the 21st century. Am J Clin Nutr 60:619-630, 1994.
2. Holick MF: Primer on the Metabolic Bone Diseases and Disorders of Mineral Metabolism, Fourth Edition, Favus, M.J., editor. Philadelphia: Lippincott-Raven. 92-98, 1999.
3. Stumpf WE, Sar M, Reid FA, Tanaka Y, DeLuca HF: Target cells for 1,25-dihydroxyvitamin D3 in intestinal tract, stomach, kidney, skin, pituitary, and parathyroid. Science 206: 1188-1190, 1979.
4. Schuessler M, Astecker N, Herzig G, Vorisek G, and Schuster I: Skin is an autonomous organ in synthesis, two-step activation and degradation of vitamin D_3: CYP27 in epidermis completes the set of essential vitamin D3-hydroxylases Steroids, 66:3-5:399-408 2001.
5. M.F. Holick, E. Smith and S. Pincus, Skin as a site of vitamin D synthesis and target tissue for 1,25-dihydroxyvitamin D3. Arch Dermatol 123:1677-1683, 1987.
6. Chen, TC, Persons K, Liu WW, Chen ML, Holick MF: The antiproliferative and differentiative activities of 1,25-dihydroxyvitamin D3 are potentiated by epidermal growth factor and attenuated by insulin in cultured human keratinocytes. J. Invest. Dermatol. 104: 113-117, 1995.
7. Hosomi J, Hosoi J, Abe E, Suda T, Kuroki T: Regulation of terminal differentiation of cultured mouse epidermal cells by 1α,25-dihydroxyvitamin D_3. Endocrinology 113: 1950-1957, 1983.
8. Lehmann B, Rudolph T, Pietzsch J, Meurer M. Conversion of vitamin D_3 to 1alpha,25-dihydroxyvitamin D_3 in human skin equivalents. Exp Dermatol Apr;9(2):97-103, 2000.

Biologic Clocks

LIGHT AND THE REGULATION OF MAMMALIAN CIRCADIAN CLOCK GENES

Michael H. Hastings, Verdun M. King and Elizabeth S. Maywood
Department of Anatomy, University of Cambridge, CB2 3DY, U.K.

Circadian rhythms are those daily cycles of behaviour and physiology that persist with an intrinsic period of approximately 24 hours when the individual is held in temporal isolation, devoid of cyclical environmental cues (Aschoff, 1981; Pittendrigh, 1993). They are expressed across divergent phyla (Dunlap 1999) and, in mammals, at every level of tissue organisation, from cells, through tissues to the whole organism. They are a fundamental aspect of eukaryotic organisation, and consequently have the most profound biological and clinical relevance (Hastings 1998). The spontaneous expression of circadian rhythms implies the existence of an internal biological clock or oscillator. To be biologically adaptive this internal clockwork has to be synchronised (entrained) to the solar day, so that internal time matches external time. This article will review recent developments in the analysis of the neural and molecular basis to circadian timing in mammals, and then consider how this core clockwork is entrained by light and by other environmental cues.

The suprachiasmatic nuclei as the circadian clockwork of mammals

The principal circadian clock of mammals is contained within the paired suprachiasmatic nuclei (SCN) of the hypothalamus (Weaver 1998, Figure 1). In the intact animal this cluster of about 10,000 neurons behaves as an assemblage, expressing co-ordinated circadian rhythms of electrical firing and gene expression, both *in vitro* and *in vivo*. The nucleus does, however, have functional sub-divisions, the most notable being the distinction between the "core" that receives direct retinal innervation (Figure 1 b), and the surrounding "shell" that has reciprocal neuronal communication with efferent targets of the SCN. Despite all SCN neurons being GABA-ergic, the distinction between core and shell is reinforced by characteristic differences in neuropeptide gene expression and neural connectivity.

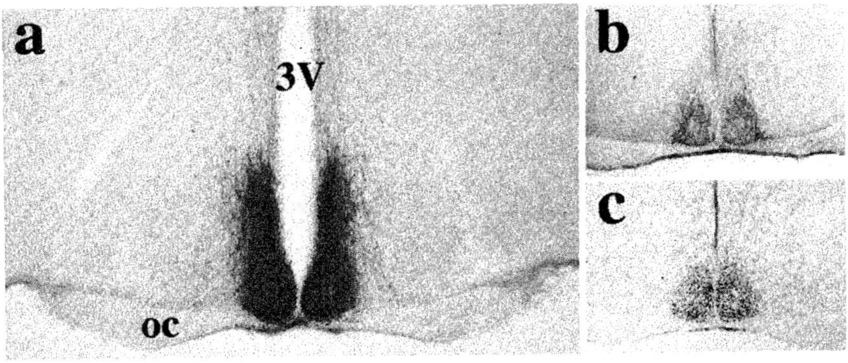

Figure 1. Coronal section of the anterior hypothalamus of a mouse, immunostained for (a) the neuropeptide PHI which is highly expressed in SCN neurons located principally in the ventral retinorecipient core. Efferent PHI-ir fibres innervate the rest of the SCN, local hypothalamic structures and more distant forebrain targets. (b) cholera toxin labelling of direct retinal innervation to SCN (RHT) revealed after intra-ocular injection. (c) mPER1 a nuclear clock antigen.. 3V: third ventricle, oc: optic chiasm.

The retinohyptohalamic tract (RHT) is both necessary and sufficient for entrainment of the SCN by light: there is no consistent evidence of a role for non-ocular photoreception by the mammalian timing system. The retinal innervation is mono-synaptic, derived from a specific population of retinal ganglion cells, and appears to facilitate luminance coding by SCN neurons i.e. the responses of the SCN to light and darkness are graded and very slowly adapting (Nelson & Takahashi 1991). They are dependent on illumination of wide receptive fields across the retina, integrating photic stimulation across time and space. Neurophysiologically, therefore, the RHT stands in contrast to the retinal pathways to the thalamus and mid-brain that subserve image formation and optokinetic reflexes. The cellular basis to circadian photoreception in mammals is also very different from the visual system, insofar as entrainment of the clockwork survives the combined loss of retinal rods and cones that otherwise renders mice visually blind (Freedman et al. 1999, Selby et al. 2000). There also remains uncertainty over the nature of the retinal photopigment mediating entrainment, and a number of proteins have been implicated including melanopsins and cryptochromes (see chapter by Provencio, this volume). On the other hand, what is clear and consistent with other retinally innervated pathways, is that the principal neurotransmitter mediating photo-entrainment of the SCN is glutamate, acting via NMDA and non-NMDA ionotropic receptors to activate neuronal firing and induce gene expression in neurons of the SCN core (Mikkelsen et al. 1995). Administration of either glutamatergic agonists or antagonists can either mimic or block, respectively, the effects of light on the SCN and the circadian system, both *in vivo* (Ebling et al. 1991, Vindlacheruvu et al. 1992) and *in vitro* (see Schurov et al. 1999 for further references).

Despite its co-ordinated activity across the SCN in the intact animal, circadian timing is not an emergent property of the neural population and its circuitry- it is generated in a cell-autonomous manner. This critical observation was made when the spontaneous firing patterns of dissociated SCN neurons were recorded on multi-electrode plates. Individual neurons exhibited high-amplitude circadian cycles of electrical firing that free-ran independently of the circadian activity of neighbouring cells (Welsh et al. 1995). Moreover, the circadian period of cell-autonomous firing rhythms is determined genetically. Mutations that alter the period length and/ or stability of the activity/ rest cycle of the whole animal (*tau* mutation in hamsters, *clock* mutation in mice) have an equivalent effect on the expression of circadian firing rhythms at the level of individual SCN neurons (Liu et al. 1997, Herzog et al. 1998). *In vitro*, the dispersed SCN neuronal population can be synchronised by exogenous GABA-ergic stimulation, presumably because this mimics the reciprocal cell-cell interactions that occur in the intact tissue (Liu & Reppert 2000). Together, these experimental observations raise two general questions. First, what is the molecular/ cellular basis to circadian timing within the SCN, and second, how are those molecular components synchronised by light and other periodic cues?

The molecular basis to the SCN clockwork

The genes encoding core elements of the circadian clockwork of mammals have been identified either on the basis of sequence homology with *Drosophila* clock genes or from a mutagenesis screen in mice (see Dunlap 1999 for review). A current model of the SCN clockwork proposes two principal families of clock gene *Period 1-3*, and *Cryptochrome 1-2*, the expression of which is driven by transcriptional complexes made up of CLOCK/ BMAL1 heterodimers acting at "E-box" regulatory sequences in the *Per* and *Cry* genes (Shearman et al. 2000, see Figure 2). A critical role of the encoded associated CRY and PER proteins is to enter the cell nucleus and suppress activation of E-boxes. This delayed negative feedback effect of the clock proteins on their cognate genes inevitably establishes an oscillatory system. The approximately 24 hour period is determined by the time courses of the component events- transcription, mRNA export and processing, translation, post-translational modification of proteins and nuclear entry. For example the *tau* mutation that shortens circadian period in the hamster to 20 hours, represents a reduction in the effectiveness of the kinase responsible for phosphorylation of PER protein (Lowrey et al. 2000).

The stability, amplitude and precision of this oscillatory loop is enhanced further by incorporation of a positive feed-forward component. The PER2 protein up-regulates expression of the *Bmal1* gene, thereby driving rhythmic availability of BMAL1 protein, the limiting factor in the transcriptional-activating complex (Shearman et al. 2000, Hastings 2000).

Negative and positive feedback loops are therefore interlocked in a relentless, self-sustaining circadian cycle. Mutations of *Clock*, *Cry*, *Bmal1* or *Per* genes generate circadian phenotypes, ranging from altered period to complete arrhythmia. In particular, gene targetting has shown that mPER2, mBMAL1 and at least one of the mCRY proteins are all necessary for the clock to function (van der Horst et al. 1999, Bunger et al. 2000, Bae et al. 2001).

Temporal information is fed out from this closed loop by the rhythmically abundant complexes of CLOCK/ BMAL and PER/ CRY which also interact to sculpt the expression of clock-controlled genes, such as neuropeptides and transcription factors (Shearman et al. 2000). The functions of the SCN neurons regulated by such clock-controlled factors ultimately determine the behaviour of the neuron, and beyond that the whole organism.

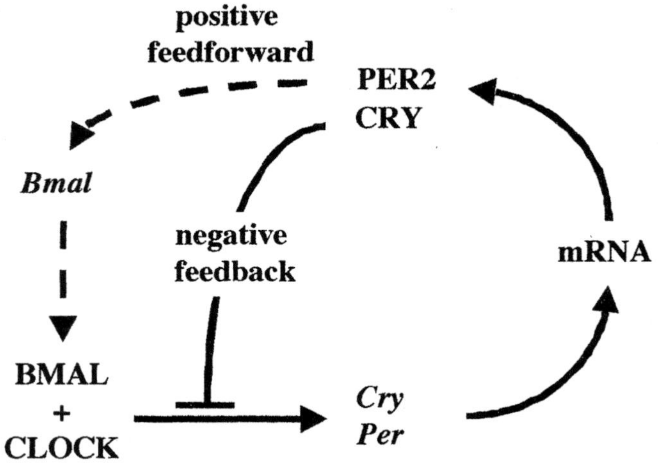

Figure 2. Schematic model of the mammalian circadian clockwork, incorporating negative feedback and positive feedforward actions of CRY and PER2 proteins respectively. See text for details. After Shearman et al. 2000.

Circadian entrainment by light and by behavioural cues in mammals

Before asking how this molecular loop can be regulated by environmental cues, it is necessary to consider the properties of entrainment in mammals. Under a stable 24 hour cycle of light and darkness, overt rhythms will be held in a characteristic phase relationship. In some species activity occurs in the day, and in others it is nocturnal, accompanied by diurnal rest/ sleep. In all species, however, secretion of melatonin by the pineal gland occurs only at night, in darkness. The temporal programme of physiology and behaviour therefore follows a species-typical pattern. An acute shift in the light/ dark cycle, the equivalent of flying between time zones in humans, is accompanied by re-entrainment of overt rhythms. This can take several days to be completed, especially where the circadian clock is asked to advance, as in eastwards flight. The condition of "jet-lag" is widely held to arise from disorganisation of the circadian programme during such re-adjustment.

The underlying basis of re-entrainment is revealed when individuals are allowed to free-run under constant environmental conditions, expressing their endogenous circadian pattern of subjective daytime and subjective night (Aschoff, 1981; Pittendrigh, 1993). Brief exposure to light during early subjective night can shift the free-running rhythms backwards, delaying subsequent cycles. Conversely an equivalent light pulse presented in late subjective night will advance the circadian system. Light presented during subjective daytime, even though subjects are held under continuous darkness, has no effect (see chapter by Czeisler this volume). The phase response curve (PRC), a plot of the phase shift as a function of when during the circadian cycle light was presented, defines this response of the endogenous oscillator to light. It is assumed that stable entrainment in life arises from the composite effects of small delays and advances arising from exposure to evening and dawn light, respectively. This sets the period of the SCN to exactly 24 hours and in so doing, holds the clockwork in a steady phase relationship to the solar day. In addition, the relative duration of circadian day and night, defined by dawn and dusk, directly represents photoperiod. Subjective day and night thereby provide an internal calendar, and in many species this is exploited to time seasonal events (see Wehr, this volume). Photic regulation of the SCN clockwork therefore imparts both time of day and time of year information (Hastings 2001).

A second category of entraining stimuli to be characterised in mammals are known collectively as non-photic. They share the common feature of initiating behavioural arousal, and have been characterised best in nocturnal rodents (Hastings et al. 1998). However, the issue of behavioural regulation of the central clock has wide applicability, not least because it is clear from observation of shift workers that sustained reversal of behavioural cycles can feedback to re-entrain the circadian timing system. The characteristic and distinguishing feature of resetting by non-photic cues is that the circadian system is most sensitive during subjective daytime, when it is not sensitive to light. In addition, non-photic cues predominantly cause phase advances- robust, large amplitude phase delays are rarely observed. The best-characterised neurochemical mediators of non-photic resetting are neuropeptide Y (NPY), which innervates the SCN from the lateral geniculate complex of the thalamus, and serotonin, derived from the median raphe nuclei of the mid-brain. Depending on the agents used, pharmacological manipulations of these systems can either mimic or block non-photic resetting, both *in vivo* and *in vitro* (Hastings et al. 1998, and see Maywood et al. 1997 and Sumova et al. 1996 for references). In addition, at both a behavioural and a neurochemical level, these non-photic systems interact with photic pathways, such that non-photic resetting can be blocked by exposure to light, and NPY can suppress the response of the SCN to glutamate (see below).

Molecular responses of the SCN to light and behavioural resetting cues

Light pulses that reset the circadian clockwork also induce gene expression in the retinorecipient SCN (Best et al. 1999). This is a rapid effect, occurring in less than one hour. A large series of studies employing the immediate-early gene *cfos* as a convenient report for SCN gene induction have characterised the signal transduction pathways mediating this transcriptional effect of light. This involves glutamatergic co-activation of NMDA and non-NMDA ionotropic glutamate receptors, leading to a rise in intracellular calcium levels and activation of calcium-dependent kinases, especially CaM kinase II/IV. This, in turn, is accompanied by rapid activation of MAP kinase (Obrietan et al. 1998) and phosphorylation of the transcriptional regulator CREB (McNulty et al. 1998, Best et al. 1999). CREB protein acts through CRE enhancer sequences to induce the expression of target genes, including *cfos*. Transgenic mice carrying a reporter gene for CRE-mediated transcription have demonstrated an increase in activity through these regulatory sequences in the SCN following exposure to nocturnal light (Obrietan et al. 1999). This transcriptional effect of light is probably wide-ranging across the SCN because it is accompanied by histone remodelling, a pre-requisite for *de novo* gene expression (Crosio et al. 2000). Consistent with the absence of behavioural resetting, there is no induction of gene expression to light pulses delivered in subjective day. The molecular response of the SCN therefore exhibits a phase-dependence equivalent to that seen in the PRC for behavioural resetting. If gene induction is causal to resetting, an intriguing question is how might both phase delays and phase advances be brought about by the same molecular response?

In contrast to the transcriptional effects of light, resetting of the circadian cycle by non-photic stimuli delivered during subjective day is not accompanied by *de novo* gene expression, nor by phosphorylation and activation of CREB (Mead et al. 1992, Sumova et al. 1994). This is consistent with the contrasting neurochemical mediation of photic and non-photic regulation (glutamate vs. NPY/ serotonin), and provides an opportunity for comparative analysis of resetting mechanisms. What is clear is that these two pathways separately addressed by photic and non-photic cues converge on the same oscillatory system. Phase advances of the activity/ rest cycle caused by a non-photic stimulus also shift the circadian rhythm of photic induction of *cfos* expression in the SCN (Mead et al. 1992).

Light and the regulation of *Period* gene expression in the SCN

Before considering resetting mechanisms, the first question to be addressed is that of nocturnality vs. diurnality. Does the cycle of clock gene expression in the SCN reflect circadian time or the activity/ rest cycle? Comparison of *Per2* expression in the SCN of diurnally active ground squirrel and nocturnally active Syrian hamster confirms that the SCN expression matches circadian time and therefore is in phase between species, despite their very different activity patterns (Mrosovksy et al. 2001, Figure 3). It is interesting to note that in the limb region of the motor cortex, expression of *Per2* is indeed oppositely phased between these species, peaking at the respective activity onset presumably because it is a reflection of local neural activation. The principal features of resetting by light, therefore, are that regardless of diurnality/ nocturnality, the clock is maximally sensitive during subjective night, that this corresponds with the nadir of *Per* expression, and that resetting is accompanied by *de novo* gene expression.

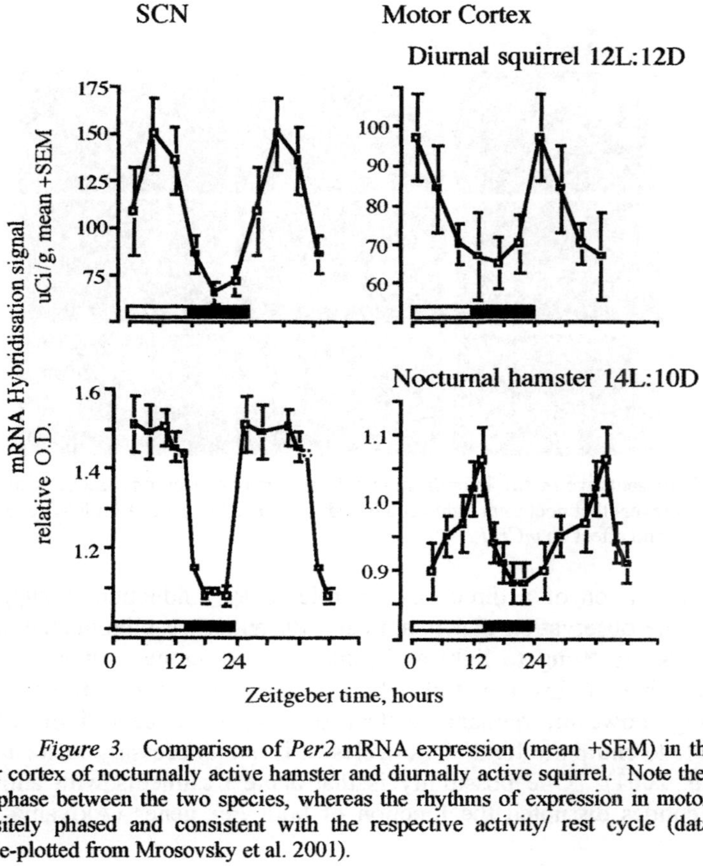

Figure 3. Comparison of *Per2* mRNA expression (mean +SEM) in the SCN and motor cortex of nocturnally active hamster and diurnally active squirrel. Note the SCN cycle is in phase between the two species, whereas the rhythms of expression in motor cortex are oppositely phased and consistent with the respective activity/ rest cycle (data re-drawn, double-plotted from Mrosovsky et al. 2001).

A synthetic model for resetting by light therefore proposes that the SCN clockwork is perturbed, and driven to a new phase in its cycle by the photic (glutamatergic) induction of *Per* genes during their nocturnal nadir. The ability of resetting light pulses to induce *Per1* and *Per2* (but not *Per3*). expression in the SCN has now been demonstrated by a number of laboratories (Shigeyoshi et al. 1997, Albrecht et al. 1997, Figure 4). Whilst induction of *mPer1* occurs with both delaying and advancing pulses, *mPer2* is induced more strongly by dusk illumination, leading to the suggestion that the molecular basis to advance and delay resetting is subtly different. The simplest interpretation of this induction is that glutamatergic activation of CREB and subsequent transcription via CRE elements in the *Per* genes is responsible for their rapid induction by nocturnal light. This is supported by the presence of several CREs in the up-stream regulatory sequences of the *Per1* gene, and the report that this region of DNA is sufficient to confer light inducibility on luciferase or fluorescent protein reporter genes expressed within the SCN of transgenic mice (Yamaguchi et al. 2000, Kuhlman et al. 2000).

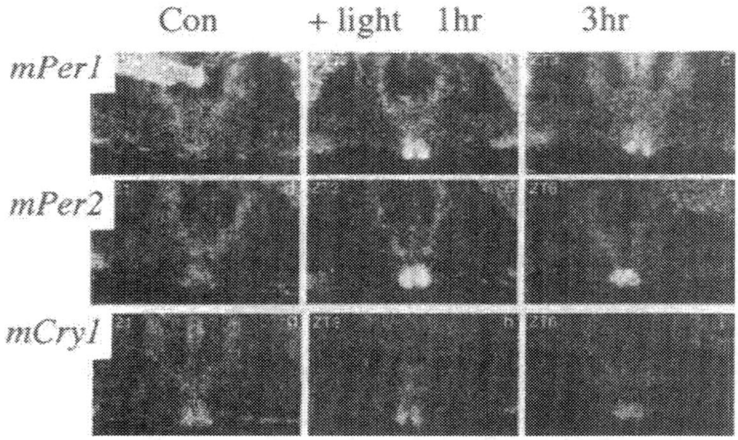

Figure 4. Representative *in situ* hybridisation autoradiographs of coronal sections of mouse hypothalamus reveal that nocturnal light acutely induces *mPer1* and *mPer2* mRNA expression in SCN, but has no effect on *mCry1*

Assignation of a direct, causal role to *Per* induction is supported further by the observation that treatment with antisense oligonucleotides to *mPer1* blocks resetting to light and glutamatergic agonists *in vivo* and *in vitro*, respectively (Akiyama et al. 1999). The precise contributions of *Per1* to resetting, however, remains unclear because mice devoid of mPER1 remain able to entrain to light/ dark cycles- it is not a necessary factor *in vivo* (Bae et al. 2001). One possibility is that acute treatments with antisense oligonucleotides disrupted the function of both *Per* genes (Akiyama et al.

1999). The primary role of PER2 protein in driving the expression of *Bmal1* to initiate the next circadian cycle would suggest that photic up-regulation of *Per2* is the key to the control of the entire clock loop.

Induction of *Per* expression by light provides an elegant, qualitative explanation for the photic PRC. Light presented in early subjective night, as spontaneous expression of mRNA is declining, will temporarily reverse that decline and thereby delay the cycle. Conversely, light encountered in late subjective night after the nocturnal nadir, will accelerate the spontaneous up-regulation of *Per* expression, thereby advancing the cycle and its dependent rhythms. Hence, the behavioural response to light (advance or delay) is phase dependent, even though at all nocturnal phases, the molecular response is the same: up-regulation of *Per* mRNA. This qualitative explanation of the PRC extends into subjective day, where spontaneous expression of *Per* is already maximal and can not be enhanced by photic cues. Consequently, subjective day represents the "dead" zone of the photic PRC (Shigeyoshi et al. 1997).

This model does not, however, account for several critical features of the circadian response to light. First, the magnitude of phase shifts is far smaller than would be expected on the basis of the massive induction of *Per* mRNA. If the levels of *Per* mRNA were the sole determinant of resetting, the clock would be pushed to a phase equivalent to the middle of subjective day, a shift of ca. 12 hours. In practice, phase shifts rarely exceed 2- 3 hours, indicating a strong inertia in the molecular loop. This can arise from several sources. First, acute exposure to light does not up-regulate *Cry* mRNA expression in the SCN (Figure 4). Therefore, only one component of the loop, the *Per1/ Per2* genes, is directly sensitive to light. Resetting of rhythmic *Cry* expression presumably depends on indirect effects mediated via *Per2/ Bmal1*. The insensitivity of *Cry* to light may have further implications for the stability of PER proteins. Under normal circumstances PER and CRY proteins form heterodimers that are important for their stabilisation and nuclear translocation (Kume et al. 1999, Field et al. 2000). Immunostaining for PER proteins after resetting light pulses reveals a very modest induction in the retinorecipient SCN, despite the high levels of mRNA (Field et al. 2000). It may well be that PER proteins newly translated from the light-induced mRNA are not stable because of the declining abundance of CRY during circadian night. Consequently the potential resetting effect of *Per* mRNA induction is dissipated.

A second important feature of resetting that is not explained by a simple model of *Per* induction is the time course of behavioural shifts. Re-entrainment to a shifted light/ dark cycle takes several days to be accomplished at the behavioural level, even though the potential of the clock to respond to light is reset within 2 hours or so (Best et al. 1999). Why should behaviour lag behind when *Per* expression is immediately up-regulated? Again, the indirect regulation of *Cry* genes appears to be the essential factor. Whereas the *Per* loop in the mouse SCN can resynchronise to an 8 hour advance of the lighting regime within 3 days, re-adjustment of the *Cry* cycle takes 8 days. Moreover, the progressive shifting of the *Cry* cycle matches directly the progressive re-adjustment of the animal's activity/ rest cycle (Field et al. in prep), suggesting that it is inertia in the response of *Cry* genes that underlies the essential features of "jet-lag".

Regulation of clock genes during resetting by non-photic cues, and the interactive effect of light.

The SCN clockwork is most sensitive to non-photic resetting cues during subjective day when the spontaneous expression of *Per* and *Cry* is at its peak. By analogy with photic resetting which perturbs the clock by induction of *Per* genes during the nadir, the simple hypothesis is that non-photic resetting involves acute suppression of *Per* gene expression during subjective day.

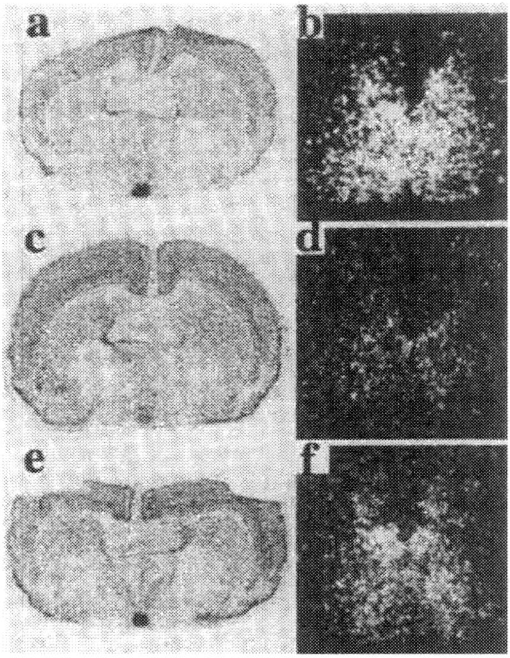

Figure 5. Representative *in situ* hybridisation images of coronal sections of hamster hypothalamus (a, c, e low power film autoradiographs, b, d, f high power emulsion) reveal convergent regulation of *Per2* gene expression in the SCN by light and by non-photic resetting cues. a, b control, c, d hamster after 3 hour of behavioural arousal (note suppression of *Per2* mRNA levels), e, f animal subjected to behavioural arousal followed by light for 0.5 hour (suppression of *Per2* mRNA reversed to control levels). (Redrawn from Maywood & Mrosovsky, 2001).

In the Syrian hamster confinement to a novel running wheel and consequent locomotor arousal is a potent behavioural cue that causes a non-photic phase advance of about 2.5 hours. Analysis of mRNA levels in the SCN reveals that this effect is accompanied by an acute (within 3 hours) suppression of both *Per1* and *Per2*, along with an earlier rise in *Per* expression on the subsequent cycle (Maywood et al. 1999, Figure 5). This supports the model that non-photic advances are brought about by an earlier decline of *Per* mRNA levels. Although it has not yet been demonstrated, the presumption is that this is followed several hours later by an accelerated decline of PER and CRY protein levels. This in turn will cause an early release of the *Per* and *Cry* genes from negative feedback, thereby advancing the initiation of the next cycle.

If suppression of *Per* expression is causal to non-photic resetting, then any manipulation that reversed the mRNA decline should also reverse the behavioural response. This is exactly what happens when hamsters are subjected to the non-photic procedure described above, but are then exposed to light for 0.5 hour. Not only is the decline in *Per* expression in the SCN halted, but their behavioural phase advance is also blocked (Maywood & Mrosovsky 2001, Figure 5). This confirms that light and non-photic cues have convergent interactive effects on the clock via opposing actions on *Per* expression, and it also reveals a cryptic sensitivity of the SCN circadian clock to light during subjective day. The "dead" zone of the photic PRC is more apparent than real: under appropriate circumstances not only can induction of *Per* gene expression be revealed, but it is also accompanied by photic resetting of the behavioural rhythm during subjective day.

Given that serotonin and NPY mediate non-photic resetting, a reasonable hypothesis is that suppression of *Per* mRNA also occurs via these neurochemical pathways. This has been tested in the hamster and mouse by treating animals during subjective day with either 8OH-DPAT (a serotonergic agonist) or native NPY respectively, and then examining the effects on behaviour and *Per* expression (Horikawa et al. 2000, Maywood unpublished data). In both cases it is clear that non-photic resetting by these neurochemical stimuli is accompanied by acute down-regulation of *Per* mRNA levels in the SCN. Furthermore, in the case of NPY infusion, both the behavioural response and the down-regulation of *Per* mRNA in the SCN can be reversed by acute exposure to light immediately afterwards (Maywood unpublished data).

Conclusion

Circadian timing by the SCN is a cell autonomous process, driven by a self-sustaining molecular loop composed of interlocked feedback mechanisms. Of the core clock genes, it appears that *Per* rather than *Cry* genes are the primary targets for resetting by both light and by non-photic cues. The contrasting phase-dependence of these stimuli on the clockwork is mirrored by their contrasting action on the *Per* genes, respectively acute up- and down-regulation of mRNA. The effects of light are mediated by glutamatergic activation of calcium-dependent transcriptional activation via CREB/ CRE. Down-regulation of *Per* mRNA can occur via NPY- and 5HT-dependent signalling, although the molecular details of the response are not yet characterised. Finally, it is clear that at the molecular level these two categories of resetting stimulus converge, exerting opposing actions on the expression of *Per* mRNA. This model of convergent, opposing regulation of *Per* expression in the SCN provides a synthetic, molecular explanation for the observed effects of light and behavioural arousal on circadian behaviour.

References

Akiyama M, Kouzu Y, Takahashi S, Wakamatsu H, Moriya T, Maetani M, Watanabe S, Tei H, Sakaki Y, Shibata S. Inhibition of light- or glutamate-induced mPer1 expression represses the phase shifts into the mouse circadian locomotor and suprachiasmatic firing rhythms. J. Neurosci. 1999; 19: 1115- 1121.

Albrecht U., Sun Z.S., Eichele G., Lee C.C. A differential response of two putative mammalian circadian regulators, mper1 and mper2, to light. Cell 1997; 91: 1055-1064.

Aschoff J, Handbook of Behavioural Neurobiology, vol. 4, Biological Rhythms. New York: Plenum, 1984.

Bae K., Xiaowei J., Maywood E.S., Hastings M.H., Reppert S.M., Weaver D.R. Differential function of *mPer1*, *mPer2* and *mPer3* in the SCN circadian clock. Neuron 2001; 30: 525- 536.

Best J.D., Maywood E.S., Smith K.L., Hastings M.H. Rapid resetting of the mammalian circadian clock. J. Neurosci. 1999; 19: 828- 835.

Bunger M.K., Wilsbacher L.D., Moran S.M., Clendenin C., Radcliffe L.A., Hogenesch J.B., Simon M.C., Takahashi J.S., Bradfield C.A. *Mop3* is an essential component of the master circadian pacemaker in mammals. Cell 2000; 103: 1009- 1017.

Crosio C., Cermakian N., Allis C.D., Sassone-Corsi P. Light induces chromatin modification in cells of the mammalian circadian clock. Nature Neurosci. 2000; 3: 1241- 1247.

Dunlap J.C. Molecular bases for circadian clocks. Cell 1999; 96: 271- 290.

Ebling, F.J.P., Maywood, E.S., Staley, K., Humby, T., Hancock, D.C., Waters, C.M., Evan, G.I., Hastings, M.H. The role of NMDA-type glutamatergic neurotransmission in the photic induction of immediate-early gene expression in the suprachiasmatic nuclei of the Syrian hamster. J. Neuroendocrinology1991; 3: 641- 652.

Field M.D., Maywood E.S., O'Brien J.A., Weaver D.R., Reppert S.M., Hastings M.H.. Analysis of clock proteins in mouse SCN demonstrates phylogenetic divergence of the circadian clockwork and resetting mechanisms. Neuron 2000; 25: 437- 447

Freedman M.S., Lucas R.J., Soni B., von Schantz M., Munoz M., David-Gray Z., Foster R. Regulation of mammalian circadian behaviour by non-rod, non-cone, ocular photoreceptors. Science 1999; 284: 502- 504.

Hastings M.H. The Brain, Circadian Rhythms and Clock Genes. Brit. Med..J 1998; 317: 1704- 1707.

Hastings M.H. Circadian clockwork: two loops are better than one. Nature Rev. Neurosci. 2000; 1: 143- 146.

Hastings M.H. Modelling the molecular calendar. J. biol. Rhythms 2001; 16: 117- 123.

Hastings M.H., Duffield G.E., Smith E.J.D., Maywood E.S., Ebling F.J.P. Entrainment of the circadian system of mammals by nonphotic cues. Chronobiol. International 1998; 15: 425-455

Herzog E.D., Takahashi J.S., Block G.D. *Clock* controls circadian period in isolated suprachiasmatic nucleus neurons. Nat. Neurosci. 1998; 1: 708- 713.

Horikawa K., Yokota S., Fuji K., Akiyama M., Moriya T., Okamura H., Shibata S. Nonphotic entrainment by $5HT_{1A/7}$ receptor agonists accompanied by reduced *Per1* and *Per2* mRNA levels in the suprachiasmatic nuclei. J. Neurosci. 2000: 20: 5867- 5873.

Kuhlman S.J., Quintero J.E., McMahon D.G. GFP fluorescence reports *Period1* circadian gene regulation in the mammalin biological clock. NeuroReport 2000; 11: 1479- 1482.

Kume K., Zylka M.J., Sriram S., Shearman L.P., Weaver D.R., Jin X., Maywood E.S., Hastings M.H., Reppert S.M. mCRY1 and mCRY2 are essential components of the negative limb of the circadian clock feedback loop. Cell 1999; 98: 193- 205.

Liu C., Weaver D.R., Strogatz S.H., Reppert S.M. Cellular construction of a circadian clock: Period determination in the suprachiasmatic nuclei. Cell 1997; 91: 855-860.

Liu C., Reppert S.M. GABA synchronises clock cells within the suprachiasmatic circadian clock. Neuron 2000; 25: 123- 128.

Lowrey P.L., Shimomura K., Antoch M.P., Yamazaki S., Zemenides P.D., Ralph M.R., Menaker M., Takahashi J.S. Positional syntenic cloning and functional characterisation of the mammalian circadian mutation *tau*. Science 2000; 288: 483- 491.

McNulty S., Schurov I.L., Sloper P.J., Hastings M.H.. Stimuli which entrain the circadian clock of the neonatal Syrian hamster *in vivo* regulate the phosphorylation of the transcription factor CREB in the suprachiasmatic nucleus *in vitro*. Eur. J. Neurosci. 1998; 10: 1063- 1072.

Maywood E.S., Mrosovsky N. A molecular explanation of interactions between photic and non-photic circadian clock-resetting stimuli. Gene Expression Patterns (Brain Res. Interactive) 2001; 1: in press

Maywood E.S., Mrosovsky N., Field M.D., Hastings M.H. Rapid down-regulation of mammalian *Period* genes during behavioural resetting of the circadian clock. Proc. Natl. Acad. Sci. USA. 1999; 96: 15211-15216.

Maywood E.S., Smith E., Hall S.J., Hastings M.H. A thalamic contribution to arousal-induced, non-photic entrainment of the circadian clock. Eur. J. Neurosci. 1997; 9: 1739- 1747.

Mead S.M., Ebling F.J.P., Maywood E.S., Humby T., Herbert J., Hastings M.H. A non-photic stimulus causes instantaneous phase advances of the light-entrainable circadian oscillator of the Syrian hamster, but does not induce the expression of c-fos in the suprachiasmatic nuclei. J. Neurosci. 1992; 12: 2516- 2522.

Mikkelsen J.D., Larsen P.J., Mick G., Vrang N., Ebling F.J.P., Maywood E.S., Hastings M.H., Moller, M. Gating of retinal inputs through the suprachiasmatic nucleus: role of excitatory neurotransmission. Neurochem. Inter. 1995; 3: 263- 272.

Mrosovsky N., Edelstein K., Hastings M.H., Maywood E.S. Cycle of *period* gene expression in a diurnal mammal: implications for nonphotic phase shifting. J. biol. Rhythms 2001: (in press).

Nelson D.E., Takahashi J.S. Sensitivity and integration in a visual pathway for circadian entrainment in the hamster (*Mesocricetus auratus*). J. Physiol. (Lond.) 1991; 439: 115- 145.

Obrietan K., Impey S., Storm D.R. Light and circadian rhythmicity regulate MAP kinase activation in the suprachiasmatic nuclei. Nature Neurosci. 1998; 1: 693- 700.

Obrietan K., Impey S., Smith D., Athos J., Storm D.R. Circadian regulation of cAMP response element-mediated gene expression in the suprachiasmatic nuclei. J. biol. Chem. 1999; 274: 17748- 17756.

Pittendrigh C.S. Temporal organisation: reflections of a Darwinian clock-watcher. Ann. Rev. Physiol. 1993; 55: 16-54.

Schurov I.L., McNulty S., Best J.D., Sloper P.J., Hastings M.H. Glutamatergic induction of CREB phosphorylation and Fos expression in primary cultures of the suprachiasmatic hypothalamus in vitro is mediated by co-ordinate activity of NMDA- and non-NMDA-receptors. J. Neuroendocrinology 1999; 11: 43- 51.

Selby C.P., Thompson C., Schmitz T.M., van Gelder R.N., Sancar A. Functional redundancy of cryptochromes and classical photoreceptors for nonvisual ocular photoreception in mice. Proc. Natl. Acad. Sci. U.S.A. 2000; 97: 14697- 14702.

Shearman LP, Sriram S., Weaver D.R., Maywood E.S., Chaves I., Zheng B., Kume K., Lee C.C., van der Horst G.T.J., Hastings M.H., Reppert S.M. Interacting molecular loops in the mammalian circadian clock. Science 2000; 288: 1013- 1019.

Shigeyoshi Y., Taguchi K., Yamamoto S., Takekida S., Yan L., Tei H., Moriya T., Shibata S., Loros J.L., Dunlap, J.C., Okamura, H. Light-induced resetting of a mammalian circadian clock is associated with rapid induction of the mper1 transcript. Cell, 1997; 91, 1043-1053.

Sumova A., Maywood E.S., Selvage D., Ebling F.J.P., Hastings M.H. Serotonergic antagonists impair arousal-induced phase shifts of the circadian system of the Syrian hamster. Brain Research 1996; 709: 88- 96.

Sumova A., Ebling F.J.P., Maywood E.S., Herbert J., Hastings M.H. Non-photic circadian entrainment in the Syrian hamster is not associated with phosphorylation of the transcriptional regulator CREB within the suprachiasmatic nucleus, but is associated with adrenocortical activation. Neuroendocrinology 1994; 59: 579- 589.

Vindlacheruvu R.R., Ebling F.J.P., Maywood E.S., Hastings M.H. Blockade of glutamatergic neurotransmission in the suprachiasmatic nucleus prevents cellular and behavioural responses of the circadian system to light. Eur. J. Neurosci. 1992; 4: 673-679.

Van der Horst G.T.J. Muijtjens M., Kobayashi K., Takano R., Kanno S., Takao M., de Wit J., Verkerk A., Eker A.P.M., van Leenen D., Buijs R., Bootsma D., Hoejimakers J.H.J., Yasui A. Mammalian Cry1 and Cry2 are essential for maintenance of circadian rhythms. Nature 1999; 398, 627- 630.

Weaver D.R. The suprachiasmatic nucleus: a 25-year retrospective. J. biol. Rhythms 1998; 13, 100- 112.

Welsh D.K., Logothetis D.E., Meister M., Reppert S.M. Individual neurons dissociated from rat suprachiasmatic nucleus express independently phased circadian firing rhythms. Neuron 1995; 14, 697-706.

Yamaguchi S., Mitsui S., Miyake S., Yan L., Onishi H., Yagita K., Suzuki M., Shibata S., Kobayashi M., Okamura H. The 5' upstream region of *mPer1* gene contains two promoters and is responsible for circadian oscillation. Curr. Biol. 2000; 10: 873- 876.

Entrainment of the Drosophila circadian clock by light

Fang-Ju Lin and Amita Sehgal, Howard Hughes Medical Institute, University of Pennsylvania Medical School, Philadelphia, PA 19104

Introduction

The phenomenon of circadian (~24 h) rhythms has been conserved from single cell (e.g. cyanobacteria) to complex organisms including plants and mammals. Although the output varies from one species to another, ranging from the circadian production of asexual spores in *Neurospora* and the circadian movement of leaves in *Arabidopsis* to the rest:activity rhythm in *Drosophila* and mammals, all circadian rhythms share some basic properties 1) They can be synchronized or entrained to environmental cues such as light or temperature. 2) They freerun in constant conditions. 3) the periodicity is temperature compensated, i.e. the period does not change over a wide range of temperature. In addition, they are all thought to be produced by a circadian system that consists of an input pathway, a central clock and an output pathway. The clock is the timekeeping component that transmits its signals to the rest of the organism through the output pathway. The input pathway serves to connect the clock to the environment, conveying signals from external stimuli such as light and temperature. Much of our current understanding of how clocks are generated in mammals is derived from research done in the fruit fly *Drosophila melanogaster*. This chapter will focus on the molecular mechanisms that entrain the clock to light in this organism.

Mechanisms underlying clock function

The two major parameters of circadian rhythms in the fruit fly *Drosophila melanogaster* are eclosion and locomotor acitivity. Fully developed adult flies are held in their pupal cases and do not emerge (eclose) until the following dawn. This process is gated and controlled by the circadian clock. Locomotor activity shows a bimodal distribution of activity in a light:dark cycle and in constant darkness, although the bimodality usually disappears, a circadian rhythm of rest:activity persists. Through genetic screens for clock mutants, the *period* (*per*) and *timeless* (*tim*) [1, 2] were identified and provided a basic mechanism for how a clock is generated. RNA and protein products of both these genes cycle with a circadian rhythm and the two proteins negatively regulate the synthesis of their own mRNAs, generating an autoregulatory loop that appears to constitute the basic timekeeping mechanism [3]; (see also Figure 1).

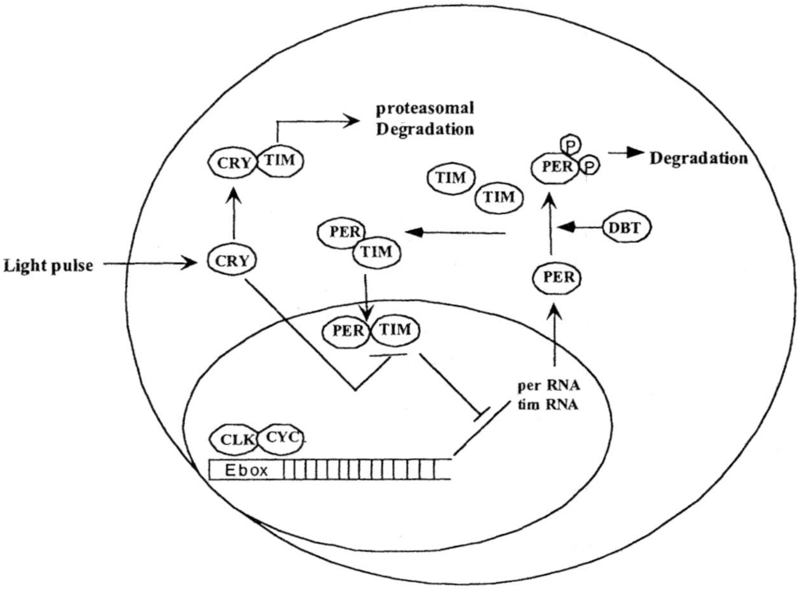

Figure 1. The feedback loop that constitutes the clock in Drosophila. The photoreceptor CRY is shown here and is discussed in the text under "The phototransduction pathway involved in circadian entrainment"

The transcription of the *per* and *tim* genes starts in the late day and peaks around beginning of the night. However, the proteins (PER and TIM) levels do not peak until 6 h later. The transcription of these two genes is controlled by a heterodimer composed of two transcription factors, CLOCK (CLK) and CYCLE (CYC), each of which contains a PAS (per, single-minded and aryl hydrocarbon receptor nuclear translocator) as well as a bHLH domain. An E-box sequence in the promoter region of *per*, *tim* and other circadianly regulated genes is known to be the target for these transcription factors. PER monomers are very unstable in the cytoplasm, due to their phosphorylation by a casein kinase called *doubletime* (*dbt*). At least one function of TIM is to stabilize PER by forming a heterodimer, which is then transported into nucleus and sequesters and inhibits the activity of the CLK-CYC heterodimer. Turnover of PER and TIM in the late night-early day releases the CLK-CYC complex and transcription of two genes resumes.

Mutations in *per* and *tim* affect the feedback loop in ways that are consistent with their effects on overt rhythms. For instance, the per^{long} (per^{l}) mutation decreases the affinity of the PER-TIM interaction, thereby delaying nuclear entry of the heterodimer and extending the cycle [4, 5]. Presumably this accounts for the longer behavioral rhythms observed in these flies. The per^{short} (per^{s}) mutation accelerates the daily disappearance of PER, most likely by decreasing its stability, thereby truncating the phase of feedback inhibition and shortening the cycle [6]. Alteration of PER phosphorylation by *dbt* mutations results in shorter or longer periods [7]. The tim^{ul} mutation appears to lengthen period by increasing the stability of TIM such that its expression in the nucleus is prolonged [8]. Constitutive expression of either *per* or *tim* or of both RNAs still results in the cyclic expression of PER and TIM proteins and the restoration of behavioral rhythms in per^{0} and tim^{0} single or double mutants respectively [9]. This suggests that the cycling of both proteins is controlled independently of the cycling of their mRNAs and that the proteins are the timekeepers in this system. Levels of the proteins are presumably indicators of the time-of-day. Although the mechanisms that drive protein cycling are not known, cyclic control of protein stability is a distinct possibility.

Resetting of the clock by light

As mentioned above, although circadian rhythms can freerun they are usually synchronized to the day:night cycle. In addition to their entrainment to cycles of light and dark (parametric entrainment), circadian rhythms also respond to pulses of light by resetting their phases (nonparametric entrainment). Light pulses in the early part of the night delay the phase of the rhythm whereas those in the second half of the night advance it. In recent years the molecular mechanisms underlying these effects of light have begun to be elucidated.

If levels of the PER and TIM proteins are measures of time-of-day it follows that a signal that resets the clock would do so by changing levels of these proteins. In fact, levels of the TIM protein are decreased rapidly upon light treatment at both the early as well as the later part of the night [10-13]. The differential behavioral response is apparently due to the delay in forming PER-TIM heterodimers in the early night and the early termination of the PER-TIM inhibition on CLK-CYC transcription activity in the late night. Evidence that the TIM response to light mediates behavioral resetting came from dose-response studies in which the effects of various doses of light on the TIM light response and on behavioral entrainment were compared [14, 15]. The levels of TIM in fly head extracts were measured by western analysis. They were unaffected by a 1 minute pulse of 10 lux delivered at zeitgeber time (ZT) 20 {ZT0=lights on and ZT12=lights off}, but they were reduced by a 25 lux pulse of equal duration. Increasing doses of light led to greater degradation and the response saturated rapidly thereafter. The behavioral rhythm advanced slightly with a 1-minute pulse of 25 lux at ZT20, but this advance was not statistically significant. A 1-minute pulse of 50 lux produced a significant shift of 35 minutes in one experiment, although the shift was not significant in another experiment indicating that this dose is close to the threshold for shifting behavioral rhythms. A pulse of 100 lux resulted in a ~1.5 hour advance in the phase of activity rhythms in both experiments. The behavioral response at ZT20 saturated with a 5-minute pulse of 200 lux, which is approximately the point at which the TIM response saturates. Thus, these experiments indicated a correlation between the TIM response to light and behavioral resetting. More importantly, the behavioral response was not more sensitive to light than the TIM response. In addition, assays using different wavelengths of light showed that both the TIM response as well as the behavioral response are most sensitive to the blue light region of the spectrum [16].

The role of TIM in light-induced resetting was also supported by analysis of phototransduction mutants that affected both the TIM response as well as the behavioral response to light. In the *Drosophila* visual system [17], rhodopsin is activated upon absorption of a light photon and it then activates the *norpA*-encoded phospholipase C. Intracellular messengers are generated that trigger depolarization of photoreceptor membranes via opening of the transient receptor potential (TRP) or TRP-like (TRPL) ion channels . To determine whether the visual system is also involved in circadian entrainment, *trp* and *trp-like* (*trpl*) were tested for their TIM response and for shifts in their behavioral rhythms in response to light [15]. We found that both responses were reduced but not eliminated. The effect of the *trp* mutants on the TIM response was particularly noticeable in lateral neurons, the clock cells that drive the rest:activity rhythm.

The phototransduction pathway involved in circadian entrainment

As mentioned above, circadian rhythms are most sensitive to the blue light region of the spectrum. Since the action spectrum does not match that of the known opsins , the role of visual photopigments in circadian entrainment has always been a subject of great debate. Earlier studies showed that mutations that affected vision either morphologically or physiologically did not affect entrainment of behavioral rhythms [18]. The discovery of the TIM response to light provided us with a molecular assay for entrainment of the clock. We reasoned that if the TIM response represented the first response of the clock to light, then by using this assay we could distinguish any effects that various mutants might have on entrainment from those on other aspects of the circadian system. Using the TIM assay, together with the behavioral response to pulses of light, we tested various visual transduction mutants [15].

We found that flies that lack eyes entrained normally to light cues, in terms of their TIM response as well as their behavioral response [15]. Although rhythms in these flies were somewhat weaker, which may be due to the morphological defects associated with them, they were able to synchronize to light:dark cycles and reset their rhythm in response to pulses. Thus, although both PER and TIM are expressed in the eye, their expression in the eye is not required for maintenance or entrainment of behavioral rhythms. Most likely, they are part of an eye-specific clock that is independent of the brain clock and mediates a rhythm of visual sensitivity [19].

As mentioned above, we found that the visual transduction mutants *trp*, and in particular the double mutant *trp;trpl* showed deficits in their response to light [15]. However, although these flies are visually blind, they were still able to entrain to light:dark cycles and also, albeit to a lesser extent than wild type, to pulses of light. Thus, they were not circadian-blind. Moreover, *norpA* mutants that lack the phospholipase C, norpA appeared to be normal for circadian entrainment. Based upon these data, we proposed that the clock in Drosophila uses a dedicated photoreceptor, most likely a protein similar to the blue light photoreceptor, cryptochrome, which had been found in plants. Given the *trp* mutant phenotype, we proposed that the visual system plays a role in circadian entrainment when it is present, but it is not required for this process.

The discovery of cryptochrome (cry) in *Drosophila*, *cryptochrome* (*dcry*), supported our prediction [20-23]. *Drosophila cryptochrome* (dCRY) is closely related to mammalian CRY, as well as to the *Drosophila* 6-4 DNA photolyase that repairs UV-damaged DNA. *cry* RNA levels cycle and highest levels are found in the early part of the day [21]. This cycling persists in constant darkness and is dependent on PER, TIM, CLK, and

CYC. In *per* and *tim* null mutants, *cry* RNA levels are constantly low and in *Clk* and cyc mutants the levels are constantly high, indicating positive regulation by the former and negative regulation by the latter two components. CRY protein cycles only in the presence of light:dark cycles. Apparently this cycling is driven by light-sensitivity of the protein because DD (constant dark) conditions lead to constant accumulation of *CRY.*

Through a genetic screen, a *cry* mutation cry^b was also found [23]. The mutant carries a one amino acid substitution in the conserved flavin-binding site and is presumably a null for photoentrainment. Although the oscillation of PER and TIM in the eyes is abolished in cry^b flies, robust cycling of these two in lateral neuron maintains behavioral rhythms both in LD and DD. In addition, these flies entrain normally to light:dark cycles [23]. However, cry^b shows no phase shift in response to light pulses, indicating a profound effect on nonparametric entrainment. In addition, cry^b flies are rhythmic in constant light, a condition that renders wild type flies arrhythmic, further supporting a role for CRY in the light input pathway [24]. Overexpression of CRY affects photosensitivity of the clock in different ways, based upon the promoter used to drive expression. When the *tim* promoter is used, photosensitivity is increased while use of the actin promoter results in reduced responses to light [21, 22]. Thus, levels of the photoreceptor may need to be tightly regulated for appropriate photic responses. Consistent with our data indicating that the eyes are dispensable for entrainment, CRY expressed in lateral neurons of transgenic flies can rescue the entrainment phenotype of cry^b flies, but fails to do so when its expression is restricted to the eyes [25].

norpA; cry^b double mutants show poor entrainment, or rather reentrainment when moved from one light:dark cycle to another, suggesting that both visual and CRY-mediated photoreception are important for circadian entrainment [23]. However, even these flies are not circadian blind, as PER and TIM continue to cycle in synchrony with the environment in the lateral neurons. Entrainment appears to be completely abolished in a *glass:*cry^b double mutant in which CRY as well as all inputs from the compound eyes, the ocelli and the Hofbauer-Buchner eyelet are eliminated [26]. The latter is a structure in the optic lobe that expresses rhodopsin and has referred to as an extra eye[27]. Together these data support our hypothesis that circadian photoreception uses a dedicated photoreceptor, but that the visual system also plays a role. It should be mentioned that even the *glass:*cry^b mutants will be driven by light:dark cycles to display rhythmic behavior although these rhythms do not show anticipation of light-to-dark or dark-to-light transitions. This driven behavior called "masking", is also seen in clock mutants and lacks the anticipatory component indicative of a clock [28]. Although not related to the clock, the persistence of masking behavior in *glass:cry^b* mutants suggests that there are additional, perhaps as yet undiscovered photoreceptors, which are relevant for the control of an organism's rest:activity.

Mechanisms that reduce TIM levels in response to light

To gain insight of the mechanism involved in TIM degradation, we first developed an *in vitro* assay [29]. We immunoprecipitated TIM from adult heads and treated the immunoprecipitates with head extracts derived either from flies kept in the dark or those pulsed with light. Light-treated extracts from either wild type or *tim^0* flies can reduce the levels of TIM, indicating a light-induced activity that degrades TIM. Pharmacological manipulations showed that this activity was blocked by inhibitors of the proteasome. To confirm these data *in vivo,* we dissected the central nervous systems out of the third-instar larvae, maintained them in culture medium and assayed them for the TIM light response. Fortunately, larvae contain lateral neurons that express TIM, although the role of clock proteins at this stage of development is not known. We found that TIM was degraded in lateral neurons in response to a 20-min light pulse, and that the degradation was blocked in the presence of proteasome inhibitors such as ALLN, MG115 and MG132. It was also blocked by an inhibitor of tyrosine kinase activity, but not by inhibitors of serine threonine kinases [29].

To determine whether TIM was the direct target of the tyrosine kinase, we immunopreciptated TIM from "dark" and "light-treated" fly heads and probed western blots of the immunoprecipitates with an anti-phosphotyrosine antibody [29]. A light-dependent increase of phosphorylation was observed. To determine whether TIM was the direct target of the proteasome we assayed TIM immunoprecipitates for the presence of conjugated ubiquitin. We failed to get consistent results with fly heads probably because ubiquitination is usually very transient and at any given time only a very small fraction (~4%) of a target protein may be ubiquitinated. However, co-expression of TIM and an HA-tagged ubiquitin octamer in Drosophila S2 cells allowed us to observe light-dependent TIM ubiquitination (Figure 2). We are currently following up on the S2 cell assay to dissect some of the early events involved in photic resetting including the role of CRY in this process.

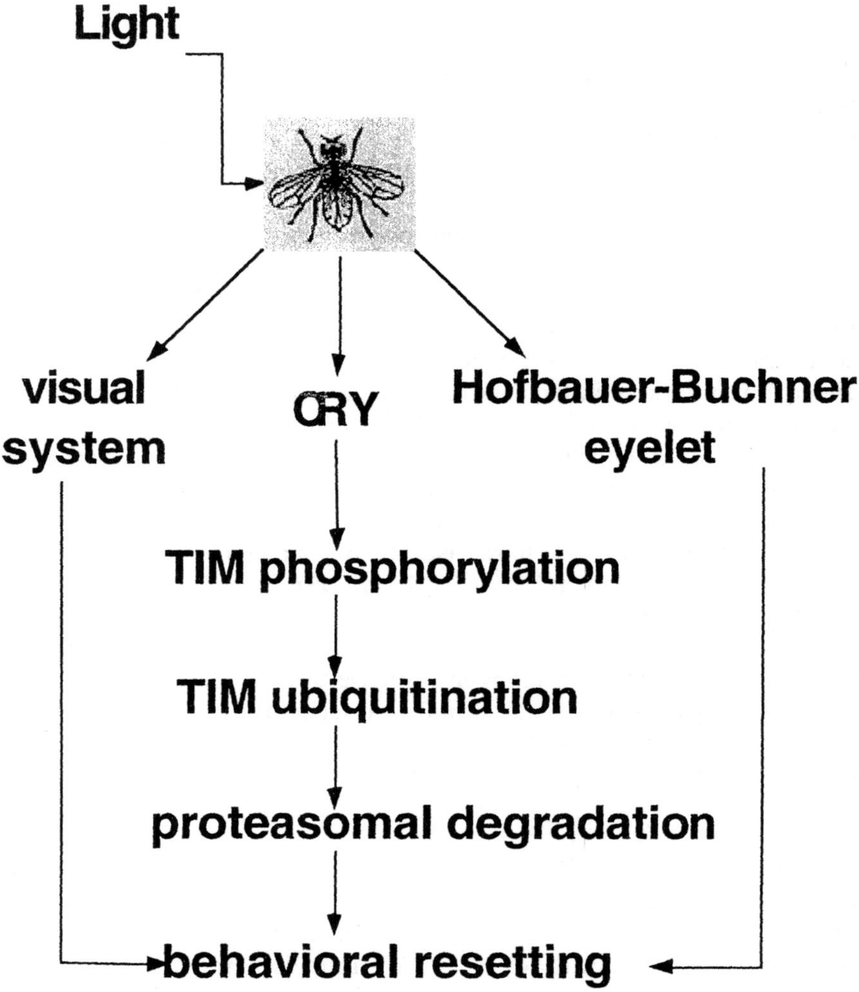

Figure 2. Model for entrainment in Drosophila. Although CRY most likely resets the clock by affecting TIM levels in response to light the mechanisms that mediate effects of the other pathways are not known.

The role of CRY is somewhat complicated by the fact that in some fly organs, as in mammals, it appears to be part of the clock [30, 31]. However, the acute response of TIM to light is abolished in different tissues in cry^b mutants, consistent with the idea that CRY mediates the circadian response to pulses of light [30]. In addition, Ceriani et al. (1999) showed that CRY and TIM interact with each other in a light-dependent manner in the yeast two-hybrid system whereas no interaction is detected between CRY^b and TIM [32]. This would explain the entrainment phenotype of cry^b mutants.

They also found that CRY attenuates the inhibition of CLK-CYC transcriptional activity by PER-TIM. Presumably, it does so by removing PER-TIM from the complex. Again, the CRY^b protein is ineffective in this assay. It is likely that the effect of CRY on PER-TIM feedback activity is the first clock-relevant event that occurs in response to light. However, this block in feedback can only persist as long as the light signal remains since it is dependent on a direct interaction between PER and TIM. On the other hand, a one-minute pulse of light can reset the clock by several hours indicating that it produces a lasting change in a clock component [15]. This lasting change is most likely the degradation of TIM that resets the cycle to a different time-of-day. We propose that CRY is involved in light-dependent TIM degradation, possibly transmitting a light signal for TIM phosphorylation, ubiquitination and eventual degradation.

REFERENCES

1. Konopka RJ, Benzer S. Clock mutants of *Drosophila melanogaster*. Proceedings of the National Academy of Sciences USA 1971; 68:2112-2116.
2. Sehgal A, Price JL, Man B, Young MW. Loss of circadian behavioral rhythms and *per* RNA oscillations in the *Drosophila* mutant *timeless*. Science 1994; 263:1603-1606.
3. Williams JA, Sehgal A. Molecular components of the Drosophila circadian clock. Annual Review of Physiology 2001; 63:729-755.
4. Curtin KD, Huang ZJ, Rosbash M. Temporally regulated nuclear entry of the *Drosophila* period protein contributes to the circadian clock. Neuron 1995; 14:365-72.
5. Gekakis N, Saez L, Delahaye BA, et al. Isolation of *timeless* by PER protein interaction: defective interaction between *timeless* protein and long-period mutant PERL. Science 1995; 270:811-815.
6. Marrus SB, Zeng H, Rosbash M. Effect of constant light and circadian entrainment of *perS* flies: evidence for light-mediated delay of the negative feedback loop in *Drosophila*. EMBO Journal 1996; 15:6877-6886.
7. Price JL, Blau J, Rothenflugh A, Abodeely M, Kloss B, Young MW. *double-time* is a novel *Drosophila* clock gene that regulates PERIOD protein accumulation. Cell 1998; 94:83-95.

8. Rutila JE, Zeng H, Le M, Curtin KD, Hall JC, Rosbash M. The tim^{SL} mutant of the Drosophila rhythm gene *timeless* manifests allele-specific interactions with *period* gene mutants. Neuron 1996; 17:921-929.
9. Yang Z, Sehgal A. Role of molecular oscillations in the Drosophila circadian clock. Neuron 2001; 29: 453-467.
10. Hunter-Ensor M, Ousley A, Sehgal A. Regulation of the *Drosophila* protein timeless suggests a mechanism for resetting the circadian clock by light. Cell 1996; 84:677-86.
11. Lee C, Parikh V, Itsukaichi T, Bae K, Edery I. Resetting the *Drosophila* clock by photic regulation of PER and a PER-TIM complex. Science 1996; 271:1740-44.
12. Myers MP, Wager-Smith K, Rothenflugh A, Young MW. Light-induced degradation of TIMELESS and entrainment of the *Drosophila* circadian clock. Science 1996; 271:1736-40.
13. Zeng H, Qian Z, Myers MP, Rosbash M. A light entrainment mechanism for the *Drosophila* circadian clock. Nature 1996; 380:129-135.
14. Suri V, Zuwei Q, Hall JC, Rosbash M. Evidence that the TIM light response is relevant to light-induced phase shifts in *Drosophila melanogaster*. Neuron 1998; 21:225-234.
15. Yang Z, Emerson M, Su HS, Sehgal A. Response of the timeless protein to light correlates with behavioral entrainment and suggests a non-visual pathway for circadian photoreception. Neuron 1998; 21:215-223.
16. Frank KD, Zimmerman WF. Action spectra for phase shifts of a Drosophila circadian rhythm. Science 1969; 163:688-689.
17. Zuker CS. The biology of vision in *Drosophila*. Proc. Natl. Acad. Sci. USA 1996; 93:571-6.
18. Helfrich C. Role of the optic lobes in the regulation of the locomotor activity rhythm of *Drosophila melanogaster*-Behavioral analysis of neural mutants. Journal of Neurogenetics 1986; 3:321-43.
19. Chen DM, Christianson JS, Sapp RJ, Stark WS. Visual receptor cycle in normal and *period* mutant Drosophila: microspectrophotometry, electrophysiology, and ultrastructural morphometry. Visual Neuroscience 1992; 9:125-35.
20. Egan E, Franklin T, Hilderbrand-Chae M, et al. An extraretinally expressed insect cryptochrome with similarity to the blue light photoreceptors of mammals and plants. The Journal of Neuroscience 1999; 19:3665-3673.

21. Emery P, So W, Kaneko M, Hall JC, Rosbash M. CRY, a Drosophila Clock and Light-Regulated Cryptochrome, Is a Major Contributor to Circadian Rhythm Resetting and Photosensitivity. Cell 1998; 95:669-679.
22. Ishikawa T, Matsumoto A, Kato T, Jr., et al. DCRY is a Drosophila photoreceptor protein implicated in light entrainment of circadian rhythm. Genes to Cells 1999; 4:57-65.
23. Stanewsky R, Kaneko M, Emery P, et al. The cry^b Mutation Identifies Cryptochrome as a Circadian Photoreceptor in Drosophila. Cell 1998; 95:681-692.
24. Emery P, Stanewsky R, Rosbash M, Hall JC. dCRY is a unique *Drosophila* circadian photoreceptor. Nature 2000; 404:456-457.
25. Emery P, Stanewsky R, Helfrich-Forster C, Emery-Le M, Hall JC, Rosbash M. Drosophila CRY Is a Deep Brain Circadian Photoreceptor. Neuron 2000; 26:493-504.
26. Helfrich-Forster C, Winter C, Hofbauer A, Hall JC, Stanewsky R. The circadian clock of fruit flies is blind after elimination of all known photoreceptors. Neuron 2001; 30:249-261.
27. Hofbauer A, Buchner E. Does Drosophila have seven eyes? Naturwissenschaften 1989; 76:335-336.
28. Hamblen-Coyle MJ, Wheeler DA, Rutila JE, Rosbash M, Hall JC. Behavior of period-altered circadian rhythm mutants of Drosophila in light:dark cycles (Diptera: Drosophilidae). Journal of Insect Behavior 1992; 5:417-445.
29. Naidoo N, Song W, Hunter-Ensor M, Sehgal A. A role for the proteasome in the light response of the timeless clock protein. Science 1999; 285:1737-1741.
30. Ivachenko M, Stanewsky R, Giebultowicz JM. Circadian photoreception in *Drosophila*: Functions of cryptochrome in peripheral and central clocks. Journal of Biological Rhythms 2001; 16:205-215.
31. Krishnan B, Levine JD, Lynch MKS, et al. A new role for cryptochrome in a *Drosophila* circadian oscillator. Nature 2001; 411:313-317.
32. Ceriani MF, Darlington TK, Staknis D, et al. Light-Dependent Sequestration of Timeless by Cryptochrome. Science 1999; 285:553-556.

MOLECULAR REGULATION OF THE *XENOPUS* CIRCADIAN CLOCK

Carla B. Green
Department of Biology, NSF Center for Biological Timing, University of Virginia, Charlottesville, VA 22904

1. INTRODUCTION

The vertebrate retina must be able to respond to ambient light levels that range over ten to twelve orders of magnitude over the course of the 24 hour day. This incredibly wide range of sensitivity is the result of many adaptive mechanisms that occur within the retina at both the intra-and inter-cellular levels, including changes in visual sensitivity, rhythmic production of neuromodulators such as melatonin and dopamine, changes in synaptic structure and function, rhythms in turnover of photosensitive membrane and many others (for reviews, see [1, 2]). Although many of these adaptive processes can be acutely affected by light, it is now clear that there is an endogenous retinal circadian clock controlling these responses in most (if not all) classes of vertebrates. These clocks maintain the correct temporal organization of the retina, allowing it to anticipate daily changes in light and to segregate certain processes (e.g. disk turnover) to the proper time of day. Disruption of this temporal organization may result in retinal dysfunction and degeneration [3, 4].

The best characterized retinal circadian system is *Xenopus laevis*, an amphibian widely used for retinal studies because the retinas are well-suited to experimental manipulation. They have large photoreceptor cells and contain approximately equal numbers of rods and cones. The retinas from these animals continue to exhibit robust circadian rhythms in culture for many days and these rhythms can be reset by light, indicating that the entire circadian system is present in this piece of tissue [5]. This system has been used for many cellular and molecular studies of the retinal clock, including the identification of several genes that are rhythmically expressed in *Xenopus* retina [6-8].

Many (but not all) of the rhythmic physiologies that have been studied are photoreceptor cell events and it is known that in *Xenopus laevis*, a clock exists somewhere within this cell layer [9]. It is not known which cell types within this layer (rods or cones or subpopulations) contain the clock or if other clocks exist in the other cell layers of the retina. The isolated photoreceptor layer synthesizes and releases melatonin under the control of this endogenous circadian clock with high levels at night and low levels during the day [9]. Dopamine is synthesized by a subclass of amacrine cells in the inner retina and, at least in some species, is synthesized with a rhythm in the opposite phase [10]. The dopamine synthesis is inhibited by high melatonin at night [11], but it is not known whether the dopamine rhythm is driven entirely by the melatonin rhythm (from the photoreceptor cells) or is also controlled by a clock within the dopaminergic amacrine cell. Circadian clocks also control the time at which photoreceptor cells shed their photosensitive membranous disks at the tip of the outer segment (for review, see [1]). These disks are phagocytized by another cell type, the opposing epithelial cell layer called the retinal pigment epithelium (RPE). It is not known whether this rhythmic shedding is controlled by the photoreceptors, the RPE, or some extracellular signal coming from another cell type. Likewise, in *Xenopus* and other lower vertebrates, the cone photoreceptor cells change the position of their outer segments throughout the day. These retinomotor movements are controlled both by light and by a circadian clock. In *Xenopus*, light control of cone contraction has been shown to be mediated through the rod cells [12], but it is not known whether the clocks controlling these events are in the cones, rods, or another cell type. Horizontal cells in fish have a circadian rhythm in rod-cone dominance [13], but the localization of the clock controlling this rhythm is unknown.

The usefulness of the *Xenopus* system has recently been expanded by the development of techniques for the generation of transgenic animals[14]. In this method, foreign DNA is introduced directly into the sperm genome before fertilization by a method called restriction enzyme-mediated integration (REMI). The transgenic sperm nuclei are then injected into oocytes, producing transgenic embryos that are non-mosaic and therefore do not have to be bred before analysis. Additionally, this method is inexpensive and yields dozens of transgenic embryos with only a few hours work. The transgenic embryos carrying promoter-reporter constructs show spatially and temporally appropriate reporter-gene expression patterns[14, 15] (and our unpublished results). We have used specific retinal promoters to demonstrate that transgene expression can be targeted to specific cell types within the retina[15] (and our unpublished results).

Part of the advantage of this system is that *Xenopus* retinas develop very quickly after fertilization. By stage 47 (5-6 days post-fertilization) they are fully differentiated with all layers and cell types, intact neurotransmitter systems and normal light-evoked responses [16-20]. By this stage, these retinas also have intact circadian systems which can control rhythmic melatonin release [21]. Therefore, transgenic retinas can be assayed for rhythmicity or other phenotypes only a few days after the embryos are generated

2. CLOCK GENES IN *XENOPUS* RETINA.

2.1. *Xenopus Clock* and *bmal1*.

As described above, the *Xenopus* retina has many advantages for studies of the central circadian clock. For these reasons, we set out to clone *Xenopus* homologs of the genes that were implicated as central clock components in other animals. The first gene that we cloned was the *Xenopus* homolog of *Clock* (*XClock*), because of its potential use for constructing dominant negative mutants (see below).

We synthesized PCR primers that flanked the PAS-B domain of mouse *Clock* because this region is highly conserved between *Clock* genes in other species. RT-PCR was performed, using *Xenopus* retinal RNA as template. An appropriately-sized band was produced and this was cloned and sequenced and shown to be nearly identical to mouse *Clock* (87% identity at the nucleotide level, 98% identity at the amino acid level). This PCR product was then used to screen a *Xenopus* retinal cDNA library and 2 positive clones were identified and sequenced. One of these clones, XClock12, contained the complete *XClock* open reading frame. Comparison of the deduced amino acid sequence of *XClock* with that of *mClock* revealed that these two proteins are 77% identical (86% similar). Similarity was even higher (94%) over the regions that correspond to the bHLH, and PAS A and B domains.

In situ hybridization of sections of adult *Xenopus* retinas showed intense staining of photoreceptor cell bodies, indicating that rod and cone photoreceptors synthesize *XClock* mRNA [22]. Other cells also show lower levels of staining that are closer to background. Scattered cells within both the inner nuclear layer and in the ganglion cell layer may also express low levels of *XClock*. Steady state mRNA levels at different times of day were measured by northern blot. The XClock12 probe recognizes a single band of about 7 Kb that does not vary in intensity throughout the day. This non-rhythmic expression pattern is similar to that observed in mouse SCN [23]. *XClock* mRNA is also expressed in many other *Xenopus* tissues [22].

We also used northern blot analysis to determine the developmental profile of *XClock* expression. RNA was isolated from whole embryos at various developmental stages and probed with radiolabeled XClock12. *XClock* mRNA can be detected by stage 25 (about 2 days post-fertilization; data not shown). At this stage, optic vesicles are present, but the retinas are largely undifferentiated.

Recent work in our lab has also resulted in the cloning and characterization of the full-length *bmal1* cDNA from *Xenopus* retina. It is highly conserved with the mouse *bmal1* with 86% amino acid identity and 91.5% similarity. We have shown that *bmal1* is a 3 kb mRNA that is expressed with a high amplitude rhythm in *Xenopus* retina. The peak mRNA levels occur near the day to night transition and this rhythm is observed in both cyclic light and constant dark conditions. *In situ* hybridization indicates that *bmal1*, like *Clock*, is expressed at high levels in the photoreceptor cells in the *Xenopus* retina.

In *Drosophila* and mammals, the CLOCK and BMAL1 proteins are transcription factors that form heterodimers, bind to a DNA element called an E-box, and activate transcription of the downstream gene (reviewed in [24]). These proteins play an important role in activating the expression of the *period* (*per*) gene in *Drosophila* and the *per1* gene in mouse. This CLOCK/BMAL1 mediated activation is critical for the circadian clock mechanism and disruption of either of these protein's normal functions disrupts normal rhythmicity [25, 26]. We tested the *Xenopus* CLOCK and BMAL1 proteins to determine if this role was conserved using a transient transfection assay. In this experiment, we transfected COS7 cells with a reporter gene construct in which three repeats of an E-box element were inserted upstream of a basal promoter driving the firefly luciferase gene. These cells were cotransfected with plasmids that resulted in the overexpression of the CLOCK and BMAL1 proteins. Using this system, we were able to demonstrate that these proteins together can cause a significant activation of the transcription of the reporter gene, but expression of CLOCK or BMAL1 alone can not. Therefore, this aspect of the circadian clock mechanism is conserved in *Xenopus*.

2.2. *Xenopus cryptochromes.*

Xenopus homologs of the *cry* genes were identified by RT-PCR from *Xenopus* retinal RNA using primers designed by the CODEHOP method [27].

The resulting PCR product was cloned and sequenced and Blast searches of the Genbank database [28] revealed a high similarity to mouse and human *crys*. The PCR product was radiolabeled and used to screen a *Xenopus* retinal cDNA library. Positive clones from this screen corresponded to 3 distinct cDNAs. One of these showed high similarity (92% identity, 97% similarity at amino acid level) to mouse *cry1* and was named *Xcry1* accordingly. The other 2 cDNAs were very similar to each other and also very similar to mouse *cry2* (85% identity, 93% similarity at amino acid level). Therefore, these cDNAs were named *Xcry2a* and *Xcry2b*.

In situ hybridization of retinal sections shows that all three *cry* genes are expressed primarily in the photoreceptor layer, with patterns similar to *Xclock* and *Xbmal1*. Northern blot analysis of *cry* expression demonstrates that *Xcry1* is rhythmic in both cyclic light and constant darkness with peak mRNA levels at ZT 16. Both *Xcry2a and Xcry2b* are also rhythmic but with lower amplitude and slightly later phases (*Xcry2a* peaks near ZT 0 and *Xcry2b* peaks between ZT 16 and ZT20). The amplitude and phase of *Xcry2* rhythms are very similar to the *cry* rhythms observed in *Drosophila* [29]. However, in mammalian retina, neither *cry1* nor *cry2* are rhythmic, and in the mouse SCN, both *crys* are rhythmic with a peak in mid to late afternoon[30].

In *Drosophila*, the CRY protein is thought to act as a photoreceptor and acts to relieve repression of CLOCK/BMAL1 mediated transcription of *per* transcription [31, 32]. In mammals, it has the opposite effect and acts to cause the repression of CLOCK/BMAL1 mediated transcription and this process does not seem to be directly light-sensitive [33-35]. We tested our *Xenopus* CRYs in the transient transfection system described above and found that they caused full repression of the CLOCK-BMAL1 mediated transcription in both light and dark, similar to the results from mammals.

2.3. *Xenopus period* genes.

Three *per* gene homologs, similar to mouse *per1*, *per2*, and *per3* have been cloned from *Xenopus* retina [36, 37] (and our unpublished results). The mRNAs from all these genes are robustly rhythmic with peaks during the early (*xper1*) to mid-day (*xper2* and *xper3*) in a light/dark cycle. In constant darkness, *xper2* is constitutively low and appears to be under direct control of the light with little or no circadian control at the level of mRNA accumulation [36].

In mouse SCN, *per1* and *per2* transcription are acutely and transiently induced by light exposure at night and this is thought to contribute to resetting of the circadian clock [38, 39]. In *Xenopus laevis* retina, *per1* expression is not acutely induced by light exposure. *xper2*, however, is acutely induced by both light and dopamine (another clock input) at all phases of the circadian cycle [36].

3. CLOCK CONTROLLED GENES IN *XENOPUS* RETINA.

In addition to the genes involved in the central clock mechanism, we have also identified other "clock-controlled genes." We have identified these genes using both a "candidate gene approach" in which the daily expression patterns of known photoreceptor genes were analyzed, and by a broad, systematic screening approach using differential display analysis.

3.1. Tryptophan Hydroxylase

Because melatonin synthesis was well known to be under circadian control in the *Xenopus* retina, we examined the gene encoding tryptophan hydroxylase (TPH), the rate-limiting enzyme in melatonin synthesis in this tissue [40]. The levels of TPH mRNA showed a distinct rhythm, with lowest levels early in the day, rising throughout the day to a peak about midnight. Peak levels were two to three times higher than trough values [6]. This rhythm persisted in both cyclic light and in constant darkness, indicating that it was under circadian control. Run-on transcription assays showed that these circadian changes in steady state TPH mRNA levels were due to actual changes in transcription of the TPH gene. *In situ* hybridization showed that the TPH mRNA is expressed in the photoreceptor layer in both rods and cones [41].

3.2. Nocturnin

In order to identify other mRNAs that exhibited rhythmic expression, we carried out a differential display screen [7]. The screen was designed to compare retinal RNAs at four different times of day (dawn, noon, dusk, and midnight) for two consecutive days in culture. The retinas were cultured in constant darkness so that only truly circadian regulated messages would be identified. One of the genes that was identified in this screen was a novel gene with a night-time peak of expression that we named nocturnin [8].

Nocturnin mRNA shows a very high amplitude rhythm in both light/dark cycles and in constant conditions. mRNA levels are undetectable during the day, begin to rise about 1-2 hours before dusk, peak at high levels about 3-4 hours after normal dark onset and by midnight have again fallen to nearly undetectable levels [8]. Like TPH, this change in mRNA levels is controlled at the level of transcription of the gene. Unlike TPH mRNA, which exhibits a low amplitude rhythm (2-3 fold) with a broad peak, nocturnin mRNA levels change by 15-20 fold and the expression time is very short (4-5 hours). The nocturnin mRNA is localized specifically to the photoreceptor cells in the adult *Xenopus* retina.

Characterization of the nocturnin cDNA revealed an open reading frame that encodes a novel protein of 388 amino acids. The deduced sequence for the nocturnin protein contains a leucine zipper-like motif near the amino-terminal end and the carboxy-terminal two thirds of its sequence shows similarity to the carboxy-terminal sequence of a yeast transcriptional activator protein called CCR4 [42, 43]. Further work is underway to discern the role that nocturnin is playing in nighttime photoreceptor physiology.

4. MANIPULATION OF THE CLOCK IN TRANSGENIC *XENOPUS* RETINAS.

Molecular analyses of circadian clocks in *Drosophila* and mammals have resulted in the development of a model for the central clock mechanism (reviewed in [24, 44]). In short, the central clock appears to be composed of a transcription/translation negative feedback loop. One of the key steps in this process is the daily activation of *period* (*per)* gene transcription by the binding of a heterodimeric transcription factor complex composed of the CLOCK and BMAL1 proteins. The CLOCK and BMAL1 proteins are members of a family of bHLH PAS transcription factors and the heterodimer binds a DNA element, called an E-box, in the *per* promoter. Mutational analyses in both mouse and fly have shown that removing all or part of the C-terminal "activation" domain of the CLOCK protein results in a dominant negative form that blocks transcriptional activation of the *per* gene and prevents clock function [45-48]. The dominant negative effect is due to the fact that the truncated CLOCK can still bind to BMAL1 and can still bind to the E-box DNA element, but can not activate transcription, therefore stopping the feedback loop.

As described above, we have cloned *Xenopus* homologs of most of the known clock genes. These highly conserved sequences support the idea that the *Xenopus* clock is similar in molecular detail to the clocks described in other organisms. These data also support the idea that data from the mouse and fly can be used to construct a tool for perturbing clock function in *Xenopus*. We have done that by generating a construct that contains the amino-terminal half of the *XClock* gene, fused in-frame with GFP. We attached GFP sequence so that we could easily determine the expression pattern of our transgene. The presence of GFP within the retina has been used in other studies, and embryos with high levels of GFP in their photoreceptors have normal ERGs, indicating that this fluorescent protein does not produce obvious deleterious effects [49]. This construct is missing the glutamine-rich activation domain and is therefore designated XClockΔQ-GFP.

This construct, called XClockΔQ-GFP should act as a dominant negative, such that over-expression should destroy the central circadian clock as predicted by the current model. We tested XClockΔQ-GFP for general translation, stability and correct nuclear localization, by cloning it in front of the *Xenopus* rod opsin promoter and producing transgenic embryos. Sections through the retinas of these embryos verified that XClockΔQ-GFP was expressed at high levels specifically in rods and was localized in the nucleus as predicted.

We tested the XClockΔQ-GFP construct for its ability to act as a dominant negative in transient transfections of COS7 cells as described above. We transfected COS7 cells with a reporter plasmid containing three repeats of the mouse *per1* E-box cloned upstream of a basal promoter and the luciferase reporter gene [47]. These cells were cotransfected with various combinations of expression plasmids with the CMV promoter driving *Clock*, *bmal1*, or XClockΔQ-GFP. All transfections included a control plasmid for normalization of transfection efficiencies. In each case, the total amount of DNA transfected was kept constant by the addition of empty vector DNA where needed. We showed that the *per* E-box luciferase construct alone produced some stimulation of luciferase activity, probably due to low levels of endogenous CLOCK and BMAL1 present in COS7 cells. Addition of exogenous CLOCK and BMAL1 activated transcription several-fold over basal levels. The XClockΔQ-GFP construct did not activate luciferase transcription at all when cotransfected with *bmal1*, verifying that this mutant version is not active. Furthermore, when XClockΔQ-GFP is added in excess, it is able to negate the normal activation of luciferase by CLOCK and BMAL1 in a dose-dependent manner, indicating that this construct is competing with wild-type CLOCK for interaction with BMAL1 and/or E-box binding sites.

We are now using this dominant-negative mutant to specifically ablate the circadian clock in specific cells within the retina. This is being done using cell-type specific promoters that allow us to target transgene expression to different retinal cell types. We are using this tool to begin to examine the cellular locations of the clocks that control the many different types of rhythms that occur in the *Xenopus* retina.

5. REFERENCES

1. Cahill GM, Besharse JC. Circadian rhythmicity in vertebrate retinas: regulation by a photoreceptor oscillator. Progress in Retinal and Eye Research 1995; 14:267-291.

2. Cahill GM, Hasegawa M. Circadian Oscillators in Vertebrate Retinal Photoreceptor Cells. Biol. Signals 1997; 6:191-200.

3. Fain GL, Lisman JE. Light, Ca2+, and photoreceptor death: new evidence for the equivalent- light hypothesis from arrestin knockout mice [comment]. Invest Ophthalmol Vis Sci 1999; 40:2770-2.

4. Fain GL, Lisman JE. Photoreceptor degeneration in vitamin A deprivation and retinitis pigmentosa: the equivalent light hypothesis. Exp Eye Res. 1993; 57:335-340.

5. Besharse JC, Iuvone PM. Circadian clock in *Xenopus* eye controlling retinal serotonin N-acetyltransferase. Nature 1983; 305:133-135.

6. Green CB, Besharse JC. Tryptophan hydroxylase expression is regulated by a circadian clock in *Xenopus laevis* retina. Journal of Neurochemistry 1994; 62:2420-2428.

7. Green CB, Besharse JC. Use of a high stringency differential display screen for identification of retinal mRNAs that are regulated by a circadian clock. Molecular Brain Research 1996; 37:157-165.

8. Green CB, Besharse JB. Identification of a novel vertebrate circadian clock-regulated gene encoding the protein nocturnin. Proc. Natl. Acad. Sci. USA 1996; 93:14884-14888.

9. Cahill GM, Besharse JC. Circadian clock functions localized in *Xenopus* retinal photoreceptors. Neuron 1993; 10:573-577.

10. McCormack CA, Burnside B. Light and circadian modulation of teleost retinal tyrosine hydroxylase activity. Investigative Ophthalmology and Visual Science 1993; 34:1853-1860.

11. Boatright JH, Iuvone PM. Meiatonin suppresses the light-evoked release of endogenous dopamine from retinas of frogs (*Xenopus laevis*). Society for Neuroscience Abstracts 1989; 15:1395.

12. Besharse JC, Witkovsky P. Light-evoked contraction of red absorbing cones in the *Xenopus* retina is maximally sensitive to green light. Visual Neurosci. 1992; 8:243-249.

13. Wang Y, Mangel SC. A circadian clock regulates rod and cone input to fish retinal cone horizontal cells. Proceedings of the National Academy of Sciences of the United States of America 1996; 93:4655-4660.

14. Kroll KL, Amaya E. Transgenic *Xenopus* embryos from sperm nuclear transplantations reveal FGF signaling requirements during gastrulation. Development 1996; 122:3173-3183.

15. Knox BE, Schlueter C, Sanger BM, Green CB, Besharse JC. Transgene expression in *Xenopus* rods. FEBS Letters 1998; 423:117-121.

16. Hollyfield JG, Rayborn ME, Sarthy PV, Lam DM-K. The emergence, localization and maturation of neurotransmitter systems during development of the retina in *Xenopus laevis*. Journal of Comparative Neurology 1979; 188(4):587-598.

17. Rayborn ME, Vijay Sarthy P, Lam DMK, Hollyfield JG. The emergence, localization, and maturation of neurotransmitter systems during development of the retina in *Xenopus laevis:* II.glycine. Journal of Comparative Neurology 1981; 195:585-593.

18. Sarthy PV, Rayborn ME, Hollyfield JG, Lam DMK. The emergence, localization, and maturation of neurotransmitter systems during development of the retina in Xenopus laevis. III. Dopamine. Journal of Comparative Neurology 1981; 195:595-602.

19. Stiemke MM, Landers RA, Al-Ubaidi MR, Rayborn ME, Hollyfield JG. Photoreceptor outer segment development in *Xenopus laevis*: Influence of the pigment epithelium. Developmental Biology 1994; 162:169-180.

20. Witkovsky P, Gallin E, Hollyfield JG, Ripps H, Bridges CDB. Photoreceptor thresholds and visual pigment levels in normal and vitamin A-deprived *Xenopus* tadpoles. J. Neurophysiol. 1976; 39:1272-1287.

21. Green CB, Liang M-Y, Steenhard BM, Besharse JC. Ontogeny of circadian and light regulation of melatonin release in *Xenopus laevis* embryos. Developmental Brain Research 1999; 117:109-116.

22. Zhu H, LaRue S, Whiteley A, Steeves TDL, Takahashi JS, Green CB. The *Xenopus Clock* gene is constitutively expressed in retinal photoreceptors. Molecular Brain Research 2000; 75:303-308.

23. Tei H, Okamura H, Shigeyoshi Y, et al. Circadian oscillation of a mammalian homologue of the *Drosophila period* gene. Nature 1997; 389:512-516.

24. Dunlap JC. Molecular bases for circadian clocks. Cell 1999; 96:271-290.

25. Vitaterna MH, King DP, Chang A-M, et al. Mutagenesis and mapping of a mouse gene, *Clock*, essential for circadian behavior. Science 1994; 264:719-725.

26. Bunger MK, Wilsbacher LD, Moran SM, et al. Mop3 is an essential component of the master circadian pacemaker in mammals. Cell 2000; 103:1009-17.

27. Rose TM, Schultz ER, Henikoff JG, Pietrokovski S, McCallum CM, Henikoff S. Consensus-degenerate hybrid oligonucleotide primers for amplification of distantly related sequences. Nucleic Acids Res. 1998; 26:1628-1635.

28. Altschul SF, Gish W, Miller W, Myers EW, Lipman DJ. Basic local alignment search tool. J. Mol. Biol. 1990; 215:403-410.

29. Emery P, So V, Kaneko M, Hall JC, Rosbash M. CRY, a *Drosophila* clock and light-regulated cryptochrome, is a major contributor to circadian rhythm resetting and photosensitivity. Cell 1998; 95:669-679.

30. Miyamoto Y, Sancar A. Vitamin B2-based blue-light photoreceptors in the retinohypothalamic tract as the photoactive pigments for setting the circadian clock in mammals. Proceedings of the National Academy of Sciences of the United States of America 1998; 95:6097-6102.

31. Stanewsky R, Kaneko M, Emery P, et al. The cry^b mutation identifies cryptochrome as a circadian photoreceptor in *Drosophila*. Cell 1998; 95:681-692.

32. Ceriani MF, Darlington TK, Staknis D, et al. Light-dependent sequestration of TIMELESS by CRYPTOCHROME. Science 1999; 285:553-6.

33. Griffin EA, Jr., Staknis D, Weitz CJ. Light-Independent Role of CRY1 and CRY2 in the Mammalian Circadian Clock. Science 1999; 286:768-771.

34. Kume K, Zylka MJ, Sriram S, et al. mCRY1 and mCRY2 are essential components of the negative limb of the circadian clock feedback loop. Cell 1999; 98:193-205.

35. Shearman LP, Sriram S, Weaver DR, et al. Interacting molecular loops in the mammalian circadian clock. Science 2000; 288:1013-9.

36. Steenhard BM, Besharse JC. Phase shifting the retinal circadian clock: xPer2 mRNA induction by light and dopamine. J Neurosci 2000; 20:8572-7.

37. Zhuang M, Wang Y, Steenhard BM, Besharse JC. Differential regulation of two period genes in the Xenopus eye. Brain Res Mol Brain Res 2000; 82:52-64.

38. Albrecht U, Sun Z, Eichele G, Lee C. A differential response of two putative mammalian circadian regulators, *mper1* and *mper2*, to light. Cell 1997; 91:1055-1064.

39. Shearman L, Zylka M, Weaver D, Kolakowski L, Reppert S. Two *period* homologs: circadian expression and photic regulation in the suprachiasmatic nucleus. Neuron 1997; 19:1261-1269.

40. Cahill GM, Besharse JC. Circadian regulation of melatonin in the retina of *Xenopus laevis*: limitation by serotonin availability. Journal of Neurochemistry 1990; 54:716-719.

41. Green CB, Cahill GM, Besharse JC. Tryptophan hydroxylase is expressed by photoreceptors in Xenopus laevis retina. Visual Neuroscience 1995; 12:663-670.

42. Malvar T, Biron RW, Kaback DB, Denis CL. The CCR4 protein from *Saccharomyces cerevisiae* contains a leucine-rich repeat region which is required for its control of ADH2 gene expression. Genetics 1992; 132:951-962.

43. Draper MP, Liu H-Y, Nelsbach AH, Mosley SP, Denis CL. CCR4 is a glucose-regulated transcription factor whose leucine-rich repeat binds several proteins important for placing CCR4 in its proper promoter context. Molecular and Cellular Biology 1994; 14:4522-4531.

44. Green CB. How cells tell time. Trends in Cell Biology 1998; 8:224-230.

45. King DP, Vitaterna MH, Chang AM, et al. The mouse Clock mutation behaves as an antimorph and maps within the W19H deletion, distal of Kit. Genetics 1997; 146:1049-1060.

46. King DP, Zhao Y, Sangoram AM, et al. Positional cloning of the mouse circadian *Clock* gene. Cell 1997; 89:641-653.

47. Gekakis N, Staknis D, Nguyen HB, et al. Role of the CLOCK protein in the mammalian circadian mechanism. Science 1998; 280:1564-1569.

48. Allada R, White NE, So WV, Hall JC, Rosbash M. A mutant Drosophila homolog of mammalian *Clock* disrupts circadian rhythms and transcription of *period* and *timeless*. Cell 1998; 93:791-804.

49. Engbretson GA, Kefalov VJ, Cameron DA, Cornwall MC, Knox BE, Barlow RB. Responses and microspectrophotometry of *Xenopus* rods expressing GFP and mutant bovine rhodopsin. Investigative Ophthalmology & Visual Science 1999; 40:S372.

THE ROLE OF MELANOPSIN AND OTHER OPSINS IN CIRCADIAN CLOCK RESETTING

Ignacio Provencio
Uniformed Services University, Bethesda, Maryland, USA

INTRODUCTION

The profound effects of the circadian clock on physiology and behavior are frequently ignored. Differences in the time of day during which drugs are administered can result in dramatic differences in efficacy and tolerance of side effects. Chronic fatigue and gastrointestinal irritability experienced by shift workers, and the general malaise of jet lagged travelers are manifestations of our reluctance to accept the impact of the body's 24-hour pacemaker. It has recently been reported that chronic jet lag produces temporal lobe atrophy and spatial cognitive deficits.[1] Several sleep disorders such as advanced sleep phase syndrome have been attributed to a dysfunctional circadian system.[2]

While much has been learned about the molecular components of the circadian clock, the mechanism by which light resets the mammalian clock remains largely unknown. Recently discovered photopigments are emerging as candidate "circadian photopigments" responsible for the resetting of the circadian clock.

PHOTOREGULATION OF CIRCADIAN RHYTHMS

Circadian photoreception in mammals has presented investigators with an apparent paradox. Although the eyes are absolutely required for the photic regulation of circadian rhythms (photoentrainment), the only known photoreceptors of the retina, the rods and cones, are not required.[3] Bilateral removal of the eyes renders rodents incapable of photoentrainment.[4] Specific ablation of the rods and cones, however, has no effect on the sensitivity of the rodent circadian system to light.[3] Mice homozygous for the *retinal degeneration* (*rd*) mutation experience a degeneration of rods within two weeks of postnatal development. The loss of rods is followed a protracted degeneration of cones.[5] Inner layers of the retina remain morphologically normal. Despite this attrition of the canonical photoreceptors, the sensitivity of the circadian system in these visually blind rodents is identical to sighted wild-type controls. In essence, the *rd* mutation has no effect on circadian photosensitivity.[3] Other retinally degenerate rodent models, including transgenic mice that lack rods and cones, also show no attenuation of circadian photoresponses.[6] To date, no animal model of retinal degeneration has shown a decreased sensitivity of the circadian system. Similarly, some humans lacking cognitive vision remain capable of photoentrainment.[7]

Within the last decade, new putative ocular photopigments have been identified.[8] Their localization to non-rod, non-cone cells has suggested a potential role in non-visual photoreceptive tasks such as the regulation of circadian rhythms. The aim of this review is to outline the known and putative mammalian photopigments and describe their potential role in circadian rhythm resetting.

CANDIDATE PHOTOPIGMENTS

Cryptochromes

Cryptochromes were initially discovered in plants. Homologs were subsequently discovered in *Drosophila* and mammals. Each cryptochrome photopigment contains a flavin and pterin chromophore. A study in mice localized cryptochromes to nonphotoreceptor cells of the retina and cells of the suprachiasmatic nuclei (SCN), suggesting that cryptochromes may be involved in photoentrainment.[9] This logic led to some unexpected findings.

Mice have two cryptochrome homologs, CRY1 and CRY2.[10] Mice null for either cryptochrome show aberrant circadian periods, but remain capable of regulating their circadian rhythms by light.[11, 12] Mice null for both cryptochromes exhibit a disrupted circadian clock.[12] This finding implies that cryptochromes are actually integral parts of the circadian machinery rather than simply being involved in regulating those clockworks in response to environmental signals. These two functions, however, are not exclusive. It remains possible that cryptochromes are light sensitive gears of the clock. This appears to be the case in *Drosophila*.[13] In mammals, however, conclusive evidence proving a photoreceptive function for cryptochromes has remained elusive. This lack of data is primarily due to the inherent difficulty of experimentally manipulating the potential photoreceptive function of cryptochromes without disrupting their clock function.

The identification of cryptochromes as molecular components of the clock was largely serendipitous. Their localization to anatomical structures known to be involved in the mammalian circadian system encouraged geneticists to assess their function in the timing system of *Drosophila*.[13, 14] These investigators established cryptochromes as elements of the circadian clock that are inherently photosensitive. When these studies were extended to mammals, the role of cryptochromes as critical clock components was confirmed,[15] although their capacity to detect light remains in question.[16]

Opsins

Opsins belong to the rhodopsin superfamily of G protein-coupled receptors. Opsin-based photopigments are the best-studied photopigments in the animal kingdom. They consist of an opsin protein covalently linked to a retinaldehyde chromophore.[17] While their function in vision is well-established, their role in the photoregulation of circadian rhythms remains correlative. Action spectra for phase shifting of circadian locomotor activity rhythms implicate opsin-based photopigments.[18-20] Other non-visual responses to light such as pupillary constriction[21] and acute suppression of pineal melatonin[22] are also thought to be mediated by opsin-based pigments. The recent completion of the consensus human genome indicates that at least eight opsins are expressed in humans: rhodopsin (rod opsin), red cone opsin, green cone opsin, blue cone opsin, RGR, peropsin, melanopsin, and encephalopsin.[23] (Table 1) Their potential roles in circadian regulation are discussed below.

Table 1. Mammalian opsins

opsin	human chromosome	sites of retinal expression	known or predicted primary function	refs.
rhodopsin	3q21.1	rods	night vision	24
red cone opsin	Xq28	red cones	color vision	24
green cone opsin	Xq28	green cones	color vision	24
blue cone opsin	7q32.3	blue cones	color vision	24
RGR	10q23	Müller and RPE cells	photoisomerase	25
peropsin	4q24-25	microvilli of RPE cells	unknown; putative photoisomerase	26
melanopsin	10q22	retinal ganglion cells	unknown; putative nonvisual photopigment	27
encephalopsin	1q43	unknown	unknown	28, 29

Rod and Cone Opsins

The opsins of rod and cone photoreceptors are required for vision. However, the degeneration of rods and cones through naturally occurring mutations or transgenically induced ablation has no effect on the photosensitivity of the circadian system.[6] While these photoreceptors are not required for circadian photoregulation, their involvement in the modulation of circadian outputs cannot be discounted. Electrophysiological recordings of light-induced activity in rodent SCN strongly implicate an input from rods and cones.[30] Rods and cones also are likely to play a role in the acute inhibition of activity in rodents. This phenomenon, know as "masking", is best demonstrated in CRY1/CRY2 null mice that lack a functional circadian system but are capable of restricting locomotor activity to the dark portion of the light:dark cycle.[12] The diurnal organization of activity in these "clockless" animals is presumably mediated through the rods and cones and their attendant photopigments. CRY1/CRY2 null mice homozygous for the *rd* mutation fail to exhibit masking and show diminished but constant activity levels throughout the light:dark cycle.[31]

Encephalopsin

Encephalopsin derives its name from its expression in the mammalian brain.[29] It was initially reported that encephalopsin was not expressed in the retina. This is in conflict with recent work claiming expression in the retina and a multitude of other tissues.[28] Like most of the newly discovered non-rod, non-cone opsins, encephalopsin is problematic to heterologously express and reconstitute, thus making the generation of an absorbance spectrum impossible. This difficulty, in combination with the broad pattern of expression, has made the determination of encephalopsin's function complicated. If encephalopsin is shown to be expressed in inner retinal neurons, it will prove to be a viable circadian photopigment candidate.

RGR (RPE-retinal G protein coupled receptor)

RGR was the first of the mammalian opsins to share greater homology with the invertebrate opsins relative to the opsins of vertebrates.[32] Specifically, RGR shares highest homology with retinochrome, a class of cephalopod opsin involved in photoisomerizing the all-*trans*-retinaldehyde chromophore to the 11-*cis* isomer. RGR's localization to Müller cells and the retinal pigment epithelium, a known site of chromophore regeneration, suggests that RGR may indeed be involved in generating 11-*cis*-retinoids. This is supported by the fact that the preferred chromophore of RGR is all-*trans*-retinaldehyde and exposure to blue light results in isomerization of bound all-

trans-retinaldehyde to the 11-*cis* configuration. Other mechanisms for generating 11-*cis*-retinoids exist within the RPE, [33] however, the activity of RGR may account for the majority of 11-*cis* chromophore in the light adapted state.[34] Taken together, the data show that RGR is not a typical signaling opsin that initiates a phototransduction cascade but rather is a functional homolog of cephalopod retinochrome. RGR's expression in non-neural ocular cells and its role as a photoisomerase makes it difficult to reconcile an alternative function in circadian photoentrainment.

Peropsin

Peropsin shares many characteristics with RGR. Like RGR, peropsin is expressed in the retinal pigment epithelium and shares greater homology with invertebrate opsins than those of vertebrates.[26] Phylogenetic analysis places peropsin among RGR and the retinochromes.[35] Again, the localization and potential function as a photoisomerase makes this putative photopigment an unlikely circadian photopigment candidate.

Melanopsin

The dermal melanophores of *Xenopus laevis* are directly light sensitive.[36] Cultured melanophores respond to illumination by dispersing their melanin-containing organelles (melanosomes) to the periphery of the cell. Melanopsin, the first member of a novel family of opsins was identified by screening a melanophore λgt10 cDNA library.[35] The library was screened with a cocktail of probes corresponding to fragments of *Xenopus* rhodopsin and violet cone opsin. The deduced amino acid sequence predicts seven transmembrane domains and a long cytoplasmic tail with multiple putative phosphorylation sites. Homologs or fragments of homologs have been identified in several vertebrate classes including amphibians, fishes, reptiles, birds, and mammals. (Figure 1)

The mammalian homologs share 55 – 60% identity with frog melanopsin at the amino acid level (excluding extracellular and cytoplasmic tails).[27] Conserved among all melanopsins is a lysine in the seventh transmembrane domain. This residue is a hallmark of all opsins and is required for the Schiff's base linkage of the retinoid chromophore to the protein. Another feature of all opsins including melanopsin is a predicted disulfide bridge connecting the first and second extracellular loops. Interestingly, despite its vertebrate origin, melanopsin like peropsin and RGR, shows greater similarity to the invertebrate opsins than those of the vertebrates.[27, 35] However, phylogenetic analysis does not place melanopsin in the same branch as RGR and peropsin, but rather positions melanopsin among the invertebrate signaling opsins, *i.e.* opsins that initiate a phototransduction cascade. The results of this analysis coupled with melanopsin's exclusive expression in neurons suggests that melanopsin is likely to initiate a phototransduction cascade.

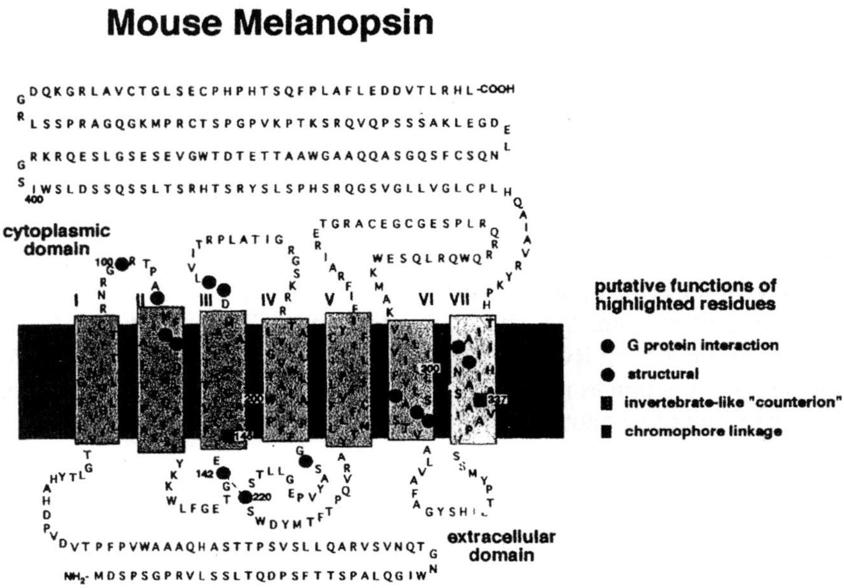

Figure 1. *Deduced amino acid sequence and predicted secondary structure and features of mouse melanopsin.*

Determination of sites of expression has been accomplished by *in situ* hybridization histochemistry. The extent and pattern of melanopsin expression varies among the vertebrate classes. For example, in amphibians melanopsin is expressed in the horizontal and ganglion cells of the retina, iridial myocytes, neurons within the brain, pinealocytes, and melanophores.[35] In mammals, however, melanopsin is solely expressed in retinal ganglion cells.[27]

This localization of melanopsin to the mammalian inner retina makes melanopsin a viable candidate circadian photopigment.[27] This candidacy is strengthened by fact that the number and distribution of melanopsin-containing retinal ganglion cells are identical to the retinal ganglion cells known to project to the murine SCN.[37] Expression of melanopsin within these cells would present the exciting possibility that SCN-projecting retinal ganglion cells are directly photoreceptive and capable of communicating photic information from the eye to the master circadian clock.

FUNCTIONAL CONSIDERATIONS

The anatomical separation of the master mammalian circadian clock in the hypothalamus and its sensory apparatus in the eye implies that the hypothalamic clock proteins probably are not directly photoreceptive. Rather, SCN cells are likely to receive information about the ambient light level via the eyes. This sensory information would subsequently be communicated through the retinohypothalamic tract to the SCN.

The previously mentioned studies in retinally degenerate models would dictate that circadian photoreceptors be situated in the inner retina.[3, 6] Melanopsin-containing cells fulfill this criterion, but the possibility of directly light-sensitive retinal ganglion cells is likely to be perceived as heretical in the field of vision science. Recent evidence, however, indicates that the retinal ganglion cells that project to the SCN are directly photosensitive.[38] Furthermore, the spectral sensitivity of the response suggests that an opsin-based photopigment mediates the observed light-induced tonic depolarization. Although melanopsin has been implicated in initiating this response, it still remains to be seen if melanopsin is indeed expressed in these noncanonical photoreceptors.

Two features of the light responsiveness of SCN-projecting retinal ganglion cells are consistent with the invertebrate-like character of melanopsin. First, most vertebrate photoreceptors *hyperpolarize* in response to photostimulation. Photoreceptive retinal ganglion cells are similar to invertebrate photoreceptors in that they *depolarize* in response to light.[38] Melanopsin's homology to the invertebrate photopigments suggests that it may have the capacity to initiate a transduction cascade culminating in a depolarization of the cell membrane. Second, invertebrate photopigments do not require anatomical juxtaposition to a chromophore-regenerating tissue such as the retinal pigment epithelium.[39] The "spent" chromophore of invertebrate photopigments is regenerated *in situ* by absorption of a photon of appropriate energy. Such a strategy would be advantageous in photoreceptors such as retinal ganglion cells that are not in apposition with the retinal pigment epithelium, the primary source of photopigment chromophore in the retina. Again, melanopsin's invertebrate-like character suggests that it may use an *in situ* chromophore regenerating mechanism similar to that of the invertebrate opsin-based photopigments.

Photoentrainment analysis of melanopsin-null mice will provide the most convincing assessment of melanopsin's role in the regulation of mammalian circadian rhythms. This kind of analysis will also establish whether melanopsin is simply one player in a functionally redundant system that requires the participation of one or more of the other opsin-based photopigments. Soon we shall see the light.

This work was supported by NIH grant R01 MH62405. I thank Mark Rollag and Guisen Jiang for helpful discussions.

REFERENCES

1. Cho K. Chronic 'jet lag' produces temporal lobe atrophy and spatial cognitive deficits. Nat Neurosci 2001; 4:567-8.
2. Toh KL, Jones CR, He Y, et al. An hPer2 phosphorylation site mutation in familial advanced sleep phase syndrome. Science 2001; 291:1040-3.
3. Foster RG, Provencio I, Hudson D, Fiske S, De Grip W, Menaker M. Circadian photoreception in the retinally degenerate mouse (rd/rd). J Comp Physiol [A] 1991; 169:39-50.
4. Nelson RJ, Zucker I. Absence of extraocular photoreception in diurnal and nocturnal rodents exposed to direct sunlight. Comp. Biochem. Physiol. 1981; 69A:145-148.
5. Carter-Dawson LD, LaVail MM, Sidman RL. Differential effect of the rd mutation on rods and cones in the mouse retina. Invest Ophthalmol Vis Sci 1978; 17:489-98.
6. Freedman MS, Lucas RJ, Soni B, et al. Regulation of mammalian circadian behavior by non-rod, non-cone, ocular photoreceptors. Science 1999; 284:502-4.
7. Czeisler CA, Shanahan TL, Klerman EB, et al. Suppression of melatonin secretion in some blind patients by exposure to bright light. N Engl J Med 1995; 332:6-11.
8. von Schantz M, Provencio I, Foster RG. Recent developments in circadian photoreception: more than meets the eye. Invest Ophthalmol Vis Sci 2000; 41:1605-7.
9. Sancar A. Cryptochrome: the second photoactive pigment in the eye and its role in circadian photoreception. Annu Rev Biochem 2000; 69:31-67.
10. Miyamoto Y, Sancar A. Vitamin B2-based blue-light photoreceptors in the retinohypothalamic tract as the photoactive pigments for setting the circadian clock in mammals. Proc Natl Acad Sci U S A 1998; 95:6097-102.
11. Thresher RJ, Vitaterna MH, Miyamoto Y, et al. Role of mouse cryptochrome blue-light photoreceptor in circadian photoresponses. Science 1998; 282:1490-4.
12. van der Horst GT, Muijtjens M, Kobayashi K, et al. Mammalian Cry1 and Cry2 are essential for maintenance of circadian rhythms. Nature 1999; 398:627-30.
13. Ceriani MF, Darlington TK, Staknis D, et al. Light-dependent sequestration of TIMELESS by CRYPTOCHROME. Science 1999; 285:553-6.
14. Emery P, So WV, Kaneko M, Hall JC, Rosbash M. CRY, a Drosophila clock and light-regulated cryptochrome, is a major contributor to circadian rhythm resetting and photosensitivity. Cell 1998; 95:669-79.
15. Kume K, Zylka MJ, Sriram S, et al. mCRY1 and mCRY2 are essential components of the negative limb of the circadian clock feedback loop. Cell 1999; 98:193-205.
16. Griffin EA, Jr., Staknis D, Weitz CJ. Light-independent role of CRY1 and CRY2 in the mammalian circadian clock. Science 1999; 286:768-71.
17. Rodieck RW. The First Steps in Seeing. Sunderland, MA: Sinauer Associates, Inc., 1998:562 pp.
18. Provencio I, Foster RG. Circadian rhythms in mice can be regulated by photoreceptors with cone- like characteristics. Brain Res 1995; 694:183-90.
19. Takahashi JS, DeCoursey PJ, Bauman L, Menaker M. Spectral sensitivity of a novel photoreceptive system mediating entrainment of mammalian circadian rhythms. Nature 1984; 308:186-8.
20. Yoshimura T, Ebihara S. Spectral sensitivity of photoreceptors mediating phase-shifts of circadian rhythms in retinally degenerate CBA/J (rd/rd) and normal CBA/N (+/+)mice. J Comp Physiol [A] 1996; 178:797-802.

21. Lucas RJ, Douglas RH, Foster RG. Characterization of an ocular photopigment capable of driving pupillary constriction in mice. Nat Neurosci 2001; 4:621-6.
22. Brainard GC, Hanifin JP, Greeson JM, et al. Action spectrum for melatonin regulation in humans: Evidence for a novel circadian photoreceptor. J Neurosci 2001; in press.
23. Lander ES, Linton LM, Birren B, et al. Initial sequencing and analysis of the human genome. Nature 2001; 409:860-921.
24. Nathans J, Thomas D, Hogness DS. Molecular genetics of human color vision: the genes encoding blue, green, and red pigments. Science 1986; 232:193-202.
25. Chen XN, Korenberg JR, Jiang M, Shen D, Fong HK. Localization of the human RGR opsin gene to chromosome 10q23. Hum Genet 1996; 97:720-2.
26. Sun H, Gilbert DJ, Copeland NG, Jenkins NA, Nathans J. Peropsin, a novel visual pigment-like protein located in the apical microvilli of the retinal pigment epithelium. Proc Natl Acad Sci U S A 1997; 94:9893-8.
27. Provencio I, Rodriguez IR, Jiang G, Hayes WP, Moreira EF, Rollag MD. A novel human opsin in the inner retina. J Neurosci 2000; 20:600-5.
28. Halford S, Freedman MS, Bellingham J, et al. Characterization of a novel human opsin gene with wide tissue expression and identification of embedded and flanking genes on chromosome 1q43. Genomics 2001; 72:203-8.
29. Blackshaw S, Snyder SH. Encephalopsin: a novel mammalian extraretinal opsin discretely localized in the brain. J Neurosci 1999; 19:3681-90.
30. Aggelopoulos NC, Meissl H. Responses of neurones of the rat suprachiasmatic nucleus to retinal illumination under photopic and scotopic conditions. J Physiol 2000; 523 Pt 1:211-22.
31. Selby CP, Thompson C, Schmitz TM, Van Gelder RN, Sancar A. Functional redundancy of cryptochromes and classical photoreceptors for nonvisual ocular photoreception in mice. Proc Natl Acad Sci U S A 2000; 97:14697-702.
32. Pandey S, Blanks JC, Spee C, Jiang M, Fong HK. Cytoplasmic retinal localization of an evolutionary homolog of the visual pigments. Exp Eye Res 1994; 58:605-13.
33. Hao W, Fong HK. The endogenous chromophore of retinal G protein-coupled receptor opsin from the pigment epithelium. J Biol Chem 1999; 274:6085-90.
34. Chen P, Hao W, Rife L, et al. A photic visual cycle of rhodopsin regeneration is dependent on Rgr. Nat Genet 2001; 28:256-60.
35. Provencio I, Jiang G, De Grip WJ, Hayes WP, Rollag MD. Melanopsin: An opsin in melanophores, brain, and eye. Proc Natl Acad Sci U S A 1998; 95:340-5.
36. Rollag MD, Provencio I, Sugden D, Green CB. Cultured amphibian melanophores: a model system to study melanopsin photobiology. Methods Enzymol 2000; 316:291-309.
37. Provencio I, Cooper HM, Foster RG. Retinal projections in mice with inherited retinal degeneration: implications for circadian photoentrainment. J Comp Neurol 1998; 395:417-39.
38. Berson DM, Dunn FA, Takao M. Phototransduction by ganglion cells innervating the circadian pacemaker [ARVO abstract]. Invest Ophthalmol Vis Sci 2001; 42:S113 Abstract nr 613.
39. Montell C. Visual transduction in Drosophila. Annu Rev Cell Dev Biol 1999; 15:231-68.

Circadian Rhythms

ACTION SPECTRUM FOR MELATONIN SUPPRESSION: EVIDENCE FOR A NOVEL CIRCADIAN PHOTORECEPTOR IN THE HUMAN EYE

George C. Brainard and John P. Hanifin
Department of Neurology, Jefferson Medical College, Thomas Jefferson University, Philadelphia, Pennsylvania 19107

Introduction

Light is a potent regulator of systemic physiology. Specifically, environmental light is the primary stimulus for regulating circadian rhythms, seasonal cycles, and neuroendocrine responses in many mammalian species including humans (1, 2, 3, 4). Shortly after the regulatory capacity of light was demonstrated in humans, light was tested as a therapeutic intervention for treating winter depression or Seasonal Affective Disorder (5, 6, 7). Since then, twenty years of clinical studies have confirmed that light therapy is effective for treating winter depression (8, 9, 10). In addition, light therapy has been studied as a treatment for selected circadian sleep disorders, nonseasonal depression, menstrual disturbances, eating disorders as well as re-entraining circadian physiology relative to the challenges of shift work or intercontinental air travel (11, 12).

During the past two centuries, extensive research has elucidated the underlying neuroanatomy and neurophysiology which supports vision. In humans and most mammalian species, light enters the eyes and stimulates the retinal rod and cone cells which initiate nerve signals to the visual cortex and other visual centers of the brain. More recently, a separate neural pathway, the retinohypothalamic tract, has been shown to project from the retina into the hypothalamus, a non-visual part of the brain (2, 13, 14). This non-visual pathway is responsible for the regulation of the circadian and neuroendocrine systems by light including the rhythmic production and secretion of the hormone melatonin.

In addition to controlling circadian rhythms of melatonin production, light of sufficient intensity, wavelength, timing and duration can acutely suppress nocturnal melatonin secretion. Acute light-induced suppression of melatonin is well documented across different mammalian species and has been used extensively to help determine the ocular and neural physiology related to melatonin regulation in humans (11, 15, 16, 17, 18). Observing melatonin responses to light has been an effective approach for clarifying the photic regulation of the circadian and neuroendocrine systems (1, 2, 11).

Until recently, the ocular photoreceptors which support circadian regulation or the therapeutic benefits of light therapy had not been identified. Initial work on the action spectrum for melatonin regulation in humans confirmed that the three cone photopic visual system is not the primary mediator of light regulation of melatonin (19). An action spectrum study with nine monochromatic wavelengths demonstrated that the phototransduction of light for melatonin regulation appears to be mediated primarily by a single photopigment with peak sensitivity in the 446-477 nm wavelength portion of the visible spectrum (20). Those data fit a vitamin A_1 retinaldehyde opsin template and indicated that the photopigment for melatonin regulation is distinct from the classical photopigments of rod and cone visual photoreceptors. The aim of this paper is to review this new research in the context of other ongoing work on circadian phototransduction and explore its implications for the therapeutic applications of light in treating circadian and affective disorders.

Studies with blind rodents suggest that rods and cones do not mediate circadian physiology

It is important to note that there is considerable diversity in the cellular structure and function of the retina across mammalian species (21). Early studies with blind mole rats and rats with destruction of retinal photoreceptors due to prolonged light exposure raised the possibility that neither the rods nor the cones of the visual system participate in regulating the circadian and neuroendocrine systems (22, 23). Despite profound loss of photoreceptors and vision, light detection for circadian and photoperiodic regulation was preserved.

Studies in mice with hereditary retinal disorders (*rd/rd* and *rds/rds*) have shown that these animals still exhibit normal light-induced melatonin suppression and circadian locomotor phase-shifts despite a nearly total loss of classical visual photoreceptors. The data support the conclusion that circadian photoreception is maintained either by 1) a very small number of rod or cone cells, or 2) an unrecognized class of retinal photoreceptors (24, 25, 26, 27). Further work with *rd* mice suggested that middle-wavelength sensitive cones (M-cones) and/or short-wavelength sensitive cones (S-cones) may be responsible for circadian photoreception (28, 29, 30). However, recent studies with transgenic coneless (*cl*) mice which have extensive loss of M-cones and S-cones show that these mice exhibit normal sensitivity for light-induced melatonin suppression and circadian phase-shifting of locomotion (31, 32). Similarly, coneless, rodless mice (*rd/rd cl*) also appeared to exhibit normal sensitivity for light-induced melatonin suppression and phase-shifting of wheel-running behavior. These results indicate that rods, M-cones and S-cones are not required for circadian photoreception in mice. Removal of the eyes however, abolished light-induced circadian phase-shifting (32). Overall, the results suggest that the mouse eye contains specific photoreceptors for circadian regulation which are different from the visual photoreceptors.

Photopigment candidates which may provide photic input to the circadian system

If the visual rods and cones in animals are not the primary photoreceptors for circadian regulation, what is the main photopigment for transducing photic stimuli to the suprachiasmatic nuclei? Studies with various species have identified several new molecules which may serve as circadian photopigments. These proposed photopigments include both opsin-based molecules such as peropsin (33), vertebrate ancient (VA) opsin (34), melanopsin (35), as well as non-opsin molecules like bilirubin (36) and cryptochrome (37). Among the newly identified opsin photopigments, the gene for melanopsin has been shown to be expressed in the human neural retina. Furthermore, melanopsin has been specifically localized in the ganglion cell and inner nuclear cell layers of both primates and mice (38). Among the non-opsin photopigment candidates for circadian phototransduction, cryptochrome has been localized to the mouse neural retina. Specifically, mouse cryptochromes (*mCRY1* and *mCRY2*) have been expressed in the ganglion cell and inner nuclear cell layers of the mouse retina (37). Cryptochromes have been studied extensively as circadian photopigments in plants and insects (39, 40). Investigators are divided, however, on the contention that cryptochromes serve as circadian photopigments in mammals (37, 41, 42, 43, 44, 45, 46).

Traditionally, the ganglion cell layer and the inner nuclear cell layers are not thought to be retinal sites for phototransduction. Recently, however, a preliminary report provided evidence that ganglion cells in the rat retina are directly photosensitive (47). Perhaps circadian phototransduction occurs independently of the retinal layer containing the rods and cones that support visual function. Certainly, the localization of melanopsin and cryptochrome in the retina, make them attractive photopigment candidates for circadian regulation. More functional evidence is needed to confirm their direct role in circadian photoreception. Furthermore, caution should be exercised in generalizing results from plants, insects, fish, amphibians and rodents to humans.

Studies with blind humans suggest that rods and cones do not mediate circadian physiology

As with studies on rodents with degenerate or atrophied retinas, some of the data on visually blind humans suggest that light detection for the human circadian and neuroendocrine systems may not rely on the photoreceptors used for vision. Acute light-induced melatonin suppression has been demonstrated in some humans with complete visual blindness (48). In that study, some blind subjects with loss of pupillary reflex, no conscious perception of the light stimuli, and no outer retinal functioning as determined by ERG testing, still exhibited acute light-induced melatonin suppression and circadian entrainment (48). A field study on 67 blind humans, however, showed that individuals who had no visual light perception had abnormal circadian rhythms and a higher incidence of free-running rhythms in urinary 6-sulfatoxymelatonin (aMT6s) compared to blind subjects who retained a degree of visual light perception (49, 50). In addition, all subjects with bilateral enucleation of the eyes had free-running aMT6s rhythms which underscores the importance of ocular photic transduction for circadian regulation.

A different study on humans with color vision deficiencies suggested that a normal trichromatic photopic system is not necessary for light regulation of melatonin secretion. Protanopic and deuteranopic observers who lacked functioning long-wavelength sensitive (L cones) or M cone photoreceptors, exhibited normal white light-induced melatonin suppression and entrainment of the melatonin rhythm (51, 52). The results from human and animal circadian studies on different forms of visual blindness suggest the possibility that melatonin regulation by light is mediated, at least in part, by a photopigment which differs from the photopigments involved in vision. Together, the results from human and animal circadian studies on different forms of visual blindness suggest that melatonin regulation by light is controlled, at least in part, by photoreceptors which differ from the known photoreceptors for vision.

Direct evidence of a novel photoreceptor providing input to the human circadian system

The elucidation of an action spectrum for acute light-induced melatonin suppression in humans provides the first direct evidence that a novel photopigment provides input to the human circadian system. There are refined guidelines for determining analytical action spectra which are applicable to light-sensitive responses in all organisms from algae to humans (53, 54, 55). Specifically, action spectra are formed by determining dose-response curves (fluence-response curves) at different monochromatic wavelengths for a specific biological response. This technique is very useful for 1) clarifying the relative effectiveness of different wavelengths for eliciting a biological response, and 2) identifying the specific photopigment involved in that response.

The first demonstration of a fluence-response relationship between light and melatonin regulation in humans showed that monochromatic 509 nm light suppressed nocturnal melatonin in a graded fashion; brighter photic stimuli elicited progressively greater suppression of melatonin (56). Preliminary results from that study also suggested that the blue-green portion of the visible spectrum provided the most potent input to the human pineal gland and that the three-cone photopic visual system was not the main photosensory system for this part of the circadian system. That concept has been supported in a recent study which showed that monochromatic light at 505 nm is approximately four times stronger than 555 nm in suppressing melatonin in healthy humans (19). Those data suggest that the ocular photoreceptor primarily responsible for human pineal melatonin suppression is not the cone system which mediates photopic vision.

The completion of a set of fluence response curves at eight monochromatic wavelengths between 440 nm and 600 nm has now provided an action spectrum which identifies the wavelength sensitivity of the photosensory system responsible for melatonin regulation in humans (20). Each fluence-response curve for each wavelength was comprised of data gathered from a complete, within-subjects design involving eight subjects in a series of seven or more nighttime melatonin suppression tests. Data from all eight fluence-response curves were fit to univariant four-parameter sigmoidal curves with high coefficients of correlation (R^2 = 0.81-0.95). The hypothesis that human melatonin suppression is modulated by a single photoreceptor type is supported by the univariance of these curves. Using the technique of Partridge and De Grip (57), an action spectrum was formed from the fluence-response data by plotting the reciprocal of incident photons required to produce the half-saturation melatonin suppression versus wavelength. As shown in Figure 1, these action spectrum data fit a vitamin A_1 retinaldehyde opsin template which predict the peak spectral absorbance of this photopigment (λ max) to be 464 nm (R^2 = 0.91). Based on its variance (± SD), a more conservative estimate of the peak wavelength region providing circadian input for regulating melatonin is 446–477 nm.

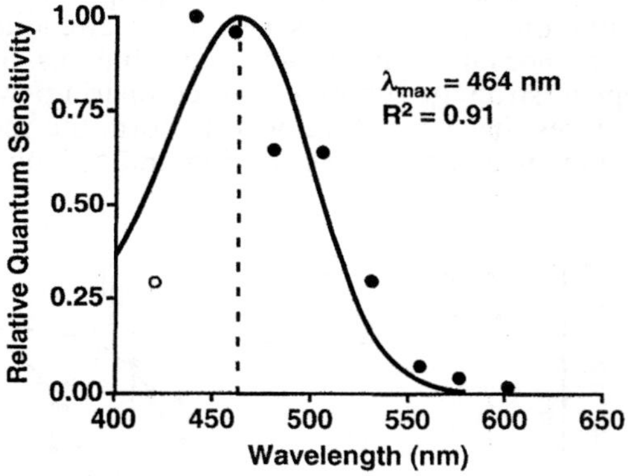

Figure 1. This graph illustrates the action spectrum for percent control-adjusted melatonin suppression in 72 healthy human subjects (20). The solid circles indicate the half-saturation constants of eight wavelengths from 440 to 600 nm which were normalized to the maximum response and plotted as log relative sensitivity. The solid line in the graph demonstrates the best fit template (R^2 = 0.91) for vitamin A_1 retinaldehyde photopigments (57). The template for these melatonin suppression data predicts a maximal spectral absorbance (λ max) of 464 nm. To probe for sensitivity to monochromatic light at 420 nm, a group of 8 subjects were exposed to a single night of no light exposure and a single night of exposure to 420 nm light at 5.58 x 10^{13} photons/cm^2. That 420 nm stimulus induced a significant melatonin suppression ($p<0.003$). The open circle represents the estimated half-saturation constant (1.83 x 10^{13} photons/cm^2) derived from the 420 nm data. This figure is from Brainard et al. (20) and is reprinted with permission (Copyright 2001 by the Society for Neuroscience).

The data shown in Figure 1 suggest that a single photopigment may be primarily responsible for melatonin suppression in humans. Furthermore, the data suggest that a novel opsin photopigment mediates melatonin regulation since the maximum spectral response is distinct from the maximal spectral responses of visual rod and cone cell photoreceptors (58). A recent study has developed a smaller melatonin suppression action spectrum which tested six monochromatic wavelengths on 22 male and female volunteers (59). The calculated λ max from that study (459 nm, $R^2 = 0.73$) was similar to the action spectrum peak shown in Figure 1, and the authors also interpreted their data to support the existence of a non-rod, non-cone photoreceptor system in humans. That study, however, showed a much higher sensitivity to 424 nm monochromatic light than the study reported here (20). The difference in response to short wavelength light between the two studies will need to be clarified with further experiments.

Comparison of fluence-response curves at 505 nm and 555 nm

Recent studies indicate that melatonin regulation is controlled by a novel photoreceptor in the human eye (19, 20, 59). That discovery reinforces an earlier proposal that it may be inappropriate to quantify light for melatonin regulation - and perhaps for all of circadian regulation - with a meter which measures lux, footcandles, lumens or footlamberts (60). This concept is strongly supported by comparing two of the eight fluence-response curves that were developed to form the melatonin action spectrum. The fluence-response curves for 505 and 555 nm are illustrated in Figure 2.

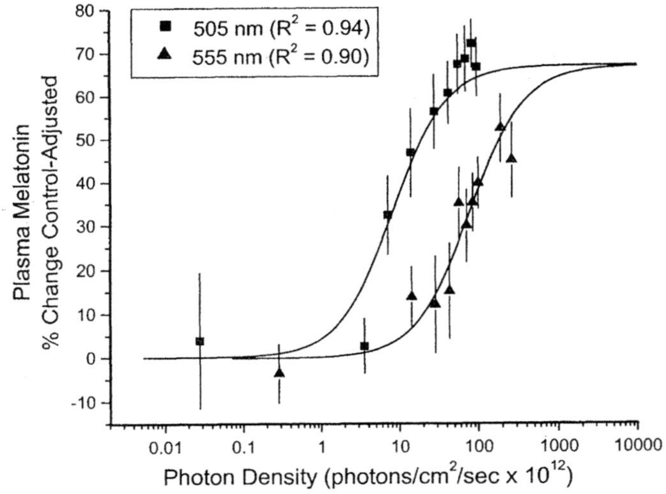

Figure 2. This figure illustrates the fitted univariant fluence-response curves for 505 nm and 555 nm monochromatic light exposures and percent control-adjusted melatonin suppression. Each fluence-response curve is derived from a separate set eight healthy volunteers who participated in a complete, within-subjects experimental design. For each graph, the data points represent group means ± SEM. The coefficient of correlations for the 505 nm and 555 nm fitted curves are 0.94 and 0.90, respectively (20).

The data shown in Figure 2 clearly illustrate that monochromatic light at 505 nm is more potent than monochromatic light at 555 nm at all photon densities which elicited melatonin suppression. Table 1 provides a comparison of both radiometric and photometric values for a single photon density used to suppress melatonin at each of these wavelengths.

Table 1. This chart shows the radiometric and photometric values of equal photon stimuli at two wavelengths along with the melatonin suppression responses they evoked in sets of eight healthy human subjects (19, 20).

WAVELENGTH λ	QUANTA photons/cm²/sec	IRRADIANCE μW/cm²	ILLUMINANCE lux		% MELATONIN SUPPRESSION
			photopic	scotopic	
505 nm	4.2 x 10^{13}	16.5	48.3	281.0	60.6
555 nm	4.2 x 10^{13}	15.0	100.4	103.6	15.2

The potential need for a specific light measurement system for circadian regulation

It is well-established that higher levels of illumination are required for stimulating the circadian system than for supporting vision in humans and other mammalian species (5, 56, 61, 62, 63). As a result, some investigators initially considered the three cone photopic visual system in humans to be responsible for mediating circadian and neuroendocrine responses since this portion of the visual system is responsive to "bright" daytime levels of illumination. Since the first demonstration of light-induced melatonin suppression in humans (5), most of the published literature on light-induced circadian responses in humans report light levels in terms of illuminance (lux, lumens) which are measures based on the standardized sensitivity curve of the photopic visual system. The peak wavelength sensitivity of photopic vision is 555 nm (21, 64). If photic input for melatonin regulation is principally provided by the three cone photopic visual system, then monochromatic light at 555 nm should be the most potent stimulus for suppressing melatonin. The data simply do not support this hypothesis. In sharp contrast, Figure 2 shows that across a range of light irradiances 505 nm is significantly stronger, photon for photon, than 555 nm for light-induced melatonin suppression. Similarly, Table 1 emphasizes that with an equal photon exposure of 505 nm versus 555 nm, a lower lux value of 505 nm was capable of inducing a four-fold greater melatonin suppression than the higher lux value at 555 nm.

These data illustrate the problem of using a photometer for quantifying light for melatonin suppression. Furthermore, there are potential clinical implications of this finding. The current standard practice is to use photometric values (lux) for quantifying light used therapeutically in treating patients with affective disorders, circadian disruption, or certain sleep disorders (65, 66). Ultimately, that measurement technique may not be optimum (60). It is important to stress, however, that the data presented here involve only the melatonin suppression response in healthy humans. Until it is determined that there is a similar wavelength sensitivity for circadian phase-shifting or other therapeutic benefits of light, there is not a compelling reason to abandon the use of photopic photometers in light therapy. In the future, however, light might best be quantified in "circadian lux" – a new metric based on the action spectrum which applies to the broad regulation of the circadian system versus the photopic visual system. Further studies of the human circadian action spectrum are likely to yield not only improvements in light measurement but also enhancements of light therapy equipment and general architectural lighting (19, 20, 60).

Conclusion

Light is a potent regulator of circadian rhythms, seasonal cycles, and neuroendocrine responses in mammals. Earlier research on rodents and humans with different forms of blindness suggested that there may be a novel photosensory system in the mammalian eye that mediated these nonvisual effects of light. The development of an action spectrum for melatonin regulation in healthy human subjects now provides the first direct evidence of a novel photopigment providing input to the human circadian system. The most potent wavelength region providing input for melatonin suppression is 446–477 nm which is distinct from the peak sensitivities of the classical visual rod and cone photoreceptors. Ultimately, this finding opens the door for redefining how light should be delivered and measured relative to circadian regulation and the therapeutic benefits of light.

Acknowledgments

The writing of this chapter was supported, in part, by grants from NIH RO1NS36590 and National Space Biomedical Research Institute under NASA Cooperative Agreement NCC 9-58 (to GCB). The authors deeply appreciate the assistance from Gena Glickman, Samar Jasser and James Gardner for reviewing the manuscript. This chapter review is principally derived from three published works (60, 19, 20). For reprints of those publications or the present chapter please contact George C. Brainard, Ph.D., Department of Neurology, Thomas Jefferson University, 1025 Walnut Street, Suite 310, Philadelphia, Pennsylvania 19107. Phone: 215-955-7644, Fax: 215-923-7588, Email: george.brainard@mail.tju.edu

References

1. Aschoff, J., ed. Handbook of Behavioral Neurobiology, Biological Rhythms. New York: Plenum Press, 1981.

2. Klein, DC, Moore, RY, Reppert, SM, eds. Suprachiasmatic Nucleus: The Mind's Clock. Oxford: Oxford University Press, 1991.

3. Wehr, TA. The durations of human melatonin secretion and sleep respond to changes in daylength (photoperiod). J Clin Endocrinol Metab 1991;73:1276-80.

4. Arendt, J. Melatonin and the pineal gland: influence on mammalian seasonal and circadian physiology. Rev Reprod 1998;3:13-22.

5. Lewy, AJ, Wehr, TA, Goodwin, FK, Newsome, DA, Markey, SP. Light suppresses melatonin secretion in humans. Science 1980;210:1267-9.

6. Lewy, AJ, Kern, HE, Rosenthal, NE, Wehr, TA. Bright artificial light treatment of a manic-depressive patient with a seasonal mood cycle. Am J Psychiatry 1982;139:1496-8.

7. Rosenthal, NE, Sack, DA, Gillin, JC, Lewy, AJ, Goodwin, FK, Davenport, Y, Mueller, PS, Newsome, DA, Wehr, TA. Seasonal affective disorder. A description of the syndrome and preliminary findings with light therapy. Arch Gen Psychiatry 1984;41:72-80.

8. Eastman, CI, Young, MA, Fogg, LF, Liu, L, Meaden, PM. Bright light treatment of winter depression: a placebo-controlled trial. Arch Gen Psychiatry 1998;55:883-9.

9. Terman, M, Terman, JS, Ross, DC. A controlled trial of timed bright light and negative air ionization for treatment of winter depression. Arch Gen Psychiatry 1998;55:875-82.

10. Lewy, AJ, Bauer, VK, Cutler, NL, Sack, RL, Ahmed, S, Thomas, KH, Blood, ML, Jackson, JM. Morning vs evening light treatment of patients with winter depression. Arch Gen Psychiatry 1998;55:890-896.

11. Wetterberg, L, ed. Light and Biological Rhythms in Man. Stockholm: Pergamon Press, 1993.

12. Lam, RW, ed. Seasonal Affective Disorder and Beyond: Light Treatment for SAD and Non-SAD Disorders. Washington, D.C.: American Psychiatric Press, 1998.

13. Moore, RY, Lenn, NJ. A retinohypothalamic projection in the rat. J Comp Neurol 1972;146:1-14.

14. Card, JP. Pseudorabies virus and the functional architecture of the circadian timing system. J Biol Rhythms 2000;15:453-61.

15. Gaddy, JR, Rollag, MD, Brainard, GC. Pupil size regulation of threshold of light-induced melatonin suppression. J Clin Endocrinol Metab 1993;77:1398-1401.

16. Brainard, GC, Rollag, MD, Hanifin, JP. Photic regulation of melatonin in humans: ocular and neural signal transduction. J Biol Rhythms 1997;12:537-46.

17. Visser, EK, Beersma, DGM, Daan, S. Melatonin suppression by light in humans is maximal when the nasal part of the retina is illuminated. J Biol Rhythms 1999;14:116-21.

18. Lasko, TA, Kripke, DF, Elliott, JA. Melatonin suppression by illumination of upper and lower visual fields. J Biol Rhythms 1999;14:122-5.

19. Brainard, GC, Hanifin, JP, Rollag, MD, Greeson, J, Byrne, B, Glickman, G, Gerner, E, Sanford, B. Human melatonin regulation is not mediated by the three cone photopic visual system. J Clin Endocrinol Metab 2001;86:433-436.

20. Brainard, GC, Hanifin, JP, Greeson, JM, Byrne, B, Glickman, G, Gerner, E, Rollag, MD. Action spectrum for melatonin regulation in humans: evidence for a novel circadian photoreceptor. J Neurosci 2001;21:6405-6412.

21. Rodieck, RW. The First Steps in Seeing. Sunderland, Massachusetts: Sinauer Associates, Inc., 1998.

22. Pevet, P, Heth, G, Hiam, A, Nevo, E. Photoperiod perception in the blind mole rat (*Spalax ehrenbergi, Nehring*): involvement of the Harderian gland, atrophied eyes, and melatonin. J Exp Zool 1984;232:41-50.

23. Webb, SM, Champney, TH, Lewinski, AK, Reiter, RJ. Photoreceptor damage and eye pigmentation: influence on the sensitivity of rat pineal N-acetyltransferase activity and melatonin levels to light at night. Neuroendocrinology 1985;40:205-9.

24. Goto, M, Ebihara, S. The influence of different light intensities on pineal melatonin content in the retinal degenerate C3H mouse and the normal CBA mouse. Neurosci Lett 1990;108:267-72.

25. Foster, RG, Provencio, I, Hudson, D, Fiske, S, DeGrip, W, Menaker, M. Circadian photoreception in the retinally degenerate mouse (rd/rd). J Comp Physiol [A] 1991;169:39-50.

26. Provencio, I, Wong, S, Lederman, AB, Argamaso, SM, Foster, RG. Visual and circadian responses to light in aged retinally degenerate mice. Vision Res 1994;34:1799-1806.

27. Yoshimura, T, Nishio, M, Goto, M, Ebihara, S. Differences in circadian photosensitivity between retinally degenerate CBA/J mice (*rd/rd*) and normal CBA/N mice (+/+). J Biol Rhythms 1994;9:51-60.

28. García-Fernández, JM, Jiminez, AJ, Foster, RG. The persistence of cone photoreceptors within the dorsal retina of aged retinally degenerate mice (*rd/rd*): implications for circadian organization. Neurosci Lett 1995;187:33-6.

29. Provencio, I, Foster, RG. Circadian rhythms in mice can be regulated by photoreceptors with cone-like characteristics. Brain Res 1995;694:183-90.

30. Yoshimura, T, Ebihara, S. Spectral sensitivity of photoreceptors mediating phase-shifts of circadian rhythms in retinally degenerate CBA/J (rd/rd) and normal CBA/N (+/+) mice. J Comp Physiol [A] 1996;178:797-802.

31. Lucas, RJ, Foster, RG. Neither functional rod photoreceptors nor rod or cone outer segments are required for the photic inhibition of pineal melatonin. Endocrinology 1999;140:1520-4.

32. Freedman, MS, Lucas, RJ, Soni, B, von Schantz, M, Munoz, M, David-Gray, Z, Foster, RG. Regulation of mammalian circadian behavior by non-rod, non-cone, ocular photoreceptors. Science 1999;284:502-504.

33. Sun, H, Gilbert, DJ, Copeland, NG, Jenkins, NA, Nathans, J. Peropsin, a novel visual pigment-like protein located in the apical microvilli of the retinal pigment epithelium. Proc Natl Acad Sci U S A 1997;94:9893-8.

34. Soni, BG, Foster, RG. A novel and ancient vertebrate opsin. FEBS Lett 1997;406:279-83.

35. Provencio, I, Jiang, G, De Grip, WJ, Hayes, WP, Rollag, MD. Melanopsin: an opsin in melanophores, brain, and eye. Proc Natl Acad Sci U S A 1998;95:340-5.

36. Oren, DA. Humoral phototransduction: blood is a messenger. The Neuroscientist 1996;2:207-10.

37. Miyamoto, Y, Sancar, A. Vitamin B2-based blue-light photoreceptors in the retinohypothalamic tract as the photoactive pigments for setting the circadian clock in mammals. Proc Natl Acad Sci U S A 1998;95:6097-6102.

38. Provencio, I, Rodriguez, IR, Jiang, G, Hayes, WP, Moreira, EF, Rollag, MD. A novel human opsin in the inner retina. J Neurosci 2000;20:600-5.

39. Ahmad, M, Cashmore, AR. HY4 gene of A. thaliana encodes a protein with characteristics of a blue-light photoreceptor. Nature 1993;366:162-6.

40. Stanewsky, R, Kaneko, M, Emery, P, et al. The cryb mutation identifies cryptochrome as a circadian photoreceptor in Drosophila. Cell 1998;95:681-92.

41. Thresher, RJ, Vitaterna, MH, Miyamoto, Y, Kazantsev, A, Hsu, DS, Petit, C, Selby, CP, Dawut, L, Smithies, O, Takahashi, JS, Sancar, A. Role of mouse cryptochrome blue-light photoreceptor in circadian photoresponses. Science 1998;282:1490-1494.

42. Griffin, EA, Staknis, D, Weitz, CJ. Light-independent role of Cry1 and Cry2 in the mammalian circadian clock. Science 1999;286:768-71.

43. van der Horst, GTJ, Muijtjens, M, Kobayashi, K, Takano, R, Kanno, S, Takao, M, de Wit, J, Verkerk, A, Eker, APM, van Leenen, D, Buijs, R, Bootsma, D, Hoeijmakers, JHJ, Yasui, A. Mammalian Cry1 and Cry2 are essential for maintenance of circadian rhythms. Nature 1999;398:627-630.

44. von Schantz, M, Provencio, I, Foster, RG. Recent developments in circadian photoreception: more than meets the eye. Invest Ophthalmol Vis Sci 2000;41:1605-7.

45. Reppert, SM, Weaver, DR. Comparing clockworks: mouse versus fly. J Biol Rhythms 2000;15:357-64.

46. Selby, CP, Thompson, C, Schmitz, TM, van Gelder, RN, Sancar, A. Functional redundancy of cryptochromes and classical photoreceptors for nonvisual ocular photoreception in mice. Proc Natl Acad Sci U S A 2000;97:14697-702.

47. Berson, DM, Dunn, FA, Takao, M. Phototransduction by ganglion cells innervating the circadian pacemaker. San Diego, California, November 10-15: 31st Annual Meeting of the Society for Neuroscience, 2001. 27. abstract.

48. Czeisler, CA, Shanahan, TL, Klerman, EB, Martens, H, Brotman, DJ, Emens, JS, Klein, T, Rizzo, JF, III. Suppression of melatonin secretion in some blind patients by exposure to bright light. N Engl J Med 1995;332:6-11.

49. Lockley, SW, Skene, DJ, Arendt, J, Tabandeh, H, Bird, AC, Defrace, R. Relationship between melatonin rhythms and visual loss in the blind. J Clin Endocrinol Metab 1997;82:3763-70.

50. Skene, DJ, Lockley, SW, Thapan, K, Arendt, J. Effects of light on human circadian rhythms. Reprod Nutr Dev 1999;39:295-304.

51. Ruberg, FL, Skene, DJ, Hanifin, JP, Rollag, MD, English, J, Arendt, J, Brainard, GC. Melatonin regulation in humans with color vision deficiencies. J Clin Endocrinol Metab 1996;81:2980-2985.

52. Brainard, GC, Ruberg, FL, Hanifin, JP, Rollag, MD. The relationship of melatonin regulation and color vision in humans. In: Webb, SM, Puig-Domingo, M, Møller, M, Pevét, P, eds. Pineal update: from molecular mechanisms to clinical implications. Westbury, New York: PJD Publications Ltd., 1997:339-48.

53. Grossweiner, LI. Photophysics. In: Smith, KC, ed. The science of photobiology. New York: Plenum Press, 1989:1-45.

54. Coohill, TP. Action spectra again? Photochem Photobiol 1991;54:859-70.

55. Lipson, ED. Action spectroscopy: methodology. In: Horspool, WM, Song, P-S, eds. Organic photochemistry and photobiology. New York: CRC Press, 1994:1257-66.

56. Brainard, GC, Lewy, AJ, Menaker, M, Miller, LS, Fredrickson, RH, Weleber, RG, Cassone, V, Hudson, D. Dose-response relationship between light irradiance and the suppression of melatonin in human volunteers. Brain Res 1988;454:212-218.

57. Partridge, JC, De Grip, WJ. A new template for rhodopsin (vitamin A1 based) visual pigments. Vision Res 1991;31:619-30.

58. Stockman, A, Sharpe, LT. Cone spectral sensitivities and color matching. In: Gegenfurtner, KR, Sharpe, LT, eds. Color vision: from genes to perception. Cambridge, United Kingdom: Cambridge University Press, 1999:53-87.

59. Thapan, K, Arendt, J, Skene, DJ. An action spectrum for melatonin suppression: evidence for a novel non-rod, non-cone photoreceptor system in humans. J Physiol 2001;535:261-7.

60. Brainard, GC. The healing light: interface of physics and biology. In: Lam, RW, ed. Seasonal Affective Disorder and beyond: light treatment for SAD and non-SAD conditions. Washington, D.C.: American Psychiatric Press, Inc., 1998:1-44.

61. Czeisler, CA, Allan, JS, Strogatz, SH, Ronda, JM, Sanchez, R, Rios, CD, Freitag, WO, Richardson, GS, Kronauer, RE. Bright light resets the human circadian pacemaker independent of the timing of the sleep-wake cycle. Science 1986;233:667-671.

62. Nelson, DE, Takahashi, JS. Comparison of visual sensitivity for suppression of pineal melatonin and circadian phase-shifting in the golden hamster. Brain Res 1991;554:272-7.

63. Zeitzer, JM, Dijk, D-J, Kronauer, RE, Brown, EN, Czeisler, CA. Sensitivity of the human circadian pacemaker to nocturnal light: melatonin phase resetting and suppression. J Physiol 2000;526:695-702.

64. Rea, MS, ed. Lighting Handbook: Reference & Application. New York: Illuminating Engineering Society of North America, 2000.

65. Society for Research on Biological Rhythms. Special Issue: Task force report on light treatment for sleep disorder. J Biol Rhythms 1995;10:99-176.

66. Lam, RW, Tam, EM, Gorman, CP, Blashko, C, Bowen, RC, Morehouse, RL, Hasey, GM. Light treatment. In: Lam, RW, Levitt, AJ, eds. Canadian consensus guidelines for the treatment of Seasonal Affective Disorder. Vancouver, British Columbia: Clinical and Academic Publishing, 1999:64-88.

Entrainment of the Non-24-hour Circadian Period of the Human Biological Clock to the 24-hour Day

Kenneth P. Wright Jr. and Charles A. Czeisler
Division of Sleep Medicine, Department of Medicine,
Harvard Medical School, Brigham and Women's Hospital,
221 Longwood Avenue, Boston, MA 02115, USA

ABSTRACT

It has long been hypothesized that exposure to bright light was required to entrain the non-24-h period of the human biological clock to the Earth's 24.0-h day. We report here supplemental data from a previously published study (1) showing that even when exposed to light levels during wakefulness that are equivalent to candlelight (averaging only ~1.5 lux in the angle of gaze), imposition of a periodic schedule of activity, sleep, meals and other routine daily activities is usually sufficient to maintain circadian entrainment to the 24.0-h day, but not to a 23.5-h or 24.6-h day. Our results indicate that either very dim light is biologically active in humans with regard to circadian photoreception and/or that the sleep-wake/rest-activity cycle itself can exert a greater influence on circadian entrainment than previously recognized. As we reported previously (1), our results also demonstrate functionally that in normally entrained, sighted adults the average intrinsic circadian period must be very near to 24 h. These results have major implications for understanding the mechanism by which the human biological clock is synchronized to environmental time and the pathophysiology of disorders of circadian regulation.

INTRODUCTION

Twenty years ago, it was reported that bright light can suppress secretion of the pineal hormone melatonin (2), leading to the hypothesis that exposure to illuminance levels similar to outside daylight was required to reset the human biological clock (2-6). The concept that circadian entrainment in humans required exposure to bright light (6) and that the human clock was "unperturbed by the use of ordinary indoor light" (4) initiated the successful use of bright light as a treatment for a host of clinical conditions thought to involve disturbances of circadian entrainment, ranging from seasonal affective disorder, delayed sleep phase syndrome during winter months in latitudes far from the equator and sleep disturbances in older people confined to indoor activities (7-10). While much has since been learned about the effects of the timing and intensity of light exposure on resetting of circadian phase and amplitude in humans (reviewed in (11)), the nature and strength of the environmental synchronizers required to maintain entrainment of the human biological clock to the 24-h day have not been previously characterized.

Light is a powerful stimulus that can reset and thereby entrain the circadian system in most species. In mammals, photic information is communicated to the master biological clock located in the suprachiasmatic nuclei (SCN) of the hypothalamus. Photic information is communicated to the SCN from specialized retinal ganglion cells via a monosynaptic retinohypothalamic tract and a multi-synaptic pathway involving the geniculo-hypothalamic tract (12). Clock gene regulatory mechanisms have been reported to underlie the self-sustained near-24.0-h periodicity of biological clocks (13-16), and these gene mechanisms within the SCN can be reset by exposure to environmental synchronizers (17-22). A number of issues, however, regarding circadian entrainment in humans remain unresolved. These include: 1) the photoreceptor responsible for circadian photoreception (23-26); 2) the minimum light intensity necessary to adjust for the daily drift due to the on-average longer-than-24-hour period (27) of the human biological clock; 3) the influence of putative non-photic synchronizers in humans, such as stimuli that cause arousal, including activity, ambient temperature, food availability and emotional stimuli (28-35), that may act on serotoninergic mechanisms in the brain or on clock genes (36-39).

Another unresolved issue in human circadian biology is the nature of the intrinsic period of the human biological clock (40;41). Based on experiments in which environmental lighting or the timing of wakefulness and sleep were not experimentally controlled, the range of entrainment of the human biological clock was reported to be centered near 25 h, which was the period observed under self-selected wake-sleep light-dark cycles (reviewed in

(27;42)). More recently, several independent laboratories have reported that the intrinsic circadian period of the human biological clock as estimated during a forced desynchrony protocol, is much closer to 24 hours (27;43-46). The circadian period assessed during forced desynchrony has been interpreted to represent the best approximation of a fundamental intrinsic periodicity of the human biological clock (41;47). However this interpretation has been challenged (40).

We hypothesized that the ability of the human biological clock to adapt to different day lengths would be dependent upon the intrinsic circadian period of the clock. We distinguish between the intrinsic circadian period, the actual period emanating from within the biological clock at a given time, from observed circadian periods, which are influenced by extrinsic resetting stimuli acting on the clock during the time of observation. Based upon the assumption that the near-24-hour circadian period estimate during forced desynchrony (27) is representative of a fundamental property of the clock, we hypothesized that more individuals would entrain to a 24.0 h but not a 23.5 or 24.6 h day in the presence of a weak environmental synchronizer. If on the other hand, the near-24-hour period estimate during forced desynchrony (27) was as argued (40), merely a result of the conditions under which the clock was studied and not representative of a fundamental property of the clock then we assumed the human biological clock would easily adapt to all near-24-hour day lengths studied. The following report provides additional analyses and examples of individual subject data expanding upon results previously published (1).

METHODS

We studied twelve healthy men and three healthy women (1). Participants gave written informed consent and the Brigham and Women's Hospital/Partners Health Care Human Research Committee approved the procedures for the protocol. Participants passed a rigorous health screening, including medical history, physical exam, electrocardiogram, blood and urine chemistries, toxicology screen for drug use, psychological tests and an interview with a clinical psychologist. None reported regular night work or rotating shift work within the past three years or crossing more than one time zone in the previous three months. Women began the study during the week of menses so that the assessment of the core body temperature rhythms for constant routine (CR) 1 and CR2 would occur during the follicular phase of their menstrual cycle since the luteal phase is associated with blunted temperature amplitude that can result in an inability to accurately assess temperature phase (48). Participants maintained a regular routine of 8-h of scheduled sleep and 16-h of scheduled wakefulness for a minimum three weeks while living at home before the start of laboratory portions of the protocol, as verified by sleep logs and call-in times to a time stamped voice recorder. Actigraphy with concurrent light exposure measurement occurred for at least one week prior to laboratory admission. Upon admission, the timing of daily activities was scheduled. Exercise and napping were proscribed. Performance testing began upon awakening and performance sessions of ~30 min in duration occurred every 2-h thereafter. Research staff had to pass through two doors, including an anteroom to enter the research suite. Staff was trained not to provide subjects with information about time. Participants were tested individually while living for up to 55 days in our laboratory in an environment free of time cues. Ambient light, room temperature, sleep-wake opportunities, nutrition intake and activity were strictly controlled. Participants were maintained on a 24.0-h schedule and initially scheduled to sleep for 8-h at their habitual bedtime for 3-6 baseline days and nights. This was followed by a 40.0-h CR that was used to assess circadian rhythms of plasma melatonin and core body temperature (27). The rhythms of plasma melatonin and core body temperature are well accepted markers reflecting the output of the internal biological clock in humans (11;27) and these markers maintain stable phase relationships to the light-dark/wakefulness-sleep cycle during entrainment. The rising phase/onset of the secretion of melatonin occurs on average shortly prior to habitual sleep time (49) whereas, the minimum of the circadian temperature minimum occurs on average shortly prior to habitual wake time (11). Given the large inter-individual differences in melatonin levels, melatonin onset was defined as the linear interpolated point in time at which melatonin levels reached 25% of the fitted peak to trough amplitude of the 3-harmonic of each individual's data from the first constant routine. This 25% melatonin onset threshold ($DLMO_{25\%}$), as determined on the first constant routine, was applied to data at all other portions of the protocol. The upward and downward mean crossing for melatonin were also calculated (53). Core body temperature phase was estimated with a dual-harmonic regression model (50). Following this first CR, individuals were scheduled for several weeks to a 23.5-h, 24.0-h or 24.6-h day. Light-dark cycles of ~1.5 lux in the angle of gaze during scheduled wakefulness and darkness during scheduled sleep were 15.83:7.66 h for the 23.5-h day, 16:8 h for the 24.0-h day and 16.4:8.2 h for the 24.6-h day. Ambient light intensity as measured with an IL-1400 photometer (International Light, Inc. Newburyport, MA, USA) with the sensor on a table top at ~76 cm aimed in the direction of the light fixtures was < 3 lux; the maximum light intensity in the room at ~183 cm with the sensor aimed in the direction of the light fixtures was < 8 lux during the 25 day schedule. A second CR was used to assess circadian phase after exposure to these scheduled day lengths. Subjects in the 24.0-h and 24.6-h conditions were then scheduled to a forced desynchrony protocol using a 28.0-h dim light (< 8 lux max)-dark activity-rest cycle, which was known to be outside the range of entrainment of the human biological clock under such conditions (27). This forced desynchrony protocol was used to estimate the intrinsic circadian period of individuals by assessing their plasma melatonin and core body temperature rhythms. Circadian melatonin and temperature period estimates were computed by a nonorthogonal spectral analysis technique with an exact maximum likelihood fitting procedure (1;27).

RESULTS

As reported (1), five of six subjects studied on the 24.0-h day (T=24.0-h) maintained phase relationships between the timing of their melatonin and core body temperature rhythms and the timing of the scheduled sleep-wake cycle that were within the normal range (49). Figures 1A-1B provide representative examples of melatonin levels above the daily mean together with scheduled

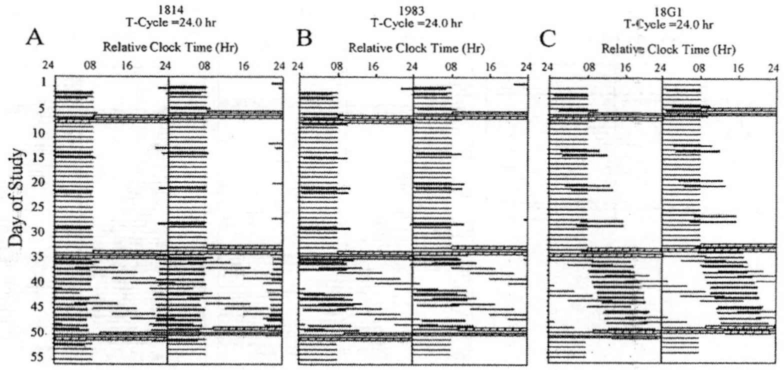

Figure 1. Plasma melatonin levels above the mean for subjects scheduled to the 24.0-h day and subsequent 28.0-h forced desynchrony. Data are double plotted with each subsequent day next to and beneath the other. Bars represent scheduled sleep episodes and CRs are boxes filled with downward diagonal bands. Thick lines represent melatonin secretion above the 24.0-h mean. **(A)** Example from a subject who remained entrained to the 24.0-h day. The forced desynchrony revealed his intrinsic period to be shorter than 24.0 h. **(B)** Example from a subject who remained entrained to the 24.0-h day. The forced desynchrony revealed her intrinsic period to be longer than 24.0 h. **(C)** Results for the subject who failed to entrain to the 24.0-h day whose circadian period was longer than and furthest from 24.0 h.

sleep episodes. During the T=24.0-h protocol, the melatonin rhythm maintained a normal phase angle of entrainment with high melatonin levels during scheduled sleep. As hypothesized, the intrinsic circadian periods of subjects who remained entrained to the 24.0-h day were limited to a narrow range around 24.0-h (23.88 to 24.12 h). Figure 1C illustrates the results for the one subject who failed to entrain to the 24.0-h day. During the T=24.0-h protocol, his melatonin rhythm drifted in a delay direction such that the phase relationship between the melatonin rhythm and the scheduled sleep-wake cycle was outside the normal range (49); and as a consequence, his melatonin levels were abnormally high during the scheduled wakefulness and low during scheduled sleep. The forced desynchrony segment of the protocol revealed that this subject had an intrinsic circadian period furthest from 24.0 h.

In contrast, none of the subjects scheduled to the 23.5-h or the 24.6-h days were able to achieve the daily phase shift required to entrain to those schedules in dim light(1). Instead, all subjects tested on the 23.5-h day (n=3)

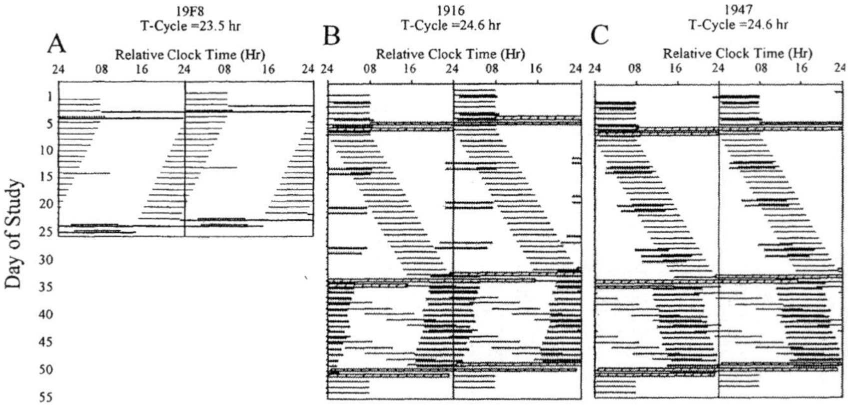

Figure 2. Plasma melatonin levels above the mean for subjects scheduled to the 24.6-h day. Symbols are the same as in Figure 1. Melatonin levels were high during scheduled sleep for the 24.0-h baseline days, whereas during the 23.5 and 24.6-h days, the human biological clock was unable to keep up with either the 30-minute daily advance or the 36-minute daily delay. Melatonin levels rose progressively later or earlier relative to the scheduled 23.5 or 24.6 h daily sleep episode respectively, and as a consequence were abnormally high during the scheduled wakefulness. **(A)** Results for a typical subject who failed to entrain to the 23.5-h day; **(B)** Results for the subject with the intrinsic period farthest from 24.6-h; **(C)** Results for the subject with the intrinsic period closest to 24.6-h.

exhibited abnormally delayed phase angles of melatonin onset relative to the timing of the scheduled sleep episode. Subjects on the 24.6-h day (n=6) showed abnormally advanced phase angles of melatonin onset relative to the timing of the scheduled sleep episode. Figure 2A shows plasma melatonin data above the 24-h mean and scheduled sleep episodes for a typical subject scheduled to the 23.5-h day. Figures 2B-2C show plasma melatonin data above the 24-h mean and scheduled sleep episodes for the subject with the circadian period furthest and closest to the 24.6-h day, respectively. Melatonin levels during 24.0-h baseline conditions were high during scheduled sleep. However, during the imposed 23.5 and 24.6-h T-cycles, the wakefulness-sleep schedules were insufficient to synchronize their biological clock in ~1.5 lux and, as a consequence, there were high melatonin levels during scheduled wakefulness and low melatonin levels during scheduled sleep. Next, we examined the absolute change in phase angle between the melatonin and temperature rhythms and the scheduled 24.0 and 24.6-h sleep-wake schedules. The changes in phase angle were calculated as the deviation in phase angle on CR2 compared to CR1 (Figure 3A). We found that the observed change in circadian phase for subjects scheduled to the 24.0-h day was significantly smaller than that for subjects scheduled to the 24.6-h day ($P < .05$). This difference in phase angle was observed even though subjects

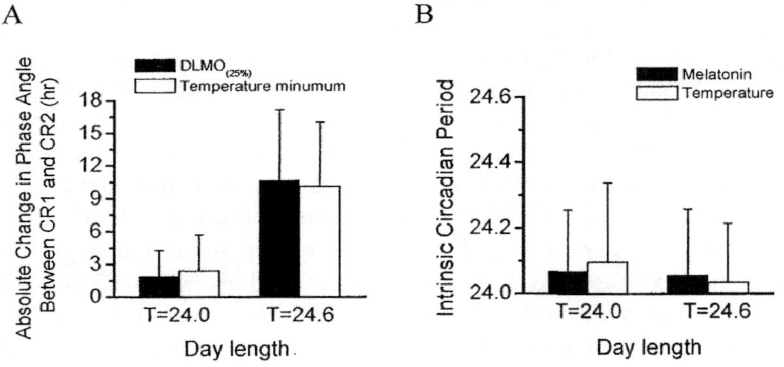

Figure 3. <u>(A) Absolute changes in the phase of melatonin onset and the temperature minimum between CR1 and CR2 **(B)** Average intrinsic circadian period.</u> The absolute deviation in phase angle was significantly less for subjects scheduled to the 24.0 h compared to the 24.6 h day ($P < 0.05$), regardless of whether subjects entrained to the day length. Forced desynchrony data were not collected for one subject in each condition. The average intrinsic period estimated during forced desynchrony was similar for subjects scheduled to the T=24.0 and T=24.6 h days. Error bars are SD.

scheduled to the 24.0 and 24.6-h day exhibited similar average intrinsic circadian periods during forced desynchrony (Figure 3B).

DISCUSSION

Our current finding that most of the human subjects tested maintained entrainment at an appropriate phase angle to a scheduled 24.0-h wakefulness-sleep cycle in as little as ~1.5 lux in the angle of gaze suggests that humans are exquisitely sensitive to photic and/or nonphotic synchronizers. It is thus clear that exposure to outside daylight or equivalent artificial light is not required to entrain the human biological clock to the 24.0-h day in most people, contrary to an earlier report (51). This finding further supports the idea that the use of artificial lighting by humans can alter the timing of the biological clock. This concept is consistent with results from resetting experiments that have reported that indoor light of ~180 lux is sufficient to reset the human biological clock (52). In addition, very recent resetting experiments reported that 50% of the weak (type I) phase delay resetting response to 9,000 lux can be achieved at ~1% of this intensity (53). These findings support the hypothesis that exposure to artificial light may underlie or exacerbate circadian sleep disorders in humans.

Simulations from the most recent mathematical model of the effects of light on the human biological clock by Kronauer and colleagues (54), that was refined using an independent dataset, show results similar to the current findings. Model simulations indicate that a 16:8 h light-dark wakefulness-sleep cycle of 1.5 lux during wakefulness and 0 lux during sleep should entrain individuals with periods of 23.90 to 24.15 h—a result remarkably consistent with the experimental results observed of entrainment in subjects with intrinsic periods of 23.88 to 24.12 h and failure to entrain in a subject with an intrinsic period of ≅ 24.36 h. However, given the larger range of intrinsic periods observed in the current study and elsewhere (27;43-46;55), it is evident that a stronger circadian synchronizer would be necessary to maintain circadian entrainment in all humans to the 24.0-h day. Furthermore, the minimum strength of the environmental synchronizer necessary to re-entrain or capture the human biological clock after a perturbation such as an abrupt phase shift of the sleep-wake and light-dark cycle is unknown and is likely to be greater than that used to maintain entrainment in the present study. The quantitative model used to predict limits of entrainment to a 16:8 LD cycle was based on phase-shifting experiments using episodes of light typically 5-6 h in duration and with intensities of light > 10 lux (56;57). Thus, the application of this model to light exposure of 1.5 lux represents an extrapolation out of this range. The slight asymmetry in the spacing of the limits relative to the 24.0-h period of the entraining cycle is due to incorporation in the model of an effect of light to shorten the oscillator period proportional to (but independent of) entraining effects, as described elsewhere

(58). This hypothesized effect of light intensity on circadian period has yet to be experimentally validated in humans.

The potential importance of non-photic synchronizers such as periodic scheduling of sleep and/or activity is illustrated by the lack of entrainment in subjects studied in a similar illuminance level whose activity-rest and light-dark cycles were not scheduled (55) and the reported entrainment to a 23.8-h day of a circadian blind human subject living in near darkness (< 0.03 lux) (59). In blind humans, pharmacological doses of melatonin have also been reported to entrain the circadian clock (60;61). In sighted humans, nonphotic stimuli have been reported to shift the timing of the circadian system (31-33;62;63). These studies as well as those from non-humans, suggest that potential nonphotic synchronizers, such as sleep and wakefulness (39) exercise (31;33;62;64), nutrition (65) and pharmacological agents (60;61) contribute to circadian entrainment. Since exercise and pharmacological agents were not allowed in the current study, the current results suggest that a periodic schedule of sleep, meals or other routine daily activities may have larger effects on the human biological clock than previously recognized.

As reported (1), the inability of subjects to entrain to the scheduled rest-activity cycle of both 23.5-h and 24.6-h when studied in ~1.5 lux demonstrates that the range of entrainment of the biological clock in sighted humans is centered near 24.0-h and that stronger environmental synchronizers would be necessary to expand the range of entrainment to include these day lengths. Previous research studies have reported that exposure to bright light (> 1,770 lux) can increase the range of entrainment of the human biological clock (3;6;66).

While we show that the biological clock is able to maintain a normal phase relationship to a weak environmental synchronizer, possible negative consequences of small changes in the phase angle of entrainment for human physiology and behavior are not well understood. In young and older humans, changes in the phase angle between the biological clock and the scheduled rest-activity cycle of as small as 2-4 hours can result in severe disruption of sleep efficiency and of waking performance (46;67).

The current findings improve our understanding about neurophysiologic mechanisms of circadian entrainment in humans showing that entrainment to the 24.0-h day can be maintained by a very dim light-dark wakefulness-sleep schedule. The near-24-hour entrainment limits of the human biological clock have important implications for the entrainment of astronauts during long-duration exploration-class space missions, For example, due to orbital mechanics, astronauts are commonly scheduled to the near-equivalent of a shorter-than-24-hour day length in Earth orbit on space shuttle missions and will be scheduled to ~24.65-h solar day of Mars on an exploration class mission to Mars. Even the ground control crew were required to adapt to such a schedule at NASA's Jet Propulsion Laboratory during the month-long operation of the solar powered unmanned Martian Rover Pathfinder in 1997. This requires adaptation of the near-24-hour period of the human biological clock to non-24-hour day lengths. Current power constraints limit the intensity of light exposure on the middeck of the space shuttle to levels similar to that used in the current studies (68). Therefore it was important, to determine whether these light levels were capable of entraining the human biological clock to the 24.0 and non-24-hour day lengths in dim light. Furthermore, the present results suggest that effective countermeasures are required to maintain entrainment of the human biological clock under these conditions. The current findings also have important implications for submariners and other Navy personnel who are scheduled to work on an 18-hour day (69). The 18-hour day is far from the period of the human biological clock and schedules such as this are likely to result in human performance failures due to the disruption of circadian rhythms and sleep that occur when the human biological clock is not properly entrained. The possibility that non-photic time cues underlie the current entrainment results maybe important for understanding how some blind individuals, who do not appear to have photic input to the biological clock and who have intrinsic circadian periods within a limited range near 24 hours, maintain entrainment to the 24 hour day (59-61). The findings that some humans have intrinsic circadian periods shorter-than-24-hours and others have circadian periods longer-than-24-hours, and the possibility that scheduled sleep-wake cycles can influence entrainment, may be important for understanding the mechanisms underlying circadian rhythm sleep disorders, such as advanced and delayed sleep phase syndrome, non-24-hour sleep-wake syndrome as well as individual vulnerability to rotating shift work schedules.

ACKNOWLEDGEMENTS

We thank the subject volunteers; research staff; subject recruiters, S. Ma, and N. Gonzalez. Thanks to R.E. Kronauer and D.B. Forger III for discussions on the mathematical simulations of circadian entrainment. Thanks to D.J. Dijk, R.J. Hughes, J.T. Hull and S.L. Lockley for scientific discussion. Supported by NASA Cooperative Agreement NCC 9-58 with the National Space Biomedical Research Institute and NASA NAG9-1035. The studies were performed in a GCRC supported by NIH MO1-RR02635. KPW was supported by fellowships from NIH T32-DK07529 and The Medical Foundation and The Harold Whitworth Pierce Charitable Trust.

REFERENCES

(1) Wright Jr. KP, Hughes RJ, Kronauer RE, Dijk D-J, Czeisler CA. Intrinsic near-24-hour pacemaker period determines limits of circadian entrainment to a weak synchronizer in humans. Proc Natl Acad Sci USA In Press.

(2) Lewy AJ, Wehr TA, Goodwin FK, Newsome DA, Markey SP. Light suppresses melatonin secretion in humans. Science 1980; 210:1267-1269.

(3) Wever RA, Polasek J, Wildgruber CM. Bright light affects human circadian rhythms. Pflügers Arch 1983; 396:85-87.

(4) Lewy AJ, Sack RL, Fredrickson RH, Reaves M, Denney D, Zielske DR. The use of bright light in the treatment of chronobiologic sleep and mood disorders: the phase-response curve. Psychopharmacol Bull 1983; 19:523-525.

(5) Daan S, Lewy AJ. Scheduled exposure to daylight: A potential strategy to reduce "jet lag" following transmeridian flight. Psychopharmacol Bull 1984; 20:566-568.

(6) Wever RA. Light effects on human circadian rhythms: A review of recent Andechs experiments. J Biol Rhythms 1989; 4:161-185.

(7) Campbell SS, Terman M, Lewy AJ, Dijk D-J, Eastman CI, Boulos Z. Light treatment for sleep disorders: Consensus Report. V. Age-related disturbances. J Biol Rhythms 1995; 10(2):151-154.

(8) Terman M, Lewy AJ, Dijk D-J, Boulos Z, Eastman CI, Campbell SS. Light treatment for sleep disorders: Consensus Report. IV. Sleep phase and duration disturbances. J Biol Rhythms 1995; 10(2):135-147.

(9) Eastman CI, Young MA, Fogg LF, Liu L, Meaden PM. Bright Light Treatment of Winter Depression. Arch Gen Psychiatry 1998; 55:883-889.

(10) Lewy AJ, Bauer VK, Cutler NL, Sack RL, Ahmed S, Thomas KH et al. Morning vs evening light treatment of patients with winter depression. Arch Gen Psychiatry 1998; 55:890-896.

(11) Czeisler CA, Wright Jr. KP. Influence of light on circadian rhythmicity in humans. In: Turek FW, Zee PC, editors. Regulation of Sleep and Circadian Rhythms. New York: Marcel Dekker, Inc., 1999: 149-180.

(12) Moore RY. Organization of the mammalian circadian system. In: Waterhouse JM, editor. Circadian clocks and their adjustment. Chichester (Ciba Foundation Symp 183): John Wiley and Sons, Inc., 1994: 88-99.

(13) Tei H, Okamura H, Shigeyoshi Y, Fukahara C, Ozawa R, Hirose M et al. Circadian oscillation of a mammalian homologue of the *Drosophila period* gene. Nature 1997; 389:512-516.

(14) Gekakis N, Staknis D, Nguyen HB, Davis FC, Wilsbacher LD, King DP et al. Role of CLOCK protein in the mammalian circadian mechanism. Science 1998; 280:1564-1569.

(15) Shearman LP, Sriram S, Weaver DR, Maywood ES, Chaves I, Zheng B et al. Interacting molecular loops in the mammalian circadian clock. Science 2000; 288:1013-1019.

(16) Lowrey PL, Shimomura K, Antoch MP, Yamazaki S, Zemenides PD, Ralph MR et al. Positional syntenic cloning and functional characterization of the mammalian circadian mutation *tau*. Science 2000; 288:483-491.

(17) Rusak B, Robertson HA, Wisden W, Hunt SP. Light pulses that shift rhythms induce gene expression in the suprachiasmatic nucleus. Science 1990; 248:1237-1240.

(18) Myers MP, Wager-Smith K, Rothenfluh-Hilfiker A, Young MW. Light-induced degradation of TIMELESS and entrainment of the *Drosophila* circadian clock. Science 1996; 271:1736-1740.

(19) Zeng H, Qian Z, Myers MP, Rosbash M. A light-entrainment mechanism for the *Drosophila* circadian clock. Nature 1996; 380:129-135.

(20) Morris ME, Viswanathan M, Kuhlman S, Davis FC, Weitz CJ. A screen for genes induced in the suprachiasmatic nucleus by light. Science 1998; 279:1544-1547.

(21) Zylka MJ, Shearman LP, Weaver DR, Reppert SM. Three period homologs in mammals: differential light responses in the suprachiasmatic circadian clock and oscillating transcripts outside the brain. Neuron 1998; 20(6):1103-1110.

(22) Ceriani MF, Darlington TK, Staknis D, Más P, Petti AA, Weitz CJ et al. Light-dependent sequestration of TIMELESS by CRYPTOCHROME. Science 1999; 285:553-556.

(23) Freedman MS, Lucas RJ, Soni B, von schantz M, Muñoz M, David-Gray Z et al. Regulation of mammalian circadian behavior by non-rod, non-cone, ocular photoreceptors. Science 1999; 284:502-504.

(24) Brainard GC, Hanifin JP, Rollag MD, Greeson J, Byrne B, Glickman G et al. Human melatonin regulation is not mediated by the three cone photopic visual system. J Clin Endocrinol Metab 2001; 86(1):433-436.

(25) Provencio I, Rodriguez IR, Jiang G, Hayes WP, Moreira EF, Rollag MD. A novel human opsin in the inner retina. J Neurosci 2000; 20(2):600-605.

(26) Lucas RJ, Freedman MS, Muñoz M, Garcia-Fernandez JM, Foster RG. Regulation of the mammalian pineal by non-rod, non-cone, ocular photoreceptors. Science 1999; 284:505-507.

(27) Czeisler CA, Duffy JF, Shanahan TL, Brown EN, Mitchell JF, Rimmer DW et al. Stability, precision, and near-24-hour period of the human circadian pacemaker. Science 1999; 284:2177-2181.

(28) Sweeney BM, Hastings JW. Effects of temperature upon diurnal rhythms. Cold Spring Harb Symp Quant Biol 1960; 25:87-103.

(29) Van Reeth O, Turek FW. Stimulated activity mediates phase shifts in the hamster circadian clock induced by dark pulses or benzodiazepines. Nature 1989; 339(6219):49-51.

(30) Mistlberger RE. Nonphotic Entrainment of Circadian Activity Rhythms in Suprachiasmatic Nuclei-Ablated Hamsters. Behav Neurosci 1992; 106(1):192-202.

(31) Van Reeth O, Sturis J, Byrne MM, Blackman JD, L'Hermite-Balériaux M, Leproult R et al. Nocturnal exercise phase delays circadian rhythms of melatonin and thyrotropin secretion in normal men. Am J Physiol 1994; 266:E964-E974.

(32) Eastman CI, Hoese EK, Youngstedt SD, Liu L. Phase-Shifting Human Circadian Rhythms With Exercise During the Night Shift. Physiol Behav 1995; 58(6):1287-1291.

(33) Buxton OM, Frank SA, L'Hermite-Balériaux M, Leproult R, Turek FW, Van Cauter E. Roles of intensity and duration of nocturnal exercise in causing phase delays of human circadian rhythms. Am J Physiol 1997; 273:E536-E542.

(34) Bobrzynska KJ, Mrosovsky N. Phase shifting by novelty-induced running: activity dose-response curves at different circadian times. J Comp Physiol [A] 1998; 182:251-258.

(35) Rajaratnam SM, Redman JR. Social contact synchronizes free-running activity rhythms of diurnal palm squirrels. Physiol Behav 1999; 66(1):21-26.

(36) Hastings MH, Duffield GE, Smith EJD, Maywood ES, Ebling FJP. Entrainment of the circadian system of mammals by nonphotic cues. Chronobiol Int 1998; 15(5):425-445.

(37) Maywood ES, Mrosovsky N, Field MD, Hastings MH. Rapid down-regulation of mammalian *Period* genes during behavioral resetting of the circadian clock. Proc Natl Acad Sci USA 1999; 96(26):15211-15216.

(38) Horikawa K, Yokota S, Fuji K, Akiyama M, Moriya T, Okamura H et al. Nonphotic entrainment by 5-$HT_{1A/7}$ receptor agonists accompanied by reduced *Per1* and *Per2* mRNA levels in the suprachiasmatic nucleus. J Neurosci 2000; 20(15):5867-5873.

(39) Grossman GH, Mistlberger RE, Antle MC, Ehlen JC, Glass JD. Sleep deprivation stimulates serotonin release in the suprachiasmatic nucleus. Neuroreport 2000; 11(9):1929-1932.

(40) Campbell S. Is there an intrinsic period of the clock? Science 2000; 288(5469):1174-1175.

(41) Czeisler CA, Dijk D-J, Kronauer RE, Brown EM, Duffy JF, Allan JS et al. Author's Response to letter entitled "Is there an intrinsic period of the clock?". Science 2000; 288(5469):1174-1175.

(42) Wever RA. The circadian system of man: Results of experiments under temporal isolation. New York: Springer-Verlag, 1979.

(43) Hiddinga AE, Beersma DGM, van den Hoofdakker RH. Endogenous and exogenous components in the circadian variation of core body temperature in humans. J Sleep Res 1997; 6(3):156-163.

(44) Carskadon MA, Labyak SE, Acebo C, Seifer R. Intrinsic circadian period of adolescent humans measured in conditions of forced desynchrony. Neurosci Lett 1999; 260:129-132.

(45) Waterhouse J, Minors D, Folkard S, Owens D, Atkinson G, MacDonald I et al. Lack of evidence that feedback from lifestyle alters the amplitude of the circadian pacemaker in humans. Chronobiol Int 1999; 16(1):93-107.

(46) Wyatt JK, Ritz-De Cecco A, Czeisler CA, Dijk D-J. Circadian temperature and melatonin rhythms, sleep, and neurobehavioral function in humans living on a 20-h day. Am J Physiol 1999; 277:R1152-R1163.

(47) Kronauer RE, Czeisler CA, Pilato SF, Moore-Ede MC, Weitzman ED. Mathematical model of the human circadian system with two interacting oscillators. Am J Physiol 1982; 242:R3-R17.

(48) Wright Jr. KP, Badia P. Effects of menstrual cycle phase and oral contraceptives on alertness, cognitive performance, and circadian rhythms during sleep deprivation. Behav Brain Res 1999; 103:185-194.

(49) Lewy AJ, Cutler NL, Sack RL. The endogenous melatonin profile as a marker for circadian phase position. J Biol Rhythms 1999; 14(3):227-236.

(50) Brown EN, Czeisler CA. The statistical analysis of circadian phase and amplitude in constant-routine core-temperature data. J Biol Rhythms 1992; 7:177-202.

(51) Nakamura K. Non-photic entrainment of human circadian clock -- effects of forced sleep-wake schedule on the circadian rhythm in plasma melatonin. Hokkaido Igaku Zasshi 1996; 71(3):403-422.

(52) Boivin DB, Czeisler CA. Resetting of circadian melatonin and cortisol rhythms in humans by ordinary room light. Neuroreport 1998; 9(5):779-782.

(53) Zeitzer JM, Dijk D-J, Kronauer RE, Brown EN, Czeisler CA. Sensitivity of the human circadian pacemaker to nocturnal light: Melatonin phase resetting and suppression. J Physiol (Lond) 2000; 526(3):695-702.

(54) Jewett ME, Forger III DB, Kronauer RE. Revised limit cycle oscillator model of human circadian pacemaker. J Biol Rhythms 1999; 14(6):493-499.

(55) Middleton B, Arendt J, Stone BM. Human circadian rhythms in constant dim light (8 lux) with knowledge of clock time. J Sleep Res 1996; 5:69-76.

(56) Kronauer RE, Forger D, Jewett ME. Quantifying human circadian pacemaker response to brief, extended, and repeated light stimuli over the photopic range. J Biol Rhythms 1999; 14(6):500-515.

(57) Forger III DB, Jewett ME, Kronauer RE. A simpler model of the human circadian clock. J Biol Rhythms 1999; 14(6):532-537.

(58) Jewett ME, Kronauer RE. Refinement of a limit cycle oscillator model of the effects of light on the human circadian pacemaker. J Theor Biol 1998; 192:455-465.

(59) Klerman EB, Rimmer DW, Dijk D-J, Kronauer RE, Rizzo JF, III, Czeisler CA. Nonphotic entrainment of the human circadian pacemaker. Am J Physiol 1998; 274:R991-R996.

(60) Lockley SW, Skene DJ, James K, Thapan K, Wright J, Arendt J. Melatonin administration can entrain the free-running circadian system of blind subjects. J Endocrinol 2000; 164:R1-R6.

(61) Sack RL, Brandes RW, Kendall AR, Lewy AJ. Entrainment of free-running circadian rhythms by melatonin in blind people. N Engl J Med 2000; 343:1070-1077.

(62) Baehr EK, Fogg LF, Eastman CI. Intermittent bright light and exercise to entrain human circadian rhythms to night work. Am J Physiol 1999; 277(6 Pt 2):R1598-R1604.

(63) Lewy AJ, Bauer VK, Ahmed S, Thomas KH, Cutler NL, Singer CL et al. The human phase response curve (PRC) to melatonin is about 12 hours out of phase with the PRC to light. Chronobiol Int 1998; 15(1):71-83.

(64) Mistlberger RE, Holmes MM. Behavioral feedback regulation of circadian rhythm phase angle in light-dark entrained mice. Am J Physiol Regulatory Integrative Comp Physiol 2000; 279:R813-R821.

(65) Kräuchi K, Werth E, Cajochen C, Weber JM, Renz C, Graw P et al. Are carbohydrate-rich meals a zeitgeber in humans? 10th Annual Meeting of the Society for Light Treatment and Biological Rhythms (SLTBR) May 6-10, 123. 1998.

(66) Eastman CI, Miescke KJ. Entrainment of circadian rhythms with 26-h bright light and sleep-wake schedules. Am J Physiol 1990; 259:R1189-R1197.

(67) Dijk D-J, Duffy JF, Riel E, Shanahan TL, Czeisler CA. Ageing and the circadian and homeostatic regulation of human sleep during forced desynchrony of rest, melatonin and temperature rhythms. J Physiol (Lond) 1999; 516.2:611-627.

(68) Dijk D-J, Neri DF, Wyatt JK, Ronda JM, Riel E, Ritz-De Cecco A et al. Sleep, performance, circadian rhythms and light-dark cycles during space shuttle missions. Am J Physio. In press.

(69) Kelly TL, Neri DF, Grill JT, Ryman D, Hunt PD, Dijk D-J et al. Nonentrained circadian rhythms of melatonin in submariners scheduled to an 18-hour day. J Biol Rhythms 1999; 14(3):190-196.

INDEX

9 780792 37669

GPSR Compliance

The European Union's (EU) General Product Safety Regulation (GPSR) is a set of rules that requires consumer products to be safe and our obligations to ensure this.

If you have any concerns about our products, you can contact us on ProductSafety@springernature.com

In case Publisher is established outside the EU, the EU authorized representative is:

Springer Nature Customer Service Center GmbH
Europaplatz 3
69115 Heidelberg, Germany

Batch number: 10755248

Printed by Printforce, the Netherlands